Miehe

Biointelligence

The present book is a habilitation thesis submitted to the University of Stuttgart in January 2021. With the publication of this work, the author fulfills the requirements of the habilitation regulations of the University of Stuttgart.

Robert Miehe

Biointelligence

Basic considerations for sustainable production

HANSER

About the author:
Dr.-Ing. Robert Miehe is head of department at Fraunhofer Institute for Manufacturing, Engineering and Automation IPA in Stuttgart, Germany.

MIX
Paper from responsible sources
FSC® C083411

Print-ISBN: 978-1-56.990-191-5
E-Book-ISBN: 978-1-56.990-268-4
ePub-ISBN: 978-1-56990-412-1

Bibliographic information of the German National Library:
The German National Library lists this publication in the German National Bibliography; detailed bibliographic data are available on the Internet at http://dnb.d-nb.de.

www.hanserpublications.com
www.hanser-fachbuch.de
Editor: Dr. Philippa Söldenwagner-Koch
Production Management: Melanie Zinsler
Cover concept: Marc Müller-Bremer, *www.rebranding.de*, Munich
Cover design: Max Kostopoulos
Cover picture: © shutterstock.com/O-IAHI
Typesetting: le-tex publishing services GmbH, Leipzig
Printed and bound by: CPI Books GmbH, Leck
Printed in Germany

Abstract

Sustainable production must remain within planetary boundaries, while being highly resilient and adaptable. Since socio-economic patterns of thought and production technologies prevailing today cannot achieve this, a reorientation in science and practice is necessary in the sense of Schumpeter's creative destruction. In this context, the concept of biointelligence has recently emerged in the scientific community. However, a comprehensive implementation is currently facing several challenges, among which are primarily the dissent about the definition of the term and its theoretical position in science. As a result, currently neither a congruent method of scientific knowledge retrieval exists nor a collection of methodological foundations for the development and management of biointelligent systems. The author thus presents basic considerations for a reorientation of the production sciences from a sustainability-oriented technology management perspective by adopting the concept of biointelligence, summarizing currently prevailing scientific trends, translating them into a coherent theory and examining them for congruence with current models and approaches. In doing so, the author sketches the picture of a normative, holistic scientific discipline that incorporates elements from engineering, information and life sciences and combines them for the purpose of sustainable production. The theories and methods developed and/or modified by the author on this basis are not to be understood as an ultimate solution to a societal challenge, but primarily serve to stimulate scientific discourse and the education of young scientists in the increasingly important field of biointelligence. Hence, the work is structured in form of a university textbook aiming to raise awareness among students and young researchers, particularly in production-oriented scientific fields, who have little experience in the scientific fields of life sciences and sustainability. Likewise, the publication is suitable for people who are committed to a reorientation of classical ways of thinking in industrial production.

Abbreviations

AAS	Asset Administration Shell
ADF	Abiotic Depletion Factor
AEL	Alkaline Electrolysis
AFC	Alkaline Fuel Cell
AFD	Allotrope Data Format
AHP	Analytical Hierarchy Process
AI	Artificial Intelligence
AM	Additive Manufacturing
AP	Acidification Potential
BDA	Big Data Analytics
BDC	Biointelligent Design Cube
BFC	Biological Fuel Cell
BII	Biointelligence index
BIOTRAIN	Preliminary survey on the biological transformation of industrial value creation in Germany (German: Voruntersuchung zur biologischen Transformation der industriellen Wertschöpfung)
BIS	Biointelligent System
BM	Business Model
BT	Biological Transformation (of Technology)
BTI	Biology Technology Interface

C	Carbon
CAD	Computer Aided Design
CBM	Condition Based Monitoring
CC	Cloud Computing
CED	Cumulative Energy Demand
CML	Centrum voor Milieukunde at the University of Leiden
CNN	Convolutional Neural Network
CPS	Cyber-physical Systems
CRISPR/Cas9	Clustered Regularly Interspaced Short Palindromic Repeats/Cas9 system
CRM	Customer Relationship Management
CS	Cognitive System
DFB	Design for Biointelligence
DFD	Design for Disassembly
DFE	Design for Environment
DFR	Design for Recycling
DFS	Design for Sustainability
DL	Deep Learning
DM	Digital Model
DMFC	Direct Methanol Fuel Cell
DNA	Deoxyribonucleic Acid
DOE	Design of Experiments
DS	Digital Shadow
DSB	Double-strand Breaks
DSS	Decision Support System
DT	Digital Twin
E-FMEA	Environmental Failure Mode and Effect Analysis
ECA	Ecological Classification Factor for Aquatic Ecosystems
ECT	Ecological Classification Factor for Terrestrial Ecosystems
EFC	Enzymatic Fuel Cell

ELM	Eco Lean Management
EL-SCP	Eco Lean Sustainability Control Panel
EM	Environmental Management
EMS	Environmental Management System
ERP	Enterprise Resource Planning
FEM	Finite Element Method
FDM	Fused Deposition Modeling
FET	Field Effect Transistor
FMEA	Failure Mode and Effect Analysis
FTA	Fault Tree Analysis
GE	Genome Editors
GBR	Green Biorefinery
gRNA	Guide RNA
GWP	Global Warming Potential
H	Hydrogen
H-FMEA	Human Failure Mode and Effect Analysis
HC	Human Toxicological Classification Factor
HDR	Homology-directed Repair
HHI	Herfindahl-Hirschman Index
1HPLC	High-performance Liquid Chromatography
HR	Homologous Recombination
HTEL	High Temperature Electrolysis
IPCC	Intergovernmental Panel on Climate Change
I4.0	Industry 4.0
ICT	Information and Communication Technology
IoE	Internet of Everything
IoMS	Internet of Manufacturing Services
IoP	Internet of People
IoS	Internet of Service

IoT	Internet of Things
ISM	In-situ Microscopy
JIT	Just-in-Time
JIS	Just-in-Sequence
JSON	JavaScript Object Notation
LCA	Life Cycle Assessment
LCC	Life Cycle Costing
LCD	Life Cycle Design
LCI	Life Cycle Inventory
LCIA	Life Cycle Impact Assessment
LCSA	Life Cycle Sustainability Assessment
LDW	Laser-based Direct Writing
LFBR	Lignocellulose Feedstock Biorefineries
LIFT	Laser-induced Forward Transfer
LM	Lean Management
MAPLE DW	Matrix-assisted Pulsed Laser Evaporation Direct Writing
MES	Method of Ecological Scarcity
MCFC	Molten Carbonate Fuel Cell
MFC	Microbial Fuel Cell
MIPS	Material Intensity per Service Unit
MIR	Mid-infrared
ML	Machine Learning
mRNA	Messenger RNA
MVDA	Multivariate Data Analysis
N	Nitrogen
NHEJ	Non-homologous End Connection
NIR	Near Infrared
NN	Neural Network
NP	Nutrification Potential

O	Oxygen
ODP	Stratospheric Ozone Depletion Potential
OECD	Organisation for Economic Co-operation and Development
OPC UA	Open Platform Communications United Architecture
PAFC	Phosphoric Acid Fuel Cell
PAT	Process Analytical Technology
PCB	Polychlorinated Biphenyl
PDCA	Plan-Do-Check-Act
PEMEL	Polymer Electrolyte Membrane Electrolysis
PEMFC	Polymer Electrolyte Membrane Fuel Cell
PHM	Prognosis and Health System
POCP	Photochemical Ozone Creation Potential
QbD	Quality by Design
QFD	Quality Function Deployment
R&D&I	Research, Development and Innovation
RAMI4.0	Reference Architecture Model Industry 4.0
RCIM	Resource-constrained Innovation Method
RCS	Resilient Control System
RDF	Resource Description Framework
REPA	Resource Environmental Profile Analysis
RGB	Red, Green and Blue
RNA	Ribonucleic Acid
SBMD	Smart Biomanufacturing Devices
SC	Special Characteristics
SCNT	Somatic Cell Nuclear Transfer
SD	System Dynamics
SLCA	Social Life Cycle Assessment
SMCA	Substance and Material Criticality Assessment
SME	Small and Medium-sized Enterprise

SMRA	Substance and Material Risk Assessment
SOFC	Solid Oxide Fuel Cell
SOP	Start of Production
ssODN	Single-stranded Oligonucleotide
STIA	System-theoretical Interdependency Analysis
TALEN	Transcription Activator-like Endonucleases
TCO	Total Costs of Ownership
TIMWOOD	Transport, Inventory, Motion, Waiting, Over-engineering, Over-processing, Defects (classic seven types of lean waste)
TPBR	Two Platform Biorefinery
TPS	Toyota Production System
TRIZ	Theory of Inventive Problem Solving
tRNA	Transfer RNA
TVT	Threshold Value Transmission
UML	Unified Modeling Language
VSM	Viable System Model
WCBR	Whole Crop Biorefinery
WGI	Worldwide Governance Indicator
XML	Extensible Markup Language
ZFN	Zinc Finger Nucleases
ZVEI	German Electrical and Electronic Manufacturers' Association
3D	Three-dimensional
4D	Four-dimensional

Contents

1 Preface

1.1 Introduction

The manufacturing industry accounts for about one quarter of the German gross domestic product (Statistisches Bundesamt, 2019b). Yet, this indicator fails to adequately reflect the economic importance of industry for the German economy. In fact, the current economic conditions of the country represent historically grown interrelationships around an industrial core. Accordingly, a large part of gross value added in the service sector is still based on industry-related services. Whereas economists long assumed a 'natural' evolution of national economies, from rural to industrial to service-dominated economies, a robust industry is now regarded as essential for the resilience of the economy (Statistisches Bundesamt, 2019b). Reasons for this are complex. However, two arguments can be listed as essential. First, manufactured goods account for over 80% of the country's exports (Bundesministerium für Wirtschaft und Energie, 2018). Industry thus decisively contributes to the fact that Germany has had a positive current account balance for decades. Apart from China, Germany is the largest exporter of goods (Bundesministerium für Wirtschaft und Energie, 2018). Second, the industry is the driving force behind innovation in Germany, accounting for almost 70% of investment in research and development (Statistisches Bundesamt, 2019a). Albeit companies in Germany are facing a number of challenges that are difficult to predict in their dynamics (e. g., shortage of skilled workers, urbanization, individualization, growing competition from emerging markets) (Hoorens et al., 2013; European Commission, 2018), they have so far maintained an excellent reputation in terms of quality and reliability. However, the Covid crisis recently revealed the fragility of the current economic model by particularly impacting global supply chains. The resulting process of rethinking industrial practices promotes the vision of a resilient and adaptable production.

Although industrial production has undoubtedly contributed significantly to today's prosperity, particularly in the 'western world', its prevalent practice is responsible for immense environmental pollution and massive socio-technological consequences. Since the first industrial revolution, thoughtless economic practices resulted in various severe consequences to the ecosystem, including climate change, resource scarcity, human- and eco-toxicity, acidification, eutrophication, noise, radiation, loss of biodiversity, particulate matter, and chemical contamination (Rockström et al., 2009a, 2009b; Fava et al., 1994; Klöpffer and Grahl, 2014; Steffen et al., 2015). Global resource consumption has more than doubled in the last 30 years (United Nations Environment Programme, 2011). To date, human activities have already caused a global warming of about 1 °C above pre-industrial levels (Intergovernmental Panel on Climate Change, 2018). Yet, forecasts predict a further acceleration of causal effects (population growth and consumption). By 2050, the world population is expected to have increased to around 10.8 billion people (United Nations, 2015). By 2100, up to 16.6 billion people could inhabit the planet. Meanwhile, the proportion of global population actively involved in the consumption of goods grows. This consumer class includes people who have more than 10 euros per day (purchasing power parity) at disposal (Manyika et al., 2012). While this class accounted for around 13% of the global population in the 1950s, its share increased to 36% in 2010. For the year 2025, with an assumed world population of 7.9 billion people, a consumer class of 53% is expected (Manyika et al., 2012). Consequently, a further doubling of the global consumption of resources can be assumed by the year 2050. Likewise, anthropogenic CO_2 emissions will continue to rise considerably based on current political and economic efforts (International Energy Agency, 2019). If it continues to increase at the current rate, global warming is likely to reach the 1.5 °C limit commonly agreed upon in the Paris accords between 2030 and 2052 (European Environment Agency, 2017).

Albeit conventional economic practice was challenged as early as the 1970s and 1980s by the 'limits-to-growth' report (Meadows et al., 1972) and the sustainability concept (World Commission on Environment and Development, 1987), outlining a fundamental change, the expected effect has largely failed to materialize. Roughly half of the cumulative anthropogenic CO_2 emissions between 1750 and 2010 were emitted in the last 40 years, i. e. to a large extent in the period after the sustainability concept was published (Intergovernmental Panel on Climate Change, 2014b). Although a consensus exists today that sustainable production is only possible within planetary boundaries (Rockström et al., 2009a, 2009b; Steffen et al., 2015), a contemporary realization appears problematic. Traditional patterns of thought and behavior, which originate in early human history and early industrialization, are too deeply rooted in decision-making at all levels.

Since present production practices apparently fail to meet the needs of future generations in an equitable manner, innovative technologies are required in addition to adapting prevailing socio-economic patterns of thought (Bauernhansl et al., 2019;

Miehe et al., 2020a). In this context, several experts consider the biological transformation of technology (BT) to have great potential, causing fundamental changes in industrial production (Byrne et al., 2018; Patermann and Aguilar, 2018; Miehe et al., 2018). Recent successes in biotechnology (e. g., CRISPR/Cas, biocomputing, enzyme technology, bioreactors, biosensorics, and bioactors) and artificial intelligence (e. g., deep neural networks, adversarial learning techniques, deep reinforcement learning) offer great opportunities for innovation leaps. A continuation of this trend may result in self-reinforcing effects, which accelerate the generation of knowledge and application strategies for biological systems (e. g., advances in computing, bioinformatics, and artificial intelligence, advancing the analysis of omics[1] data and/or increasing possibilities of reprogramming DNA). According to Chui et al. (2020), with regard to biological systems, four essential fields of innovation emerge: [1] the mapping, measuring, and engineering of molecules, [2] the engineering of cells, tissues, and organs; [3] the interface between biology and machines, and [4] the use of cells or molecules such as DNA for computation.

With roughly 60% of global physical inputs to the economy that could be produced biologically and about 45% of the current global disease burden that could be addressed, the (direct and indirect) economic impact of biological innovations appears substantial (Chui et al., 2020). According to Miehe et al. (2018; 2020a), Bauernhansl et al. (2019), and Chui et al. (2020), fundamental human needs such as health, nutrition, energy, consumption, and housing are already beginning to be transformed by biological innovations. Prominent examples are CAR T-cell therapies, offering great potential for the treatment of liquid (short term) and solid tumors (medium term), as well as cultured meat, replacing entire value-added chains (Chui et al., 2020). The transformative potential of biotechnology also became apparent in the Covid crisis with the development of highly effective mRNA vaccines at a record pace. This example, however, also illustrates the current dilemma of these innovative approaches. Although the necessary knowledge for the development of vaccines has been gathered quickly, an equally efficient supply of the population was limited by the lack of local production capacities and expertise as well as the fact that existing regulations do not stimulate licensing models (i. e. 'the sale of blueprints'). Then again, biotechnological advancements are associated with potentially cascading and long-lasting effects on entire ecosystems or species. The most significant risks result from the creation of novel organisms by DNA engineering, to which current species and natural ecosystems are not adapted. Due to the interconnection of ecosystems, this may cause effects that cannot be foreseen today. In addition, access is becoming easier and cheaper, potentially enabling anyone to interfere with genetic material. Consequently,

[1] The term 'omics' refers to several scientific research streams in biology (e. g., genomics, epigenomics, transcriptomics, proteomics) that collectively characterize and quantify pools of biological molecules at different levels and processes of living systems.

extensive social reservations prevail. Apparently, humanity possesses more technologies than it is able to use in a commonly accepted manner. Yet, it is precisely this appropriate deployment that is decisive for the future viability of the socio-economic model. As prevailing socio-political conditions in form of existing individual, social, and political value systems largely determine technological development, a solution to a fundamental reorientation such as the sustainability issue can never be solved on a microeconomic basis only.

In this context, Miehe et al. (2018; 2020a) and Bauernhansl et al. (2019) present the concept of biointelligence, seeking to reinvent production by appropriately using the above sketched self-reinforcing moments of information and biotechnology. Based on developments in various scientific disciplines, a new form of application-oriented research is to be developed, which follows a normative, holistic maxim. This will be achieved by linking biological and technical systems via information and communication technology (ICT), enabling regional, small-scale, recyclable, resilient, demand-oriented, and highly automated production technologies and value-added chains, which in turn will allow the purposeful management of biological resources in the form of a technology-oriented demand economy. Crucial to the development of such approaches is a highly interdisciplinary research approach (Bauernhansl et al., 2019; Miehe et al., 2020a). Due to its recent emergence, the concept of biointelligence has not yet been sufficiently established in the scientific community, which hinders both its urgently required further academic advancement as well as a comprehensive dissemination and awareness of future decision-makers in higher education.

1.2 Objective and approach

The objective of this work is to establish the concept of biointelligence in science and practice. Rather than providing prefabricated solutions, this work presents basic theoretical considerations to stimulate scientific discourse and equip the reader with a specific mind-set. The here introduced normative framework of thought and design is intended to provide a systematic basis for a sustainability-oriented technology management with respect to recent findings in production science. The framework is supplemented by a review of essential approaches to developing and managing biointelligent systems (BIS). This work thus contributes to the structuring of an emerging field of research and promotes the necessary awareness of several generations of young scientists for a reorientation of science and practice in the context of sustainable production. The intention is not only to describe current trends in technology and economy, but to rigorously pursue a normative, holistic design logic.

The work represents a summary of selected research and teaching activities of the author. Its scientific approach follows the research process according to Ulrich and

Hill (1976), distinguishing between discovery, reasoning and application, whose respective requirements for insight generation are integrated into the chapter structure (Figure 1.1).

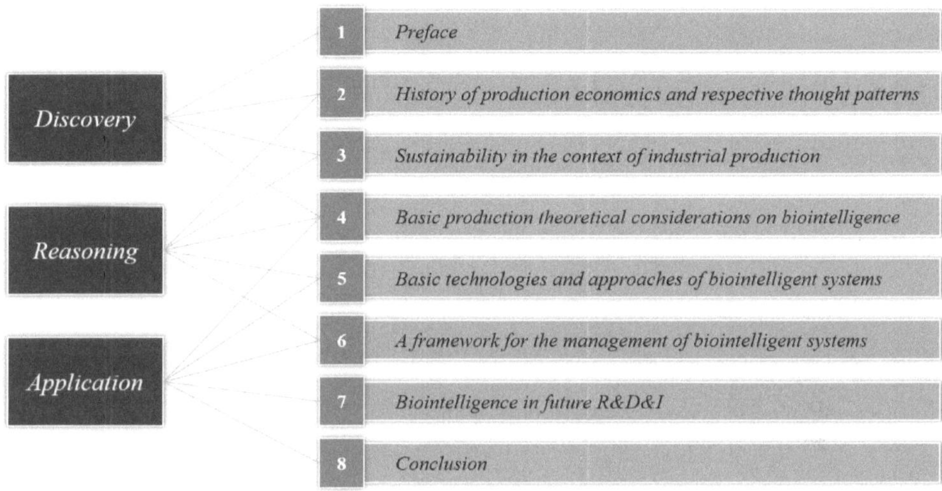

Figure 1.1 Chapter structure and research process according to Ulrich and Hill (1976)

The work is structured in form of a university textbook, which is intended to support the transfer of knowledge in academic teaching. The target group includes students and young researchers, particularly in production-oriented scientific fields (especially production engineers and managers), who have little experience in the scientific fields of life sciences and sustainability. Likewise, the publication is suitable for people who are committed to a reorientation of classical ways of thinking in industrial production. The document is structured as a coherent text. Each section closes with a conclusion, a bullet-point summary of the lessons learned, and a number of follow-up questions that facilitate the transition to the subsequent section. This will provide the reader with an understanding of essential interrelationships, challenges, basic technologies, and methods that will allow him or her to decisively support the necessary change to sustainable production.

After the introduction in Chapter 1, Chapter 2 provides a brief summary of currently dominant patterns of thought, which are commonly identified as the root cause of the sustainability problem. Based on the insight of a necessary fundamental reorientation, Chapter 3 introduces fundamentals of the concept of sustainability, discusses its implications for theory formation, and outlines existing models of future economies. Chapter 4 then presents basic production-theoretical considerations on biointelligence by summarizing the origin of the term, locating it in scientific theory, and reviewing the consistency of the current doctrine. On this basis, Chapter 5 outlines basic technologies and selected approaches for the development of BIS, while Chapter 6 summarizes basic thoughts on the management of BIS with regard to a holistic benefit criterion. Subsequently, Chapter 7 provides an overview of selected research and development issues, before the work concludes with a summary, a critical review and an outlook (Chapter 8).

2 History of production economics and respective thought patterns

In order to understand, classify, and change today's production methods and thought patterns, it is necessary to understand their origin. Thereby, it is not enough to consider the emergence of production technology or science exclusively. Rather, it is important to place it in its socio-cultural context and illustrate its impact on the environment. Hence, this section outlines the history of production on the basis of four essential characteristics: economic form, dominant economic theory, technological basis, and impact on the natural environment. The section, which is classified in a pre-industrial and industrial era, focusses on the major socio-economic upheavals in human history. The section does not claim to be complete in the sense of a historical scientific discourse, but rather serves the iterative understanding and awareness for the development of today's thought patterns in order to create the basis for a reorientation in the following section.

2.1 Pre-industrial era

Economic behavior is based on the cooperation of people in large groups. Crucial to this is the creation of so-called intersubjective realities, imaginary entities that only become reality because a large number of people believe in them (Harari, 2014). Examples of this are gods, states, companies, money and hierarchies. The mental abilities necessary to create these constructs developed comparatively late in human history and are closely linked to the creation of the first more complex tools and technologies (boats, needles, oil lamps, bows and arrows), the emergence of religions and trade, and the establishment of social classes. This epoch of human history is called the *cognitive revolution* and, as a process of intellectual and social transformation, is dated to a period of about 70,000 to 15,000 years ago. During this time, people lived in nomadic tribes (Spier, 1998; Harari, 2014; Parzinger, 2015). Their livelihood was based on hunting and gathering. Naturally, a consensus on an economic system of that time did not exist. Since the production of goods followed the current demand, the economic form can be described as an *early demand economy*. The technological basis consisted of a comparatively small number of tools, which almost everyone had to master equally in order to ensure daily survival. The only existing form of trade was barter (goods for goods) (Schenk, 1991; Rudolph, 2009). In contrast to the often expressed notion that people of this era lived 'in harmony with nature', extensive impacts on the ecosystem are documented. While the release of produced goods into the environment following their use without hesitation had no consequences due to the small number and complexity of the products, systematic hunting led to a substantial loss of biodiversity. The effective cutting of prey, the preservation of food and the development of thermal insulation from animal skins enabled the spread of humans across the entire globe, which in turn resulted in an extinction of about 50% of all land mammals over 50kg body weight by the time of the agricultural revolution. Compared to current standards, however, the total utilization of natural sources and sinks can be classified as marginal due to a small world population and the quantity and complexity of goods (Spier, 1998; Harari, 2014; Parzinger, 2015).

A first significant change in economic-social structures and lifestyles occurred shortly after the end of the last ice age about 15,000 years ago with the emergence of systematic production methods in agriculture (e. g., ploughing a field with the help of livestock). The groups, previously organized in nomadic communities, increasingly developed into a sedentary community characterized by agriculture and animal breeding. Animals and plants were no longer hunted or picked, but bred and harvested. This upheaval, which began about 11,500 years ago in eastern Turkey, is today called the *Neolithic revolution* (also agricultural revolution) (Harari, 2014). Similar to the cognitive revolution, the transformation was not a short-term global change, but a fundamental development over thousands of years, which continued to be affected by the parallel existence of nomadic tribes in different parts of the world (Spier, 1998). At the same time, it is regarded as a fact today that the Neolithic revolution did not have a single point of origin and spread from there, but started completely independently of each other in different parts of the world. Figure 2.1 illustrates the global development of the agricultural revolution according to Bellwood (2004) and Harari (2014).

For humanity, however, the change did not initially imply a comprehensive improvement in living conditions, since the transformation of the form of work, the restriction of nutrition to a few calorie sources and the hygienic conditions of settling down represented elementary challenges that first had to be mastered (Harari, 2014). At the same time, the systematization of production processes led to an increase in output. As the population began to grow, a locally heterogeneous dynamic developed, which resulted in new social and technical innovations. Elementary developments were the artificial irrigation (about 9000 years ago), the invention of the wheel and the training of farm animals (about 6000 years ago) (Parzinger, 2015).

Figure 2.1 Regions with own agricultural revolution according to Bellwood (2004) and Harari (2014)

Yet, the transition to sedentariness did not automatically produce complex societies and states, as there was simply a lack of any possibility of state organization. This changed with the emergence of the first state structures in the region of today's middle east in the 4th millennium BC (Spier, 1998). A decisive factor in this was the development of writing systems that allowed extensive amounts of data to be stored and processed (Harari, 2014). The first writing system known today is considered to be the partial writing system of the Sumerians, which originated around 3100 BC and was transformed into a complete writing system by the middle of the 3rd millennium BC. In a similar period of time a comparable writing system developed in Egypt with the hieroglyphics. Other complete scripts emerged in China (ca. 1200 BC) and Central America (ca. 1000 BC) (Harari, 2014).

Initially a consensus on an economic system did not exist. This form of economy, which dominated from the Middle Ages to the industrial revolution and increasingly appeared in monarchical forms of government, can today be described as an *agricultural-based demand economy*. In addition to the recycling of various resources (e.g., animal excrements for agriculture), increasingly complex goods were developed, which in turn required new raw materials. In the course of time, the products were further adapted to the needs of the individual and manufactured in local workshops by qualified specialists who passed on their experience and skills by qualifying young people. In this way, craft guilds and technological centers were formed, e.g., in places where ores were processed (e.g., the German Harz mountains) (Bauernhansl and Miehe, 2020). In addition to goods-for-goods exchanges, from about 1000 BC goods with established exchange values (e.g., shells) and from about 100 BC money was established as a form of payment (Schenk, 1991; Rudolph, 2009). In comparison to the previous epoch in human history, the Neolithic revolution and the increasingly complex societies that followed it resulted in far more extensive impacts on the natural ecosystem. The phenomena of natural species extinction, which had already occurred in ancient times and which was triggered by larger groups of people staying in one place for a longer period of time, was intensified by mass settlement. Similarly, the conversion of many areas of land into agricultural land, often in the form of slash and burn land clearance, led to a further decline in biodiversity. Due to the small world population and the still comparatively small quantity and complexity of goods, the utilization of natural sources and sinks was still limited (Harari, 2014; Bauernhansl and Miehe, 2020).

Until the first half of the 15th century AD only few people were able to learn, develop and document knowledge about natural, spiritual and religious issues due to their social position. This changed abruptly with the modern letterpress printing technology developed by Gutenberg (Eisenstein, 1997; Giesecke, 2006). Subsequently, a wealth of theories and approaches rapidly developed and spread, which continues to influence our thinking today. The most elementary approaches are likely to be *humanism* and *rationalism*.

Since the cognitive revolution, a multitude of intersubjective realities had developed in the form of gods and religions, serving to explain various unknown phenomena in order to give meaning and significance to human existence. In contrast, humanism emphasized the individuality of man (Campana, 1946; Kristeller, 1974). The development of this further intersubjective interpretation (of man as unique), raised man further above nature in philosophic theory. The natural environment as well as all its forms of life serve exclusively the well-being of man.[1)] The doctrine, formulated as early as the 14th century AD, spread throughout Europe in the following two centuries, driven by the new possibilities of reproduction (Böhme, 1986). According to Harari (2016), the early school of humanism later formed the basis for three different interpretations which decisively shaped the recent history of mankind: [1] liberalism (liberal humanism), [2] communism (social/Marxist humanism) and [3] nationalism (evolutionary humanism).

On the other hand, Descartes' rationalism (Descartes, 2011) elevated rationalist thinking ('cogito ergo sum', i.e., 'I think, therefore I am') to the most important building block of knowledge. While fragments of faith and culture indoctrinated up to this point (so-called *memes*[2)]) were accepted almost without contradiction, an epistemol-

---

1 In sustainability research, this interpretation is referred to as an anthropocentric perspective.

2 Meme theory describes the formation of human consciousness based on biological processes. The idea is that the contents of consciousness in the form of fragments of faith and culture (so-called memes) are passed on through communication and shaped by environmental influences (Blackmore 2000). On the basis of the genetic preconditions of the single individual, coherent constructs of different memes (so-called memplexes) are formed, which in turn can be passed on and changed (Lenzen, 2003). Meme theory is used to describe socio-cultural developments, e.g., cultures, religions.

ogy developed that only accepted what could be verified as plausible by step-by-step analysis and logical reflection (Cottingham, 1988; Peacocke, 2002). Rationalism not only formed the basis for elementary scientific knowledge, such as Newton's mechanics (Newton, 1999), and for philosophical breakthroughs, such as enlightenment, but is still regarded today as a foundation of scientific epistemology (rationalism is partly considered as the origin of the so-called *scientific revolution*).

From the 16th century onwards, increasingly coherent theories also emerged in economics, accompanying the transition from feudalist to early capitalist forms of economy. Although these theories were not holistic, but rather individual concepts, the predominant economic school of outgoing European feudalism and incipient early capitalism between the 16th and 18th century AD is today referred to in the economic sciences as *mercantilism* or *pre-classicism* (Schaefer, 1993; Gömmel, 1998). In order to meet the demand of the still absolutist governed states of this epoch for strong armies and a growing bureaucracy, some theories developed which propagated the greatest possible promotion of productive domestic forces and the generation of surpluses in foreign trade. In economic practice, exports of manufactured goods were actively promoted, while imports were limited. The economic theory of mercantilism is characterized by its focus on the domestic market and interventionism as a structural tool (Bofinger, 2011).

Table 2.1 summarizes the socio-economic development of the pre-industrial era on the basis of its essential characteristics.

Table 2.1 Socio-economic development of the pre-industrial age based on various characteristics [own summary][3]

	Cognitive Revolution	Neolithic Revolution	
Period	about 70,000 to 15,000 years ago	about 15,000 years ago to the 15th century	16th–18th century
Characteristics of the society or social upheaval	living in nomadic tribes	sedentary society based on agriculture and livestock	dissemination of knowledge through printing, humanism, enlightenment, rationalism, colonialism, strengthening of armies, establishment of large civil service apparatuses
Economic form	early demand economy	agricultural-based demand economy	

3 Population data are based on Roser et al. (2020).

Table 2.1 Socio-economic development of the pre-industrial age based on various characteristics [own summary] (*continued*)

	Cognitive Revolution	Neolithic Revolution	
Form of trade	goods-for-goods	goods-for-goods (until 1000 BC) goods-for-goods with recognized value (from 1000 BC) goods-for-money (from 100 BC)	
Technological basis	simple tools and technologies (e. g., fire, boats, needles, oil lamps, bow and arrow, grinding)	more complex tools, technologies and goods, the production of which required extensive experience and skill, use of new raw materials (e. g., ores), development of local workshops, technological centers and craft guilds	
Prevailing economic theory	none	none	mercantilism
Global population	about 15,000 to 4 million	about 4 to 500 million	about 500 to 800 million

2.2 Industrial era

The industrial age can roughly be divided into three phases (also known as industrial revolutions), which were decisively shaped by social and technological transformations.

2.2.1 First phase of industrialization (around 1750 to 1850)

The emphasis on the uniqueness of the human being (humanism) and the turn towards natural sciences (rationalism), which was already initiated in the 14th century AD, led to the development of a wide range of philosophical concepts and constitutional ideas in the following centuries, marking the transition from the Middle Ages to modernity. In interaction with the Protestant ethics of the Calvinists, Puritans and Quakers, which not only legalized the unbiased pursuit of property and the acquisition of goods, but regarded it as God's will (Weber, 1992), a dynamic developed which manifested itself in profound social, philosophical, economic and technological changes. Compared to medieval ideals of atonement and punishment, societies were from then on increasingly shaped by a belief in progress and self-determination (Bauernhansl and Miehe, 2020).

A decisive catalyst for this development was the rapid technological progress in Europe, which around the middle of the 18th century marked the transition to the industrial age. Three major breakthroughs were instrumental in what is today referred to as the first industrial revolution (primarily occurring in Great Britain): [1] the mechanization of work previously performed by humans and animals by machines (e. g., spinning machines, mechanical looms), [2] the mechanical generation and conversion of energy by the steam engine developed by Newcomen and Watt, and [3] the mass use of mineral raw materials (especially iron and coal) (Landes, 1986). The former provided the basis for the emergence of factories as separate socio-technical production units, a completely new form of production of goods that was already based on a partial internal division of labor. The importance of the steam engine for the mechanization of energy production and work only slowly increased, finally replacing regenerative generation technologies (especially hydroelectric power) as a primary energy source around the middle of the 19th century, i. e., a little more than 100 years after its invention (Mende, 1993).

The first industrial revolution enabled the entire population to be supplied with clothing and food at comparatively low cost for the first time, as the transport system (steam navigation, railways) was considerably improved in addition to production (Bauernhansl and Miehe, 2020). With the help of these working and power machines, structurally caused famines could be avoided in the newly industrialized countries, which in turn led to an exponential increase of population. On the other hand, the technological transformation had considerable effects on society: increasing productivity in the production of basic goods, e. g., in agriculture, led to a significant reduction in employment in areas of traditional crafts and agriculture. While the pre-industrial age was characterized by three social classes (clergy, nobility, citizens/farmer), the industrial revolution created two new classes: the factory workers and the factory owners. Through skillful economic activity, more and more people were able to achieve wealth. This perspective attracted many people to the cities, where they hoped to find well-paid work in one of the many newly established factories. Dramatic population growth, coupled with increasing urbanization, often led to catastrophic hygienic conditions in the cities of the early industrial era. Diseases and epidemics were frequent causes of death. In addition, adverse working conditions in the factories were a further factor. Child labor, high levels of pollution, and little or no occupational safety resulted in high mortality rates among factory workers. The exploitation of factory workers and the structural poverty of the workforce that accompanied it are referred to as pauperism (Bauernhansl and Miehe, 2020).

Fundamental social concepts of what is now known as the age of European enlightenment were the rejection of prejudice, religious tolerance and orientation towards so-called natural rights (Peter-André Alt, 2007; Ehrhard Bahr, 2008; Annette Meyer, 2009). These basic (intersubjective) assumptions, understood as universally valid principles of order, form the basis of today's conception of human rights, which in turn provided the foundation for the two major social transformations of the modern era shaped by the global 'west', the American revolution of 1776 and the French revolution of 1789. While the Declaration of Independence of the United States of America defined human rights as life, freedom, and the pursuit of happiness, the slogan of the French revolution emphasized freedom, equality, and fraternity. In comparison to the US-American interpretation, which accentuates the individuality of the human being, the ideals of freedom and equality formulated in the course of the French revolution, today a foundation of western European democracies, are considered to be an elementary contradiction. The incompatibility of these two ideals continues in different forms throughout the history of economics and politics, culminating in the sustainability debate to this day (Felber, 2010; Harari, 2014).

With the first industrial revolution, capitalism developed into the leading economic and social order in Europe. Elementary characteristics of this economic system are the control of production and consumption by the market, private ownership of production resources, accumulation of capital, and the pursuit of profit (Willke, 2006; Hüther, 2006; Weber, 2015). A characteristic feature of capitalism was the reinvestment of profits in order to increase productivity and generate even more profits. Figure 2.2 illustrates the core principle of capitalism in a highly simplified form according to Harari (2014).

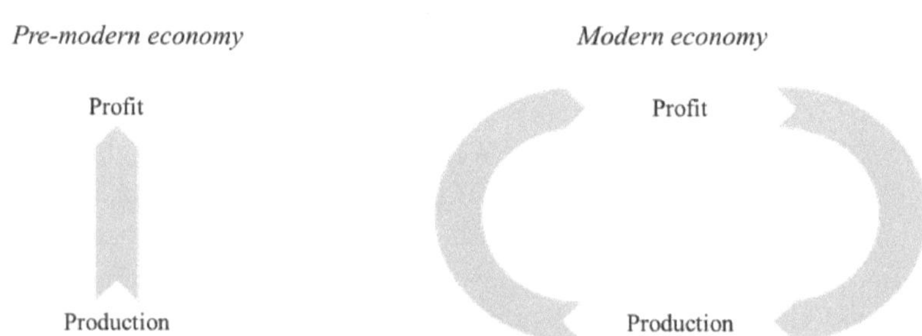

Figure 2.2 Basic principle of capitalism in comparison to pre-modern economic forms according to Harari (2014)

While the term capitalism is used consistently in English, the term market economy serves as a synonym in Germany (Woll, 2008).[4] Consequently, the loose economic theories of mercantilism and pre-classism were replaced by holistic theories that claimed to optimally represent the functional logic of 'markets' (Hüther, 2006). Economic and political theories, which from then on were renewed in an ever more rapid succes-

[4] In German, the term capitalism has a rather negative connotation.

sion, mostly emerged as a reaction to socio-political uncertainties and, depending on the socio-political conditions, formed the basis of the economic policy of various states for a limited time (Harari, 2014).

The first theoretical construct known as a holistic economic school was *physiocracy*, whose emergence can be explained by misguided developments in the mercantilist economic policy of late feudal France. The concept formulated by Quesnay in 1758 postulated land as the sole source of a country's wealth (Quesnay, 1971). While still under the impression of the strong expansion of bureaucratic apparatuses and loss-making state intervention in the economy, which led to the decline of French agriculture, Quesnay (1971) shifted the creation of added value to the production phase. While mercantilism attributed the wealth of a country to trade, the physiocracy emphasized that only agriculture produces surpluses (Priddat, 2001; Holub, 2010; Bach, 2018). The contribution of nature was seen as a 'gift', which is equivalent to a value that should not be assigned to it in any other economic school until the second half of the 20th century. The accumulation of monetary capital was rejected and instead a continuous circulation of money was emphasized. In the course of this, Quesnay developed the economic circuit model of macroeconomics[5)], which is still fundamental today and explains the creation of added value through the interaction of three classes (today: households, companies, and capital providers). Figure 2.3 illustrates the economic circuit model in simplified form according to Quesnay (1971).

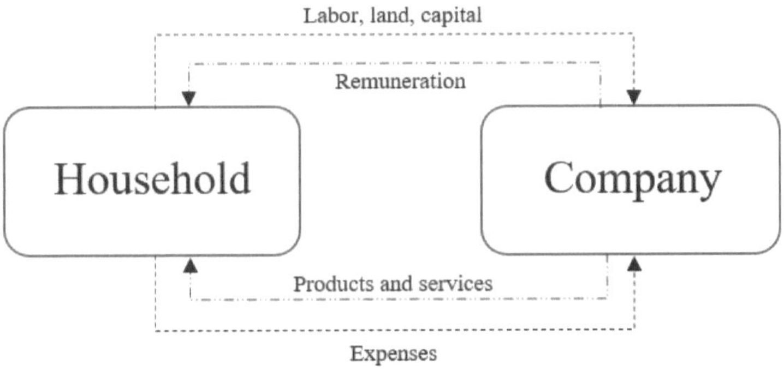

Figure 2.3 Economic circuit model according to Quesnay (1971)

Decisive for the physiocratic school was the notion of a natural law established by the enlightenment, which had to be harmonized with the law created by man in order to

[5] Today, empirical sciences essentially distinguish two perspectives for the analysis of socio-technical systems, the macro and micro level. While the macro level examines holistic system interrelationships, the micro level considers exchange relationships between related individual systems. Applied to concrete scientific questions, problems of economic relations and political options are to be assigned to macro or political economy. The analysis of possible actions of individual economic units (companies, households), on the other hand, takes place in microeconomics, the most common form of which is business administration (Miehe, 2018).

achieve the greatest possible welfare for all people and to allow a free economic activity. Free pricing, free trade, and a capitalist lease system were seen as prerequisites. Correspondingly, the physiocrats demanded a limitation of state intervention in the economy and in private ownership of the means of production ('laissez-faire policy') (Priddat, 2001; Hüther, 2006; Holub, 2010; Bach, 2018). Elements of physiocratic economic policy were implemented in the final stages of absolutist France, although without ever exerting great influence.

Far more significant influence continues to this day with the second major economic theory of this period, *classical national economy* or *classical economics*, which shaped European economic policy from the late 18th century onwards. The approach first formulated by Smith (2013) in 1776 in his work 'The wealth of nations' represents a transfer of the principles of liberalism to the economy. Influenced to a similar extent by the failures of mercantilism as Quesnay, Smith transfers the ideal of individual freedom, already emphasized at the time in both the US Declaration of Independence and the motto of the French revolution, to economic interaction between economic entities. The central concern was to promote the prosperity of the nation as a whole by creating a self-regulating system (Eagly, 1974; Aspromourgos, 1995; Ziegler, 2008). Smith (2013) assumed that the self-interest of the individual would bring the entire system into a state of equilibrium like an 'invisible hand'. To justify the principle of self-interest, Smith adopted the views of the Protestant ethics of the Calvinists, Puritans, and Quakers, praising the wealth of the individual as moral and outlining a vision of alternating advantage (win-win). As the representatives of classical economics partly contradict each other in certain points, the following basic principles are considered characteristic of classical national economy today: [1] the principle of self-interest, [2] the division of labor, [3] the equilibrium thesis, [4] the theory of labor value (i. e., the interpretation of the value of a good as the labor value contained in it), [5] the exchange of goods on a market (the greater, the better), [6] the formation of market prices through the agreement of suppliers and consumers, [7] the demand for freedom of economic activity (i. e., avoidance of intervention in economic activity). Classical economics further emphasized the concept of capital, already introduced by Quesnay, as that part of wealth that is used to generate profit (Hüther, 2006). Nature was regarded as free in the awareness of its infinity. At the same time, in 1798, Malthus already pointed out the techno-economic challenges of the emerging problem of overpopulation (Malthus, 2007). The end of the first phase of industrialization marked the first emergence of worker councils, which would provide the basis for a further upheaval of political and economic conditions in parts of the world during the second phase of industrialization (Schönhoven, 2017).

2.2.2 Second phase of industrialization (around 1850 to 1950)

While the first phase of industrialization is inseparably linked with enlightenment, liberalism, and mechanization as well as the development of the first coherent economic theories, the second phase of industrialization is characterized by electrification, industrial mass production, the chemical revolution, the diversification of economic theories, and the associated socio-cultural transformations.

Around the middle of the 19th century AD, technical innovations were increasingly developed, which enabled new products as well as a more effective production of goods both in unit and continuous production. These included the assembly line, electrical power engineering, the combustion engine, and chemical synthesis. After first mechanical applications as semi-continuous assembly or conveyor belts in the production of ship's biscuits in England, in slaughterhouses and automobile production in the USA, as well as in the production of baked goods in Germany between 1830 and 1910, Ford developed the first electrically operated, permanent assembly line for the production of the T-model in 1913 (Banham, 2003; Ford, 2008). After 1920, the continuous assembly line was applied in automobile production in Germany (Herlyn, 2011).

Technological progress in unit-load production was accompanied by the emergence of *operational production science or factory management theory*, decisively influenced by the work of Taylor (2001), who transferred the principle of the division of labor of classical national economy to production as an instrument of business organization. According to Bauernhansl and Miehe (2020), the essential approaches of the scientific management formulated here include the separation of executive and planning work (work preparation), the division of labor and the sequencing of processes, the implementation of time studies for process improvement and the determination of target times. Further elements are the introduction of a differential wage system to create incentives, guidelines for the daily workload, the introduction of a function master system, and the efficient employment of low- and unskilled workers for repetitive tasks in the production process. The practical implementation of the internal division of labor was thus based on the categorization of the production process into small, simple activities, which were to be carried out by different employees. Ford is also considered a pioneer in this respect, as it was the first company to implement Taylor's principles, realizing unimagined productivity increases. In addition, Ford recognized that the differences in income prevailing at that time would prevent the success of the unit-intensive production. In addition to reinvesting the profits in research and development as well as in machinery and equipment, Ford increased the income of its employees. In this way, the company compensated its workforce for the reduced work content, developed them as new customers and thus stabilized the sales of its products (see Figure 2.2). This principle was soon copied by others, including the Robert Bosch corporation in Germany (Bauernhansl and Miehe, 2020).

In a similar period, a number of groundbreaking developments emerged in chemistry, which led to the mass production of new basic substances and materials that continues to this day.[6] Since more complex molecules had previously only been found in nature and could only be used as such, the urea synthesis developed by Wöhler in 1828 enabled the artificial production of such a substance for the first time (Fischer, 2012; Wächter, 2018). Although the first organic substances that could be synthetically produced (urea and acetic acid) were initially not available on the market, Runge discovered in 1834 that aniline contained in coal tar could be used to produce dyes by adding other chemicals. Perkin developed the necessary process for this in 1857 (Bensaude-Vincent and Stengers, 1997; Fischer, 2012). An abundance of new dyes could be developed quickly, which in turn led to completely new possibilities for the fashion industry, among others. The possibilities of dye synthesis heralded the period of promoterism ('Gründerzeit') in the now *fossil-based chemical industry*[7], from which, among others, the companies BASF, Bayer, Hoechst, and Agfa emerged (Osteroth, 1985). The industry quickly entered new fields of application, including explosives and synthetic rubber for vehicle tires (Mortimer and Müller, 2015). At its inglorious peak, the conglomerate of companies then known as Interessengemeinschaft Farbenindustrie AG (I. G. Farben), consisting of the above-mentioned firms, employed a large number of forced laborers and established the first privately financed concentration camp (Auschwitz III Monowitz Concentration Camp) (Osteroth, 1985; Bensaude-Vincent and Stengers, 1997; Fischer, 2012; Mortimer and Müller, 2015; Wächter, 2018).

The second phase of industrialization is accompanied by the increasing importance of crude oil as a basic material for the chemical industry and as fuel for mobile systems (Bauernhansl and Miehe, 2020). After Gesner was granted approval for a production process for petroleum from crude oil in the USA in 1855, a worldwide large-scale mining of oil deposits quickly began, heralding the founding of the petroleum industry, which is still the largest branch of industry in the world today (Campbell et al., 2007). The impact on people and the environment, especially in the chemical industry, soon became apparent. Already at the end of the 19th and the beginning of the 20th century,

6 Excursus: the Chemical Abstracts Service (CAS) register, in which every existing chemical substance must be registered, currently contains over 100 million substances (Chemical Abstract Service, 2015). About 75% of these have been added in the last 10 years alone, i. e., over a period of 10 years a new substance was entered every 2.5 minutes. Consequently, the number of substances classified as potentially harmful to humans and the environment is increasing. Since the introduction of the REACH regulation in October 2008, the number of entries in the list of substances of very high concern (SVHC) has increased more than tenfold (European Chemicals Agency, 2015). In the same period of time, the entries in the 'Substitute It Now' (SIN) list, which is an indicator of new substance regulations, have quadrupled (International Chemical Secretariat, 2015).

7 Still 90% of the global chemical industry's raw material demand is covered by fossil sources (in particular crude oil and natural gas), which represents a significantly higher share of demand than the 80% fossil fuels used to generate the global energy supply (Fischer, 2012). The pollution of the fossil-based chemical industry is increased by the various toxic effects during production and use as well as by residues (including after disposal).

institutes and offices for the research and monitoring of the environmental properties of chemical substances were established, among others the 'Preußische Landesanstalt für Wasser-, Boden-, Lufthygiene' (today: Federal Environmental Agency) and the 'Kaiserliches Gesundheitsamt' (today: Federal Health Agency).

The rapid technological development and its resulting social changes also influenced the development of new economic theories. Under the impression of the massive contradictions of early capitalism, which produced enormous wealth as well as extreme poverty and pollution, Marx and Engels formulated 'The Communist Manifesto' in 1848, which later evolved into Marxist economic theory (Marx and Engels, 2017; März, 1985). The basic assumption here is the interpretation of human history as class struggle (among others, slave owners against slaves, feudal lords against servants), which continued in the modern age. Here, however, it was the capitalists and the proletarians (workers) who were hostile to each other as oppressors and oppressed. According to Marx and Engels, the resulting unequal relations of property and production led to enormous social disruptions, culminating in a revolution of the proletarians (Marx, 2014). This forecast became reality with the Russian October revolution in 1917 and the spread of communism or socialism in large parts of the world, peaking in the second half of the 20th century. The ideal of Marxist economic logic was the abolition of private property to dissolve class antagonisms. In comparison to physiocracy and classical national economy, which each propagated a liberal economic order, the idea of equality is emphasized here for the first time. This was to be realized by a co-operative central administrative economy in the hands of the state. Its essential tasks should be the anticipation of demand, the coordination of production and the organization and distribution of goods (März, 1985; Nutzinger and Wolfstetter, 2008). In this context, Marx and Engels demanded a centralized credit system by a monopolistic national bank, a centralized expansion of communal production capacities (factories, estates) and a centralization of the transport system (Baßeler et al., 2006; Devine and Devine, 2010). After the attempt to establish a 'pure form' of a communist state failed with the collapse of the Soviet Union at the beginning of the 1990s, a large number of economic systems and economic theories today show parallels to the philosophy formulated by Marx and Engels. In addition to various people's republics and democracies, such as China or North Korea, which can be characterized as semi-socialist political systems, Marxist economic theory is regarded not least as a basic element of the social market economy prevailing in Germany and Austria (Koenen, 2017).

Around 1850 List and von Nitterdorf also criticized classical national economy by rejecting universally valid economic laws and propagating a reaction to current circumstances (e. g., technological maturity of an economy) in their place (Baßeler et al., 2006; Heertje and Wenzel, 2008; Menger, 2006). Likewise, the principle of self-interest was criticized by attributing human activity primarily to changing cultural factors. Consequently, the *historical school of economics* did not adopt an economic order based on natural law, but rather explained the social science identification of the developmental stages of an economy with the help of empirical research and historical observations as a basic building block of dynamic economic theory (Bofinger, 2011; Mankiw and Taylor, 2012). Since the theory tended to reject judgements per se, the historical school of economics provided less concrete approaches to political action. At the same time, the representatives of the *new historical school* in particular outlined a different role of the state compared to the classical national economy. Instead of economic liberalism, the responsibility of the state was always accorded decisive importance, a view that was later adopted in US influenced institutional economics (Mankiw and Taylor, 2012; Menger, 2006; Ziegler, 2008). In addition, the stimulation of methodological discourse in German-speaking economics is regarded as an essential merit of this theory, culminating in the controversy over value judgements (discussion point: quality and binding nature of normative statements for politics) and methods (discussion point: admissibility of value judgements, i. e., the significance of inductive and deductive research procedures) of the late 19th and early 20th centuries (Lachmann, 2004; Schröder, 2015).

The most prominent economic theory of this period, which from the late 19th century to the present day occupies the status of classical national economy, was *neoclassical theory* or neoclassicism. After the classical period shifted the focus from trade (in mercantilism) to the creation of wealth in production, the neoclassical period revised this by focusing more closely on the emergence of economic players in markets (Felderer and Homburg, 2005; Vilks, 1991). Like the historical school of national economics, the representatives of neoclassical theory also focused on the analysis of individual phenomena and their methodological-mathematical representation. Although today the theory is not regarded as a unified concept (Hampicke, 1992), its main representatives including Jevons (1871), Pareto (2007), Menger (2006), and Walras (2003) promote some common features, which are essentially a continuation of the liberal economic logic. These include the emphasis on markets as an allocation mechanism in the sense of Smith's 'invisible hand', the assumption of a rational behavior of individuals (homo oeconomicus), the thesis of individualistic-utilitarian tendencies of action as well as the rejection of regulatory intervention by the state to the greatest possible extent (Bofinger, 2011; Baßeler et al., 2006; Heertje and Wenzel, 2008). Although the pioneers of this school, who like Marx and Engels experienced the turmoil of early industrial capitalism, were also seeking ways to fight poverty among the population, they came to diametrically opposed conclusions (Mankiw and Taylor, 2012; Menger, 2006; Ziegler, 2008). Starting point was Thünen's marginal principle, which for the first time tried to solve economic questions with the help of differential calculus (Thünen, 1990). Here, cost and benefit[8)] functions, which are supposed to repre-

---

[8] Today's understanding of the term benefit is mainly influenced by the school of marginal utility according to Gossen (2010), which in turn has its origins in the utilitarianism of Bentham (1823) and Mill (2006; 2014). According to the principle of maximum happiness, preference is to be given to that action which promises the greatest possible degree of well-being for all concerned.

sent certain individual preferences, were developed by applying partial differentials. On this basis, Cournot developed the price-demand function in 1838 (Cournot, 1924). Also based on Thünen, Gossen outlined two laws of need satisfaction in 1854, the saturation law and the marginal utility compensation rule (Gossen, 2010). The former stated that an increase in the consumption of a good inevitably leads to a decreasing additional benefit (marginal utility). The second Gossen law justified the tendency of an individual to allocate his or her resources in the desire for the highest possible satisfaction of needs (total utility). The decisive factor here was the acceptance of scarce resources, i. e., goods for which costs are incurred. Pareto (2007) concluded that the poverty of the population could only be effectively counteracted if the gross domestic product grew faster than the population. Similar to classical economics, the assumption of a natural law led Pareto (2007) to develop a highly questionable social Darwinian theory ('cycle of elites') in addition to advocating the superiority of the mathematical and scientific method (Hüther, 2006). A trend in neoclassical theory that is still renowned today, albeit less based on mathematical models of economic relationships, is the *Austrian school* (Baßeler et al., 2006; Ziegler, 2008). A characteristic feature of this theory is the focus on the actions of actual individuals and their subjective needs (methodological subjectivity) (Blumenthal, 2007; Gloria-Palermo, 1999). The majority of the representatives of this school around Menger (2006) and von Mises (2005) support a price system free from state intervention as the main control mechanism of the market. Although from an economic point of view some differences to pure neoclassicism exist (e. g., criticism of market equilibria), the Austrian school emphasizes very similar liberal economic positions (e. g., advocacy of free markets, marginal utility, ideology partly legitimized by natural law). For further understanding, it is not necessary to deepen the discussion at this point. Further literature can be found in Heertje and Wenzel (2008), Ziegler (2008), Bofinger (2011), and Mankiw and Taylor (2012).

Another economic theory developed during this period, that – in contrast to the neoclassical and Austrian school – emphasizes the idea of equity, is the free economy. Influenced by the economic crisis in his home country Argentina, its main representative Gesell (2018) adopted essential characteristics of physiocratic and Marxist economic theory in order to realize a socially compatible market economy. The principle of ownership of land, which was elementary in classical and neoclassical times, was rejected and a transfer of ownership to public ownership was demanded (Werner, 1989; Huber, 1998). However, in order to enable a functioning market economy at the same time, it was intended that private and cooperative entrepreneurs should be able to lease the land and cultivate it freely (open land). While the lease could be provided to the general public (e. g., in the form of health care), a specially developed form of payment, free money, was meant to stimulate the trade of goods. In order to enable this, the free money should, in contrast to common currencies, be subject to a constant decline in value (Wirth, 2003). Once again, this demonstrates how an economic model is based on a 'natural' order, in which the flow of nominal goods, similar to the flow of real goods, is controlled by its own impermanence. A money creation[9], as it is common practice in today's neoclassical economic order, would thus not be possible. Although the free economy has so far only been applied in the form of a few local money experiments, the basic concept has recently received increased attention in the course of modern challenges (environmental problems, modern complementary currencies, loss of importance of human labor etc.) (Märkt, 2005; Kendel, 2007).

From the first half of the 20th century onwards, the thought processes developed in economics, increasingly revolving around the ideals of freedom and equality formulated in the motto of the French revolution, received greater attention in microeconomic matters (Felber, 2010). While the emerging business management science was primarily oriented towards the idea of freedom by considerably improving the methodological basis for corporate management, factory management science provided approaches for optimized production control and performance of employees (e. g., through performance-related remuneration) as well as for the ergonomic design of workplaces (e. g., according to aspects of ergonomics, physical stress, noise and health protection) (Taylor, 2001). Likewise, the early phase of the 20th century marked the development of standards for internal co-determination by employees, whose ideal of constructive cooperation between employer and employee representatives (worker council) is still anchored in the majority of German companies today (Bauernhansl and Miehe, 2020).

2.2.3 Third phase of industrialization (around 1950 until today)

The third phase of industrialization, ranging from post-war Germany to the present day, was initially characterized by reconstruction and systemic competition between capitalism and communism. From the late half of the 20th century onwards, the increasing worldwide integration of economic structures (globalization) and the emergence of the environmental crisis followed. With the collapse of the Soviet Union at

[9] While in the early days of money-for-goods trading, the global quantity of money was determined by the existing stock of gold and silver, the introduction of paper money and online trading led to the possibility of a potentially infinite multiplication. If a bank grants a loan to private individuals or companies today, it does not finance it via deposits from its savers or its own gold reserves, but borrows it from the central bank itself. This so-called giral money does not exist in a physical form, only as a digital figure, i. e., as a number on the borrower's account. Since the borrower has to pay interest on the borrowed capital in any case, which is usually higher than the risk reserve that the issuing bank has to deposit with the central bank, the issuing bank generates real money (money creation) from fictitious money. On the basis of the division of labor and globally interconnected economic relations, money once lent theoretically enables an infinitely high creation of money (Binswanger (1991; 2016). Considers this decoupling of globally available money and its equivalent value to be the cause of the current environmental problem. In connection with the interest on lending transactions, debt is forced, which results in a growth spiral. In this context, Binswanger (1991; 2016) pleads for a moderation of the exploitation of nature through institutional reforms in the monetary system.

the beginning of the 1990s, concepts of neoclassical economic logic gained even greater importance in the short term, causing a stagnation of consistent economic theory formation (Priddat, 2005). From a technological point of view, the third phase of industrialization was primarily characterized by the rise of ICT, allowing an increased robot-supported automation. In addition to digitalization, which is expected to lead to a complete autonomous networking of all means of production, the early phase of the 21st century also marks the emergence of industrial biotechnology (Bauernhansl and Miehe, 2020).

Under the impression of the nationalist turmoil in the first half of the 20th century and the emerging systemic competition between communism and capitalism, the first years of the third phase of industrialization were dominated by economic theory, which postulated both the market economy and a comparatively strong state. The most important schools of this period are Keynesianism and ordoliberalism. Although Keynes (2002) published his ideas by 1936, it was not until after World War 2 and until the mid-1970s that his doctrine became the dominant economic logic in all western countries. The concept postulated the active control of demand, if necessary with the help of government spending and adequate monetary policy (Hüther, 2006). In this way, it should be possible to mitigate crises and achieve low unemployment. Although Keynesianism is hardly present in today's economy, the labor market continues to determine economic policy decisions as a central measure alongside gross domestic product. The approach of ordoliberalism, which was published by Eucken in 1937 (Eucken, 2013), distinguishes between the political-economic framework (ordo), which is determined by the state, and direct economic activity, which would function without state intervention (van Suntum, 2005). This was intended to effectively unite the social idea and the principle of performance (Kruber, 2002). On the basis of ordoliberalism, Müller-Armack (1976) developed the concept of the social market economy, which still characterizes the economic order of Germany and Austria today. Compared to ordoliberalism, however, the social market economy provides for considerably more extensive state intervention in the economy and economic structure (Ptak, 2013). The last independent economic theory is monetarism, primarily influenced by Friedman and Schwartz (2015). In contrast to Keynes' demand-oriented theory, it postulates a supply-oriented economic policy whose central control variable is not interest rates but money supply (Hüther, 2006; Bofinger, 2011; Mankiw and Taylor, 2012). Money supply should be kept at a constant level as far as possible, since a too large money supply in the market would lead to inflation and a too small to deflation. Although the concept gained more attention under the impact of the global oil crises in 1973 and 1978, resulting in massive price fluctuations, the experience with monetarist economic policy was disappointing (Bofinger, 2011). From the second half of the 20th century, a growing number of approaches addressed the issue of distributive equality. Particularly noteworthy are the theory of justice by Rawls (1999), which today forms a moral-philosophical foundation of environmental economics, and the book 'Poverty and Famines' by Sen (1990). The latter clearly rejected the thesis that poverty and famines are primarily caused by scarce resources. Sen's prominent emphasis on distributive justice provided a basis for the sustainability concept developed in the late 1980s and paved the way for the concept of welfare economics (Hüther, 2006).

After Germany's productive capacity had already reached the pre-war level at the beginning of the 'economic miracle' (around 1950) and real wages had even exceeded it, further growth-driven expansion continued into the early 1960s on the basis of the existing technological infrastructure (Zündorf, 2006). With the end of the 'boom', manufacturing companies increasingly focused on technological renewal. After the first companies began using computers for data processing in the 1960s, their large-scale application began in the 1970s, driven primarily by the invention of the microchip and its continuous performance improvement. From then on, companies used computers to support various human activities, including design, work scheduling, manufacturing, and production planning. This led to the collective term CAx (computer aided x) (Paul, 1991; Waldner, 1992). With the further development of computer technology, industrial robot technology emerged (Maier, 2016; Kreuzer et al., 1994; Weber, 2007b), which was first transferred to practical applications in Japan and Germany at the end of the 1960s. With the help of these programmable machines for handling, assembling or processing workpieces, many activities previously performed by humans were replaced. With the first wave of automation, lasting until the early 2000s, companies were able to produce both more cost efficient and with more variants (Maier, 2016; Kreuzer et al., 1994; Weber, 2007b).

In the 1980s, however, western companies faced increasing pressure from Japanese car manufacturers (especially Toyota) in terms of price, quality and flexibility.[10] Based on the principles of the Toyota production system (TPS) (Ohno, 2005), Womack et al. (1990; 2003; 2015) presented the concept of *lean management* (LM). By implementing the postulated slimming of production processes with the help of autonomous automation (jp. *Jidōka*) and the principle of continuous improvement (jp. *Kaizen*), factory management (scientific management), which had been shaped by Taylor (2001) until then, was raised to a new level. Today, LM is considered a standard in manufacturing companies. Many large and medium-sized companies (including VW, Audi, Daimler, BMW, Siemens, Festo, TRUMPF, ZF) have established an in-house production system (Röhm et al., 2020).

While LM was aimed at organizational improvement, the parallel rapid development of ICT (e.g., machine learning and artificial intelligence) and the increasing possibili-

[10] While western car manufacturers produced on inventory, i.e., accumulated stocks of semi-finished and finished goods in order to be able to react quickly to a sudden increase in sales, Toyota aligned its system with the customer's wishes. While European and U.S. companies faced problems when the customer did not want exactly what had already been produced, Toyota was able to react flexibly to any changes in demand and requirements by reducing the lead time of the produced goods to a minimum. The central lever of this approach is the avoidance of any activities for which the customer is not willing to pay.

ties of networking (e. g., Internet of Things) (Weiser, 1991; Ashton, 2009) resulted in a further wave of technological automation since the early 2010s, widely referred to as Industry 4.0 (I4.0) (Kagermann et al., 2011; Bauernhansl et al., 2014).[11] The technologies, some of which are still in the development stage, are intended both to support people in physically stressful activities (e. g., through exoskeletons or wearables) and to autonomize the entire production system with the aid of suitable sensor and actuator technology, artificial intelligence, the Internet of Things, big data analytics and cloud technology (Bauernhansl et al., 2014; Lee et al., 2015; Qin et al., 2016; Seif et al., 2019; Stock and Seliger, 2016; Uhlemann et al., 2017a; Vaidya et al., 2018; Wollschlaeger et al., 2017; Zhong et al., 2017). The baseline notion of Industry 4.0 is the perception of production as a cyber-physical system, i. e., a deep interlinking between the physical and digital sphere. As a result, companies are increasingly adopting digital platforms as new business models (BM) in addition to their core technological business.[12] Likewise, consumers' participation is gradually evolving from a mere consumer to a prosumer, i. e., a consumer who is an active part of the production process. While ICT is already being used in every sector, biotechnology is increasingly developing into an economically viable technology.[13] In addition to recent scientific achievements in areas such as biocomputing, enzyme technology and biosensor technology, the CRISPR/Cas9 method (see Section 5.3.7), first published in 2012, deserves special attention (Jinek et al., 2012). Compared to other methods of genome editing, the ease of use and low costs of this method quickly led to its widespread use in research and industry (Mahfouz et al., 2014; Pennisi, 2013).

After the predominant economic theory and the resulting actions (until the end of the 1980s in the 'western' world, then worldwide) were shaped by the notion of freedom, a partial global rethinking took place from the 1970s onwards. The limits of the resources available on earth, which were first pointed out at the beginning of the decade by Meadows et al. (1972), led to a critical questioning of the traditional economic system. This triggered a controversial discussion in business, research and politics, which culminated in the concept of sustainability in 1987 (World Commission on Environment and Development, 1987). Compared to traditional economic models, this approach represented a fundamental change. Instead of the explicit emphasis on growth and welfare, a notion of an equity-oriented primarily regulatory pattern of

---

[11] Section 5.4.1 provides a detailed summary of the Industry 4.0 concept.

[12] The increasing focus on digital platforms as a business model is often referred to as the platform economy.

[13] Excursus: one of the greatest achievements in molecular biology in recent years was the decoding of the human genome in the context of the US Human Genome Project. The project started in 1990, and since 2003 the human genome has been officially considered to have been decoded. Since then, the costs for decoding have dropped dramatically, driven by the increased computing capacity. While it was possible to sequence DNA base pairs for US$383.35 in 2003, by 2015 this figure had risen to 33.32 million, an increase of 87,000 times in only 12 years (Roser and Ritchie, 2020). If the decoding of a human-like being with about 3 billion base pairs had cost more than US$7.8 million in 2003, it would have cost US$90 only a little more than a decade later.

thought was emphasized. At the same time, the linear, fossil-based approach to value creation (products are produced on the basis of fossil raw materials and disposed of after use without hesitation) was identified as the main problem (Miehe, 2018).

Table 2.2 summarizes the socio-economic development of the industrial era on the basis of the three phases of industrialization outlined above.

Table 2.2 Socio-economic development of the industrial era based on various characteristics [own summary][14)]

	First phase	Second phase	Third phase
Period	About 1750 to 1850 AD	About 1850 to 1950 AD	About 1950 AD until today
Characteristics of the society or social upheaval	Sedentary society, increasingly characterized by industrial production	Increasing industrialization, increase in productivity, social disruptions	Economic miracle, system competition, urbanization, emerging environmental challenges
Economic form	Early fossil-based capitalism ('linear' value creation)	Fossil-based capitalism and communism (partly nationalism, characterized by protectionism)	Fossil-based capitalism (partially communism)
Technological basis	Mechanical loom, steam engine, railway, water power	Electrical power engineering, assembly line, chemical synthesis, scientific management	Information and communication technology, robots, lean management, Internet (of Things)
Prevailing economic theory	Classical economic theory (further: physiocracy)	Neoclassical theory (further: Austrian school Marxist economic theory, historical school, free economy)	Neoclassical theory, Keynesianism, social market economy (further: ordoliberalism, monetarism)
Global population	About 800 million to 1.3 billion	About 1.3 to 2.5 billion	About 2.5 to 7.5 billion

[14] Population data according to Roser et al. (2020)

2.3 Chapter review and conclusions

Today's production methods and thought patterns are the result of complex developments in the history of mankind. Decisive stages for today's socio-economic order were: the formation of intersubjective realities (cognitive revolution), the development of writing systems, the emergence of systematic modes of production in agriculture (Neolithic revolution), the invention of modern printing, the emergence of the guiding ideas of humanism and rationalism (scientific revolution), the systematization of industrial production processes (industrial revolution), the methodological theory evolution in economics and politics, as well as the emergence of ICT and the Internet. The diffusion of the technological basis represents a history of continuous socio-technical paradigm shifts, known today as Kondratieff cycles (Grinin et al., 2017). The starting point is always a technical innovation that leads to an improvement of the existing scope of design and thus creates a demand. This is followed by massive investments, which in turn favor the market penetration of a certain technology. As a result of the general establishment of the technology, a decline in investment occurs, culminating in an overall economic downturn. This continues until a new technical innovation is developed. Today, the Kondratieff cycles are often used to illustrate the continuous development of the technological basis of economies. To date, consensus exists on five cycles, while Grinin et al. (2017) predict a sixth cycle, which is determined by the diffusion of self-regulating systems. The technical basis for this is provided by the so-called *MANBRIC* technologies (*medical, additive, nano-, bio-, robo-, info-, and cogno-technologies*). The sixth cycle, which will begin in the period between 2020 and 2040, is primarily associated with the need for holistic health and slowing down the human ageing process (Grinin et al., 2017). Figure 2.4 illustrates the diffusion of the technical basis in the form of Kondratieff cycles based on Grinin et al. (2017).

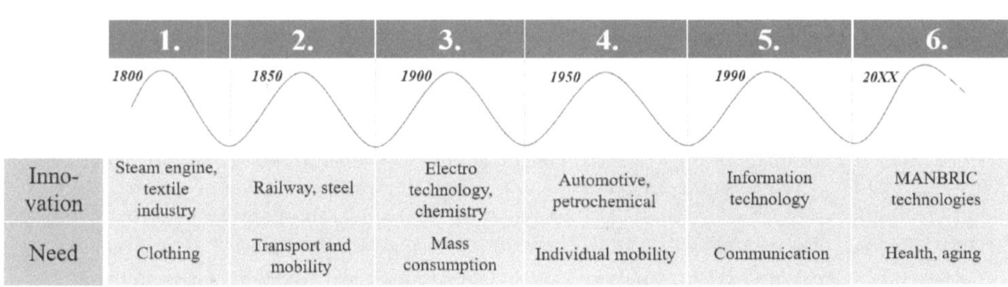

Figure 2.4 Diffusion of technological basis in form of six Kondratieff cycles according to Grinin et al. (2017)

The result of continuous technological renewal has been enormous progress in the fields of welfare, health and the ability to cooperate, which is manifested in large parts of the world by a significant increase in life expectancy and basic income as well as long phases of peaceful coexistence.[15] In contrast, sensitive effects of humanity on the natural ecosystem occurred. On one hand, the history of mankind shows that the idea of a 'life in harmony with nature' never existed. The ecosystem that mankind found at the time of the agricultural revolution was already heavily influenced by human intervention. On the other hand, industrialization initiated a process that permanently shifted the limits of human influence on the ecosystem. The rise in prosperity resulting from the increase in productivity led to an almost exponential worldwide population growth. After the dominant economic theories propagated unrestricted economic growth and, believing in its inexhaustibility, regarded nature as a free good, prices, e. g., for fossil fuels, never reflected their real value (Jarass et al., 2009). The transfer of these macroeconomic thought patterns to everyday microeconomic actions in companies and households resulted in consumption exceeding the level of natural regeneration capacity. This trend is accompanied by considerable resource depletion and serious anthropogenic environmental impacts.[16] Figure 2.5 illustrates the qualitative course of essential socio-economic and ecological indicators since the first industrial revolution according to Steffen et al. (2007).

Consequently, geologists describe industrialization today as the transition from the Holocene to the Anthropocene, the geological epoch in which mankind became the most important factor affecting the earth's geosystem (Zalasiewicz et al., 2008; Crutzen, 2011; Steffen et al., 2007).[17] If the impacts of individual distributed groups on nature in early economic forms were easily absorbed, today's world population and its globally networked societies are stressing it beyond its ability to regenerate.[18] A critical examination of the existing theoretical framework is necessary to overcome this. In doing so, it is important not to commit the mistake of evaluating a single theory as unrestrictedly bad or good, since it never singularly influences the thinking

[15] Another consequence, which will not be elaborated here, is the emergence of a deeply rooted individual and societal belief in progress and technology. Although this leads to the preservation of essential system components by, e. g., encouraging the granting of loans, it restricts the scope for a reorientation in the social-political debate.

[16] To name only a few examples: since 1970 more resources per year have been consumed than the earth can provide in the form of its biocapacity (World Wide Fund For Nature, 2014). Driven by the combustion of fossil fuels, the greenhouse gas concentration in the atmosphere is rapidly approaching the critical level of 450 ppm (Intergovernmental Panel on Climate Change, 2014a). In Germany alone, a three-digit billion euro amount is outsourced to uninvolved third parties every year (Lorenz, 2000). A study commissioned by the European Commission (2013) estimates costs of around one trillion Euro per year (€23 billion directly and up to €940 billion indirectly), solely for damage caused by air pollution.

[17] This becomes evident, for example, through global meat consumption. Today, humans and their domesticated farm animals account for 97% of the body weight of all terrestrial vertebrates. The mass of all wild vertebrates is merely 3% (World Society for the Protection of Animals, 2008). Around 30% of the world's available farmland is used solely to produce food for farm animals and therefore cannot be used to satisfy basic human needs (Steinfeld et al., 2006). Meat consumption is also responsible for 14 to 18% of total anthropogenic greenhouse gas emissions (Gerber et al., 2013). Meanwhile, around 30% of food in German households is disposed of without being used.

[18] In comparison, today's world population exceeds that of the phase of the agricultural revolution by a factor of about ten. In comparison to the time before settlement, the world population grew by a factor of 2000.

and actions of actors in politics, economics and science. Today's economic forms and thought patterns in macro- and microeconomics are much more likely to represent the result of human history, which is available and individually adapted by the acting protagonists in the form of different memeplexes. The social market economy, for example, clearly represents a construct of previously formulated theories (classical, neoclassical, Marxist economic theory, ordoliberalism). A production manager rarely seeks exclusively a technical solution for optimization problems, but rather relies equally on elements of scientific management.

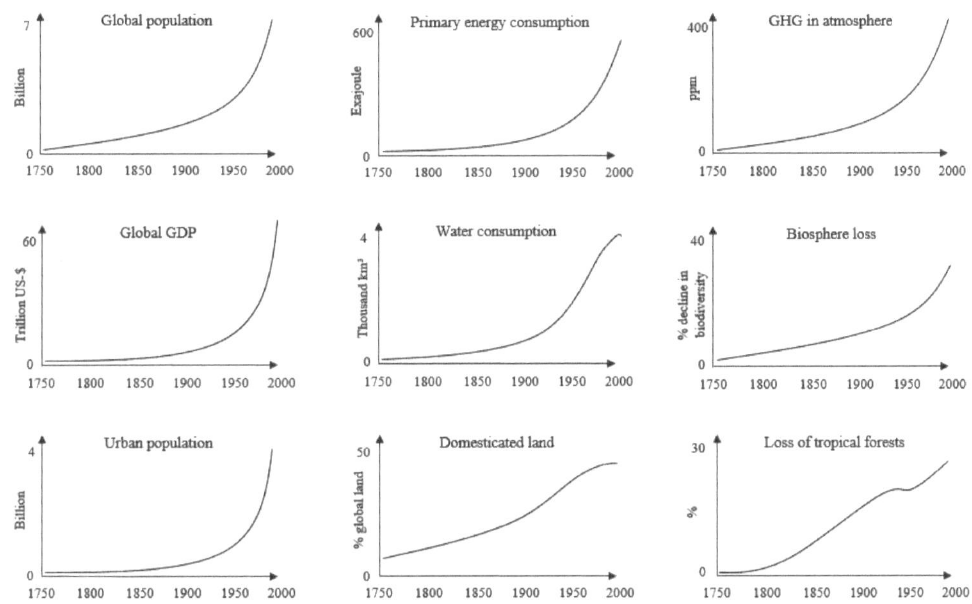

Figure 2.5 Qualitative development of essential socio-economic and ecological indicators since the first industrial revolution according to Steffen et al. (2007)

Equally erroneous as the defamation of individual theories, is their use for the singular legitimation of today's economic practices. The majority of economic theories that still dominate today were formulated at a time when the socio-ecological consequences of their actions were not apparent to anyone. Correspondingly, their advocates assumed completely different preconditions than those that apply today.[19] Weizsäcker and Wijkman (2017) describe this discrepancy of basic conditions as an

[19] The basis of Smith's expression of the 'invisible hand' was the balance between market and legislation, which was clearly established in the 18th century. At that time, it was simply not conceivable that a market could reach a scale or produce effects that exceeded the legislative, executive and judicial power of a single state. Today this is the case in two particular areas. First, the financial sector has reached a size that no state in the world can regulate individually. This becomes clear from the global financial flows, of which only 2% are used to pay for real goods, 98% are speculative (Weizsäcker and Wijkman, 2017). On the other hand, environmental protection requires a globally coordinated approach, otherwise it remains ineffective in many cases.

empty and full world. In an empty world, people could do business without hesitation, but in a full world this is no longer possible due to the limited resources. Consequently, Jonas (1993) advocated a reorientation of the philosophical perspective already in the 1990s. While the approaches developed up to this point primarily sought solutions for the here and now (in this context particularly the compatibility of the ideals of freedom and equality), Jonas (1993) emphasizes the responsibility for individual action. From an ethical perspective, an action can only be considered moral, if it is in harmony with the permanence of human life on earth. Although the hypothesis has a purely philosophical background, its transfer to economic and production science is inevitable. Figure 2.6 illustrates the difference between an 'empty' and a 'full' world according to Weizsäcker and Wijkman (2017).

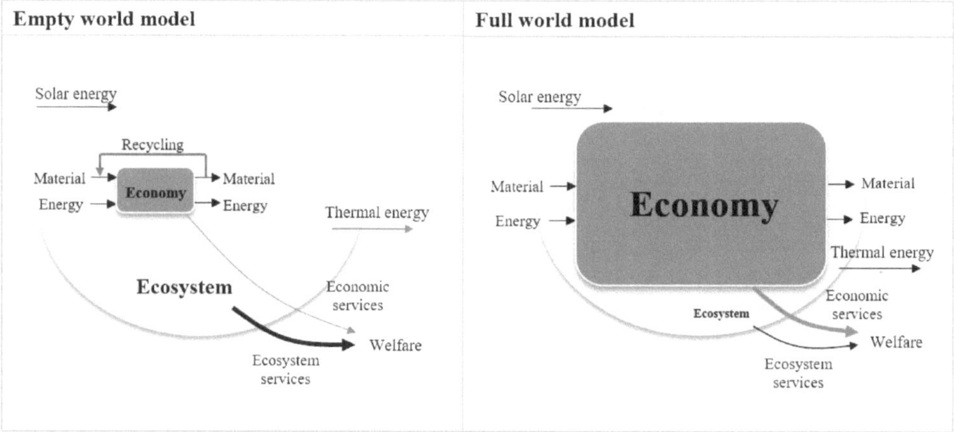

Figure 2.6 Difference between an 'empty' and a 'full' world according to Weizsäcker and Wijkman (2017)

In the context of a reorientation of science, Weizsäcker and Wijkman (2017) also stress the need for a modern enlightenment in order to eliminate misconceptions in dealing with existing theories. In addition, they advocate an adaptation of the 'eastern' philosophy of equilibrium (yin and yang) in 'western' theories. In the sense of Jonas' principle of responsibility, the role of science as a whole must also be questioned. For example, it should be clarified whether application-oriented scientific disciplines may in future continue to rely solely on the economic proof of functionality. In this respect, it should be noted that free scientific knowledge does not exist (Harari, 2014). Rather, research shows a symbiotic relationship with the prevailing ideology. While science legitimates ideology, ideology justifies the costs of science. Consequently, a consideration or adaptation of the dominant ideology is necessary to achieve a Kuhnian paradigm shift towards responsible scientific practice (Kuhn, 1996b). In this context, the 'networked thinking' of biocybernetics, outlined by Vester

(2011), is decisive. In contrast to the prevailing 'linear thinking' of a constructivist-reductionistic world view, it aims to consider all possible interconnections of a decision with its surroundings. If the decision-making space is perceived as a complex system, dogmatic thought patterns of existing theories lose their validity and are replaced by contemporary approaches.

Lessons learned

- The history of production distinguishes two epochs (pre-industrial, industrial), each of which being characterized by strong social and cultural upheavals that have led to today's technologies, life styles, societies, and economic practices.
- Human actions have always had an impact on the environment. Due to accumulation in the course of massive population growth since the first industrial revolution humanity is currently stressing the earth's biocapacity largely beyond its capability to regenerate.
- Compared to earlier centuries, in which far fewer people inhabited an area of land (empty world), the dramatically increased population and its resulting global consumption of goods has resulted in a full world.
- Generally, the formation and change of societies, ideals, scientific theories, and technology is always closely interrelated.
- Western societies in particular have incorporated two ideals into their constitutions, which are partly contradictory: freedom and equality. Recently another ideal is emerging: responsibility.
- Existing theories, especially in economics, have to be examined in the context of their emergence, i. e., if a theory has been developed for an empty world (e. g., putting the idea of freedom above the idea of equality), its suitability for a full world cannot be assumed per se.
- Technology development until today has been running in (Kondratieff) cycles. Next in line are so-called MANBRIC technologies (medical, additive, nano-, bio-, robo-, info-, and cogno-technologies).
- In order to allow future generations to encounter possibilities similar to our own, a radical rethinking in theory and practice is necessary. For this purpose, the sustainability concept was formulated.

Follow-up questions

- What is the concept of sustainability?
- What are basic strategic options of a reorientation?
- How can it be implemented at macro and micro level?
- What are the implications for theory formulation?
- How does a sustainable economic system differ from the current one?
- What are the challenges at macro and micro level?
- What is the current status of the implementation of the sustainability concept?

3 Sustainability in the context of industrial production

After the previous section demonstrated that today's production methods and thought patterns are hardly suitable for ensuring long-term permanence of human life on earth, this section discusses the current status and perspectives of a reorientation of basic theoretical concepts. Therefore, the notion of sustainability is fundamental. The following sections thus serve as an introduction to the concept in order to examine its interpretations, obstacles, and perspectives for implementation in various scientific disciplines and society.

3.1 Fundamentals of the sustainability concept

Thinking and acting in terms of sustainability has posed enormous challenges to science and practice for about half a century. The need for a fundamental change in human behavior is often cited as a cause. However, this is precisely a question which the concept is unable to answer adequately due to the lack of social consensus to date.

The term sustainability was introduced in its current literal sense in German-language literature at the beginning of the 18th century (Carlowitz, 1713), yet the roots of the term can already be found in early antiquity (Grober, 2010). However, an ongoing discourse on sustainability only began in the early 1970s, largely initiated by Meadows et al. (1972). In their 'limits to growth' report, the finiteness of globally available resources became accessible to a broad public for the first time. On this basis, Meadows et al. (1972) sketched an end to the traditional economic system. This led to a controversial discussion in business, research and politics, culminating in the concept of sustainability in 1987 (World Commission on Environment and Development, 1987). According to the definition of the Brundtland commission, sustainable development is defined as a process that enables the present generation to meet its needs without depriving future generations of this possibility. Thereby, the concept of needs is understood as basic human needs. Accordingly, the guiding principle of sustainability is essentially built on two basic principles, intergenerational (future generations) and intragenerational (global) justice. Compared to concepts of liberal economic theories, sustainability thus primarily emphasizes the ideal of equality.

Although the concept represents a coherent implementation of the principle of responsibility postulated by Jonas (1993), the vague definition of the guiding principle by the Brundtland commission (World Commission on Environment and Development, 1987) has led to an ongoing discourse on an appropriate regulatory design (Kopfmüller, 2007). Daly (1990) derives the need for a rethinking in two key areas: [1] ensuring a fair distribution of goods and [2] limiting the utilization of the natural environment to an optimal quantity. Particularly in the context of optimal use of nature, two, partly contradictory, interpretations have been developed: strong and weak sustainability (Daly, 1990; Norton and Toman, 1997; Neumayer, 2003). The basis of dissent is the legitimation of the exchange of natural, human and physical capital, for which the Brundtland report does not mention specific regulations (Pearce, 1988; Hartwick, 1977; Costanza and Daly, 1992). Strong sustainability is characterized by the demand not only to preserve, but to revitalize the remaining fund of natural capital (Ott and Döring, 2008; Döring and Ott, 2001). The substitution of natural, human and physical capital is strongly limited in this interpretation. Another characteristic of strong sustainability is the complementarity between natural and physical capital (Daly, 1990). The economy is perceived as a subsystem of the biosphere. The aim must be to ensure a reasonable degree of utilization of the ecological system by the economic system. To preserve the economic system, it is necessary to invest in the limiting production factor: nature. This interpretation is also known as the priority model, as its primary goal is environmental protection. In contrast, the second interpretation, weak sustainability, assumes a general substitutability of capital, which is reflected in the triple-bottom-line model (economy, ecology, society) (Kopfmüller, 2007). The term sustainability is interpreted as the maintenance of the sum of all capital. With the help of integrative optimization, an ideal value between the three pillars is supposed to be realized. In Germany, this interpretation was primarily shaped by the Enquete commission of the German Bundestag (1998), Hüther et al. (1999, p. 74) and the Hans-Böckler-Foundation (2000). Although the interpretation has become particularly widespread in company practice, it is often criticized from a scientific perspective. On one hand, it allows different actors to assert their positions and interests at will (Brand and Jochum, 2000, pp. 79–80). On the other hand, the simultaneous consideration of the three pillars leads to a complexity that can hardly be controlled (Kopfmüller, 2007, p. 17). In this context, both a weak and a strong characteristic can be considered generally critical (Meyer-Abich, 2001; Kopfmüller, 2007). Weak sustainability does not suffice to preserve nature, whereas strong sustainability is too restrictive to be practicable. Figure 3.1 outlines the difference between priority and triple-bottom-line model according to Brand and Jochum (2000) and Behlau (2012).

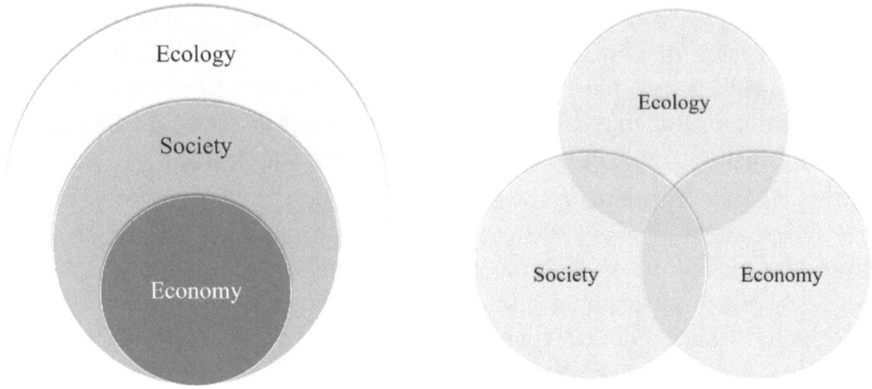

Figure 3.1 Priority and triple-bottom-line model according to Brand and Jochum (2000) and Behlau (2012)

Instead of capital substitution, other interpretations focus on inter- and intragenerative justice. Jörissen et al. (1999, p. 48) express the goal of sustainability as the preservation of earth's productive capacity over time. Kopfmüller et al. (2001) propose the determination of critical limits for each type of natural capital, which must not be exceeded. This concept has recently been adopted by Hauschild et al. (2017; 2020) under the term 'absolute sustainability'. The authors outline an alignment of the economy with the planetary boundaries (see Section 3.3.2).

3.2 Basic strategies for a sustainable economy

In order to create a sustainable economy, generally, three not always clearly distinguishable strategies are differed: efficiency, consistency, and sufficiency (Huber, 1995). The former has its origins in the economic principle and is traditionally understood as the ratio of benefit and costs. Consistency or effectiveness, on the other hand, describes the ability to achieve an intended effect in terms of causality. While the term efficiency describes how an action should be carried out correctly, effectiveness questions the correctness of an action in itself (Drucker, 1963). In contrast, representatives of the sufficiency strategy plead for a rationalization of consumption and production.

3.2.1 Efficiency

The concept of efficiency plays a particularly important role in the sustainability debate and exists in combination with various dimensions, including energy, material, resource, eco, socio- and ultra-efficiency (Schaltegger and Sturm, 1999; 2000; Rogall,

2008; Braungart and McDonough, 2014; Lentes et al., 2017; Miehe et al., 2017b). In comparison to the traditional business sense of efficiency, the system boundary of 'costs' is extended by the dimension to be examined. However, a clear distinction between the individual dimensions (e. g., eco- and resource efficiency), which could serve as justification for the divergent terms, does not exist. While eco- and resource efficiency seek to determine the most holistic possible use of nature in relation to an intended benefit, energy and material efficiency cover sub-areas (Patterson, 1996; Erdmann and Zweifel, 2008). Similarly, no clear differentiation of efficiency and productivity in the context of sustainability has been presented to date. While Rogall (2008) views them as synonyms, both Schütz and Bringezu (2008) and Coelli et al. (2005) distinguish them from each other. According to Schütz and Bringezu (2008), the term productivity is primarily determined by the choice of an economic reference value (e. g., value added). In contrast, efficiency is characterized by a physical reference (functional unit). Based on the principle of rationality, the concept of efficiency is a manifestation of the traditional way of economic thinking. The difficulty of transferring this traditional way of thinking to the concept of sustainability is illustrated by various deficits. Several authors show that increasing the efficiency of any technical system usually induces a more feudal consumption of the same or other goods (Greening et al., 2000; Berkhout et al., 2000; Sorrell and Dimitropoulos, 2008). In literature, this phenomenon, first described in 1865 by Jevons (1906), is called the rebound effect. Another effect that is associated with efficiency in the context of sustainability is the failure to consider the entire life cycle and/or the interrelationships with other systems. Decisions based on the assumption of a too narrow system boundary or temporal frame of reference usually leads to a shift of the problem into precisely those areas that are not considered (phases of the life cycle, regions, time periods, etc.). In literature, this phenomenon is referred to as the leakage or spillover effect (Nertinger, 2015). Last but not least, efficiency is a relative measure. An increase does not necessarily lead to an improvement of absolute environmental impact (Müller-Christ, 2010). In the case of non-renewable resources, the strategy can only achieve a prolongation of availability, but not a permanent access. In scientific literature, this phenomenon is referred to as the quantity effect (Nertinger, 2015). Thus, a solution to the environmental problem through a microeconomic increase in efficiency is highly controversial.

3.2.2 Effectiveness

The concept of effectiveness also originates from traditional business thinking. In the context of sustainability, the term consistency is commonly used, referring to an adaptation of human actions to nature, which from a technical perspective is also the aim of biomimetics (Oertel and Grunwald, 2006). In the case of production, the consistency strategy envisages an adaptation of the industrial to natural material and energy cycles (Freimann and Walther, 2012). Accordingly, a consistent action is characterized by not exceeding the environment's natural capacity for regeneration and absorption (Carnau, 2011). Like the efficiency concept, however, this strategy cannot be accepted without contradiction as a path to a sustainable economy. On one hand, no sound criteria exist in the context of sustainability that could a priori attest an action's effectiveness in the global context. An assessment of environmental compatibility by decision-makers and/or stakeholders is always subjective and therefore difficult to reproduce (Schaltegger et al., 2003). Taghizadegan (2010) further criticizes that the mere sketching of such a possibility fuels existing technological optimism and ultimately acts as a driver of unreflected consumption. As such, the consistency strategy leads to a conflict of objectives with the sufficiency described below. In addition, doubts about the general feasibility of the concept exist (Basel, 2010), which are supported by the economic theory of entropy (Georgescu-Roegen, 1987). According to Georgescu-Roegen (1987), the production of a good is never to be seen as the creation of something completely new, but merely as a metamorphosis of something that already exists. If nature is recognized as the producer of primary goods, the annual amount of extraction can only be within the range of that which is regenerated by solar radiation. The ecological environment is therefore not a free but a scarce good. A further effect of transferring the thermodynamic principles to economic issues is the rejection of a 100% recycling quota, which would result in an infinitely high energy requirement.

3.2.3 Sufficiency

While the concepts of efficiency and effectiveness are to be seen as adapted from traditional economic thinking, the sufficiency strategy is a pure expression of the concept of sustainability. The approach introduced by Sachs (1993) identifies the rationalization of resources or needs as the key lever for sustainable development. The target state of the sufficiency strategy is a return to the satisfaction of real needs. Compared to the strategies discussed above, sufficiency is a purely behavioral approach (Schaltegger et al., 2003). Carnau (2011) outlines the concept as a central element of an ecological redesign of previous thought patterns. In literature, four measures are commonly mentioned as concrete measures for implementation: [1] reducing the quantity of products, [2] increasing self-sufficiency, [3] regionalizing supply chains and [4] decelerating consumption (Sachs, 1993; Schneidewind and Palzkill, 2011; Baumast, 2013). Representatives of the sufficiency strategy are correspondingly critical of the demand for unrestricted growth of national economies. Instead, the use of resources and the environmental impact should be minimized by means of the lowest possible consumption. In the context of the dependency of economies, industries and companies on raw material imports, lower consumption or increased self-sufficiency represents a significant lever for minimizing risks for all stakeholders. While this strategic option was long considered unsuitable for industrial production, the Covid crisis has recently prompted a reappraisal. Although from a purely ecological perspective the sufficiency concept provides the most logical arguments for sustainable development, it is not free of deficits. First, the approach contradicts conventional patterns of thinking which are based on growth. Compared to the strategies outlined above, the sufficiency concept thus represents a fundamental modification of the socio-economic framework conditions (Gandenberger, 2011), whose real implementation appears to face almost insurmountable hurdles. Furthermore, the effects of rationalization and deceleration on countries in the Global South, which can have adverse effects on global distributive justice, need to be examined (Nertinger, 2015). Renunciation can only take place in regions where consumption exceeds the satisfaction of real needs (Carnau, 2011).

A continuation of the traditional economic thinking patterns of efficiency and consistency reveals various deficits. Yet, the strategies must be maintained. Both the efficiency and consistency paradigms certainly have the potential for ecological modernization of existing technical systems and serve as drivers for innovations in the context of sustainable development (Huber, 2011). At the same time, the idea of sufficiency must not be ignored. Although it represents a comparatively radical change in the traditional patterns of economic thinking, it is by far the most valid argument from an ecological perspective.

3.3 Implications for theory formulation and operationalization

The implementation of the concept of sustainability undoubtedly requires a reorientation of theory and practice. As explained above, especially this is severely compromised by the vague formulation of the guiding principle. While the concept comparatively clearly emphasizes an anthropocentric perspective[1)] (Döring and Ott, 2001), a number of elementary questions that would require a social consensus remain unanswered. The non-existence of a socio-political agreement results in far-reaching consequences, which prevent both an appropriate theory formulation and an adequate technology development. The partly related, unresolved questions are:

[1] The anthropocentric perspective is a term from sustainability research that describes the interpretation of the natural environment as a production factor. Nature has no demands out of itself. This represents a common view throughout the history of mankind in various religions, in humanism, as well as in liberal economic theories. In the context of sustainability, the protection of the environment thus only serves to ensure the needs of future generations of humankind. Although the pros and cons of the anthropocentric perspective in the context of sustainability are still controversially discussed today, they form the basis of the Brundtland report.

1. Management of the natural environment: assuming an anthropocentric perspective, it remains to be clarified whether the ecosystem as such or only its productive capacity should be preserved (Daly, 1990; Jörissen et al., 1999). While the former would inevitably imply a limitation to an optimal quantity of use, the latter would also be plausible through technological renewal.
2. Optimal level of environmental utilization: the changed basic conditions of the 'full world' essentially results in one insight: Nature is not a free good, but a scarce one (Weizsäcker and Wijkman, 2017). However, this only becomes relevant in economics when it is translated into the economic-political decision-making. Such a translation can only be achieved by systematically taking external effects into account. Suitable instruments for this are the internalization of external costs as well as adequate regulations, standards, and incentive systems.
3. Classification of the environment: according to the anthropocentric principle, nature is not assigned a right out of itself. At the same time, nature cannot express any wishes or sensitivities. The basis for the preservation of the natural environment as such, as well as for a control via external effects and/or costs, can therefore only be achieved in the form of a classification.
4. Inter- and intragenerational equity: equity between successive generations in the sense of constant natural capital could be realized on the basis of the above-mentioned classification of the environment, e.g., through intergenerational leases that would be renegotiated at regular intervals. However, the implementation of this within one generation would require a more far-reaching rethinking. This applies in particular to the questioning of the liberal economic logic, which would largely have to be replaced by centralized planning. As large parts of the environmental problem perceived today cannot be handled by individual states, a worldwide consensus and/or a global-feudalist social and economic order would be necessary in both cases. This would also require a critical examination of the different views of global justice.

The points outlined here are prerequisites for effective operationalization in politics and business. Since it is not possible to resolve the open points in this work and social consensus building is not expected in the near future, the goal of science must be to develop solutions for dealing with increasing complexity in technical, economic and political practice (Vester, 2011).

In order to provide the reader with the basis for a differentiated discourse within the framework of sustainability, the following sections summarize the state of knowledge on external effects/costs and the classification of the environment.

3.3.1 Externalities and external costs

A purely liberal economic logic leads to a phenomenon: profits are privatized, while a majority of losses are socialized, i. e., outsourced to society. In order to answer the question to what extent a society is prepared to continue this practice in the future, an elementary understanding of external effects and costs is essential. Both concepts cannot be explained without the other, as the theoretical foundation of external costs is based on the economic understanding of external effects, i. e., the consequences of actions not covered by the market. First works in this field date back to the beginning of the 20th century, when Pigou (1911), for the first time, differed between private and social costs. The basis for this was the fact that no costs are incurred for certain goods, especially the environment. While Pigou (1911) attempted to solve the economic allocation problem with the help of state intervention (Pigou tax), Coase (1960) later outlined a purely private sector solution of the environmental problem by the respective market participants. Each person affected by environmental pollution is intended to pay a certain amount to the polluter to avoid the impairment, which in turn would finance the prevention measure(s). The Coase theorem is to be understood rather as a theoretical model, as a practical application is possible in very few cases (Coase, 1960). Since social costs are not taken into account in pricing on the market, a good is offered at too low a price. Accordingly, the market demands a larger quantity than is socially desirable. The difference between social and private costs is the external effect. Figure 3.2 illustrates the emergence of external effects according to Bofinger (2011).

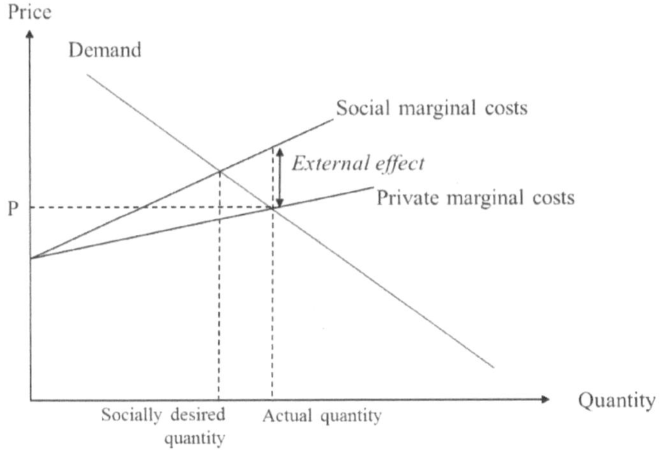

Figure 3.2 Emergence of external effects according to Bofinger (2011)

External effects can have both positive and negative consequences on the production and consumption function of third parties (Roth 1992; Bofinger 2011). Third parties can be individuals, companies, households or entire economies. A positive external effect (external income) is defined as a positive difference between social and private income. A negative external effect (external costs) occurs when the difference between social and private costs is negative (Bofinger, 2011). External revenues thus improve the economic position of third parties, while external costs worsen it. From a microeconomic or business management perspective, social costs refer to the sum of all costs incurred by a national economy through production activities (Roth, 1992; Wicke, 1993; Burschel et al., 2004).

In traditional corporate accounting, externalities are not taken into account as they are not financially relevant to the company at any time (Burschel et al., 2004; Schreiner, 1996; Hoitsch and Lingnau, 2004). This results in goods being offered on the market at prices that do not correspond to the real value consumption of input factors and the costs of eliminating negative externalities are outsourced to the general public (Wicke, 1993; Fichter et al., 1997). In literature, this is identified as one of the central problem areas of traditional economic activity and as a contrast to the desired sustainable development (Wicke, 1993; Burschel et al., 2004). In order to be relevant for operational accounting, external effects must first be determined, assigned to a specific action and monetized. While the assignment of environmental impacts to operational actions is particularly complex due to a recurring loss of information between the actual consequences and those that can be assigned to an action, various approaches exist in literature for the monetization of external effects. The choice of methodology depends on the form of the damage occurring. Generally, a distinction is made between material damage (e. g., loss of production or income), immaterial damage (aesthetic losses, reduced well-being), and damage to human health. The most common methods for monetization of external effects are damage and abatement cost approach (Adensam et al., 2002; Quinet, 2004). While the former seeks to estimate the damage caused, the latter can be understood as costs of prevention (Kesicki and Ekins, 2012; Vogt-Schilb and Hallegatte, 2014; Ward, 2014). This approach, which is sometimes highly controversial in economic theory, assesses measures on suspicion without reliable information on causal relationships or damage costs (Adensam et al., 2002). Nevertheless, the concept is widely used, as it requires little effort compared to damage cost-based approaches. Further literature on this subject is provided by Söllner (1997), Heertje and Wenzel (2008), Bofinger (2011), Mankiw and Taylor (2012), and the German Federal Environment Agency (Umweltbundesamt, 2012).

3.3.2 Categorization of the natural environment

In the context of the sustainability debate, the terms ecosystem and natural environment are often incorrectly used as synonyms. For a categorization, it is thus first necessary to define the terms system, ecosystem, environment and nature. The term system originates from the ancient Greek word sýstēma, characterizing a whole composed of individual parts (Heinrich, 2007; Erk, 2016; Harary and Batell, 1981).

Today, it has proved useful in many scientific disciplines to interpret the object of investigation as a system (Ulrich, 1970). This approach, known as systems theory, allows a wide range of research areas (e. g., biological cells, institutions, society, a state, a machine, the weather, plate theory) to be structured, described and analyzed according to the same models. A system is interpreted as a set of different elements that are related to each other in a limited space (Heinrich, 2007; Erk, 2016; Harary and Batell, 1981). The system structure or order is determined by the number (quantity) and variety (quality) of the relations between the individual components, i. e., the elements. The boundary of the space under consideration is described as the system boundary. Figure 3.3 illustrates the basic model of a system according to Harary and Batell (1981), Heinrich (2007), and Erk (2016).

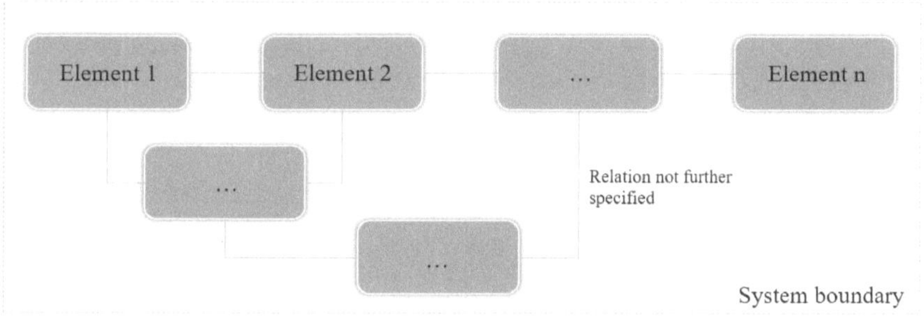

Figure 3.3 Basic model of a system according to Harary and Batell (1981), Heinrich (2007) and Erk (2016)

Systems theory, first developed around the middle of the 20th century, now forms the basis for a wide range of interdisciplinary work (Wiener, 1961; Diesner, 2015; Heinrich, 2007; Erk, 2016). The interpretation of an object of observation as a system is commonly referred to as 'systemic thinking'. As a rule, a distinction is made between different types of systems. According to the way they are formed, natural (inorganic, organic) and synthetic (logical, mechanical, social, combined) systems can be distinguished. Frequently a distinction is also made between total and partial or sub-systems. Likewise, open and closed systems can be differed according to their interaction with surrounding systems. It should be noted that closed systems are only a theoretical construct, since no real systems are known to date that do not interact with their environment in one way or another (Vester, 2011). Furthermore, systems can be distinguished with regard to their resilience. Stable (i. e., resilient) systems are able to return to a state of equilibrium on their own in case of a disturbance. Cybernetic systems use the feedback principle for this purpose. Unstable systems, on the other hand, remain in a state of imbalance in case of disturbances.

In this sense, the ecosystem describes a dynamic complex of organic life and inanimate nature in the form of various interdependencies (Jax, 2006; Chapin et al., 2011). As such, it is composed of biocenosis, i. e., the community of organisms of different species, and biotope, i. e., the natural habitat. The term environment, on the other hand, refers to the surrounding of a living being with which it interacts in a variety of ways (Trepl, 2005). Only by adding the adjective 'natural', a relation to nature is established. The term nature includes everything that was not created by man. Generally, animate (biotic) and inanimate (abiotic) nature are differed (Weber, 2007a). While the term ecosystem exists in itself, the term natural environment presupposes a reference perspective based on the individual.

While different categorization systems exist for the ecosystem and nature (e. g., the biosystematics or taxonomy of living beings in species), based on the anthropocentric principle of sustainability (see above), only a categorization of the natural environment is possible. To date, however, a generally valid system does not exist for this purpose. Nearly all approaches presented up to now, in the context of sustainability, are based on the assignment of a value for humanity. In this context, Krol et al. (1999) present a socio-economic categorization approach based on different user values. The user is understood as the person from whose perspective an evaluation is made. At a first level, the value for users and the non-user value are differed. The former represents the immediate usefulness for the viewers, while the latter is characterized by the utility for other users, e. g., future generations or nature itself. On a second level, a distinction is made between direct and indirect utility values as well as option, heritage and existence values. Direct utility values result from the use of resources (source function), whereas indirect utility values result from the use of the environment as a receiving medium (sink function). Figure 3.4 summarizes the understanding based on Krol et al. (1999).

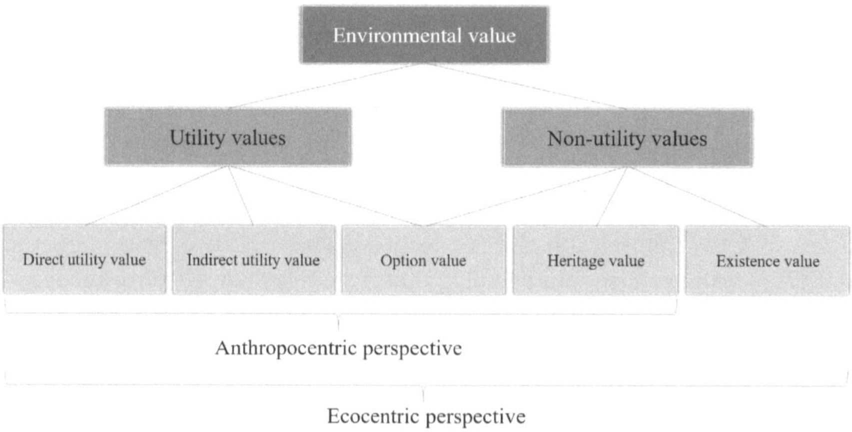

Figure 3.4 Systematization of the environmental value according to Krol et al. (1999)

However, the dominant view today is a classification based on socially perceived problem areas. A problem field is not understood as an existential threat to a certain aspect of nature, but as a consequence that is not foreseeable for mankind with regard to the currently prevailing way of life and economy. Two possibilities exist to determine the social perception with regard to the natural environment: a cross section of the perception of non-specialists with the help of survey results among the total population or the adaptation of existing classifications in relevant studies. While the former specifically documents the requirement of social perception specified above, the latter represents a far more well-founded classification. Among the environmental issues of greatest concern to society are climate change, air and water pollution, resource scarcity, health impacts, man-made and natural disasters, waste generation, and biodiversity loss (Dittmann, 2008). Further specification of individual problem areas in terms of their characteristics and/or any limit and reference values is not further elaborated in the surveys considered. Since these results do not clarify, what is specifically understood by air and water pollution, a conclusive categorization of environmental problems on this basis is not possible. Life cycle assessment (LCA) research provides methodologically and comparatively valid systematization approaches (see Section 6.2). Generally, input and output-oriented[2)] classification is used (Klöpffer and Grahl, 2014; Guinée et al., 2001; Frischknecht, 2013). Table 3.1 exemplifies a categorization scheme according to Klöpffer and Grahl (2014).

Table 3.1 Overview of socially perceived environmental problem areas based on Klöpffer and Grahl (2014)

Input-oriented environmental problem areas	Output-oriented environmental problem areas
▪ Abiotic resource consumption ▪ Biotic resource consumption ▪ Freshwater use ▪ Utilization of natural space ▪ Air use	▪ Climate change ▪ Stratospheric ozone depletion ▪ Summer smog ▪ Acidification ▪ Eutrophication ▪ Air pollution ▪ Human toxicity ▪ Ecotoxicity

Since social perception and scientific knowledge in ecosystem sciences are in a constant state of transition, the approaches outlined above represent a dynamic form of categorization. Rockström et al. (2009a; 2009b) present a similar, albeit more coherent

[2] The input-output approach, which dates back to Leontief (1986), is an elementary approach of operational system theory and is used in various scientific disciplines (e. g., for national accounts, process technology or life cycle assessment). A complete system (e. g., company, supply chain, production process) is characterized by the relationships between the input and output variables of its open subsystems. This essentially represents a conceptual thought pattern that understands a process as a black box. The approach differentiates between process input variables (input), process throughput and process output variables (output) (Zelewski 1999). Throughput is understood as the process of transformation from input to output variables: Input → Throughput → Output

approach in terms of social discourse with the concept of planetary boundaries. In addition to classification, the authors present a response to the follow-up problem resulting from systematization, the question of a desirable state of the natural environment.

A short digression: the optimization of an (existing, felt and/or predicted) state can only be achieved in relation to a reference state. In the context of sustainability, however, this is only available for individual socially perceived problem areas. In global environmental policy, it is common practice to formulate saving targets for the triggering moments within individual problem categories in relation to a specific reference year.[3] Although these are usually based on scientifically sound studies, they are often relativized under questionable circumstances, e. g., through lobbying by groups from outside the field. The concentration on individual environmental problem categories also leads to a situation where interrelationships between different problem areas receive little attention. The structural basis for a holistic improvement in terms of the sustainability concept is thus deprived.

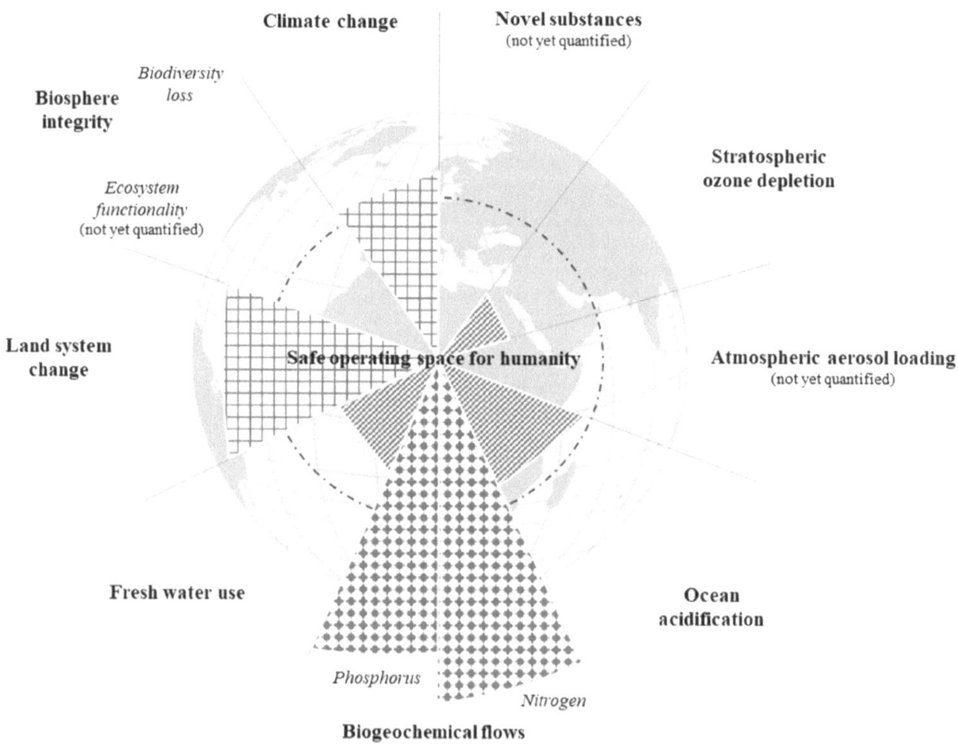

Figure 3.5 Qualitative illustration of planetary boundary concept according to Rockström et al. (2009a; 2009b)

3 For example, the majority of climate policy savings targets refer to the reference year 1990.

In this context, Rockström et al. (2009a; 2009b) postulate the orientation towards planetary boundaries, which they calculate for the first time in 2009. In addition to the ecological limits, the current level of use is calculated and set in relation to the pre-industrial state. The ten ecological problem areas identified here are climate change, loss of biodiversity, the nitrogen and phosphorus cycle, stratospheric ozone depletion, ocean acidification, global freshwater consumption, changes in land use, particulate matter, and chemical contamination. Although the study does not claim to be exact, it does highlight the problem of nature's ability to regenerate. The most urgent problems described are those for which the current status has already exceeded the ecological limits. These are the nitrogen and phosphorus cycle and the loss of biodiversity. Other highly critical issues are land system change, climate change and ocean acidification. Figure 3.5 illustrates the results according to Rockström et al. (2009a; 2009b).

3.4 Macroeconomic models of a sustainable economy

Although the foundations for an intensified consideration of nature in resource economics were already laid in the first half of the 20th century (Cassel, 1921; Hotelling, 1931), an ongoing far-reaching discourse only began in the second half. In particular, the representatives of the critical theory of the Frankfurt school surrounding Horkheimer (2007), Adorno (1992), and Marcuse (2004) advocated a greater attention to nature by criticizing the economy as the driving force of its destruction. At the economic level, Kapp (1971) and Polanyi (1973) developed the basis for today's environmental economics and industrial ecology from the 1950s onwards. Nature, in the sense of the biosphere, represents the basis of all life and thus also of all economic actions (Illge and Schwarze, 2006) (see priority model). The scarcity of a good is no longer determined exclusively by the costs incurred but by the danger to man and the environment. The basic assumption here is the insight of microeconomics and biocybernetics that any action exhibits an optimal scale, in the sense of a balance of marginal costs and revenues and/or dimension (Daly, 1990; Vester, 2011). To date, however, none of the theories presented has even begun to gain the significance that the approaches outlined above for the three phases of industrialization have developed. Rather, some basic lines of thought have developed into approaches that serve as vivid macroeconomic visions of a sustainable economy. These are essentially a transfer of the elementary strategies of sustainability described above into objectives that are comprehensible to the general public. The common concern of the concepts presented so far is the creation of a regenerative economy, i. e., an economy whose production and consumption of goods does not exceed the ecological capacity of the earth. Three approaches are currently being discussed: circular economy, bioeconomy, and demand economy. Similar to the three sustainability strategies, these cannot be completely separated from one another.

3.4.1 Circular economy

The concept of a circular economy can primarily be assigned to the consistency strategy. The term circular economy serves to distinguish a cyclical economy from today's linear economy (Geissdoerfer et al., 2017). While current economies burn the majority of raw materials used after the products have been utilized, the circular economy aims to establish a system that constantly reuses all materials and energy in a circular industrial metamorphosis. The concept represents an extension of today's idealized views on ancient agricultural economies (see Chapter 2), which were always based on a continuous exchange of elementary production factors. For example, production energy used to cultivate the fields came from animal or human muscle power that was again derived from the cultivated area, which formed the basis of nutrition. The resulting production residues (e. g., straw) and product waste (e. g., excrements) were directly fed back into production. The current understanding of the circular economy, whose origins date back to the early 1990s, is largely based on the US American industrial ecology of the late 1980s (Frosch and Gallopoulos, 1989; Andersen, 2007). Since the framework conditions have changed decisively compared to agricultural economies after more than 250 years of industrially influenced economic activity, the realization of a circular economy faces the challenge of far more complex interactions between economic units. Especially in western industrial nations, lifestyles are characterized by a one-time consumption of goods. This has resulted in a linear sequence, starting with the extraction of resources, continuing through production and use and ending with disposal. The life cycle of a product in this economic model is described as cradle-to-grave. This production model, which is primarily designed for economies of scale and manifests itself not least in complex, globally distributed supply chains, is increasingly contradicting the desired 'green economy'. In line with the consistency strategy of sustainability, the circular economy aims to close material cycles following the example of nature. As such, the term circular economy can be understood as a theoretical framework for a multitude of approaches presented to date, postulating the circularity of industrial metabolism as a core element of a future-proof economy.

The scientific foundation of the circular economy is industrial ecology (Ayres and Ayres, 2002) that postulated the requirement of an industrial system to behave like an ecosystem in which every waste in turn serves as a resource for the first time. The basic principles of industrial ecology are therefore the circular design of material flows and the cascading of energy flows. The learning of natural ecosystems is often cited as a structural instrument. However, the dominant understanding of the circular economy in Europe is based on the cradle-to-cradle concept of Braungart and McDonough (2014). The decisive factor in the formulation of the vision was the recognition that every living creature has an impact (e. g., through the production of waste and emissions) on the natural environment, but nature always finds ways to use it sensibly. In this respect, it is not the production of waste and emissions per se that is rejected. Instead, the choice of the place and time of entry into natural ecosystems is identified as the cause of impact. Braungart and McDonough (2014) then formulate a design paradigm that should enable sustainable innovations for products, industrial systems and buildings. Thereby, the differentiation of two cycles is crucial, whose substances can occur together in products, but whose circular utilization must take place in strictly separated systems. The biological cycle is used here as a synonym for the natural metabolism. It describes the cyclic path from the extraction of biological materials and substances, through production and use in production, to reuse in the sense of biodegradation of dead organisms. This contrasts with the technical cycle, which is essentially a reflection of the concept of industrial metabolism of US industrial ecology. In this concept, the technical cycle includes all substances and materials that cannot be degraded in a biological cycle by natural processes. These materials and substances have to be utilized for industrial applications in the long run by recovery and recycling. Products and processes are to be conserved in such a way that their biological and technical components can be separated from each other after use with little effort and transferred to the respective cycles. Figure 3.6 illustrates the difference between the technical and biological cycles according to Braungart and McDonough (2014).

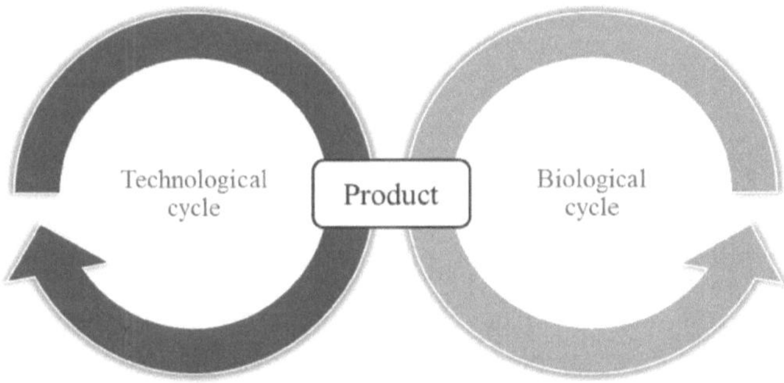

Figure 3.6 Technical and biological cycle in the cradle-to-cradle concept according to Braungart and McDonough (2014)

Another interpretation currently present in Europe is provided by the Ellen MacArthur Foundation (2015). Here, the distinction between biological and technical cycles is adopted from Braungart and McDonough (2014). According to the Ellen MacArthur Foundation (2015), the goal of the circular economy is to maintain the value or usefulness of products, components and materials at a constant level over time. This is to be achieved by cascading use without waste (zero waste) and without emissions (zero emission). The Ellen MacArthur Foundation (2015) names optimization, regeneration, recycling, sharing, virtualization, and exchange as measures on the way to realizing a circular economy.

A further vision that can be attributed to the theoretical framework of the circular economy is the concept of natural/regenerative capitalism postulated for the first time by Hawken et al. (1999). On the basis of a comparative study of various related concepts (including conscious capitalism, donut economy, sharing economy, steady state economy), Fullerton presents a series of universal patterns and principles that need to be taken into account when designing a sustainable economic system (Fullerton, 2015):

1. Appropriate relations: On the basis of the elementary insight that humanity is a constituent element of a network of life, the separation between 'us' (human beings/humanity) and 'it' (the environment), which exists in thought patterns and economic actions, must be overcome. Damage to one part of the network affects other parts. Human economic activity is always based on biosphere conditions. Accordingly, the productive capacity of the biosphere must be maintained.
2. Interpret prosperity holistically: True prosperity cannot be depicted in monetary terms. In this sense, a new definition with regard to the well-being of society as a whole should be pursued, integrating social, cultural, living, and experiential aspects.
3. Participation: The expanded understanding of prosperity should be designed in the sense of distributive equality.
4. Innovative, adaptable, responsive action: Innovations must be developed in all areas of society in order to ensure the adaptability of humanity to changing environmental conditions.
5. Recycling and dematerialization: Material and energy flows must be designed in cycles, the total material and energy throughput must be reduced.
6. Expanding the marginal effect: Natural edges (forest edge, pond banks) are particularly rich. These ecosystems must be preserved in terms of natural capital.
7. Strive for balance: The resilience of ecological systems is based on the balance principle. Accordingly, unnatural concentrations of power in economic systems should be avoided.
8. Form stable communities and regions: Healthy communities are the basis for exchange, innovation and togetherness and thus form the foundation for a sustainable economy.

Other related approaches that are not further elaborated here are performance economy, blue economy, and regenerative design (Lyle, 2008; Stahel, 2010; Pauli, 2012). Weizsäcker and Wijkman (2017) summarize a number of recent publications in order to derive policy options for the realization of a circular economy that effectively contributes to decarbonization and dematerialization. These include: increasing the recycling and reuse rate, limiting waste incineration, shifting taxes from labor to resource taxes, intensifying political measures to promote renewable energies, design requirements for products (e. g., simpler repair, maintenance, and dismantling as well as a ban on planned obsolescence), new public procurement guidelines, expanding climate policy (e. g., to include material efficiency), supporting recycling management through elementary infrastructure investments and developing new business models.

3.4.2 Demand or subsistence economy

The terms subsistence economy and demand economy are used as synonyms in this work. Traditionally, agricultural forms of economy are referred to as subsistence farming. In contrast to neoclassical patterns of thought (e. g., marginal utility, economic growth), the focus in these economies is self-sufficiency. The goal of production is therefore not the maximization of profits but the independent, local production of goods for daily use. At the beginning of the 21st century about 40% of the world population lived in traditional subsistence economies (Fankhauser, 2014). After this economic form was considered outdated for a long time, the concept recently received increased attention in the context of the sustainability debate. Consequently, Paech (Paech, 2012) and Latouche (Latouche, 2015) outline the demand economy as a form of a post-growth economy. Furthermore, various authors identify the limitation of consumption to 'real' needs and decentralized production as elementary components of a moral economy (Sachs, 1993; Schneidewind and Palzkill, 2011; Baumast, 2013; Scott, 1977). Accordingly, the concept can be primarily assigned to the sufficiency strategy. The demand economy does not in any way exclude the recycling of materials and energy. On the contrary: traditional agricultural forms of economy that have been implemented to date are the only known forms of comprehensive closed-loop recycling systems. The concept can therefore be attributed equally to the consistency strategy. Although the demand economy, similar to the sufficiency strategy, has the most rational arguments purely from the perspective of ecology, its suitability for real implementation is often doubted today. In addition to the considerations outlined in Section 3.2.3, the current, historically grown socio-political structures hinder implementation.

A further related approach, which should be mentioned at this point, is the concept of the common welfare economy. Although the approach postulated by Felber (2010) is still based on market economy principles—including the existence of private companies, money and market prices—the entrepreneurial production of services is embedded in a holistic framework that is intended to replace the pursuit of profit and competition with the pursuit of the common good and cooperation. In this understanding, companies no longer serve the mere physical satisfaction of external demand, i. e., the production and sale of goods to achieve monetary profits. Rather, successful companies are defined by their contribution to society and ecology. The idea of growth is not completely rejected, but viewed as less important. Furthermore, Felber (2010) presents a catalogue of criteria that companies report in their main balance sheet to show their contribution to social responsibility, ecological sustainability, democratic participation, and conformity with relevant regulations. In this catalogue, the generation of a profit is understood merely as a vehicle for realizing a socio-ecological contribution that is as far-reaching as possible. Although the approach is already supported by a number of companies today, a comprehensive implementation of the concept seems unlikely at present, since even the companies that support it will, in case of doubt, prioritize economic (e. g., profit) and social (e. g., jobs) over ecological aspects. A decisive factor for the implementation of the common welfare economy (similar to the demand economy) would be a transformation of the political framework conditions, which in turn would be triggered by a social need.

3.4.3 Bioeconomy

Another vision of a sustainable economy is the bioeconomy (Pfau et al., 2014; D'Amato et al., 2017). Historically, two interpretations can be distinguished: the initial holistic 'bio-cultural' understanding based on the theory of commodities and the narrow 'materialistic' understanding.

The term is based on the concept of bioeconomics formulated by Georgescu-Roegen (2014). From the 1970s onwards, the concept served as the basis for the theory of commodities. A commodity is interpreted as an instrument for satisfying human needs and, following the biological principle of order, is grouped into categories of goods according to their utility value (including type, genus and family of goods). The basis for a commodity is the commodity stock. Commodity science distinguishes between the mineral, animal and plant world. The decisive factor here is the holistic view, which does not reduce the significance of a commodity exclusively to its purchasability, but integrates, e. g., ecosystem services and other non-monetized resources as core elements into the consideration. Georgescu-Roegen (2014) describes the resulting manifold connections between economy and environment as bioeconomics. In this understanding, the bioeconomy is interpreted to be a medium for achieving a balance between society and biosphere. Subsequently, various authors presented papers addressing different elements of a bioeconomy, including the structuring of the economy according to biological principles and the use of biological resources (Mohr, 1990; Gowdy and Mesner, 1998; Witt, 1999; Kiridus-Göller, 2011). Their common ground is the rejection of a strictly material understanding of the bioeconomy and the emphasis on a bio-cultural context of human actions, which manifests itself through biomimetics/biomimicry as a structuring instrument (Gottwald and Krätzer, 2014). The approach developed by Benyus (2002), whose name is derived from the ancient Greek words life (bios) and imitation (mīmēsis), postulates the study of nature and the resulting inspiration for novel concepts and processes as an elementary method for solving human problems. A similar approach is the concept of biocracy formulated by Caldwell (1987), which postulates a renaissance of the economic approach of physiocracy (see Section 2.2.1) with the technologies available today. From a production perspective, Beer's cybernetic viable system model (Beer, 1967) and the emergence of biocybernetics can also be attributed to the original understanding of the bioeconomy (Vester, 1993).

Since the beginning of the 21st century, the term bioeconomy has received increased attention in the political debate, leading to a narrowing of the term (Vivien et al., 2019). The vision, which was decisively shaped by the former EU research commissioner Janez Potočnik, describes the concept as an economy based on renewable resources, which serves as a counter-draft to the current fossil-based economy (Potočnik, 2005). Thereby, especially biotechnology is regarded as an enabling technology. Today, this biobased economy is often seen as a model for global agriculture and food production and as a strategic element of decarbonization, which differs considerably from its original interpretation (Vivien et al., 2019). In comparison to the holistic 'bio-cultural' understanding, however, the goal formulated here is less the closing of cycles within the framework of planetary boundaries, but rather the production and use of biological resources to provide products, processes, and services in all economic sectors within the framework of a future-oriented economic system. It remains unclear, however, which sectors are included in the bioeconomy. While some sources describe the bioeconomy as a vision of agriculture and the food industry, others emphasize the opportunities in areas such as bioprocess technology, medicine, materials, and energy (Pfau et al., 2014; D'Amato et al., 2017). The political interpretation continues to emphasize sustainability, but in a less pronounced form. The use of biological principles as a structural instrument is merely considered. In contrast, far more attention is paid to the innovation potential and the development of new raw material deposits under traditional economic conditions. While the original interpretation demands nothing less than a paradigm shift in economics, the narrow 'materialistic' understanding envisages a comparatively marginal change of course within today's economic patterns. Rajan (2009) thus refers to this interpretation of bioeconomy as biocapitalism.

Figure 3.7 illustrates the difference of holistic 'bio-cultural' and narrow 'materialistic' understanding with regard to their scope, instrumental basis, and sustainability measure.

Today, the politically influenced interpretation serves as a basis for a multitude of institutions and frameworks in science, politics and society. From an economic perspective, this is justified by considerable success. The sectors of the European Union that are unofficially associated with the bioeconomy today provide jobs for around 17 million people (i. e., 8.5% of the working population) and generate more than two trillion Euro annually (El-Chichakli et al., 2016). At the same time, it becomes clear that a purely economy-oriented perspective leads to undesirable ecological and social consequences, including competition for land, adverse effects on ecosystems, incompatibilities (Cristóbal et al., 2016; Grefe, 2016; Gottwald and Krätzer, 2014; Cooper, 2014; Marazzi, 2013). As a result, a reorientation of the narrow towards the holistic concept recently experienced increased attention (Bugge et al., 2016). Yet this primarily involves the instrumental basis and the economic scope, rather than the basic economic concept.

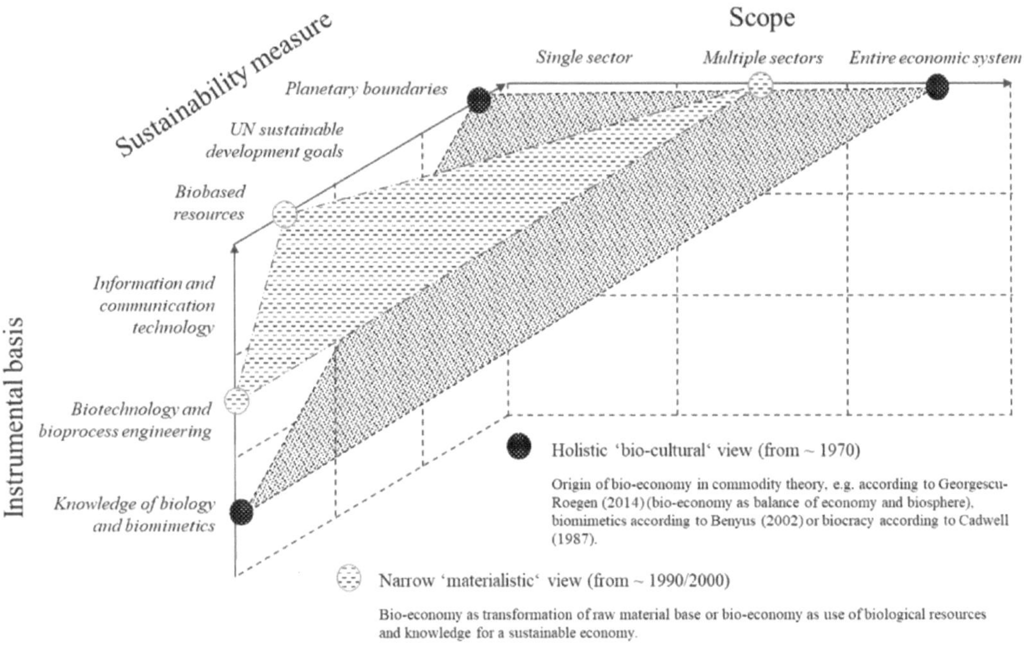

Figure 3.7 Evolution of the bioeconomy concept [own illustration]

3.5 Operationalization as a neuralgic conflict of objectives

Manufacturing companies occupy an ambivalent position in the context of sustainability. On one hand, they are one of the main causes of the problem. On the other hand, they are regarded as a major lever for solving the problem, since only they have the necessary problem-solving capabilities (Reichwald and Piller, 2009; Günther and Nowack, 2008). In order to act effectively, however, companies need planning security. To the same extent that this applies to business relationships within a supply chain, it is necessary for socio-economic framework conditions. The greater the uncertainty, the more difficult the decision-making process. In case of doubt, this can lead to a complete rejection of high-risk trade. One obstacle is the vague concept of sustainability, which cannot be considered a priori as an instrument to be implemented (Hülsmann, 2004). The question thus arises as to how a manufacturing company can optimally adapt to the diverse requirements of the concept. In this context, the (neo-)classical orientation of companies towards the short-term goals of growth and profit maximization appears to be a fundamental conflict of objectives (Nertinger, 2015), contradicting the concept of sustainability, which is designed to secure the company's long-term existence. It is precisely this safeguarding of the stock of natural capital that forms the basis of all economic activities. From a long-term perspective, sustainability is not only a social but also an entrepreneurial goal. This long-term complementarity of objectives, in the sense of a dependence of access to resources, is primarily described by representatives of the resource dependence theory (Pfeffer and Salancik, 2003; Pugh and Hickson, 2007). On this basis, Nertinger (Nertinger, 2015) characterizes sustainable entrepreneurial success by means of three dimensions. [1] The requirement of organizational preservation through the continuous achievement of an adequate return on capital employed, [2] the constant assurance of preservation of and access to the required resources, and [3] the consideration of the preservation of eco- and socio-systems. In contrast, doubts prevail about the practical compatibility of the variables outlined above, especially their multi-dimensionality. On one hand, the significant short-term orientation of western companies as outlined by Hofstede (2003) leaves little room for long-term decisions. Instead of securing the company's long-term existence, decision-makers prioritize short-term success and risk permanent damage of resource access. On the other hand, the very goal of entrepreneurial continuity proves to be problematic. The legitimation of operational actions in terms of sustainability has to lead to a market exclusion and thus to a termination of the enterprise if it is not observed in the long run. If this is not the case, the incentive for a long-term orientation is missing.

Although a large number of concepts for the implementation of sustainability in companies have been presented to date, postulating an anchoring in corporate strategy, an increase in transparency with the help of performance indicator systems or an adaptation of business processes, a coherent economic optimization is hardly possible due to the currently prevailing framework conditions. At the same time, it has become apparent that companies are increasingly being valued on how they deal with sustainability. For example, the number of global environmental regulations has grown almost exponentially for years, which poses increasingly complex challenges for globally active companies in particular. In order to ensure global market access, more and more requirements in the design, production, and packaging of the product have to be taken into account, in turn requiring extensive adjustments along the entire value chain. In addition to mandatory conformity with existing law, the social diffusion of the sustainability debate is accompanied by a projection of additional requirements that initially have little to do with the traditional production of goods and services in companies. According to Matten et al. (2012), sustainable corporate action includes, in addition to philanthropy, elements such as voluntariness, internalization of external effects, and the perception of ecological and social responsibility.

3.6 Chapter review and conclusions

The anthropocentric notion of sustainability requires a questioning of existing production methods and patterns of thought. However, the lack of a uniform interpretation of the concept of sustainability massively impedes its operationalization. To date, a number of elementary questions that would require a social consensus remain unanswered. On a macro level, the model of an economy that is aligned with planetary boundaries is undoubtedly fundamental for the sustainable design of the economy. It remains unclear, however, whether this requires the preservation of the ecosystem as such or merely its productive capacity. Furthermore, the compatibility of the macroeconomic guiding principles presented so far (circular economy, demand economy, bioeconomy) with the predominant liberal economic logic of the neoclassical period has not been clarified. On one hand, this is astonishing, since the Brundtland report already emphasizes a clear regulatory perspective, which would suggest a worldwide rethinking towards more social economic theories (e. g., ecological economics). On the other hand, it becomes clear that interventions in the production of goods and services in the individual enterprise are increasing, e. g., due to the rapid rise in the number of environmental regulations, which can be understood as a clear sign of a stronger prioritization of an equality-oriented economic logic. The same applies to the question of the desirable state of the natural environment. With the concept of planetary boundaries an approach exists which would allow a comparatively pragmatic orientation. However, no official global orientation of economic and environmental policy towards the concept has been established. Sustainable entrepreneurial activity is thus still largely based on voluntary action, which leads to a low significance of sustainability for individual economic issues. Meanwhile, the history of the terms (especially the terms sustainability and bioeconomy) shows that the contemporary interpretations differ greatly from their original understanding. Apparently, a conceptual dilution is unavoidable in the course of political consensus-building. This is not least due to elementary conflicts of objectives at different decision-making levels. In a simplified form, three different groups can be named as examples, whose obvious actions necessarily lead to conflicting decisions in terms of sustainability (especially ecological preservation), each of which influencing each other: individual, politics and companies. The majority of individuals tend to place their own well-being and that of their relatives above that of global society. Politicians are therefore not elected because they have the best rational arguments, but because they have the best 'promise of salvation' for the individual. In order to secure the favor of the voters, meaningful decisions in the sense of sustainability (e. g., the pricing of environmental goods, the adoption of regulations) are insufficiently advanced in the political discourse. As a result, market conditions and incentives for companies are already in deficit, which in turn determine technological development. Figure 3.8 illustrates the phenomenon, which is described here as a vicious circle of sustainability according to Schmidt-Bleek (2014a).

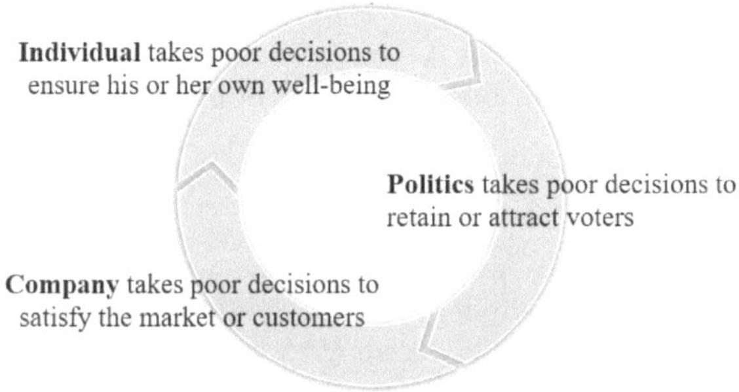

Figure 3.8 Vicious circle of sustainability according to Schmidt-Bleek (Schmidt-Bleek, 2014a)

In summary, what conclusions can be derived for manufacturing sciences at the macro- and microeconomic level?

1. Implications for manufacturing sciences at macro and micro level

Circular economy, bioeconomy and demand economy are not strictly separated visions or separate stages of a chronological sequence of economic development towards more sustainability. Rather, they are expressions of the same vision, a form of economy within the framework of planetary boundaries, each of which emphasizes different aspects. However, it is precisely the politically influenced interpretations of terms (e. g., bioeconomy) that do not sufficiently meet the requirements of a sustainable form of economy, especially with regard to ecological preservation. In the following, the bioeconomy will therefore be interpreted in the sense of its initial, holistic 'bio-cultural' understanding. Of substantial importance for the production sciences as a whole is the development and pursuit of a positive vision of sustainable value creation. The vision should be functional and presentable, link current and future technological development, and evoke positive associations in the public. In terms of sustainability, it should be a combination and specification of the above discussed approaches, which may be summarized in seven elementary design principles, each of which being derived from the already formulated approaches of circular economy, bioeconomy and demand economy:

- Complete substitution of fossil fuels by renewable raw materials and energy sources (substitution principle)
- Calibration of the prevailing liberal economic logic by aspects of a socio-ecological economic logic, i. e., stronger regulation of markets (regulatory principle) as well

as dismantling of the principle of division of labor, increase of self-sufficiency and regionalization of supply chains (decentralization principle)

- Realization of a minimum energy and material expenditure for the production of goods by adequate choice of type and quantity as well as technological renewal (renewal principle)
- Orientation towards natural principles, phenomena, and metabolisms (eco-convergence principle)
- Complete transparency over every substance, material and assembly used (transparency principle)
- Complete recycling of every substance, material, and assembly used (closed loop principle)

In order to realize the vision, four principles of action are also required, which must be implemented at both the overall and the individual economic level:

- Systemic, holistic thinking and acting, i. e., thinking in terms of interactions, life cycles and planetary boundaries (holism principle)
- Consistency in acting according to elementary principles (principle of permanence)
- Performing inter- and intragenerational responsibility in the sense of the permanence of human life on earth (principle of responsibility)
- Involvement of all parties affected by a decision (principle of participation)

2. Implications for manufacturing technology

The development of manufacturing technologies takes place almost exclusively at the micro level. State control can be exercised at most by means of regulations (including bans and subsidies). Since this work does not attempt to solve the vicious circle of sustainability, it can be assumed that the aspects of profitability, liquidity, and securing existence will remain the most important determinants of corporate decisions in the foreseeable future. Nevertheless, the goal of sustainable production technology should be based on the principles outlined above and result in a holistic value creation process involving as many actors involved in and/or affected by the production process as possible. At the same time, a core objective of production technology should be the exploration and identification of technically sustainable possibilities and their demands on society and politics. In order to adequately meet the needs of future generations, innovative solutions must be developed in areas of need that are highly critical from an environmental point of view (housing, energy, nutrition, health, consumption, mobility). The decisive factor here is the networked thinking well-known from the field of biocybernetics, the aim of which is to develop small-scale, system-compatible production modes adapted to the environment (Vester,

2011). According to Vester (Vester, 2011), the following logic applies: technology is an element/subsystem of a complex natural overall system. Each element/subsystem is connected to every other element/subsystem of the complex natural overall system via manifold relations. Each element/subsystem has an optimal size (e. g., appropriate level of input/output or amount of information, turnover), which results from the interaction with the other elements/subsystems. The natural overall system is reactive, i. e., it adapts its target values to new constellations by means of feedback loops in order to ensure a high error tolerance. Consequently, no a priori desirable or investment-worthy alternatives exist. The focus on individual, unchangeable target variables (GDP, jobs, growth, etc.) is to be rejected. The main objective of technical development is always to increase and secure the viability of the complex natural system as a whole. A technical development must therefore not take place exclusively within an individual discipline, otherwise the overall picture is lost. Rather, technical advancement requires a complete system description, which includes quantitative as well as qualitative factors (consensus, aesthetics, quality of life, motivation, etc.), their elements/subsystems and relations.

Lessons learned

- Today's understanding of the anthropocentric concept of sustainability is largely shaped by the Brundtland report of 1987, defining sustainable development as a process that enables present generations to meet their needs, while not compromising the possibilities of future generations.
- While the Brundtland definition is based on two pillars (inter- and intragenerational equality), the report does not provide any information on substitutability of natural, human, and physical capital. Hence two interpretations exist, each of which can be understood as an extremity: weak sustainability (triple-bottom-line model) and strong sustainability (priority model).
- In comparison to dominant economic theories (especially classics and neoclassics), which primarily emphasize liberal thinking (liberal humanism), the concept of sustainability clearly underlines equality (regulatory approach).
- Three basic strategies for the development of a sustainable economic practice exist, which cannot be clearly distinguished from one another: [1] reducing the input involved in the production of a given good (efficiency), [2] selecting the right input materials and technologies (effectiveness), and [3] abandoning fictitious needs (sufficiency).
- Due to prevailing empty world theories as well as a number of unanswered elementary questions (preservation of ecosystem as such or its productive capacity, internalization of externalities, etc.) that would require social consensus, theory formation represents a major challenge, while a generally acknowledged operationalization is practically impossible today.
- The hitherto existing possibility of enforcing subjective demands not least results in a shift of responsibility to other parties (e. g., politics, producers, consumers), often without realizing that responsibility is a cycle of trade (vicious circle of sustainability).

- Among others, this becomes clear by existing macroeconomic future visions (e. g., circular economy, bioeconomy, subsistence economy). Here it appears that initial theories are often materialized through public discourse, i. e., conceptual delusion occurs depriving sustainability-oriented concepts of their original purpose.
- Both production economics and technology require a reorientation towards the concept of sustainability, enabling the consideration of elements of all basic strategies and future visions. For this purpose the concept of biointelligence was developed.

Follow-up questions

- How can a transformation towards a sustainable economy succeed?
- What is the concept of biointelligence and how does it reflect the concept of sustainability?
- Does it represent a separate scientific discipline or is it a form of an existing one?
- How does it affect existing theories and approaches in production science?

4 Basic production theory considerations on biointelligence

After having challenged existing patterns of thought, this chapter discusses basic production theory concepts of biointelligence.

4.1 Notion of a reorientation of production economics and technology

The approach primarily developed by Miehe and Bauernhansl serves as a basis for the reorientation of production economics and technology (Miehe et al., 2020a; Miehe et al., 2018; Bauernhansl et al., 2019). Based on a Delphi-supported scenario-based investigation of the BT of industrial value creation in Germany (BIOTRAIN), the authors conclude the following basic assumption, which is regarded as a precondition below: 'under the given socio-economic conditions (see Chapter 2 and Chapter 3) realistic sustainable value creation implies maintaining the current standard of living while not depleting natural resources. This requires a transformation in the sense of a renaissance of sustainable production methods with innovative technologies, which must be sustained by an adequate economic theory.' Figure 4.1 qualitatively illustrates the initial assumption based on the historical development of production methods according to Koren (2010) and Miehe et al. (2020a).

Bauernhansl et al. (2019) describe the notion of a new form of production and the transformation to it in a vivid portrayal for Germany as follows.

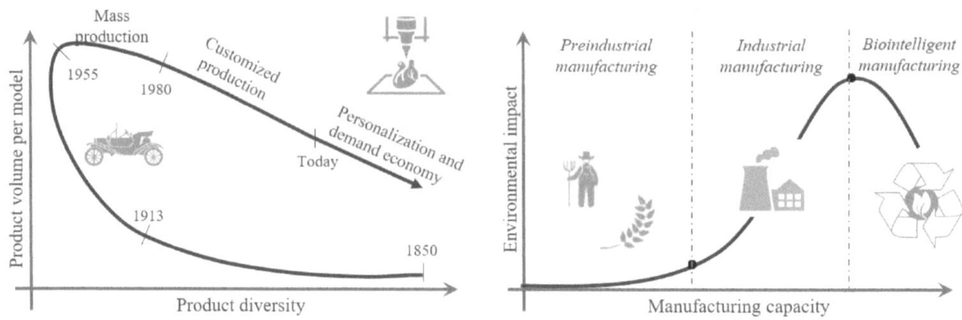

Figure 4.1 Qualitative illustration of historical and projected development of production regarding volume, diversity, capacity, and impact according to Koren (2010) and Miehe et al. (2020a)

1. Vision

Biology comprehensively influences production and enables mankind to satisfy its needs in an ecologically sustainable fashion. Smart biomanufacturing devices (intelligent, decentralized bioproduction cells, SBMD) facilitate the transition to a biobased, personalized and decentralized production of consumer goods and food as well as health care applications. Despite not all products being manufactured locally (defense, electronics, etc.), the technological renewal of industrial value creation is developing into an advanced economy that takes into account the planetary boundaries: the technology-based demand economy. Goods are only produced once they are needed (just-in-time). Waste no longer exists, since any output of one system represents the input of one or more other systems.

2. Transition

Initially, the transition is based on a profound digitalization of the industry, what is undoubtedly occurring already under I4.0. Resulting from intensive research and development as well as the unprecedented commitment of manufacturing companies, the effective application of ICTs enables the networking of machinery and equipment, which in turn allows enormous increases in productivity.

While the use of self-learning algorithms in industry initially serves merely the purpose of improving productivity, research institutions are increasingly using them to solve complex challenges of human coexistence and economic activity (climate change, scarcity of resources, preservation of social goods, etc.). The enormous progress in development as well as the price decline in the field of sensor technology simultaneously enable a massive expansion of technology deployment. The amount of information available grows exponentially. With the help of machine learning, the immense amount of data can be evaluated in a targeted manner and be made available to potential users in databases.

Meanwhile, biotechnology has experienced a dynamic rise as a result of the increasing networking of knowledge and is rapidly entering industries such as pharmaceuticals, medicine, and chemicals. The systematic application of knowledge of biology and the associated convergence of production, information, and biotechnology profoundly changes products, production processes, forms of organization, teaching and learning methods as well as entire regions and lifestyles. While optimization measures since the first industrial revolution have either concentrated solely on the technical system (e.g., realization of flow production) or aimed at connecting technical and information systems (e.g., networking of production facilities via cyber-physical systems), a synchronous linking of technical, information and biological systems becomes increasingly feasible.

While digital transformation (I4.0) has enabled extensive flexibilization potentials in manufacturing companies, it is now possible to create production units that are fully transformable. Biointelligent production systems are able to adapt their architecture autonomously and ad hoc to the optimal solution of a given production task. Production systems are no longer to be understood as delimited entities (factories, plants, etc.), but rather as regional socio-technical cells that possess all relevant information (DNA) in order to exchange and intelligently use resources (metabolism), to adapt to environmental conditions (evolution), to network spontaneously and autonomously with each other and to find specific solutions for given tasks collectively (i.e., in swarms). In case of unexpected failures of parts of the production system, it adapts itself automatically and in real time (resilience). Biointelligent production cells can occur in a wide variety of dimensions (molecule, cell, organism, community, region, etc.), are organized decentrally, operate regional resources, and have diverse exchange relationships with their surrounding units and biological resources.

SBMD are gaining in importance in industrial value creation, allowing a variety of products to be produced independently on site with the given biological resources. For this purpose, production processes based on additive manufacturing are coupled with self-learning algorithms in order to prepare and process regionally available biobased materials (e.g., bioreactors, biorefineries) and adapt them to specific requirements (e.g., use, end-of-life). The technological and biological systems are enabled to communicate with each other and to learn from each other. Small-scale cell factories, based on the principles of tissue engineering, enter households, enabling the production of meat and other nutrition. Household waste is directly utilized, e.g., using bioelectrochemical production cells, converting it into energy and/or material for new products.

The spread of SBMD enables a completely new form of consumer participation in the production process by simultaneously occupying the role as consumers (traditional value creation), designers (digitalized value creation), and suppliers of raw materials and components. As a result, material cycles are increasingly being closed.

Yet, the transformation might not only serve to decentralize and modify production. It also allows the integration of adaptive microorganisms into materials and surfaces, functionalization and adaptation of products to the conditions of use as well as refabrication. Intelligent communication interfaces enable consumers to actively intervene in the development and use process at any time by controlling essential parameters.

Increased communication capabilities between technical and biological systems offer novel potentials for human-machine interfaces. Future production systems allow humans and machines to work together intuitively. Both 'worlds' will be able to communicate with each other without barriers. Machines pass their information, e. g., via intelligent wearables (e. g., augmented reality glasses) and humans are reproduced within machines via avatars, i. e., digital twins (DT). This enables the machine to interact with each person individually, e. g., to ergonomically adjust the workstation, change the language or take into account other physical or mental states of the interacting individual. The recognition of emotions as well as physical stress conditions enables a specific adaptation of machines and systems as well as a needs-based work design. With the help of novel sensor, actuator and data processing systems (e. g., as combination of human brain interfaces and exoskeletons), people are enabled to intuitively control technological systems (e. g., machine states) and thus realize a stress-oriented and at the same time highly efficient production.

Decentralization of value creation and increasing opportunities for active participation in production also changes lifestyles. Food not produced in small-scale cell factories is grown in own or shared gardens and breeding facilities (horticulture). Urban areas are increasingly equipped with vertical and horizontal gardens on private and public buildings. Entire building complexes and quarters are becoming biointelligent, far beyond the introduction of smart grids and smart entertainment. Intelligent platform technologies enable an ideal capacity utilization of the installed SBMDs. As a result of the regionalization of value creation, transport routes are being massively reduced. Small and medium-sized companies with flexible structures as the basis for innovation processes gain importance.

The trend towards cooperative use leads to numerous social innovations. Sharing concepts are becoming accepted more widely. The open-source principle largely replaces the ownership principle. A new common 'feeling of community and self-determination' develops between residents of regions and communities. Production will be less concentrated in metropolitan areas. Regions and communities formerly affected by rural exodus will recover.

From the technical renewal of industrial production, a progressive economy develops: the technology-based demand economy. Instead of selling goods, an increasing number of construction plans are being purchased, which can be processed by intelligent (self-learning) systems and adapted to local conditions (e. g., resource deposits, energy supply). While the majority of consumer products are produced locally (in households or in local manufacturing cells), established companies continue to manufacture complex products (machines, systems, vehicles, etc.). In contrast to today's centralized linear production methods, however, this is decentralized and organized in loops.

Ultimately, decentralization, substitution of ecologically disadvantageous production processes via BIS, increased collaborative use (shared economy), and problem-solving skills of self-learning systems result in a substantial reduction of the consumption of non-regenerative materials and fewer emissions.

Loops are thoroughly closed according to the cradle-to-cradle approach. Re- and upcycling are an intended aspect of a product's life cycle, which is already considered in its early stages. Biobased products can easily be released to the biosphere and returned to production. The substitution of imported biobased materials by regionally available biobased materials allows a considerable reduction of import dependency. Standards for the valuation of sustainable production are actively promoted by governments at an early stage, allowing for specific regulatory interventions. The extension of accounting boundaries and the consistent pricing of external effects leads to a fundamental shift in corporate attitudes. As a result, stakeholder orientation finally and thoroughly replaces shareholder focus. Decision-makers increasingly prioritize the service to the community. The total individual footprint of citizens decreases to a socio-ecologically acceptable level. The ecosphere is relieved in absolute terms.

4.2 Terminology of the concept of biointelligence

After providing an overview of a reorientation of production technology and economy in the previous section, this section introduces a scientific definition of biointelligence. Yet, prior to defining the term itself, a few preliminary considerations are necessary. The following section thus initially compiles elementary terms and concepts from existing literature in order to create the basis for gradual development and recombination. For further understanding, the definition of the prefix bio and the term intelligence is crucial. The prefix bio originates from Greek and is a synonym for life (Langenscheidt, 1993). The term intelligence is derived from the Latin word intellegere, which is often translated as recognize, see or understand (Langenscheidt, 2010). While the roots of the terms are clear, their interpretation appears to be complicated. Equally important for the delimitation of biointelligence is the definition of technology as well as its BT.

4.2.1 Intelligence

A generally accepted interpretation of the term intelligence has not yet been established. Since the existing views of the term today primarily arise from the scientific disciplines of philosophy and psychology, the discussion is increasingly supplemented by contributions from the information and communication sciences in the course of the debate on so-called artificial intelligence (AI). While psychology attempts to describe the existing intelligence of an individual as accurately as possible, e. g., in order to predict his or her professional career, the philosophical approach focuses on describing the cause of intelligence. In addition to the concept of intelligence, the term emergence is of crucial importance.

4.2.1.1 The concept of intelligence in psychology

In order to describe the ability of the human mind to understand what is true, the term intellectus was already introduced in the Middle Ages as a technical term in classical philosophy for the word nous (mind). However, the understanding of human intelligence in psychology, which dominates today, was primarily influenced by psychometry at the beginning of the 20th century by Binet (2015), Thurestone (2014), and Spearman (1904a; 1904b; 1914). While Binet (2015) referred to the term as human judgement or 'common sense', Wechsler (1944) identified intelligence as the ability of an individual to act purposefully, think rationally, and deal effectively with his environment. Humphreys (1979), in contrast, characterized intelligence as a process sequence for dealing with information, consisting of gathering, storing, retrieving, combining, comparing, and using. Other proposed definitions highlight elements such as dealing with complexity, problem-solving skills, goal-oriented action and the ability to adapt to changing life situations (Sternberg, 1982; Gardner, 2011). Due to the diversity of existing definitions, a group of fifty-two researchers published a statement in the Wall Street Journal in 1994. The term intelligence is described there as a general mental ability to think, plan, solve problems, think abstractly, grasp complex ideas, learn quickly, and learn from experience (Gottfredson, 1997).

To date, various theories, models, criteria and tests of intelligence have been presented with the aim of mapping the cognitive performance of an individual in order to predict academic success and the prestige of the profession later achieved. The most prominent approaches in psychology are called factorial theories, as they are based on a factor analysis of human intelligence (Horn and Cattell, 1966; Marshalek et al., 1983; Kyllonen and Christal, 1990; Kane et al., 2005; Ackerman et al., 2005; Finkel et al., 2007; Süß and Beauducel, 2015; Stern and Neubauer, 2016). Here, the results of a large number of test items are combined into a data structure that can be described by a few factors. Well-known examples of this theory genre are the two-factor theory, the primary factor model, and the two-factor model (Spearman, 1904a; 1904b; 1914; Thurestone, 2014; Horn and Cattell, 1966). Further multidimensional approaches are the cube model, the Berlin intelligence structure model, the theory of multiple intelligences, and the triarchic intelligence theory (Gardner, 1993; 2011; Süß and Beauducel, 2015; Sternberg, 1978; 1982; 2003).

All these approaches have in common the description of intelligence on the basis of a number of characteristics that an object of investigation is more or less capable of fulfilling. For each characteristic, tasks are formulated whose success in solving them determines the intelligence of an individual. While Spearman (1904a, 1904b, 1914) distinguishes between general intelligence, i. e., intelligence that influences all areas of performance, and specific intelligence, i. e., intelligence that is only relevant in certain areas (e. g., mathematical tasks), the two-factor model differentiates between fluid and crystalline intelligence (Horn and Cattell, 1966). The former is innate or inherited and cannot be influenced by the environment, whereas the latter comprises all abilities that are learned or determined by the environment in the course of life (i. e., semantic knowledge, e. g., factual knowledge, and learned characteristics, e. g., behavior, mathematical knowledge). The primary factor model, on the other hand, describes intelligence as an area-specific composition of various individual abilities (Thurestone, 2014). These include the perception of space as well as of different objects and their relation to each other, mathematical and computational skills, memory, logical reasoning, word fluency (e. g., finding synonyms) and the 'correct' interpretation of verbal relationships. Meanwhile, the Berlin intelligence structure model distinguishes between two modalities, each of which can be described by certain bundles of abilities: the operations and the contents (Süß and Beauducel, 2015). The first comprises criteria such as the speed of processing a task, the ability to remember verbal, numerical, and figural material, the ingenuity in solving problems and the processing capacity in dealing with complex information. The modality 'content', in turn, includes the ability to acquire language-, numerical-, and view-based operations. The theory of multiple intelligences adds further psychometric features to the focus by distinguishing between seven types of intelligence (Gardner, 1993, 2011; Neisser et al., 1996):

- *Linguistic intelligence*: affinity for spoken and written words.
- *Logical-mathematical intelligence*: logical and mathematical abilities.
- *Spatial intelligence*: ability to form a mental model of a spatial world and to maneuver and operate with this model.
- *Musical intelligence*: ability to produce an excellent or absolute pitch.
- *Physical kinesthetic intelligence*: ability to solve problems or design products with the whole body or with parts of the body.
- *Interpersonal intelligence*: ability to see things from the perspective of others or to understand people in terms of empathy.
- *Intrapersonal intelligence*: ability to create a precise, accurate model of oneself and use this model to work effectively in life.

According to the expression in one or more areas of multiple intelligences, potentials arise for certain professional groups. For example, people with a high physical-kinesthetic intelligence are usually good dancers, athletes, surgeons or craftsmen, while a pronounced interpersonal intelligence offers advantages especially for teachers or politicians.

In order to describe human competence even more comprehensively, the triarchic theory or theory of successful intelligence was proposed (Sternberg, 1978, 1982, 2003). In this theory, the intelligence of a person is not defined according to absolute standards, but rather on the basis of individual standards and in the respective socio-cultural context. Accordingly, three basic aspects of intelligence are distinguished. First, analytical intelligence comprises all mental processes by which intelligence is expressed. Creative intelligence, on the other hand, describes how an individual deals with new challenges. Last but not least, practical intelligence encompasses the active adaptation, selection, and design of the environment. It is strongly dependent on the respective socio-cultural milieu of the individual. An individual is sufficiently intelligent if he or she can use the processing skills available to him or her in an appropriate way (i. e., for success) and compensate for existing weaknesses.

While the categorization of human intelligence, especially with regard to more complex abilities (e. g., logical reasoning), depends on the underlying theory, there is a consensus that processing speed and working memory are fundamental mental abilities and are therefore elementary components of intelligence. Since this does not only apply to humans, but also to living beings in general, the field of animal cognition or cognitive ethology has developed (Tolman, 1948; Premack and Woodruff, 1978; Settleworth, 2009). The focus of this discipline is the analysis of the mental abilities of non-human living beings.

4.2.1.2 The concepts of intelligence and emergence in philosophy

The field of neuroscience is concerned with the neuronal basis of human intelligence. Here, intelligence is usually reduced to the processing of signals and information in the brain (especially in the cortex). However, this interpretation does not go far enough to describe the cause of intelligence. In order to approach the philosophical goal, i. e., the investigation of the origin of an individual's intelligence, it is first necessary to introduce the term emergence (also supersummativity or fulguration) (Goldstein, 1999; Corning, 2002; Binder, 2009). Helbig (2018) sketches the term on the basis of the beginnings of the development of different fields of science. For instance, mankind began some time ago to pursue the question why certain observable phenomena occur and why the underlying processes always appear in a certain describable way. On the basis of individual observations, explanatory approaches were developed from which entire sciences arose. However, the focused approach of the disciplines, which were separated from each other by definition, led to the fact that the connections between the sciences were not completely described. To illustrate this, Helbig

(2018) presents an approach based on the division of the world into different layers of being. The individual levels can be assigned to fields of science (Table 4.1).

Table 4.1 Layers of being according to Helbig (2018)

Level	Scientific discipline
Atomic and subatomic range	Physics, cosmology
Molecules and their relationship to each other	Chemistry
Living cells	Cell biology
Living creatures (e. g., plants, animals)	Biology
Brain, individual consciousness, language	Neurology, psychology, linguistics
Man-made artifacts	Engineering and computer science (including AI)
Social groups/societies including technical and economic tools as well as history	Sociology, ecology, history
World consciousness, global emergent phenomena	Philosophy, theology

The hierarchically arranged levels are subject to a strict order of precedence, as each level is composed of units of the level mentioned above. For example, organisms consist of organs, which in turn consist of cells, which again consist of organelles. Last but not least, organelles are a construct of different molecules. A central question is how the properties and rules that apply to the units of a level are based on the properties and rules of the units of a lower level. Emergence deals with this question. It describes that the interaction of several units of a level E_n can develop properties on a higher level E_{n+1}, which are not present or observed on the level of the units E_n. Simplified: a system is considered emergent if its properties cannot be explained from its components (Helbig, 2018). Examples of emergent phenomena exist in a wide range of scientific disciplines. In neurosciences, for example, consciousness is considered an emergent attribute of the brain (Mahner and Bunge, 2000). Similarly, in biology: protein molecules have characteristics that none of their atoms have (Campbell and Reece, 2005). Life as such: the ability to reproduce is a property of a higher level, which results from the interaction of chemical and biological molecules and polymers. But molecules cannot reproduce themselves.

Generally, strong and weak emergence are distinguished (Laughlin, 2007; Kim, 2016). While strong emergence describes that properties of a phenomenon of a higher level E_{n+1} are not attributable to the properties of the units of the lower level E_n (see vitalism, Section 4.2.2.1), weak emergence is based on the assumption that all properties of phenomena of all levels are attributable to the properties of the lowest level. If this is not possible, it is because the connections cannot yet be explained with the instruments available today (Helbig, 2018).

Emergence and intelligence of systems have an inherent relationship. This becomes particularly clear by the swarm behavior in nature. For example, swarms of ants build complex structures and organize their colonies in hierarchical structures, which is often referred to as socialization. It is found that the ant, due to its limited perception and processing of information, is hardly considered intelligent, but the colony as a whole develops a collective intelligence (Helbig, 2018). However, the concept of intelligence is difficult to define here as well. Helbig (2018) lists, similar to the approaches outlined above, ten essential abilities that can serve as characteristics of intelligence, but stresses that these should be understood solely as a guide and less as a complete list:

1. Knowledge acquisition and memory
2. Use of the stored knowledge and conclusion
3. Ability to abstract and specialize
4. Ability to plan and solve problems
5. Real-time adaptability
6. Ability to learn (assess progress vs. regression)
7. Symbol processing (words, characters, ...)
8. Ability to act in incompletely described or recognized situations
9. Object/pattern recognition capability
10. Ability to communicate (speech processing)

4.2.1.3 The concept of intelligence in information technology

The concept of intelligence as a process of information processing is not only popular in the neurosciences, but also in information and communication sciences. Computers have already surpassed the computing power of the human brain by far when dealing with individual tasks. Accordingly, McCarthy developed the term artificial intelligence (AI) in 1956, aiming to design machines that act as if they are intelligent (Ertel, 2016). Today, the term AI is defined wider and is commonly associated with systems that perceive their environment and take measures to maximize their chances of achieving a given goal as well as possible (Russel and Norvig, 2010). In contrast to electronic data processing (EDP) systems, which operate according to the principle of 'input – processing – output', learning and understanding are essential attributes of AI. Generally, three types of AI systems are distinguished. First, analytical AI only exhibits characteristics that are compatible with the understanding of cognitive intelligence outlined above. In terms of technical implementation, the focus here is primarily on learning from past experience in order to be able to independently make appropriate decisions. Human-inspired AI, in contrast, contains elements of cognitive as well as emotional intelligence. The aim of these AI systems is primarily to understand human emotions and to take them into account when making decisions. Not least, humanized AI displays characteristics of cognitive, emotional and social intelligence. These AI systems are able to be aware of themselves and their interactions. Today, AI systems are elementary components of various technical systems (Russel and Norvig, 2010; Ertel, 2016). Although they are extremely well able to solve specific tasks, they are clearly inferior to humans with regard to numerous other tasks.

In the course of this work the following interpretation is implied: *The concept of intelligence can be differentiated into two interpretations, a narrow and a broad one. Intelligence in the narrow sense describes the cognitive abilities of living systems (i. e., cells, organisms, humans, swarms). Intelligent behavior cannot always be explained by the sum of its parts (emergence). In a broad sense the term is synonymous with the ability to act 'rationally', i. e., practical intelligence. Yet, rational behavior is less understood here in the sense of the economic concept of homo economicus, but as the ability to develop the best possible solution by taking into account the totality of influencing factors, i. e., to act reasonably. Artificial intelligence, in contrast, is defined as the deliberate imitation of cognitive abilities of living systems (i. e., intelligence in the narrow sense) by a technical system. A technical system can achieve different degrees of intelligence, which depend on the independence, complexity and efficiency of problem solving. The basic building blocks for such 'cognitive' performance are perception, understanding, action, learning, and the ability to adapt to new environments.*

4.2.2 Life (Bio)

For the prefix Bio, synonymous with life, a generally accepted definition has not yet been presented either. While today it seems hardly possible to trace its origin, early basic considerations, similar to the concept of intelligence, can be found in the philosophy of ancient Greece and religion. The current understanding, however, is primarily influenced by natural sciences. The following sections summarize some approaches to definitions from philosophy and natural sciences in order to derive an appropriate interpretation for this work.

4.2.2.1 The concept of life in philosophy

Ancient Greek philosophy mentioned two words for the term life, zoḗ and bíos (Bedau, 2010; Karafyllis, 2012). The former is equivalent to the word soul. In contrast to later periods, the presence of a soul in ancient Greece was attributed to both humans and animals. The term zoḗ thus described the common psychological origin of living beings. Meanwhile, the word bíos referred to the way of life dictated by reason. While this distinction shaped almost the entire philosophy of the end of the archaic age (ca. 800–500 BC) and the beginning of the classical epoch (ca. 500–336 BC) of ancient Greece, medieval philosophy summarized both terms in the word vita (Bedau, 2010; Karafyllis, 2012).

Apart from the interpretation of the term life, the ancient Greek philosophers searched for its origins. The idea that life consists of one or more basic elements was elementary. These included: water, air, fire and earth. Thales of Miletus and Anaximenes were among the first ancient philosophers of nature to postulate the thesis that all life originates from water. Anaxagoras, in contrast, identified the seed (here a synonym for earth) as the smallest element of all life. Heraclitus, in turn, regarded an eternally living fire as the very basis of existence. The first combination of the four basic elements (four-element doctrine) dates back to Empedocles, who described the origin of animate bodies in his phylogenetic theory. This theory, now considered the precursor of the theory of evolution, describes all physical (i. e., animate and inanimate) objects as mixtures of the four elements water, air, fire and earth (Geoffrey et al., 2001). Differences between animate and inanimate objects as well as between different species arise from the mixture ratio. The history of creation begins with the damp earth, out of which the first plants, animals, and humans developed from the unifying power of love. Empedocles already emphasized the importance of coincidence and individual biological functions (procreation, embryonic development, respiration, and sensory perception) for the origin of living beings. Aristotle, in turn, was the first to introduce a hierarchy of living beings according to their 'soul capacity'. Therein, he distinguished between three stages: [1] plants, whose life consists solely of nutrition and reproduction, [2] animals, which were additionally able to perceive senses and to move, and [3] humans, who are the only ones able to think.

Until the middle of the 20th century two contrary streams developed in philosophy. Mechanicism attempts to explain life solely from the presence, motion, and interaction of matter and to derive universal laws from this. Mechanical philosophy assumes that metaphysical concepts such as spirit, will, soul or consciousness can be explained exclusively by material connections (see Section 4.2.1.2) (Lüthy et al., 2001). Immaterial interpretations are strictly rejected in this mind-set. Today's predominant philosophical-physical interpretation, atomism, is regarded as a manifestation of this view, as are Newtonian mechanics or molecular biology. The concept of vitalism, on the other hand, which is shared in a similar way by a multitude of religious faiths, is based on the assumption of an essential difference between everything organic and everything inorganic. The concept assumes that life results from a primeval force, the soul or the substance of life (Gerabek et al., 2004). Due to the further development of biotechnology, which was able to demonstrate that the production of organic substances is possible without an additional life force, as well as the emergence of synthetic biology and the successes in scientific disciplines that are related to mechanism, vitalism is now considered largely obsolete. Today, the mechanical philosophy is no longer entirely questioned, but rather relativized in relation to individual questions, e. g., in systems or organism theory. The latter may be seen as an integrative approach between mechanism and vitalism from the perspective of biology. In contrast to the biological reductionism resulting from mechanism, which seeks to explore biological phenomena at the micro level, the concept emphasizes the existence of emergent properties of living beings (Laughlin, 2007). On this basis, the approach proposes a reciprocal choice of perspective in order to analyze phenomena appropriately.

4.2.2.2 The concept of life in the natural sciences

While natural sciences are still not able to explain how life developed, the following is regarded as a consensus (Schrödinger, 1967; Jeuken, 1975; Eigen et al., 1981; Koshland and Daniel, 2002; McKay, 2004; Trifonov, 2012; Görtz and Brümmer, 2012; Fritsche, 2015): the smallest form of life is the cell. A cell contains all components necessary for life and separates these from the outside world by a cell wall. If the wall is broken, the cell dies. All known forms of life consist of at least one cell. The necessary components within the cell are structured the same way in all life forms. Whether plant cell, animal cell or microorganism, the most important components consist of molecules and fulfil the same tasks everywhere. Lipids are involved in building the cell wall, carbohydrates are the basis for the metabolism and nucleic acids are building blocks for the storage of genetic material. In all life forms, the storage of the construction plan and the organization of the organism is based on the combination of just four nucleic acids, the genetic code, although from a chemical point of view a variety of nucleic acids would be possible.

In an approach similar to dataism[1)], Witzany (2019) identifies communication as an essential criterion of the concept of life. In contrast to inanimate matter, communication is found on every level of life (including cells, tissues, organs, organisms) as sign-mediated interactions. On the other hand, the currently dominant understanding of life, comparable to the concept of intelligence, is based on a list of characteristic features (Görtz and Brümmer, 2012; Fritsche, 2015). Depending on the author, the list contains more or less aspects, yet basically resulting in the same statements (Jeuken, 1975; Koshland and Daniel, 2002; McKay, 2004; Trifonov, 2012): the structuring of life in cells requires an inner state of low entropy, which can be maintained by a continuous exchange of energy. Energy is absorbed through metabolism based on the restructuring of molecules, which are more easily accessible by the cell when it can absorb and process information about its environment. If good environmental conditions allow the cell to grow to its natural limit, it can divide to spread its genetic code. The principle of reproduction is subject to evolution, in which only the best

[1] Dataism represents a not yet further specified trend in the broader field of information and communication sciences, which attributes physical, chemical, and biological phenomena to the exchange of information or the processing of data.

adapted life form survives (Görtz and Brümmer, 2012; Fritsche, 2015). The listed characteristics apply to all living beings on earth. However, even here borderline cases exist. Cryptobiotic life forms present a challenge to the definition of life. Tardigrades, for example, reduce their metabolism in poor environmental conditions to a point where it is no longer measurable. A similar mechanism can be observed in certain bacteria. Under poor environmental conditions, these bacteria form capsules in which they survive without a measurable metabolism. Besides cryptobiotic life forms, viruses are also considered borderline cases. Although they possess their own DNA, they cannot reproduce themselves but need a host cell. Therefore, viruses are not considered to be living organisms. A classification according to the criteria listed above has therefore not yet been satisfactorily completed. The decisive factor is thus not the fulfilment of individual characteristics, but the sum of all characteristics.

The following understanding will be assumed in the further course of this work: *The term life (as a synonym for the prefix bio) describes the sum of properties that must be fulfilled in their entirety in order to distinguish a system from inanimate matter. These are: [1] an internal state of low entropy, [2] a continuous exchange of energy, [3] the presence of a metabolism, [4] reproduction, [5] evolution, and [6] communication.*

4.2.3 Technology and its biological transformation

The term technology originates from the ancient Greek word téchnē (engl. art, craft) and describes the knowledge of nature and technique as well as the teaching of them. While the German language distinguishes between the terms technology and technique (Technik), the English language adopts both interpretations in the term technology. It can thus be used to refer to the entire spectrum of scientific and technical knowledge that represents the basis for all products and production processes or simply to a collection of techniques to achieve a certain purpose (Bain, 1937; Mannix and Stratton, 2005; Schatzberg, 2006; Borgmann, 2006; Brian, 2009). In the following, this work refers to the interpretation of the German VDI 3780 that understands technology as the entire quantity of man-made objects as well as the manufacturing processes, knowledge, and skills required for this. An elementary component of technology is the idea of a technical (factual) system, which is considered to be sufficiently described by the sum of four properties (Verband Deutscher Ingenieure, 2000):

1. Consisting of inanimate material-concrete matter (i.e., materials with defined properties)
2. Application orientation
3. Explicit functional principle (to convert, transport and/or store material/mass, energy and/or information)
4. Spatial structure with geometrically defined shape

Technical systems always pass a certain 'life cycle' consisting of the extraction of the necessary raw materials from the natural environment, the targeted use, conversion and utilization of materials and/or energy as well as the disposal at the end of their service life (Klöpffer and Grahl, 2014; Guinée et al., 2001; Frischknecht, 2013).

Technology today is both the basis of the prosperity of so-called industrialized countries and a driving force behind extensive environmental impacts. Recent achievements in biotechnology and AI have led various authors to conclude that BT has considerable potential for disruptive and sustainable innovation.[2] Biologicalization is occasionally used as a synonym (Byrne et al., 2018; Patermann and Aguilar, 2018). While the term BT is increasingly establishing itself in the socio-political discourse, the term biologicalization represents a gradual change that can lead to unwanted misunderstandings, since its use in the context of the social Darwinist teachings of the Third Reich (1933–1945) is sufficiently documented (Vom Bruch and Kaderas, 2002). In the course of the incorrect interpretation of 'natural' selection in the social 'struggle for existence', countless cruelties were committed in nationalist eugenics programs (Lenzen, 2003). Since this understanding in no way corresponds to the interpretation intended by the author, the term biologicalization is refrained from being used in this work. Consequently, the term BT is used exclusively. However, the term is not to be confused with the term biotransformation, which is known in biology and describes a process in the metabolism of living beings in which substances that cannot be secreted are converted by chemical processes into excretable substances (Stürmer and Breuer, 2006; Ashauer et al., 2012; Amadi et al., 2018). The word transformation is rather understood here in the Kuhnian sense of a paradigm shift that occurs primarily in the context of the microeconomic development of new technologies (Kuhn, 1996b). After some authors initially suggest the term as a systematic application of biotechnology to create a bioeconomy in the narrow 'materialistic' understanding (Kircher, 2012; Besi and McCormick, 2015; Patermann and Aguilar, 2018), Byrne et al. (Byrne et al., 2018) present a purely technical approach to production by predicting an increasing use and integration of biological and bioinspired principles, materials, functions, structures, and resources for intelligent and sustainable manufacturing technologies. Meanwhile, Miehe et al. (2018; 2020a) and Bauernhansl et al. (2019) advocate a holistic understanding of the transformation process towards a bioeconomy. Thereby, the previous understandings are adapted by describing BT as the increased use of materials, structures, processes and organisms of living nature in

[2] The term disruptive innovation is based on Schumpeter's concept of 'creative destruction' Schumpeter (2005; 2007). With creation, i. e., the recombination of production factors, destruction occurs, i. e., old structures are displaced. The basic assumption here is the entrepreneurial 'joy of creation', a characteristic which, according to Han (2002), was primarily shaped by 'western' philosophy. Through a new development that is well received by the market, the entrepreneur will soon become a monopolist (Christensen, 1997). Since successful concepts are often copied by competitors, imitation follows. Schumpeter considers this alternation between innovation and imitation to be the driving force behind competition.

technology (Bauernhansl et al., 2019; Miehe et al., 2020a). As a target state, an economic form (the technology-oriented demand economy) is outlined, which resembles the initial, holistic understanding of commodity theory, yet supplements it with essential elements of circular and demand economy. The interpretation presented here presupposes the distinction of three system levels. The technical system describes the totality of man-made objects that fulfil a value-added purpose. The biological system comprises the entire living nature, while the information system contains the totality of information. Through the integration of one or more levels, so-called integrative systems are created. Figure 4.2 illustrates the understanding of the system levels according to Satava (2002) and Miehe et al. (2018; 2020a).

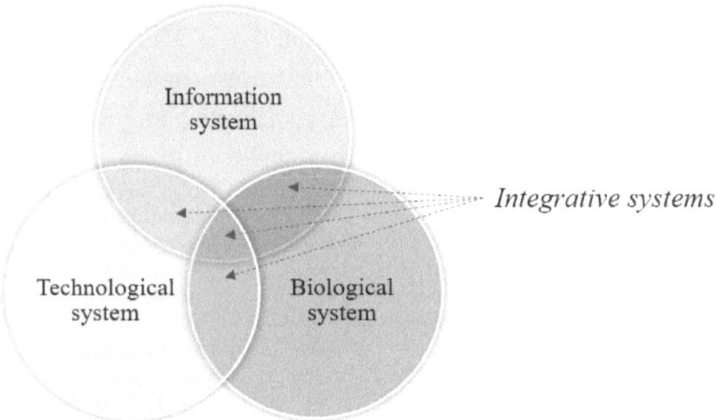

Figure 4.2 Levels of integration of BT according to Satava (2002) and Miehe et al. (2018; 2020a)

Depending on the depth of integration of technical, biological, and informational systems, Miehe et al. (2018; 2020a) further distinguish between three development modes. Inspiration describes the transfer of biological phenomena into technology, a concept derived from biomimetics/biomimicry. As a result, a bioinspired system represents a pure expression of a technical system. Biologically inspired new types of materials and structures, which are used in lightweight construction, for example, continue to consist mostly of inanimate matter. Innovative organizational and cooperation solutions in manufacturing companies, e. g., swarm-intelligent manufacturing systems based on neural networks or evolutionary algorithms, are still inanimate production systems. Hence, an integrative system is not created by inspiration. In contrast, integration is defined as the incorporation of biological systems into technology. In the context of production systems this can be, for example, the substitution of chemical processes by biological processes or the use of biological systems (microorganisms, algae, etc.) to close cycles (recovery of rare earths from magnets, production of bioplastics from CO_2 waste streams, etc.). The integration mode thus comprises a wide range of approaches from biotechnology and bioprocess engineering. In contrast to the first two modes, the comprehensive interaction of technical, informational, and biological systems represents a new way of thinking that is intended to enable the development of new, self-sufficient production technologies and structures. Integrative systems resulting from the interaction of technical, informational, and biological systems are referred to as BIS. BT thus primarily serves to create a new technological basis for an economy. Its modes of development are thus motion variables, while their results represent state variables. Figure 4.3 illustrates the understanding of BT according to Miehe et al. (2018; 2020a).

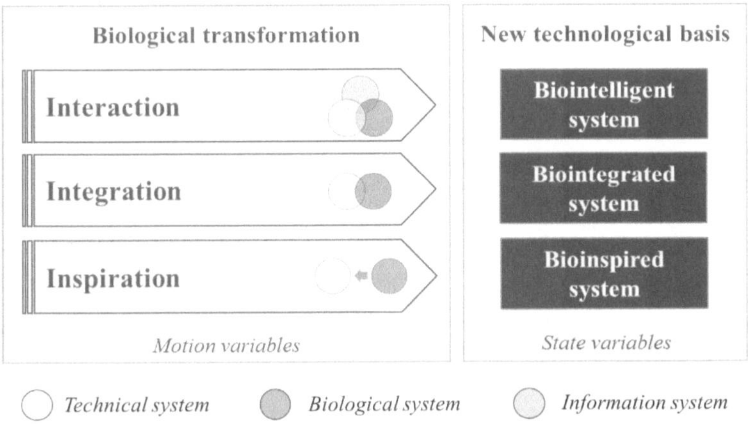

Figure 4.3 BT of technology according to Miehe et al. (2018; 2020a)

4.2.4 The sustainable transformation of society

For years, the perception of sustainability is rising in society and increasingly determines the way of thinking and acting on an economic and political level. The societal debate, which is still emerging on a comparatively small scale, is seen by various authors as a further foundation for a reorientation, in addition to the required technological renewal (Miehe et al., 2018; Miehe et al., 2020a; Bauernhansl et al., 2019; Weizsäcker and Wijkman, 2017). The goal of this transformation process is the development of a new narrative in the sense of an ideology that is supported by the majority of society and thus affects the essential theoretical constructs of science. At its core, the debate is concerned with the relative importance of the three ideals of freedom, equality, and responsibility. Similar to the modes of development of BT, three elementary building blocks of the transformation process can be distinguished. First, the enlightenment postulated by Weizsäcker and Wijkman (2017) requires the clearance of misunderstandings (especially with regard to the use of (neo)classical theories to legitimize today's economic methods) and the identification of the potentials and risks of new technology. Consequently, these must be eroded in the sense of the principle of participation (see Section 3.6) through a discourse involving society as a whole and embedded in the form of a modified framework (transition). A decisive factor in this process is to abandon 'linear thinking' in individual disciplines in favor of an integrative, systemic decision-making process that includes as many interdependencies as possible in a transparent manner (Vester, 2011). A process of creative destruction in the sense of Schumpeter (2005; 2007) can only emerge by an interaction between the BT and the sustainable transformation of society, resulting in sustainable innovations in essential fields of human need (e.g., health, housing, nutrition, energy, consumption) within a new economic structure. Figure 4.4 illustrates the interaction.

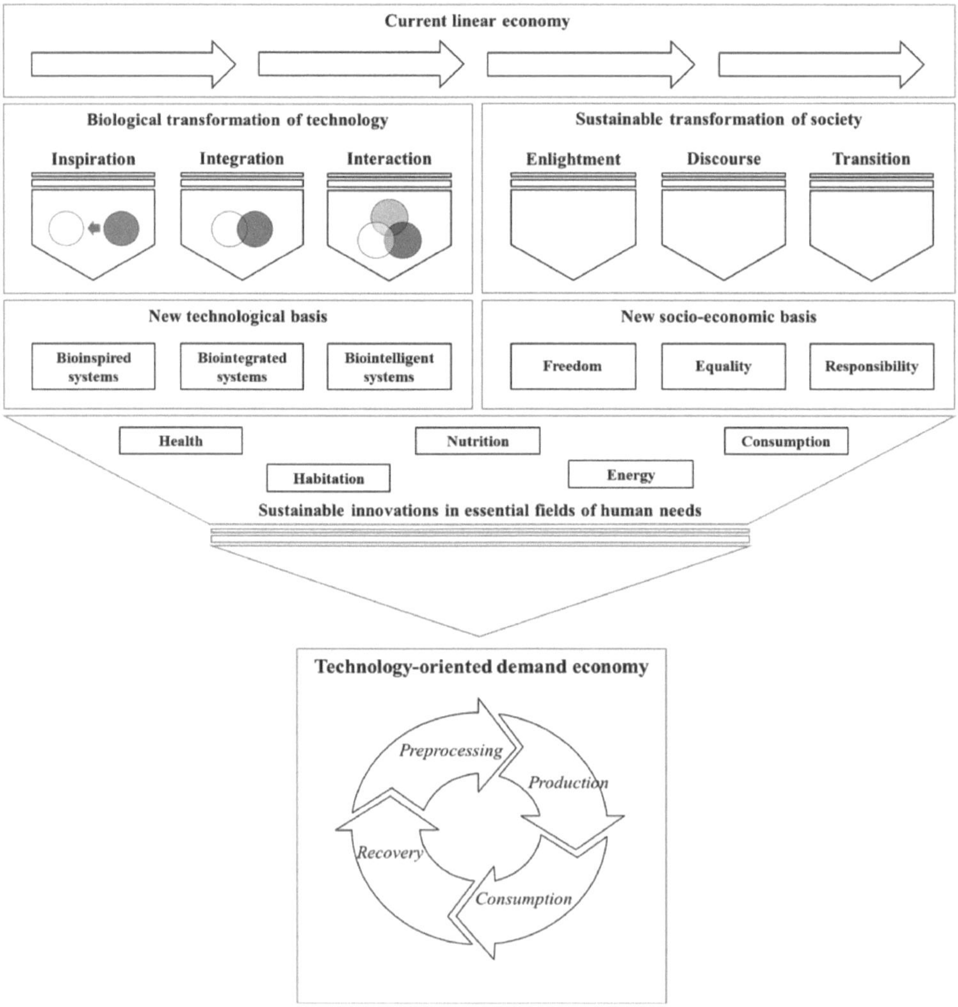

Figure 4.4 Holistic change towards technology-oriented demand economy [own illustration]

4.3 Biointelligence

After establishing biointelligence as a result of the systematic interaction of technical, biological and informational systems, a further specification of the term is required.

4.3.1 The term biointelligence

First mentions of the term biointelligence in the scientific community date back to the late 20th century. Stonier (1997) initially describes the concept in contrast to machine intelligence as the 'true' intelligence of nature, which is expressed in the complexity and balance of natural organisms and systems. Other early approaches to biointelligence are rather related to bioinformatics and systems biology in medicine. Satava (2002) outlines the concept by the example of robotics-aided or computer-aided surgery. Here, electronic information interfaces are expected to replace the surgeon's hands and brain. Mousses et al. (2008) sketch a concept of biological intelligence-based knowledge recovery that may be applied for various personal genomics in medicine. Farley et al. (2013) present a scalable computational platform for biomedical knowledge computing aiming to represent complex syntactic and semantic relationships in unstructured biomedical information. From the perspective of robotics, Pearce (2012) outlines the vision of self-improving organic robots being capable of modifying their own source code and optimize themselves independently. While the vision of self-improving organic robots already contains elements of the understanding outlined below, it fails to meet the purpose and time horizon of the proposed definition. An appropriate interpretation in the context of this work requires the consideration of both macroeconomic and microeconomic requirements. Based on the previous contributions, two basic interpretations may be distinguished: the comprehensive use of information technology interacting with biological systems, and the recognition of the uniqueness of living systems or natural ecosystems. Below, this work thus distinguishes between two interpretations of the term biointelligence, which will be pursued equally in the further course of this work. While the explicit concept of biointelligence is a purely technical one, i. e., it focuses exclusively on the micro level (individual systems such as machines, humans, etc.), the implicit concept represents a holistic view (macro level).

Both interpretations are based on the understanding of human intelligence. Depending on the fulfillment of a set of characteristics (see Chapter 5), different systems can exhibit distinct forms of biointelligence. In comparison to AI, the term is not associated with a bundle of tools that a priori guarantee biointelligence. While this might be plausible in an explicit sense, a system can only be regarded as fully biointelligent once it has provided evidence of the implicit interpretation.

1. *Explicit concept of biointelligence:* in this understanding, the term refers to the systematic information-technological networking of biological and technical systems. In this sense, a system can be described as biointelligent, if it contains a biological and a technical component and/or active principle, if these interact in any form (biology technology interface, BTI) and their interaction is regulated in the form of an autonomous control. It is irrelevant whether the control is designed intelligently as such, i. e., based on AI approaches, or whether conventional control concepts (e. g., electronic data processing according to the Input – Processing – Output principle) are used. Decisive is that the overall behavior of the system corresponds to characteristics of intelligence in the narrow sense.
2. *Implicit concept of biointelligence:* according to this definition, a system is referred to as biointelligent, if it demonstrates a responsible handling of biological resources ranging from single cells to entire ecosystems. Its target state is balance with its natural environment and surrounding systems.

4.3.2 Biointelligence as a scientific discipline

After an initial introduction to the definition of biointelligence, this section discusses the question of its scientific and theoretical implications.

4.3.2.1 Basic considerations on the scientific positioning

Sciences can be grouped into different categories depending on the object of investigation and the method of obtaining knowledge. However, a generally binding classification of the sciences does not exist (Zelewski, 1999). The most widely accepted approach to scientific systematics first differentiates between object and meta-sciences (Zelewski, 1999). The object of investigation of metascience is science itself. It includes philosophy, logic, and sociology of science. Object sciences, in contrast, deal with all questions whose object of investigation is not the sciences itself. Object sciences are traditionally distinguished into formal and real sciences (Zelewski, 1999; Ulrich and Hill, 1976). Zelewski (1999) supplements structural sciences. The formal sciences investigate formal objects. Their aim is to construct sign and language systems. Formal sciences include mathematics, logic, and philosophy. Characteristic for this category of science is the method of deductive reasoning (Zelewski, 1999). In contrast, the object of investigation in real sciences is the real object outside scientific language systems, while structural sciences focus on the analysis of general structures that can be found in both formal and real sciences (Zelewski, 1999). The structural sciences include synergetics, system and problem theory. A further differentiation of the real sciences into social and natural sciences is based on the goal of describing, explaining, and shaping empirically recognizable sections of reality (Ulrich and Hill, 1976). While natural sciences pursue the goal of generating theoretical models of explanation, social studies focus on the development of practical decision models and processes. Kuhn (1996a) identifies the set of recognized paradigms in a discipline as another distinguishing feature. These are found much more frequently in the natural sciences (Ulrich and Hill, 1976). The natural sciences include physics, chemistry and biology. Representatives of the social sciences are primarily economics and sociology. According to Zelewski (1999), the characteristic method of real sciences is the nomographic approach, i. e., the search for generally valid rules (hypotheses) and their empirical testing.

Many established applied disciplines (e. g., engineering sciences) operate at the interface between natural and social sciences by combining the laws, approaches, and knowledge of both fields in order to develop practical alternatives for action. Since the object of investigation of biointelligence is undoubtedly the real object outside of scientific language systems, i. e., not science itself, it is clearly a discipline of real sciences. Similar to engineering sciences, it also aims at the specific combination of natural and social science findings for the purpose of practical application. Decisive for the application potential is however not only the economic efficiency, but a balance of economic, ecological, and social benefits. According to the interpretation described above, this represents a mutual dependency on natural and social sciences, since in order to develop the systems it is necessary to rely on both scientific disciplines and to prove the advantages. From this perspective, biointelligence represents a new scientific discipline. Figure 4.5 illustrates the positioning of biointelligence in the systematics of science according to Ulrich and Hill and Zelewski (Zelewski, 1999; Ulrich and Hill, 1976).

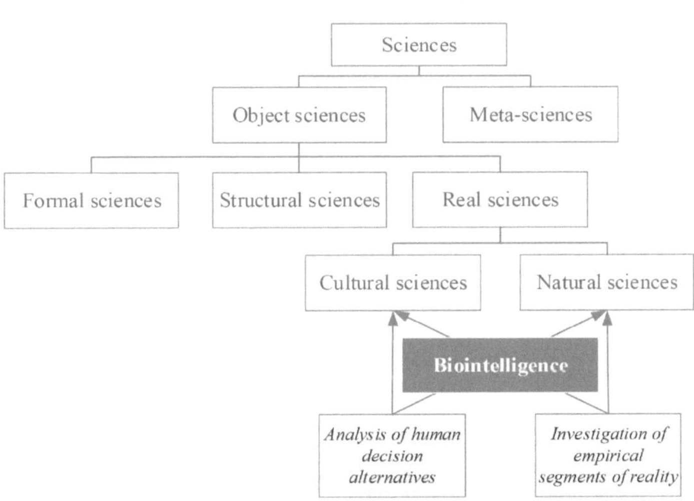

Figure 4.5 Placing biointelligence in the systematics of science according to Ulrich and Hill (1976, p. 307) and Zelewski (1999)

4.3.2.2 Method of insight generation

According to Warnecke (1992), new products and processes are not just the result of scientific advances; new scientific advances are also the result of new needs. Sections one to three have comprehensively portrayed the transformative nature of needs with regard to the imperative of reorienting established patterns of thought in the production sciences. Scientific progress therefore follows two paths: first, new products and processes are created on the basis of new scientific findings; second, new needs lead to new findings (Warnecke, 1992). For biointelligence, this results in two concepts for further scientific progress (Figure 4.6): bottom-up and top-down approach. The former designs biointelligent systems in the explicit sense on the basis of new scientific findings. The latter generates scientific findings by examining their contribution to the diverse needs described above in the implicit sense of biointelligence. In this context, the concept of needs must be oriented not only to individual needs, but equally to societal and ecological aspects.

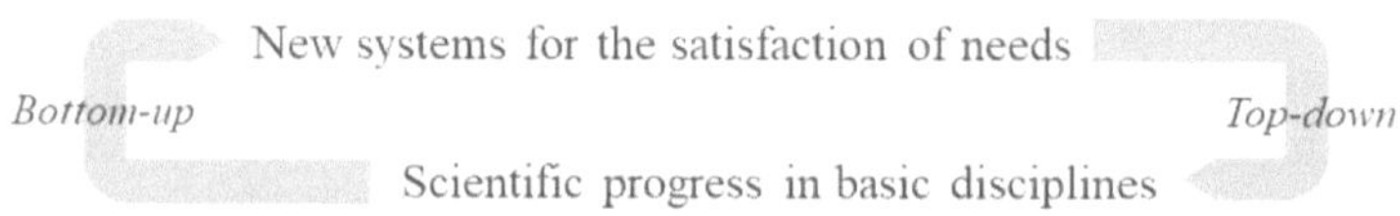

Figure 4.6 Bottom-up and top-down approach of scientific advancement in the context of biointelligence [own illustration]

The research process of biointelligence follows the standard approach of real sciences. In this context, Ulrich and Hill (1976) identify three core tasks. It is necessary to execute the initially subjectively perceived sections of reality by describing and defining concepts, to abstract on the basis of individual cases, and to develop alternative courses of action for the realization of future realities. Ulrich and Hill (1976) divide the process of insight generation into three steps: discovery, reasoning, and application (Table 4.2).

Table 4.2 Research process of real sciences according to Ulrich and Hill (1976)

	Objective	Problem field	Problem type	Utility criteria
Discovery	Development of reference framework	Science psychology	Heuristic	Suitability
Reasoning	Empirical substantiation of reference framework	Science logic	Induction	Truth
Application	Application of scientific evidence according to specific interests	Science policy	Relevance	Benefit

The focus of biointelligence as a scientific discipline is the development of a new technical basis for a full world as well as the proof of its advantageousness in terms of anthropogenic action alternatives. To this end, aspects of biology, engineering, computer science, business and economics, ecology, and ethics are combined and adapted. As a real science, the process of insight generation follows the discovery, reasoning, and application according to Ulrich and Hill (1976).

4.3.2.2.1 Discovery

The discovery process refers to the conceptual basis of a scientific research project. Here, it is important to define its purpose. This is done by specifying a field of investigation, identifying a problem, clarifying the basic concepts and developing and systematically structuring working hypotheses. The discovery process illustrates how appropriate solutions can be developed on the basis of incomplete information and limited resources. Discovery is thus considered to be a heuristic problem (Ulrich and Hill, 1976). In the context of biointelligence, systems theory generally provides the foundation for the investigation of complex scientific entities (Ulrich, 1970; Wiendahl, 2010). According to Ulrich and Hill (Ulrich and Hill, 1976), systems theory represents an interdisciplinary concept that allows the investigation of any design problem.

4.3.2.2.2 Reasoning

Reasoning describes the methodology used in the research project. The aim of the methodology is to verify the conceptual basis. Conditions have to be identified which allow the researcher to draw conclusions from singular observations to universally valid dependencies. This represents an induction problem that is to be examined for its consistency with reality. Yet, an additional need for deductive conclusions exist, especially when existing theorems are rigorously linked (Ulrich and Hill, 1976). As a basis, Ulrich and Hill (Ulrich and Hill, 1976) propose the critical rationalism of Popper (2007). Popper (2007) abandons the complete induction of empiricism and instead demands a limited induction, the so-called falsification. The logical and factual truth of hypotheses and observations is thereby tested in the sense of an inductive research process following Ulrich and Hill (Ulrich and Hill, 1976). The scientific discipline of biointelligence is based primarily on an inductive logic of argumentation, which is supplemented by deductive conclusions based on the combination of various theoretic approaches.

4.3.2.2.3 Application

Application refers to the utility of the results of a research project in terms of critical reflection. Ulrich and Hill (1976) emphasize the social function of scientific statements and introduce a social benefit criterion. In the context of biointelligence, this social benefit criterion is assigned an extended meaning derived from basic principles of a sustainable reorientation of science and practice (see Section 3.6). The benefit is not only a mere economic one, but, according to Jonas, the contribution to the permanence of human existence on earth (Jonas, 1993). In comparison to existing real scientific disciplines, biointelligence thus presupposes a holistic benefit criterion that is to be understood as normative.

4.3.2.3 Comparison to related scientific disciplines

According to the definition, biointelligence is based on three scientific fields, collectively referred to as a variety of real science disciplines, each of which represents a characteristic or intersection of natural and social sciences: life sciences, engineering, and information sciences. Life science today serves as a generic term for all scientific disciplines whose object of investigation are processes or structures of living nature (Gerber-Kreuzer, 2007). In addition to biology and its sub-disciplines, it also includes disciplines such as biotechnology and bioprocess engineering. The life sciences are largely separated from the engineering sciences, whose focus is the research, development, and construction of a primarily inanimate technology as well as its production and testing (Buchheim and Sonnemann, 1993; Spur, 1998). Information science, in contrast, deals with the flow of information and information processing in mostly digital systems (Stock and Stock, 2013).

In order to specify the positioning of biointelligence in the context of related disciplines, it is first necessary to discuss similarities and differences with key research areas in the life sciences. While a distinction between biology and biointelligence is obvious[3)], the disciplines of systems biology, bioinformatics, synthetic biology, biotechnology, and bioprocess engineering exhibit considerable similarities. All these disciplines have a certain application focus and a clear interdisciplinary orientation. Like biointelligence, they are interface disciplines between social and natural sciences. Their differentiation from biointelligence results from the dominant goals and benefit criteria of the insight generation. In contrast to the focus of biointelligence, which is the creation of a sustainable technological basis by systematically combining biological, technical and informational systems, the goal of synthetic biology is the development of novel, purely biological systems that do not exist in nature (Noireaux and Libchaber, 2004; Benner and Sismour, 2005; Gutmann, 2011; Schwille, 2011; Hodgman and Jewett, 2012; Lentini et al., 2017). A holistic benefit criterion in the sense of sustainability is not evident here. Although bioinformatics also combine ICT and biology, it serves primarily the generation of insights of biological systems with the help of computer-aided methods (Gibas and Jambeck, 2001; Xiong, 2006; Reeves et al., 2009). The same applies to systems biology, which aims to determine the principles of regulatory processes in and between biological systems with the help of systems theory methods (Alon, 2006; Klipp et al., 2016; Voit, 2012; Tavassoly et al., 2018). While bioinformatics and systems biology as entire disciplines focus more on gaining knowledge than on application, their insights are fundamental for the development of BIS. Not least, the disciplines of biotechnology, bioengineering, and bioprocess engineering exhibit distinct application orientations by transferring principles of engineering

[3] The goal of biology is to generate knowledge about living nature, but not to use this knowledge to generate practical expertise for applications (Campbell and Reece, 2005). The same applies to the sub-areas of biology (including omics, ecology, molecular and cell biology, biophysics). Biology thus provides a knowledge base for biointelligence, but is by no means to be equated with it.

and natural sciences to biological systems in order to realize industrial processes (Skalak, 1986; Panke and Wubbolts, 2002; Shuler and Kargi, 2002; Renneberg and Süßbier, 2005; Gerber-Kreuzer, 2007). While biotechnology and bioengineering are basically synonyms, bioprocess engineering can be seen as a subfield of them. Biotechnology and bioengineering focus on a wide range of applications and implementing sectors, while bioprocess engineering primarily addresses continuous process technology, processes, and methods. In both cases, neither a holistic benefit criterion nor an obligatory involvement of the information sciences is apparent. A further differentiation of biointelligence from the sub-disciplines of engineering and information sciences is not appropriate at this point, since the difference is evident solely on the basis of the non-existent integration of biological systems. Likewise, none of the sub-disciplines known here has a holistic benefit criterion.

The comparison to related scientific disciplines repeatedly proves that biointelligence represents an independent scientific discipline which, although relying on existing disciplines, generates new knowledge through a strict interdisciplinary application orientation and a holistic benefit criterion.

Accordingly, not individual enabling technology(s) or thought patterns of individual existing scientific disciplines are decisive for biointelligence, but rather systematic interdisciplinary cooperation, leading to an increasing convergence of the sciences and culminating in a sustainable technical renewal in the long term. Figure 4.7 exemplifies the convergence of science according to Miehe et al. (2019b).

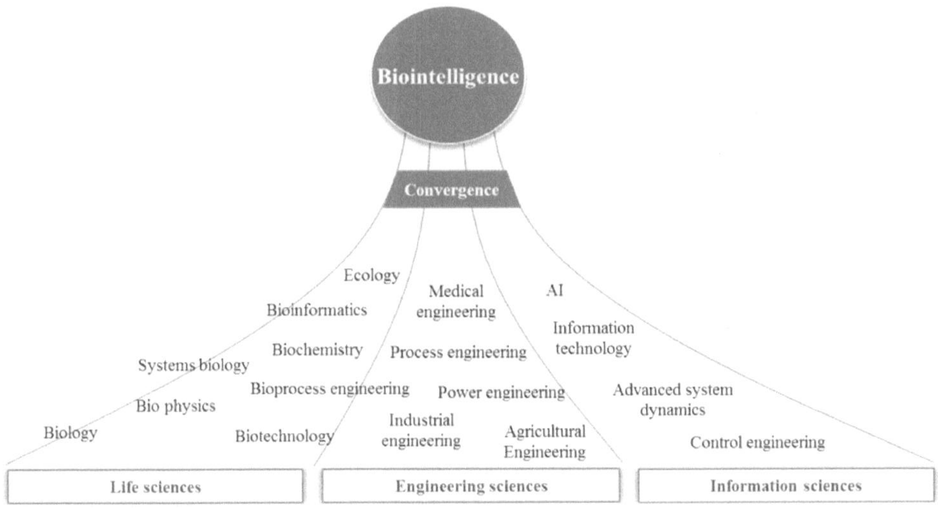

Figure 4.7 Exemplified convergence of sciences according to Miehe et al. (2019b)[4)]

[4] The scientific disciplines listed here represent a loose, exemplary selection and do not claim to be complete. The core message of the chart is the fundamental importance of the convergence of knowledge between life sciences, engineering, and information sciences for biointelligence.

4.4 Biointelligence in the context of production theory

After biointelligence has been established as a scientific discipline, this section reviews existing concepts of production theory with regard to the requirements of the intended reorientation.

4.4.1 Systems-oriented life cycle thinking in cellular units

Whether for development, management or simply the understanding of BIS in both an explicit and an implicit sense, certain thought patterns are essential, which are derived directly from the requirements for a reorientation of production science (see Section 3.6). The basic mind-set of biointelligent production may be referred to as a systems-oriented life cycle thinking in cellular units that seeks a balanced generation of benefits. The individual elements of the postulated mind-set require a thorough introduction.

Systems-oriented life cycle thinking is based on the findings of systems theory, ecology, and life cycle engineering (LCE). It represents a mixture of the ecological life cycle and the biocybernetic systems thinking approach. First, systems thinking refers to a specific interpretation of reality. The approach outlined here is based on the system dynamics (SD) approach of Forrester (1971; 1993; 1994) and the biocybernetics of Vester (1993; 2011). Fundamental to this is the insight that the behavior of complex systems corresponds less to a linear, unidirectional than to a continuous circular pattern. Consequently, the logic inherent in many studies and approaches, that a problem leads to an action that leads to a solution, has to be replaced. More suitable for describing complex problems is a systematic approach that recognizes that every action is based on current conditions, every action influences future conditions, and every changed condition becomes the basis for future actions. Figure 4.8 illustrates the difference between a linear, unidirectional and a continuous, circular logic of systems-thinking according to Forrester (1993).

However, systems thinking does not only imply that individual actions necessarily result in subsequent events, but also that any action has an impact on a fragile superordinate system (e. g., the socio- or ecosphere) (Vester, 1993; 2011). Hence, the logic of the systems thinking approach is to always seek a balance with the surrounding systems in order to avoid damaging the overall system. In an implicit sense, a system thus can only be described as biointelligent if it does not endanger the eco- and sociosphere. The philosophy of unconstrained growth of single system entities, is replaced by a holistic approach (Vester, 2011).

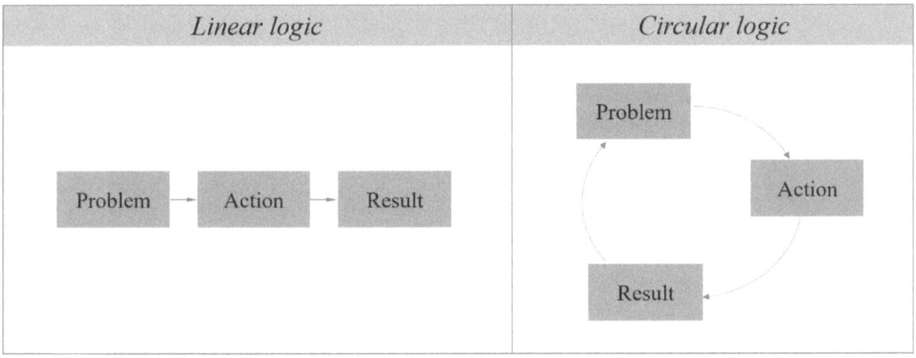

Figure 4.8 Difference between linear, unidirectional and continuous, circular logic of systems-thinking according to Forrester (1993)

In contrast, *life cycle* thinking (mostly associated with the discipline of life cycle engineering) refers to the basic insight of ecology that each entity has an impact on the environment throughout its entire life. In order to assess whether an entity (product, action, etc.) is favorable from an ecological perspective, the entire life cycle of the entity concerned has to be considered. In comparison to the economic interpretation of a product life cycle (introduction, growth, maturity, decline), in the context of biointelligence the term refers to the complete path from cradle to grave of an examined entity (Heiskanen, 2002; Thabrew et al., 2009). Figure 4.9 illustrates the ecological understanding of a life cycle according to Klöpffer and Grahl (2014).

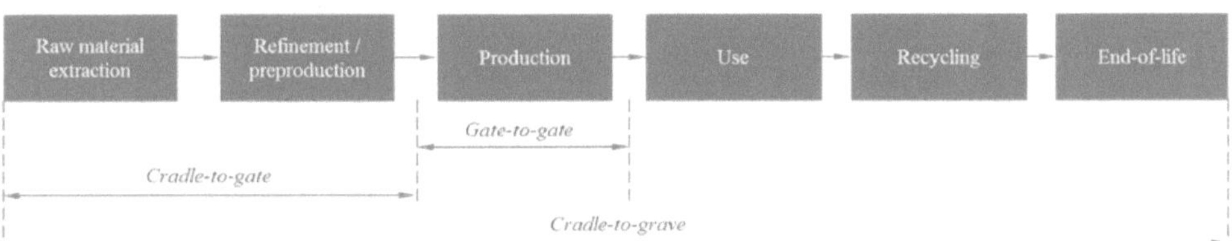

Figure 4.9 Ecology-oriented life cycle understanding according to Klöpffer and Grahl (2014)

In order to perceive reality as interconnected in each element, another thought pattern is crucial, the division of reality into cellular units. Each cell comprises a segment of reality and, in accordance with systems thinking, interacts with a large number of surrounding cells and the complex overall system in a variety of relations. Cells can occur at different levels. For instance, molecules, biological cells, organisms, sociotechnical systems, products, and entire ecosystems can be interpreted as cells. Cells are able to assemble spontaneously, establish various connections with each other and detach themselves from each other instantaneously. A cell is sufficiently described in terms of biointelligence by six factors which directly result from above characteristics of life, technology, and systems-oriented life cycle thinking:

- *Type of matter* (biotic, abiotic, miscellaneous)
- *Internal processes* (hierarchy, segments, metabolism, etc.)
- *Interrelations* with surrounding cells and overall system (physical, nominal, informational)
- *Purpose for application* (production, decomposition, cohesion, etc.)
- *Functional principle* regarding the intended application (mechanical, chemical, optical, biological, economic, etc.)
- *Conditions* to fulfil its purpose (connectivity, viability, etc.)

Figure 4.10 illustrates the systems thinking cellular approach of biointelligence. In addition to the formation of cells at different levels, the idea of complete networking in form of various interrelations is crucial. Although presented here as a single line, they can occur in three different forms: physical (mechanical, electrical. chemical, biological, etc.), nominal (monetary), and informational interaction.

Finally, biointelligence seeks a balanced generation of benefits. This refers to the basic insight of production science that customer benefit orientation is the foundation for flexible, efficient manufacturing of goods. Yet, according to systems thinking, not the individual, i. e., the customer, alone is decisive. The goal must rather be to create a benefit for the individual, society, and the ecosystem. According to Vester (2011), this requires a shift from a product to a function-oriented approach to create benefit. While the former leads to adherence to existing, outdated products, organizations and/or thought patterns, the renewal in terms of biointelligence requires simply maintaining and improving functionality. The overall system must be capable of removing hazardous products, organizations and/or behavior, whether by technology, social change and/or political instruments.

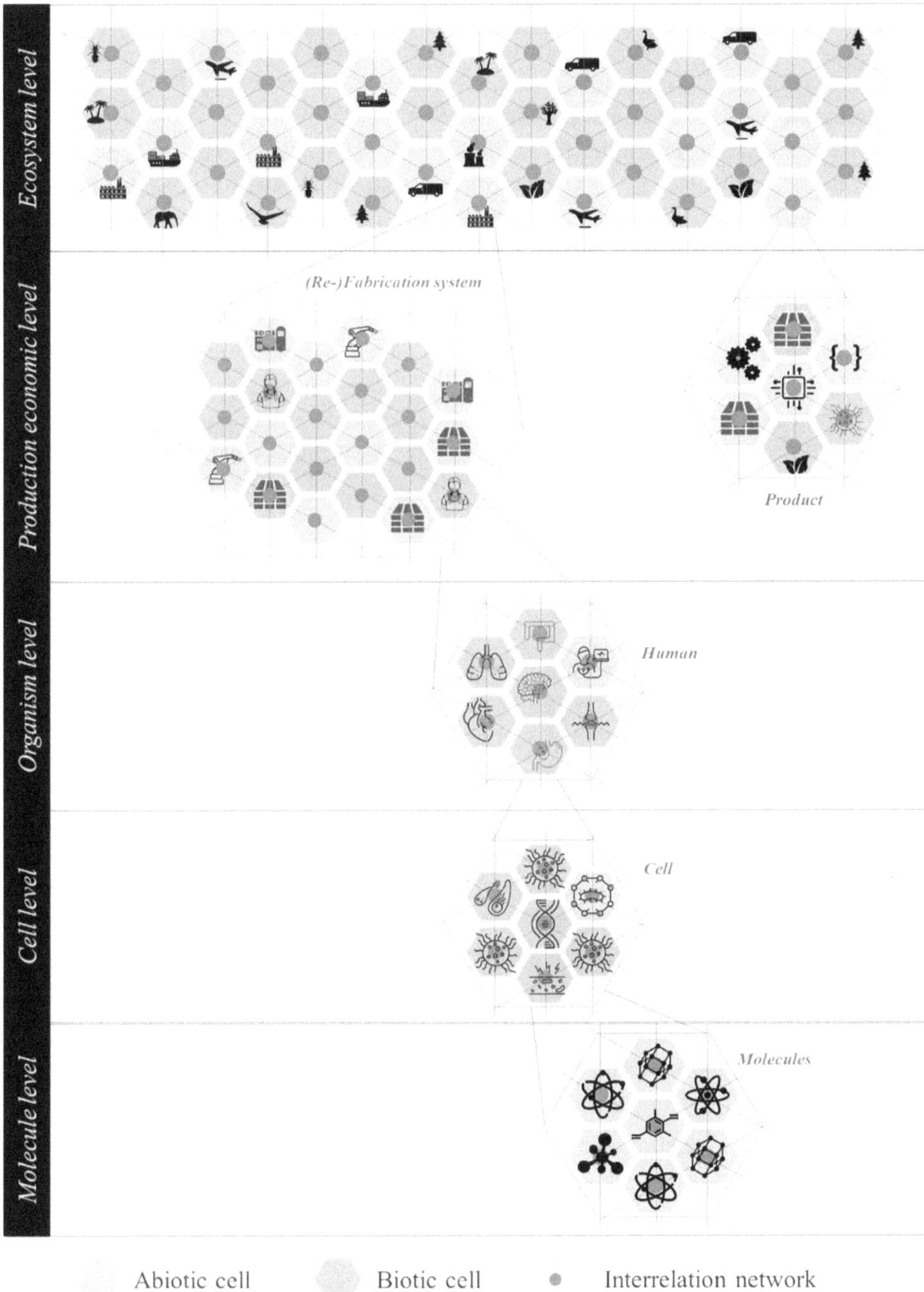

Figure 4.10 Systems thinking in cellular units [own illustration]

4.4.2 Biointelligent production

Based on the fundamental mind-set, the question arises as to what is to be regarded as biointelligent production. Before defining the term accurately, some general reflections on the current understanding of the concept are required. The term production (lat. producere = to bring out) is commonly understood in a broad sense (Schneeweiß, 2002). Its interpretation requires the understanding of a multitude of related notations, approaches, and methods. The first definitions of the concept of production originates from the beginning of the 19th century. Müller (1922) interpreted it as a combination of two components to create a third, the product, enabled by the human mind that controls the combination process in a targeted manner. On the basis of the factors of production discussed in economics (labor, soil, capital), Biermann (1904) identified five components. These are the production facilities and tools, capital, nature, including soil, raw materials, and auxiliary materials, as well as executive labor and the economic cycle. The distinction between executive and dispositive activities is evident here for the first time. On this basis, Gutenberg (1988) and Heinen (1994) develop the currently accepted system of operational production factors. According to Corsten (2007), three possible definitions of production can be distinguished in literature:

1. *Production as factor combination*: Gutenberg (1988) identifies the composition of production factors for the creation of an end product as the most important characteristic of differentiation. In this understanding, any operational action of combining factors of production is understood as production.
2. *Production as a phase of the operational process between procurement and sales*: This significantly narrowed definition, among others represented by Dyckhoff and Spengler (2010), emphasizes the transformative power of production. Transfer functions (procurement, sales, etc.), which are part of the factor combination in the above approach, are explicitly excluded.
3. *Production as value added*: This definition, which among others was adopted by Engels (1962), Zahn and Schmidt (1996) and Westkämper (2006), serves primarily to distinguish production from consumption. This is based on the assumption that values are created in the production phase, but destroyed in the consumption phase.

With regard to the holistic benefit criterion of biointelligence and its basic mind-set, these interpretations are to be rejected to a large extent. The views of production as a combination of factors or as a phase of the operational process between procurement and sales clearly neglect the holistic principle (see Section 3.6), as they represent pure expressions of a traditional linear logic. In this sense, questions concerning the remanufacturing of products would have no relevance. The same applies to production as value creation. While a sole focus on monetary value added in relation to the individual purchase decision of the consumer also does not permit a holistic perspective, a comprehensive approach to value creation is conceivable here. Furthermore, the decentralization principle (see Section 3.6) requires a dismantling of the global division of labor, which today manifests itself in highly complex global supply chains. Since the holistic benefit criterion already addresses life cycle thinking, the cellular systems thinking approach requires an interpretation of production as a cluster of cells at different levels. Thus, the following understanding is assumed in the further course:

1. *Biointelligent production in an explicit sense*: spontaneous or planned combination and recombination of cellular units, i. e., local production factors (at least one biological and one technical), under the provision of information technology principles for the purpose of regulation and autopoiesis. Its overall behavior corresponds to the requirements of intelligence in the narrow sense.
2. *Biointelligent production in an implicit sense*: recurring phase of a constant cellular merging and diverging cycle between the provision of cellular units of subordinate levels, i. e., raw materials, and the sale of intended cellular unit, i. e., product. In this context, production is not understood as an independent unit, but rather in its entirety as a self-contained organism whose natural basis (the planetary boundaries) must not be exceeded.

4.4.3 Biointelligent products

The objective of production is the satisfaction of own and/or external needs by supplying products (Zelewski, 1999). Similar to the term production, an accurate definition of the term initially requires some general reflections on the current understanding of the concept. The term product serves a similar purpose as the term good in economics and the concept of commodity in commodity science. However, a clear definition of the term does not exist. Production economy generally uses the terms product and good as synonyms to characterize the material and immaterial means of satisfying needs (Corsten, 1985; 2007; Beattie, 2009; Rasmussen, 2012). From a technical point of view, the former represents the output of a production process and may be divided into unit goods, bulk goods, and continuous goods. While unit goods are the result of the production process, bulk goods and continuous flow goods represent the results of process engineering (Hirschberg, 1999; Westkämper and Warnecke, 2010). Unit goods are self-contained objects with a defined function (e. g., machines, devices) combined from a set of parts. Examples of bulk and continuous goods include raw materials, building materials, and fertilizers. From a sales management perspective, material products are divided into final, intermediate and waste products (Corsten, 2007). End products are usually consumer or investment goods. Corsten (2007) identifies services, work, and information as intangible goods. Besides material or immaterial goods, hybrids exist, so-called system products. A further perspective is the consumer view. Assuming the satisfaction of needs as a benefit, Kotler et al. (2007) differentiates between three interpretations of the term. The substantive interpretation describes a product as a bundle of physical-technical properties to satisfy functional customer needs. The extended definition interprets the product as a service package for the comprehensive satisfaction of functional customer needs by material and immaterial means. If, in addition to functional benefits, other categories of benefits (e. g., emotional or social benefits) that can be satisfied in a material or immaterial way are also included, the generic product term is used.

Existing microeconomic interpretations of the term product are insufficient to fully describe a biointelligent product. Although the needs orientation of Kotler et al. (2007) basically corresponds to the benefit and/or function orientation of biointelligence, the exclusive consideration of customer needs eventually results in imbalance with other affected systems. Hence, biointelligent products cannot exclusively be developed for single customers, but must be understood as an integrative optimum between individual, social, and ecosystem benefits. Figure 4.11 illustrates the approach. Thereby, the individual benefit is determined by the contribution of a product to essential individual fields of need (Bauernhansl et al., 2019). The social benefit is determined by generally accepted objectives, while the ecological benefit is expressed in terms of its impact on the global and local system (Miehe, 2018).

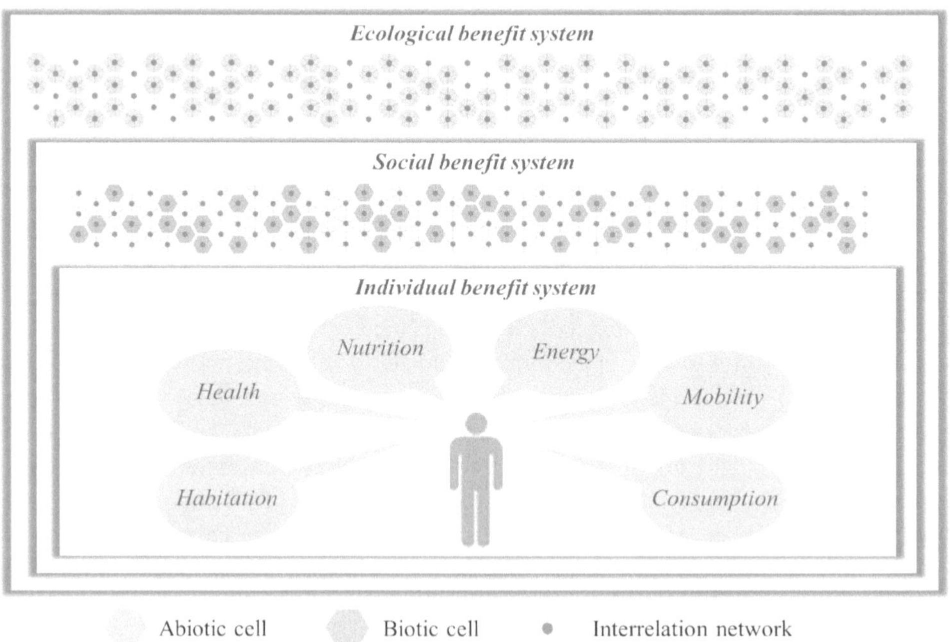

Figure 4.11 Framework of holistic needs satisfaction of a biointelligent product [own illustration]

According to the interpretation of the biointelligence concept outlined above, two perspectives of a biointelligent product can be distinguished:

1. *Biointelligent product in an explicit sense*: material result (cluster of cells) of a production process that enables the user to gain a functional benefit. It consists of at least one biological and one technical component as well as information-technological mechanisms for the purpose of regulation and autopoiesis. Its overall behavior corresponds to the requirements of intelligence in the narrow sense. A biointelligent product in an explicit sense can only be an end or intermediate product.
2. *Biointelligent product in an implicit sense*: material and/or immaterial good, which fulfils a proven socio-ecological benefit.

4.4.4 Classification of production factors

The material foundation of any production process are the production factors, a concept predominantly discussed in economics and business. The understanding in traditional production science is based on the interpretation of economics and business administration. Classical economics differentiates between labor, capital and land (Mankiw and Taylor, 2012; Heertje and Wenzel, 2008). Since this concept was considered insufficient in the microeconomic context to describe production-economic transformation processes, Gutenberg (1988) developed the currently dominant factor-combinative paradigm. This was based on the interpretation of production as a combination of factors, which had already been rejected above. The classification of the production factors is initially divided into elementary factors and dispositive factors. The characteristic of the distinction is the possibility of disposition of factor combination. While the elementary factors cover all materials and direct work on the product, the planning factors describe organizational aspects such as management, planning, organization, and monitoring. Furthermore, Gutenberg (1988) and Heinen (1994) distinguish elementary factors according to the number of factor combinations in potential and repeat or consumption factors. Only one factor combination exists for consumption factors (repetition factors). For potential factors, several factor combinations may prove to be effective. According to Gutenberg (1988) and Heinen (1994), repetition factors describe those factors that are directly consumed in a production process, i. e., raw materials and supplies. Potential factors, on the other hand, represent those factors that are used indirectly. These are both tangible and intangible operating resources as well as the work on the product itself.

To date, various authors have presented extensions of the traditional business classification to integrate additional factors such as services (Corsten, 1985) or immaterial factors such as information (Wild, 1970) and rights (Vormbaum 1967; Wittmann 1977). However, especially the integration of the environment leads to dissent. Although Kern

(1992) already includes environmental factors that are available to the company free of charge, various authors criticize the consideration of the natural environment in the conventional classifications of production factors as insufficient. Thus, Schreiner (1996) suggests integrating the environment as a separate category in the conventional factor system, while other authors strictly reject this (Steven, 1994). The traditional view of production factors, e. g., according to Gutenberg (1988) and Heinen (1994), considers only the input variables of a production process. According to Steven (1994), a consistent allocation of the production factor environment is not possible in this understanding, since only that part of the environmental goods can be considered which is used as input in a production process. The ability of the environment to absorb the undesirable outputs from a production process (sink function) would be neglected. Stirnberg (1999), in contrast, identifies inconsistencies in conventional factor systems with regard to the orientation towards the flow of goods. According to Dyckhoff (2008), potential factors can be understood simultaneously as input and output factors, since they are not directly consumed but leave the process again.

An alternative interpretation, which has not yet been transferred to operational production theory, is the concept of factors of production in ecological economics. According to Daly and Farley (2003) as well as Zencey (2012), the concept of ecological economics rejects previous classifications and replace them by three types of factors: matter, energy, and intelligent design. Taking into account the economic law of entropy, the following basic assumption is crucial: *Productivity increases are fundamentally impossible by simply combining matter and energy. The only way to realize this is to increase design intelligence.*

An extended understanding of the production factor soil in the classic economy as natural capital is provided by modern environmental economics (Rogall, 2008). Natural resources are understood here as goods that already exist in nature and serve to satisfy human needs (Schütz and Bringezu, 2008). Siebert (1982) expands the conceptual framework of a natural resource as a mere production factor, since this would not take into account many conservation goods (e. g., the protective cover of the earth). Schütz and Bringezu (2008) distinguish natural resources according to their nature into raw materials, area, soil, air, water, and biodiversity. The understanding thus includes not only the production function but also the sink function of the economic system.

From a biointelligence perspective, the microeconomic classification of production factors according to Gutenberg (1982; 1988) and Heinen (1994) is only partially appropriate. Apart from the absence of the natural environment as a separate category, in the sense of ecosystem services, the concept of elementary factors in particular represents a contradiction to the closed-loop principle (see Section 3.6). In the sense of a closed-loop system, a one-time or multiple factor combination without constant reuse must be strictly rejected. An appropriate classification in terms of biointelligence interprets all material factors as stabilizing factors. The elementary aim is to maintain their suitability for constant recycling, either within an industrial-technical cycle or by appropriate release into the natural environment. The environment is interpreted here in the sense of modern environmental economy as source and sink (ecosystem service). The basic materials can originate from primary production or from the existing product cycle. Table 4.3 summarizes the understanding.

Table 4.3 Classification of production factors in the context of biointelligence [own classification according to Gutenberg (1982; 1988) and Heinen (1994)]

Stabilizing factors					Dispositive factors					
Basic materials			Operating equipment							
Raw materials	Auxiliary materials	Operating materials	Physical operating equipment	Non-physical operating equipment	Labor on product	Ecosystem service	Administration	Planning	Organization	Monitoring

The cellular systems thinking approach requires further development at other levels. As outlined above, production in the classical sense takes place primarily at the economic level, yet it can equally apply to other levels from the perspective of biointelligence, e. g., in individual biological cells or the natural environment (ecosystem services). The development of a corresponding production factor logic at all levels represents a subject for future research.

4.4.5 Biointelligent production process

As the core of microeconomic production, the production process is primarily a manifestation of the explicit interpretation of biointelligence. Traditional production science distinguishes between a narrow and an extended definition of a production process with regard to the system boundary. The former comprises only the work on the product, while the latter integrates the entire entrepreneurial production of goods and services. In both cases, the idea of a transformation from a lower to a higher design or functionality is assumed (Fritz and Schulze, 2010). Representatives of this narrow interpretation include Schneeweiß (2002) and Westkämper (2006), among others. In contrast, Corsten (2007) and Dyckhoff and Spengler (2010) support an extended interpretation. However, with regard to the holistic benefit criterion of biointelligence, both approaches do not suffice. In order to supplement the broad understanding of Dyckhoff and Spengler (2010), the following understanding will be assumed:

A biointelligent production process is a directional quantity of cellular units that is necessary to provide a certain output within a constant product cycle.

In order to further specify the production process as a function of various parameters (e. g., industrial sector, product type, type of organization, technology), three characteristics are decisive for the biointelligent production process (Dyckhoff and Spengler, 2010).

4.4.5.1 Technology characteristic

A biointelligent production process requires the presence of biological active principles for the purpose of production. In the sense of a directed quantity of cellular productive units within the framework of a constant product cycle, these can be supplemented by physical, chemical, and/or informational active principles. According to Dyckhoff and Spengler (2010), these can be mechanical (e. g., shaping, machining, surface treatment, assembly), electrotechnical (e. g., electrolysis) or caloric processes (e. g., heat transfer, thermal cutting). A further decisive factor for a biointelligent production process is the existence of ICT principles for the purpose of control and autopoiesis (Miehe et al., 2019a).

4.4.5.2 Structural characteristic

Depending on the production process, the structure of the material flow varies greatly. Dyckhoff and Spengler (2010) distinguish between consistent, converging, diverging, and reorganizing processes. The former are characterized by continuous processing (e. g., paint shop). A converging process structure is usually found in assembly. Here, a subsequent work step is dependent on several previous ones. Diverging ones represent an inversion of converging processes. One is followed by several work steps. A process is to be classified as reorganizing if several inputs result in several different outputs. This form of process organization is particularly common in the process industry. In addition to the four basic types, which are elementary for the production sciences as a whole (assuming the system boundary factory), the sustainability-related reorientation of biointelligence requires a complementary view. Here, the consideration of a production process as joint production is fundamental (Arndt, 1997; Baumgärtner, 2000). According to the holistic benefit condition of biointelligence, a production process can never be regarded as the output of a single good. Instead, all outputs (regardless of their use) must be considered. The result of a biointelligent production process is thus the product and its derivatives. According to Arndt (1997), the latter can be further divided into five categories:

1. Co-product = co-product with the relatively most significant economic yield
2. By-products = not the actual goal of production, but can significantly influence the profitability of a company through their sales opportunities

3. Recyclable waste = recycling within the company or sale on the market
4. Non-recyclable waste = products to be disposed of at cost, e. g., tangible and intangible waste and emissions to air, water and soil
5. Pure losses = unused conductors, e. g., waste heat, cold, non-harmful exhaust gases

In addition, the closed-loop principle requires the exploitation of all possibilities for the cascade use of production factors on a microeconomic level (within and outside the microeconomic production system). A complementary view, which is primarily based on a microeconomic perspective, is a cascading production process. The holistic benefit criterion requires that the productive functionality of all input factors remains constant. While all of the processes described above can occur at the microeconomic level, a holistic-looped view in terms of biointelligence is indispensable. While the microeconomic cascading perspective has a descriptive character, the holistic-looped perspective is normative. Any consideration of a production process in terms of biointelligence must take a holistic view in addition to the microeconomic view. Assuming production as either a spontaneous or planned merging of different cells fulfilling an intended purpose, two types of biointelligent production processes may be differed: [1] stand-alone cascading sequences of individual cellular units within a higher-level cellular productive unit, and [2] versatile, reticular integral cycle of various productive cells. Figure 4.12 illustrates the understanding.

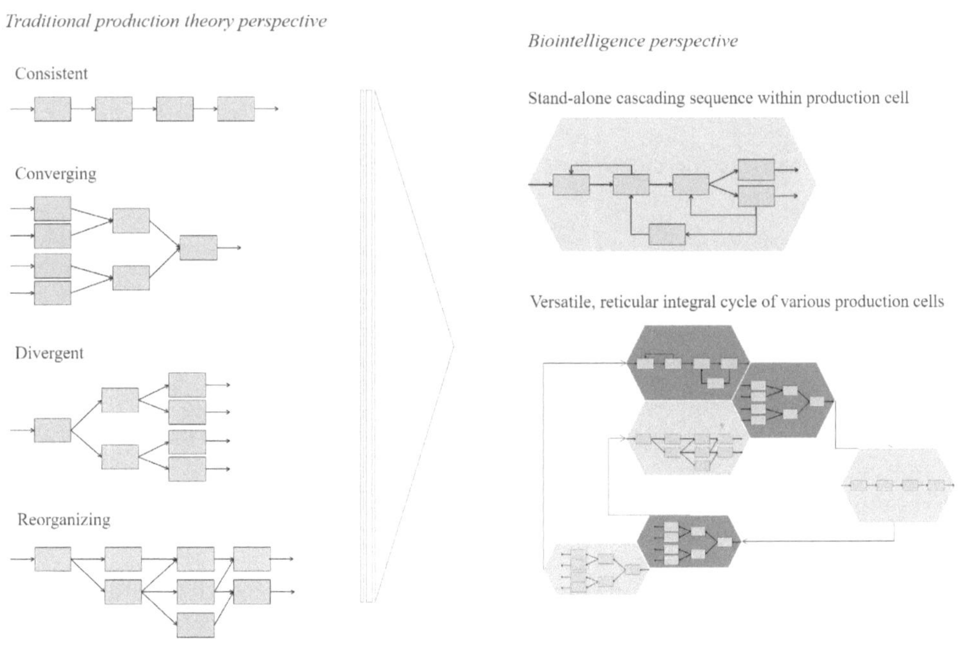

Figure 4.12 Classification of conventional and biointelligent understanding of production processes [own illustration according to Arndt (1997) and Dyckhoff and Spengler (2010)]

4.4.5.3 Frequency characteristic

The frequency characteristic indicates the number of repetitions of a constant sequence of production elements. Here, individual, series and mass production are differed. The former describes the separate, usually customer-specific manufacturing of a product (engineer-/make-to-order). Series production is defined as the production of a large quantity of products in a similar sequence over a given period of time. The differentiation from mass production is based in particular on the occurrence of changeovers. If this does not take place and the product is manufactured in almost the same design over a given period of time, it is mass production. Although series and mass production cannot be substituted in certain areas, the goal of biointelligence is personalized individual production (decentralization principle).

Figure 4.13 summarizes the understanding of a production process in the context of biointelligence. It distinguishes between the classical descriptive view of today's production sciences and the normative perspective of biointelligence.

Features	*Characteristics*					
Technology	Physical		Chemical		Biological	Informational
Structure	Consistent	Converging	Divergent	Reorganizing	Stand-alone cascading	Versatile, reticular, integral cycle
Frequency	Mass production		Serial production		Single/repetitive production	

Minimum requirement of normative view of biointelligent production process
Conventional descriptive view of production process

Figure 4.13 Features and characteristics of production processes [own classification]

4.4.6 Biointelligent production systems and enterprises

Literature essentially provides two perspectives on production systems. One definition equates it with manufacturing companies, the other focuses on the interaction of different entities for the purpose of producing an intended product. While the latter apparently matches the underlying concept of biointelligence more accurately, the first should not be ignored either as it represents a sublevel of the first. Similar to the term production and product, an accurate definition initially requires some general reflections on the current understanding of the concepts.

In classical production science, the term describes an entity whose purpose is production (Corsten, 2007; Günther and Tempelmeier, 2011). As such, it can be a factory, a production segment or a production network. In the context of biointelligence this is however insufficient, as the concept seeks an equally applicable interpretation in an explicit and implicit sense. Thus, the following definition applies in the further course:

A biointelligent production system is a cluster of cells whose interaction yields in previously disconnected cellular entities. Cluster of cells may occur at any level, ranging from molecules to ecosystems.

In the context of production theory, manufacturing enterprises represent the most prominent form of a production system. Generally, they can be categorized according to two characteristic features: [1] the purpose of the business activity and [2] the description of the entity as such. The derivation of the understanding according to the purpose is based on the classification of economic units. The characteristic feature here is the satisfaction of needs. Accordingly, consumption and production operations are differed (Wöhe and Döring, 2010). The former pursues the goal of covering their own demand and are referred to as households, the latter exists solely to cover external demand (companies, public enterprises). Table 4.4 illustrates the understanding according to Wöhe and Döring (2010).

Table 4.4 Classification of economic entities according to Wöhe and Döring (2010)

Economic entities			
Production entity (external demand coverage)		**Consumption entity (internal demand coverage)**	
Private enterprises	Public enterprises	Public authorities	Private households

According to Ulrich and Hill (1976), only three paradigms exist for the universal description of a company as such: the factor-theoretical approach according to Gutenberg (1982; 1988), the systems theory approach according to Ulrich (1970), and the decision theory approach according to Heinen (1969; 1994). Gutenberg (1982; 1988) understands a company as a place of factor combination to create an intended satisfaction of needs. Meanwhile, Ulrich (1970) describes a company as an open system, which is in diverse exchange relationships with superordinate and subordinate systems. Not least, Heinen (1969; 1994) considers the company to be a place of real phenomena relevant to decision-making. The following definition applies in the further course:

A biointelligent production enterprise is a goal-oriented, participatory socio-technical system of individual cellular units.

Crucial to its understanding are thus the system-theoretical positioning, the specification of participation, and the specification of goals.

4.4.6.1 Systems-theoretical positioning

The interpretation of a production company as an open system requires the definition of various exchange relationships with superordinate systems. The core of the enterprise is the transformation system, which is characterized by its relations between its input and output variables. The input-output approach, which dates back to Leontief (1986), is generally used to describe the transformation process of a production plant. In accordance with the three interrelation types of BIS, the exchange relations that production enterprises establish with their surrounding systems for the purpose of exchanging goods can be divided into three categories: [1] real flow of goods (exchange of real goods), [2] nominal flow of goods (exchange of financial resources) and [3] information flow (exchange of information). The ecosystem comprises the entire spectrum of environmental factors that have direct and indirect implications for economic activity in production. Thereby, an economic, an ecological, and a social ecosystem are differed (Zelewski, 1999; Wiendahl et al., 2009; Wiendahl, 2010; Dyckhoff and Spengler, 2010; Schuh, 2011; Schenk et al., 2014).

From an economic perspective, the most prominent interdependencies exist with the procurement, capital, and sales markets. Zelewski (1999) further subdivides the economic system into performance and financial elements. The exchange relationships of a company with the factor and procurement markets as well as the sales and product markets in terms of the exchange of real and nominal goods characterize the performance-related economic system. The financial system, on the other hand, is characterized by the interaction of the enterprise with the money and capital markets. Here only nominal goods are exchanged.

The social perspective can be grouped into the social and the sovereign system. The social system consists of any person or group that is not directly integrated into the internal economic actions of the company, in the broadest sense the entire public. Yet, two forms of mutual influence exist between the company and external persons or groups. On one hand, external influences on internal company actions can occur, on the other hand external persons and groups can be affected by internal actions. The sovereign system includes all institutions whose competence allows them to directly influence internal actions. This also includes the respective legal situation. A multitude of possibilities exist for influencing, e. g., directives, regulations, subsidies.

The ecological environment consists of the natural environment with which a company is always in contact through the exchange of real goods. According to Döring and Ott (2001), the use of the ecological ecosystem by manufacturing companies is determined by the availability of resources ('sources') and the capacity of natural sinks to absorb them.

The traditional interpretation of the production economy, which is represented by Zelewski (1999), Wiendahl (2010), Dyckhoff and Spengler (2010) and Schuh (2011), among others, does not weight the surrounding systems. However, the levelling of the surrounding systems does not do justice to the natural ecosystem in particular, which biointelligence understands as the basis of life. Accordingly, the systems-theoretical understanding of ecology differs from that of production theory. The European Commission (2010c) proposes a distinction between a foreground and background system. The company is understood here solely as a foreground system, which due to its actions is in interrelations with a background system. The distinction is determined by the degrees of freedom of influence. The background system thus represents those upstream and downstream processes which a company cannot influence. The company is embedded in the technosphere, the system of man-made technologies. The technosphere is ultimately a subsystem of the ecosphere, the animate and inanimate system earth.

For the understanding of a biointelligent production company, both views reveal various deficiencies. First, the understanding of ecology leaves little room for explaining economic exchange relationships. Second, the understanding of business economics neglects essential causal relationships between companies and the environment. Finally, the importance attributed to the ecological system does not reflect the understanding of ecology. In this context, Miehe (2018) presents an extended interpretation, which in the following serves as a basis for the system-theoretical understanding of a biointelligent production enterprise. First of all, the existence of certain natural preconditions for production is emphasized and specified in the priority system of the ecosphere (see priority model of sustainability). The ecosphere comprises all animate and inanimate elements of the overall system, i. e., earth. In this understanding, human existence represents a sub-area of the ecosphere which is referred to as the anthroposphere. The anthroposphere is in turn divided into two subsystems, economic and social. An enterprise or entrepreneurial activity represents a characteristic of human existence. It is thus assigned to the anthroposphere.

In terms of systems theory, a company is initially regarded as a black box whose core objective is to satisfy its own and/or other people's needs by supplying goods. In order to achieve this goal, certain input variables are required and their combination generates certain output variables. The company obtains input variables from the markets of the economic and social system and the ecosphere. Similarly, output variables are supplied to the markets of the economic and social system as well as the ecosphere. A fictitious impact system is introduced to assess the impact of corporate production on the ecosphere and anthroposphere. This is initially assigned to the ecosphere, but displays various exchange relationships with the anthroposphere. Ecological consequences of extraction (source function) or delivery (sink function) can lead to problems at global and local level. The impact system is thus further divided into global and local systems. While all forms of exchange (real and nominal flows of goods, flow of information) between the enterprise and its direct anthroposphere surrounding systems are obvious, the interaction of the enterprise with the impact system is based solely on the transfer of real goods. Yet, these each generate a nominal flow of goods, i. e., external costs that mostly do not affect the company itself but other anthropospheric systems. In the case of both input and output variables, the effects of the exchange of real goods are exclusively evident in the direction of the impact system. The use of raw, auxiliary, fuel or operating material on the input side leads to a withdrawal of resources and various emissions in the upstream chain. The internal transformation within the company results in direct emissions and waste as well as indirect effects in downstream stages. This is clearly illustrated by the example of operating resources, which are only used in the company during the utilization phase. Enterprises are generally not responsible for their own production and remanufacturing. In case of a more detailed examination, a foreground and background system is to be specified for these. Figure 4.14 summarizes sustainability-oriented systems-theoretical understanding according to Miehe (Miehe, 2018).

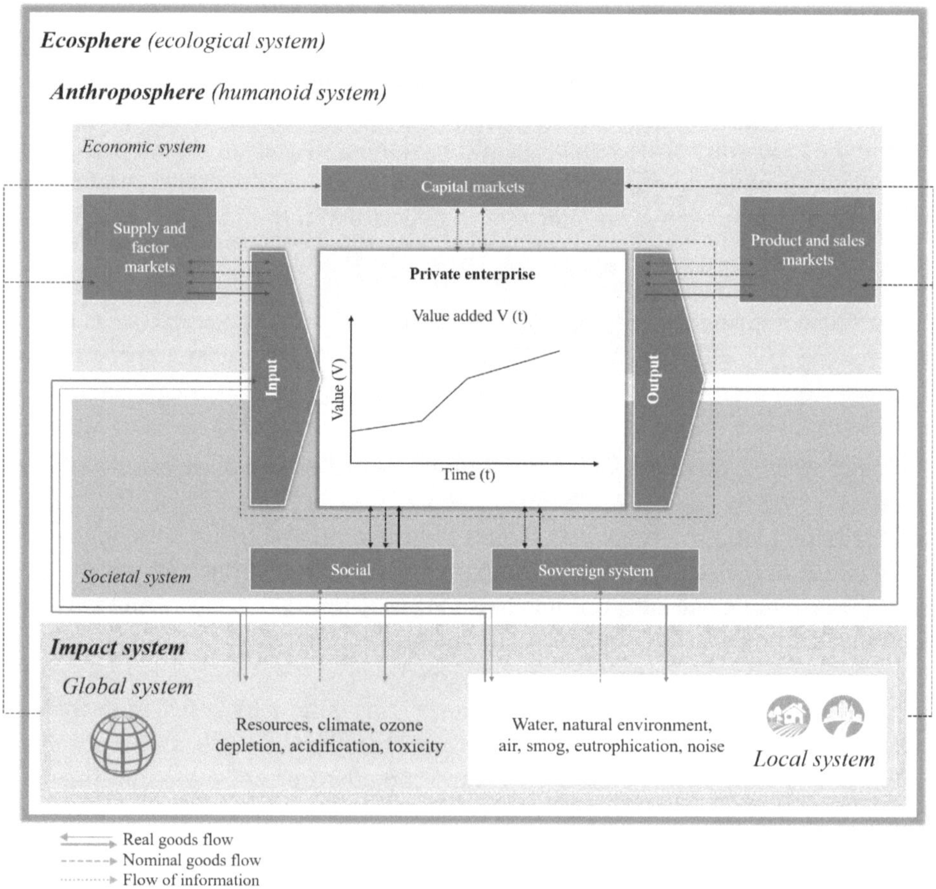

Figure 4.14 Sustainability-oriented systems-theoretical understanding of a production enterprise according to Miehe (Miehe, 2018)

While the approach sufficiently enables a consideration of a company's impact on its surrounding systems, it does not thoroughly explain its material cycles. In addition, it does not comply with the concept of random merging of individual cells for a production task. While on a microeconomic level it is advisable to concentrate on separate units such as companies and factories, the concept of a biointelligent production system must offer the possibility of transfer to different levels. For instance, production can equally occur at home and/or in communities (e. g., through the additive production of consumer products or the production of drugs in microbioreactors in hospitals). Yet, every conceivable biointelligent production system must seek to a closed-loop material cycle along its entire life. The result of production is the product and its derivatives (non-usable by-products). All by-products that cannot be recycled within the manufacturing system are transferred to a remanufacturing system, aiming to return as much of the products as possible to the state of a usable raw material. Figure 4.15 illustrates the following interpretation of a production system.

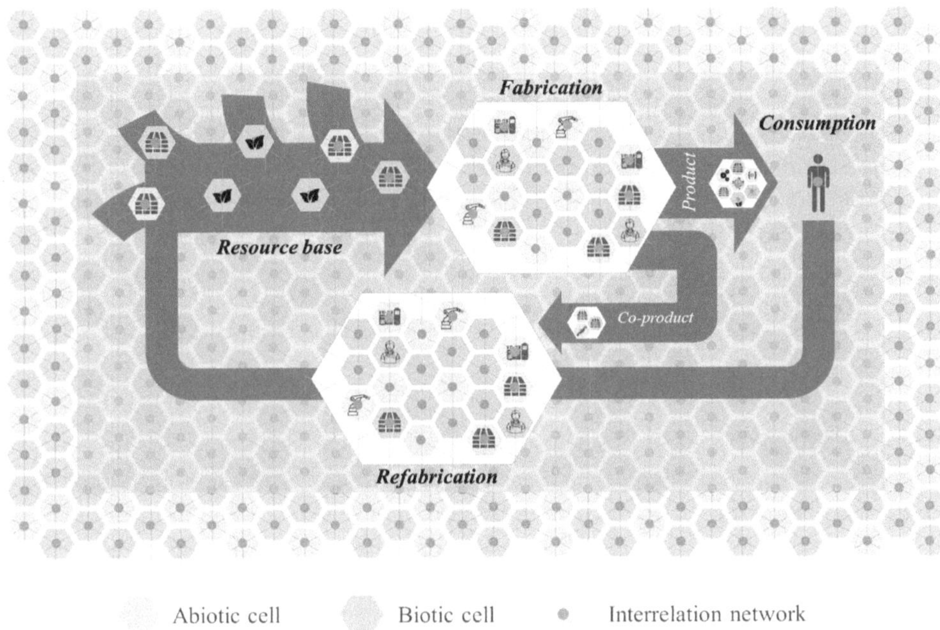

Figure 4.15 Concept of a biointelligent production system [own illustration]

4.4.6.2 Participation

In business practice two conflicting concepts of participation in companies exist today, shareholder and stakeholder value approach. The former is based on the considerations of the former General Electric CEO, Jack Welch, in the 1980s, before it was first elaborated in scientific literature by Rappaport (1986; 1998). According to this approach, the reason for the existence of a company is purely to satisfy the needs of its shareholders. The shareholder value represents the market value of the equity capital. This approach, which is often predominant in business practice, focuses on the generation of profit or free cash flow. Various authors, including Figge and Schaltegger (2000), Gleißner (2004) and Coenenberg et al. (2007), criticize the one-sided calculation of enterprise value focused on the interests of the owners and the resulting lack of target-oriented incentives for management. According to literature, a consensus exists today that equity investors are not the only stakeholders in a company. Burschel et al. (2004) state that entrepreneurial activities always have an impact on third parties, e. g., customers, the public, the environment. Although concepts exist to expand the shareholder value approach, e.g., by Figge and Hahn (2002) and Rauschenberger (2002), the failure to consider third parties led to the development of the stakeholder value approach by Freeman (1984). Thereby the stakeholder value is characterized as the benefit of all stakeholders in a business activity. Besides owners, stakeholders include employees, customers, suppliers, capital markets, the state and the public, as well as the environment. In the context of the environment, the approach is closely linked to the resource dependency theory. This theory examines the system-environment relationships of economic entities (Pugh and Hickson, 2007). Based on considerations of Pfeffer and Salancik (2003), the theory explains the behavior of an organization in relation to the environment through its dependence on a resource. Dependency results from the relative share of the resource in the inputs and outputs as well as the independence and market position of the resource owner or controlling entity.

A biointelligent production enterprise strives to achieve a benefit for all stakeholders involved in and affected by the operational performance.

4.4.6.3 Objectives

Production science distinguishes between actual and formal company targets (Zelewski, 1999), which also determine the objective system of biointelligent production companies. The actual objective is the satisfaction of own and/or external needs by supplying goods. Formal goals, however, can be further subdivided into technical (flexibility, capacity utilization, product quality, etc.), economic (profitability, productivity, profit, growth, etc.), ecological (resource consumption, environmental impact, compliance with legal limits, etc.) and social (humanity of working conditions, development of individual ideas, social responsibility, etc.) goals. Among formal objectives three types of target relationships exist: [1] goal compatibility (i. e., the achievement of one goal has a positive influence on the achievement of the other goal), [2] goal difference/neutrality (i. e., two goals can be pursued separately), and [3] goal conflict (i. e., the achievement of one goal has a negative influence on the achievement of the other goal). The resolution of formal goal conflicts is one of the core challenges for biointelligent companies and the application-oriented research of biointelligence in an implicit sense.

4.4.7 Biointelligent production technology

Classical production engineering distinguishes three main techniques: manufacturing, process, and energy technology. According to the German DIN 8580, production engineering includes all techniques for the manufacturing of geometrically defined solid bodies (Deutsches Institut für Normung e. V., 2003). The result of production engineering is therefore a unit good. The manufacturing processes include techniques for generating geometrically defined shapes from the shapeless state and for changing the shape and material properties. The most prominent classification of manufacturing processes in Germany, DIN 8580, distinguishes six main groups depending on the change in material composition (Fritz and Schulze, 2010). Original forms (main group 1) summarize processes that create material cohesion. Examples of this group are casting (master forms from liquid state), injection molding (master forms from plastic state), and pressing or sintering (master forms from granular or powdery state). Forming (main group 2) refers to the process of plastically changing the shape of a solid body. These include rolling, extrusion, deep drawing, stretching, bending, and twisting. Separation (main group 3) describes techniques for changing the shape of a solid body by locally breaking up its cohesion. Examples of this group include drilling, milling, turning, and grinding. Joining (main group 4) comprises techniques for creating cohesion (also with shapeless materials). Examples are riveting, decanting, welding, and gluing. Coating (main group 5) refers to the application of a firmly adherent layer of a formless material. This category includes painting, filling, and plastering. Finally, main group 6 describes all processes to change the properties of a material. This can be realized by diffusion, chemical reactions or quality dislocations, among others. Examples of this group are forging, annealing, hardening, and exposure. In general, the processes of the six main groups are also possible as combinations of several groups.

In contrast to manufacturing technology, process engineering comprises processes for the production and transformation of materials, mostly into a geometrically undefined form (Hirschberg, 1999). The result of process engineering is usually referred to as continuous flow material. The most common classification of process engineering is the classification according to the physical processes of the basic operations. Here a distinction is made between mechanical, thermal, chemical, and electrochemical process engineering. Mechanical process engineering primarily refers to techniques based on material conversion processes by means of mechanical impact. It includes the four main groups crushing, agglomerating, mixing, and separating. Thermal process engineering characterizes processes of thermal separation and purification. Examples for the group are distillation, rectification, and extraction. Chemical process engineering comprises techniques of material conversions by chemical reactions. A related group is electrochemical process engineering, which comprises processes for material conversion by means of electrochemical approaches. Examples of this group include the synthesis of chemicals and the electrolytic refining of metals. Battery and fuel cell technology can also be assigned to electrochemical process engineering. Last but not least, bioprocess engineering encompasses processes of material conversion with the help of biological or biotechnological processes. These include fermentation, biocatalysis, and synthesis. While process engineering, e. g., in the health industry, produces end products such as drugs directly from a raw material or substrate, the production of consumer products usually requires both process and production engineering methods. Process engineering is mostly used to provide the basic material (raw material), which is then converted to a higher degree of geometric accuracy in production engineering. Manufacturing and process engineering differ not only in the diversity of their processes, modes of operation, and produced goods, but also in their essential control criteria. While factors such as surface quality (roughness), tolerance (IT quality), and throughput time are essential for manufacturing technology, process stability, for example, is of far greater importance in process engineering.

The third main technology is energy technology, which includes the extraction, transport, storage, and use of energy in all forms. Every process in manufacturing and process engineering is dependent on the provision of energy. In addition to the main technologies, three auxiliary technologies are attributed to production technology: [1] handling technology (e. g., industrial robots), [2] conveyor technology (e. g., assembly lines, driverless transport systems) and [3] information technology (e. g., machine control).

Not assigned to production technology, although closely related to it, is waste and recycling technology. It includes all processes for the treatment of different types of waste. At the first level it is divided into waste recovery and landfill technology. The former is further divided into material recovery (recycling technology) and energy recovery (waste-to-energy technology). The processes of waste and recovery technology are divided into four categories. Biological waste and recovery technology refers to all processes whose function is based on biological mechanisms. These include composting, fermentation, and microbial leaching. Mechanical waste and recycling technology refers to techniques based on mechanical action. They include, among others, shredding and packaging. Thermal waste and recycling technology primarily comprises incineration processes, mostly for the purpose of energy generation. Examples are fluidized bed combustion or pyrolysis. Chemical-physical processes in waste and recycling technology use both chemical reactions and physical mechanisms of action. These include drying, immobilization, decanters or centrifuges. Not least, electronic waste and recycling technology includes processes whose operating principle is based on the use of electrical energy. Examples of this include electrohydraulic and electrodynamic fragmentation. Combinations of the techniques described here are possible as well.

The goal of biointelligent production technology is the development of SBMDs, which enable the manufacture and re-fabrication of personalized goods on site based on local resources. The prerequisite for this is the availability of the basic material and energy supply as well as the production and recycling technology at a single location. For the cumulative integration of biological materials, structures, and processes into the traditional production environment, an increasing convergence of the main and auxiliary techniques of production technology is necessary.

4.5 Chapter review and conclusions

The concept of biointelligence is intended to enable a reorientation in production theory and practice towards sustainability, based on BT. Its goal towards technology development is the creation and application of SBMDs, which enable the manufacture and re-fabrication of personalized goods on site based on local resources. In order to achieve this, two interpretations of the term are required: explicit and implicit. While the explicit understanding interprets the term as the systematic information-technological networking of biological and technical systems, the implicit interpretation regards the responsible handling of biological resources ranging from single cells to entire ecosystems as decisive. From a sustainability perspective, the concept thus implies a profound technological renewal, through a rigorous combination of information-driven intelligence with the development and establishment of new biological, medical or therapeutic processes, systems, and devices, under the condition of responsibility for the persistence of human life on earth (holistic benefit criterion). As such, it addresses nearly all major premises of a transition towards a sustainable economy (see Section 3.6), including the substitution of fossil fuels by renewable raw materials and energy sources (substitution principle), a regulative approach towards science and technology development (regulatory principle), a major focus on the increase of self-sufficiency and regionalization of supply chains (decentralization principle), an orientation towards natural principles, phenomena, and metabolisms (eco-convergence principle), as well as systems-oriented life cycle-thinking in cellular units (holism principle). Meanwhile, the comparison to related scientific disciplines repeatedly proves that biointelligence represents an independent scientific discipline which, although relying on existing disciplines, generates new knowledge through a strict interdisciplinary application orientation and a holistic benefit criterion. Accordingly, not an individual enabling technology or thought pattern of individual existing scientific disciplines is decisive for biointelligence, but rather systematic interdisciplinary cooperation, leading to an increasing convergence of the sciences and culminating in a sustainable technical renewal in the long term. Due to its underlying systems-oriented life cycle thinking in cellular units, biointelligence requires an adaptation of essential concepts of production theory:

- Biointelligent production:
  - Explicit: spontaneous or planned combination and recombination of cellular units, i.e., local production factors (at least one biological and one technical), under the provision of ICT principles for the purpose of regulation and autopoiesis.
  - Implicit: recurring phase of a constant cellular merging and diverging cycle between the provision of cellular units of subordinate levels, i.e., raw materials, and the sale of intended cellular unit, i.e., product.
- Biointelligent product:
  - Explicit: material result (cluster of cells) of a production process that enables the user to gain a functional benefit. It consists of at least one biological and one technical component as well as information-technological mechanisms for the purpose of regulation and autopoiesis.
  - Implicit: material and/or immaterial good, which fulfils a proven socio-ecological benefit.
- Biointelligent production process (only explicit): directional quantity of cellular units that is necessary to provide a certain output within a constant product cycle.
- Biointelligent production system (only implicit): cluster of cells whose interaction results in previously disconnected cellular entities. Cluster of cells may occur at any level, ranging from molecules to ecosystems.
- Biointelligent production enterprise (sub-concept of production system): goal-oriented, participatory socio-technical system of individual cellular units.

However, it remains clear that the vision outlined here can only become reality if it is accompanied by a sustainable transformation of society that adapts to the demands of the whole world.

Lessons learned

- Realistically sustainable value creation implies maintaining the current standard of living while not depleting natural resources. This requires a transformation in the sense of a renaissance of sustainable production methods with innovative technologies, which becomes possible with the concept of biointelligence, resulting in a technology-based demand economy.
- Biointelligence as novel scientific discipline that is based on insights of various fields of research as well as their future interdisciplinary collaboration.

- Two interpretations of biointelligence that apply equally:
  - Explicit concept: systematic information-technological networking of biological and technical systems in order to enable an overall intelligent system behavior.
  - Implicit concept: responsible handling of biological resources in form of thorough balance with natural environment and surrounding systems.
- The understanding, development, and management of BIS requires a specific mindset. This systems-oriented life cycle thinking implies:
  - Separation of reality in cellular segments, ranging from single molecules to entire ecosystems. Each cell is sufficiently described by six factors: type of matter, internal processes, relations with surrounding cells, and overall system, as well as purpose, functional principle, and conditions for and to application.
  - Assuming that each cellular unit is directly and/or indirectly interconnected with each other cellular unit, resulting in a circular logic of consequences.
  - The impact of a system on its surrounding systems and the overall system can only be thoroughly described over its entire life ('cradle to cradle').
  - Seek a balanced generation of benefits for individual, society, and ecosystem via function orientation, not technology and/or product focus.
- Biointelligence requires a rethinking of standard concepts of production theory (production, product, production process, production factors, production systems, production enterprises, production technology, etc.) that demands for further research.

Follow-up questions

- What approaches exist for the development and management of BIS?
- How do they relate to and/or complement each other?
- Which research and development fields currently exist?

5 Basic technologies and approaches of biointelligent systems

After discussing basic production-theoretical considerations on the concept of biointelligence in the previous section, this section presents methodological principles and elementary technologies for the development of BIS in an explicit sense. As discussed above, biointelligence does not constitute an entirely new field, it rather aims to (further) develop individual basic technologies by adding methods from various fields of research. In fact, it is supposed to stimulate thinking in certain structures. Yet, for the development of BIS in the explicit sense, individual basic approaches exist that emerge from the three modes of development of BT. Thereby, the key assumption is that each subsequent level continues to build upon the knowledge of the previous one. For bioinspiration, basic engineering and biological knowledge as well as profound experience of the biomimetic transfer process are of particular importance. The instrumental foundation of biointegration is the knowledge of basic biotechnology approaches, yet in addition to the prior know-how of bioinspiration. Based on both development modes, basic information technology know-how is fundamental for biointegration. This includes knowledge of basic approaches to I4.0 and AI as well as the concept of the biology-technology-interfaces (BTI). Figure 5.1 illustrates the fundamental toolbox for the development of BIS in an explicit sense.

Since it is not possible to provide a complete overview of the entire spectrum of biology, engineering, biotechnology, and computer science within this work, only selected methods are outlined in the following, which have particular significance for the development of BIS. These include the biomimetic transfer process, selected technology examples which are attributed a great transformative potential, the CRISPR/Cas9 system, AI as well as BTI. The latter also requires the elaboration of elementary relationships in online and inline analytics, soft sensors, and the concept of the digital twin (DT). Further approaches in individual scientific disciplines are only referred to in the respective sections. Accordingly, the toolbox does not claim to be complete, but initially serves to create a foundation for the establishment of biointelligence in science and to stimulate an innovation process in corporate practice.

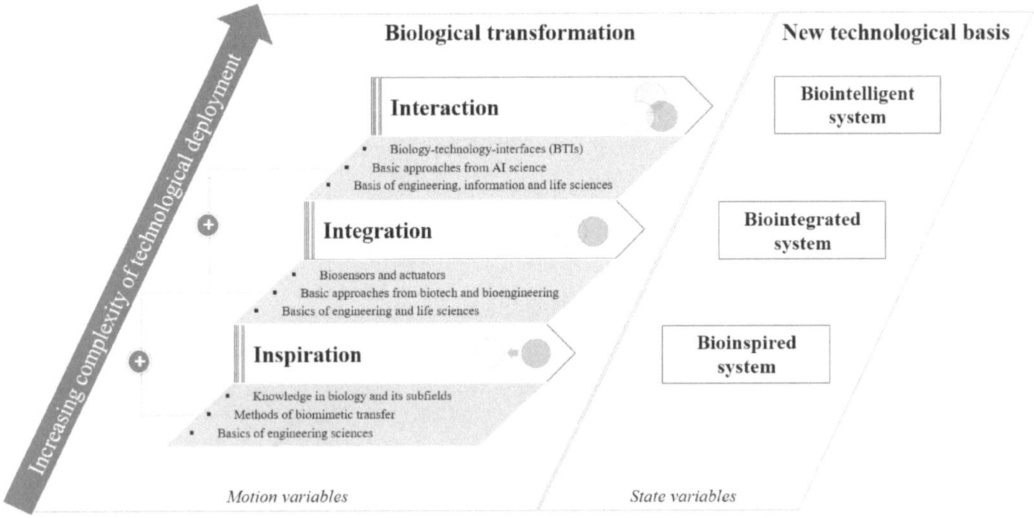

Figure 5.1 Basic toolbox for the development of BIS in an explicit sense [own illustration]

5.1 Examples of bioinspiration, biointegration and biointeraction

In order to provide a deeper understanding of the three development modes of BT as a basis for biointelligence, this section summarizes selected examples for the individual levels, which are characterized by the fact that they either contribute directly to the concept of BT or use similar approaches in a different context (Table 5.1). The examples presented here are based on preliminary works such as Bauernhansl et al. (2019) and Miehe et al. (2020a), yet further supplemented by research and subjective classification by the author. Consequently, the examples and technologies listed here do not claim to be exhaustive, but should enable the reader to gain a more in-depth understanding of BT and biointelligence. References and/or further literature are provided for each example. Especially evident is that the last mode of development is less a technological one than an integrative one. Accordingly, the approaches listed here are primarily visionary concepts that are still in an early phase of technical development and/or have not yet been implemented.

Table 5.1 Selected examples for individual development modes of BT

Description of example and/or technology	Reference/real-world example/ further information
Bioinspiration	
Efficiency improvement for hot water steam nozzle using the rules of evolution strategy, resulting in completely unexpected zigzag-shaped longitudinal geometry	(Kramer, 1960; Nachtigall and Wisser, 2015)
Automatic shape optimization of flapping wings for more efficient movement of micro air vehicles and small boats with fin drives	(Bilo, 1971; Nachtigall and Wisser, 2015)
Dolphin skin coating to reduce resistance for accelerating the floating of torpedoes and submarines under given propulsion power	(Kramer, 1960; Nachtigall and Wisser, 2015)
Biological balance as the foundation of systems theory and biocybernetics	(Vester, 1993; 2011; Nachtigall and Wisser, 2015)
Development of Velcro according to the large burdock of the thistle arctium lappa	(Nachtigall and Wisser, 2015)
Application of the gecko principle for climbing devices, allowing window cleaners to climb vertical glass surfaces	(Autumn et al., 2000; Nachtigall and Wisser, 2015)
Mirror optics in the eye of crabs as basis for X-ray telescopes	(Vogt, 1980; Nachtigall and Wisser, 2015)
Neural networks in biology (e. g., movement process of the stick insect) as basis for diverse applications in computer science (speech recognition software, learning machines, etc.)	(Cruse et al., 1997; Nachtigall and Wisser, 2015)
Building air conditioning according to the principle of a termite structure	(Nachtigall and Wisser, 2015)
Polar bear fur as basis for transparent thermal insulation of facade elements	(Tributsch et al., 1990; Koon, 1998; Nachtigall and Wisser, 2015)
Honeycomb principle as basis for lightweight structures (e. g., highly rigid membranes for large loudspeakers, lightweight bricks)	(Meretz, 1963; Nachtigall and Wisser, 2015)
Scoring films based on the principle of shark scales to reduce the frictional resistance and thus the fuel consumption of large aircrafts	(Reif and Dinkelacker, 1982; Nachtigall and Wisser, 2015)
Water production by fog condensation according to the principle of the dark beetle (onymacris unguicularis)	(Enders and Henschel, 2000; Nachtigall and Wisser, 2015)

Description of example and/or technology	Reference/real-world example/ further information
Development of multifunctional packaging according to the coconut principle	(Küppers and Tributsch, 2001; Nachtigall and Wisser, 2015)
Micro- and nanostructuring of super-hydrophobic contact surfaces of various basic materials for self-cleaning according to the lotus leaf and lotusan principle	(Barthlott and Neinhuis, 1997; Patankar, 2004; Gao and McCarthy, 2006; Nachtigall and Wisser, 2015)
Self-repairing composite materials by monomer systems embedded in the polymer matrix, using nano materials or for application in soft robots	(Lee et al., 2004; Nachtigall and Wisser, 2015; Cheng et al., 2014; Wang et al., 2015)
Colon peristalsis as basis for micro feed pumps	(Smela et al., 1999; Nachtigall and Wisser, 2015)
Self-sharpening tools based on the principle of rodent teeth to improve the energy and material efficiency of cutting processes	(Rechberger, 2012; Nachtigall and Wisser, 2015)
Application of the pneumatic principle, e. g., for the development of air domes, convertible lock constructions or pneumatic parking garages	(Nachtigall and Wisser, 2015)
Application of the tensairity principle (form stabilization via a higher internal in relation to external pressure), e. g., for emergency bridges from tubular balloons, self-repair for pneumatic systems or as stabilizer in large buildings	(Luchsinger et al., 2004; Pedretti and Luscher, 2007; Rampf et al., 2012; Nachtigall and Wisser, 2015)
Butterfly scales as basis for selective facade elements, allowing visible light to fully pass through whilst filtering UV or IR radiation	(Nachtigall and Wisser, 2015)
Bioinspired robots	(Bar-Cohen and Breazeal, 2003; Pfeifer et al., 2007; Spenko et al., 2008)
Gripping arm inspired by the trunk of an elephant	(Nachtigall and Wisser, 2015)
Streamlined car body based on the example of fish	(Nachtigall and Wisser, 2015)
Hair sensors for flow monitoring according to the principle of spider (cupiennius salei)	(Tao and Yu, 2012; Dagamseh et al., 2012; Nachtigall and Wisser, 2015)
Localization of objects in currents by sensor arrangements according to the principle of fish sidelines and/or electrosensory organs	(Nachtigall and Wisser, 2015)

Table 5.1 Selected examples for individual development modes of BT (*continued*)

Description of example and/or technology	Reference/real-world example/ further information
Measurement of minimal deformations and changes in mechanical tension, inspired by splitting organs of spiders	(Nachtigall and Wisser, 2015)
Technical reproduction of insect eyes for various applications in air traffic monitoring, e. g., panoramic imaging, safe crossing of narrow passages, terrain following overflight, altitude and landing control, hodometry	(Nachtigall and Wisser, 2015)
Development of autosensitive ('feeling') materials, e. g., based on the example of muscle length control	(Nachtigall and Wisser, 2015)
Decentralized control of snake robots, based on the leg movements of locusts to improve the momentary adaptation to roughness (small obstacles) and local inclinations of the ground	(Sato et al., 2011; Nachtigall and Wisser, 2015)
Development of artificial silicone retina based on the natural functional transmission of cochlea and retina to support severely hearing-impaired people to adapt to sound sources	(Del Priore, 2005; Nachtigall and Wisser, 2015)
Intelligent sensor implants and sensor-actuator controlled prosthetic knee joints for imitating the natural organs	(Basmer et al., 2010; Nachtigall and Wisser, 2015)
Artificial light trapping systems/antennas based on the antenna pigments of chlorophyll ('artificial leaf')	(Fetisova et al., 1987; Imahori, 2004; Aratani et al., 2009; Ziessel and Harriman, 2011; Nachtigall and Wisser, 2015)
Evolutionary algorithms for improving designs and production environments	(Eiben et al., 1999; Zitzler et al., 2000; Deb, 2009; Yu et al., 2011; Hecker et al., 2013)
Structural optimization of highly stressed working tools in raw material extraction based the large anteater (Myrmecophaga tridactyla)	(Nachtigall and Wisser, 2015)
Systems thinking and systemic behavior according to the patterns of biological systems	(Vester, 1993; 2011; Nachtigall and Wisser, 2015)
Evolutionary and systemic management according to the example of natural systems	(Beer, 1967; Vester, 1993; 2011; Küppers, 2004; 2010; Malik, 2006)
Efficient fluid transport for cooling lubricants to the cutting point in a cutting machine based on the skin structure of lizards	(Comanns et al., 2015; Nachtigall and Wisser, 2015)

Description of example and/or technology	Reference/real-world example/ further information
Mimicking the swarming behavior of different animals for various applications, e. g., collision avoidance, robot control in production, general management	(Zhang et al., 2007; Hinchey et al., 2007; Karaboga and Akay, 2009; Azevedo et al., 2010; Tan and Zheng, 2013)
Biointegration	
Biomaterials as substitute for fossil-based raw materials, e. g., biopolymers produced from cellulose, starch, proteins or peptides	(Hench and Wilson, 1984; Hubbell, 1999; Griffith, 2000; Babu et al., 2013; Hatti-Kaul et al., 2020)
Biodegradable materials for closing and/or replacing harmful material cycles, e. g., as starch- or protein-based bio-nanocomposites or based on silk fibroin and keratin	(Ozawa et al., 2002; Nair and Laurencin, 2007; Chu, 2007; Lapitsky et al., 2008; Zhao et al., 2008; Vasconcelos et al., 2008)
Electroactive/adaptive biomaterials as basis for 'smart' biomaterials, e. g., as plant or microbial biopolymers with ionic functional groups or thermoplastic starch as electroactive biopolymers	(Finkenstadt and Willett, 2005; Babensee, 2008; Balint et al., 2014; Tandon et al., 2018)
Biobased coolants and lubricants, e. g., to improve lubrication of workpieces in production machines on a non-fossil basis, improve biodegradability and lubricity	(Schneider, 2006; Honary and Richter, 2011; Zulkifli et al., 2016; Syahir et al., 2017; Zainal et al., 2018)
Biosensors (incl. organ-on-chip technologies), e. g., to improve data acquisition in complex systems or to invent entirely new features in machines and robots (smell, taste, etc.). Examples are: ▪ Measurement of analytes by electrochemical detection using potentiometry, e. g., by coating the measuring electrode with a compatible enzyme ▪ Determination of bacterial content of bathing and/or waste water by applying antibodies to an oscillating membrane ▪ Marker-free analysis of proteins with in situ silicon field effect sensors ▪ Determination of penicillin concentration in a bioreactor in which strains of fungi are cultivated using the enzyme Acylase ▪ Simulation of activities, mechanics and physiological reaction of organ systems in form of multi-channel 3D microfluidic biochips made of polydimethylsiloxane (organ-on-chip)	(Eggins, 1997; Chaubey and Malhotra, 2002; D'souza, 2001; Gerard et al., 2002; Jianrong et al., 2004; Berggren et al., 2001; Ronkainen et al., 2010; Turner, 2013; Borisov and Wolfbeis, 2008; Su et al., 2011; Selimović et al., 2013; Chapin et al., 2011; Caplin et al., 2015; Wikswo et al., 2013; Bhatia and Ingber, 2014; Zhang and Radisic, 2017; Rogal et al., 2017; Christoffersson et al., 2017; Kilic et al., 2018; Sun et al., 2019; Zhao et al., 2019; Sriram et al., 2018; Lud et al., 2006)

Table 5.1 Selected examples for individual development modes of BT (*continued*)

Description of example and/or technology	Reference/real-world example/ further information
Bio actuators for control of complex systems (bioreactors, fermenter, etc.), e. g., using insect dorsal vessels, as integrated skeletal muscle cell-based bio-actuator, as double acting sleeve muscle or for virus-based alignment of inorganic, organic, and biological nanosized materials	(Lee et al., 2003; Hayden et al., 2003; Touyama et al., 2009; Fujita et al., 2011; Suzumura et al., 2011; Zheng and Shen, 2013; Hanashi et al., 2014)
3D/4D bioprinting as basis for flexible shaping of cell tissue in a multitude of applications, e. g., printing of vascularized, hetero-generous cell-laden tissue constructs or periodontal tissue regeneration	(Hubbell, 1995; Griffith and Naughton, 2002; Kolesky et al., 2014; Murphy and Atala, 2014; Shin et al., 2003; Gao et al., 2016; Li et al., 2016; Mandrycky et al., 2016; Zhang et al., 2017; Yang et al., 2019; Carter et al., 2017)
Additive biomanufacturing, e. g., enzyme process technology for low-energy natural cohesion or printing of water-based composites and regenerated biomaterials (e. g., chitosan, cellulose or sodium alginate) as building components	(Poh et al., 2016; Singh et al., 2017; Zadpoor, 2017; Vinson et al., 2017; Zhang et al., 2014; Mogas-Soldevila and Oxman, 2015)
Additive manufacturing of biomaterials, e. g., as hybrid additive manufacturing of silicone biostructures	(Campoli et al., 2013; Harun et al., 2018; Liravi and Toyserkani, 2018; Bose et al., 2018)
Biobased medical therapy, e. g., RNA-based personalized therapy or personalized CAR-T cell immunotherapy for liquid lymphoma	(Daka and Peer, 2012; June et al., 2018; Lu et al., 2014)
Production and application of biopharmaceuticals, e. g., as biologics involving recombinant DNA technology (e. g., growth hormones, monoclonal antibodies, fusion proteins) for therapy of rheumatoid arthritis, psoriatic arthritis, or ankylosing spondylitis	(Walsh, 2003; 2004; 2005; Crommelin et al., 2003; Giddings et al., 2000; Butler, 2005; Karg and Kallio, 2009; Bailon and Won, 2009; Kerr, 2010)
Biobased energy carriers, e. g., biobutanol, algae as source for biodiesel, formic acid for hydrogen binding or microbiological, enzymatically catalyzed material transformations	(Fischer and Schrattenholzer, 2001; Lemus and Lal, 2005; Dürre, 2007; Weiss et al., 2007; Field et al., 2008; Demirbas, 2008; Sims et al., 2010)

Description of example and/or technology	Reference/real-world example/ further information
Biobased energy conversation and/or storage technologies, e. g., microbial fuel and electrolysis cells using living microorganisms that process organic substances as part of their energy metabolism to produce electricity or hydrogen or soil carbon storage by switchgrass grown bioenergy	(Katz et al., 1999; Chen et al., 2001; Katz and Willner, 2003; Moore et al., 2005; Ramanavicius et al., 2005; Minteer et al., 2007; Arechederra et al., 2007; Kamitaka et al., 2009; Sleutels et al., 2009; Zhou et al., 2019; Schiebahn et al., 2013; Liebig et al., 2008; Read and Lermit, 2005)
Biobased separation, recovery and recycling technologies, e. g., bioleaching of rare earths from end-of-life magnets, heavy metals or from sewage sludge	(Olson et al., 2003; Rohwerder et al., 2003; Pathak et al., 2009; Bosecker, 1997; Moslemizadeh et al., 2016; Mansouri et al., 2017; Tabatabaee and Kurth, 2017)
Recycling technologies for biobased materials, e. g., recycling of thermoplastic composites with wood fibers	(Soroudi and Jakubowicz, 2013; Peinado et al., 2015; Clavé et al., 2016; Alaerts et al., 2018)
Human-brain-interfaces, e. g., for activity-aware mental stress detection or real-time detection and classification of emotional states	(Sitaram et al., 2011; Sun et al., 2010; Wolpaw, 2007; Zander and Kothe, 2011; Zander et al., 2017)
Biorobots, as microscale robots for delivery of small molecules, DNA vectors, and proteins to edit cellular machinery and probe cell signaling systems or as reconfigurable organisms by using stem cells from frogs	(Guo et al., 2005; Pavone et al., 2006; Assaf et al., 2013; Wong et al., 2013; Hunter et al., 2018; Kriegman et al., 2020)
Biointeraction	
Biobased soft sensors as combination of biosensors and soft sensors for variety of applications, e. g., for online/inline measurement in bioprocesses, data gathering in biobased products and biological environments (e. g., aquaponic systems).	Among others, preliminary work documented by Assis et al. (2000), Luttmann et al. (2012), and Gustavsson and Mandenius (2013)
Thorough digital twins of biology-technology-systems (e. g., for human organs or cellular metabolism) as prerequisite for information technical linking	Among others, preliminary work documented by Ahmed et al. (2019), Anderl et al. (2018), Negri et al. (2017) and Nargund et al. (2019)

Table 5.1 Selected examples for individual development modes of BT (*continued*)

Description of example and/or technology	Reference/real-world example/ further information
Closed-loop intuitive prosthetics and implant technology with feedback function and flexible barrier-free exoskeletons for a wide range of applications, treatment of disabilities, assistance at workplace, etc.	Among others, preliminary work documented by Petrini et al. (2019)
Interactive, function adaptive, learning biomaterials, e. g., to adapt the functionality to its use or learn new features	Among others, preliminary work documented by Abramson et al. (2005), Tartarini and Mele (2016), Karanasiou et al. (2017; Tandon et al., 2018)
Multi-product fermenter, bioreactors and biorefineries, using switchable organisms	Among others, preliminary work documented by Wang et al. (2017)
Instant on-site production of personalized biopharmaceuticals/biologics	Among others, preliminary work documented by Daka and Peer (2012) and June et al. (2018)
Living and learning robots, e. g., based on reconfigurable organisms, able to redesign their own source code	Among others, preliminary work documented by Pavone et al. (2006), Pearce (2012) and Kriegman et al. (2020)
Smart biomanufacturing devices (SBMD) for on-site production of personalized food, pharmaceuticals, and consumer products, e. g., by combining multi-product micro-fermenter with 3D/4D bioprinting or additive biomanufacturing devices or bioelectrochemical processing of household wastage	Among others, preliminary work documented by Bauernhansl et al. (2019) and Miehe et al. (2020a)
Self-sustaining/-supplying apartment building which is capable of performing a task in a spontaneously emerging cellular production network, by integrating various types of SBMDs, efficient fruit and vegetable cultivation (next level urban gardening/horticulture), various sensor technologies to measure the well-being of the residents, a variety of adaptive (actuator) elements for real-time response to residential demands and overall learning control to develop new features and improve performance and/or prevent incidents	None

5.2 Basic approaches of bioinspiration – the biomimetic transfer process

Since this work cannot provide a comprehensive overview of the state of biological and technical knowledge, this section focuses exclusively on the biomimetic transfer process. The definition of terms in this section initially requires a distinction between the German and English language (Wanieck, 2019). While in German the term bionics ('Bionik') represents a combination of the words biology and technology ('Technik'), in English three terms exist which can be understood as counterparts of it, i. e., bionics, biomimetics and biomimicry. According to Nachtigall (2013), the German scientific discipline of bionics focuses on the technical application of processes and principles of biological systems. Here, the term biology is broadly defined. Thereby, the interaction of animate and inanimate nature can also serve as a basis for inspiration. Hence, its aim is to transfer biology to technology. It interprets biological phenomena as solutions to problems, inventions, and innovations and seeks options for transferring this knowledge into technical systems (Verein Deutscher Ingenieure, 2012b). The conception of flying machines by Leonardo Da Vinci, which was based on the observation of the flight behavior of birds (Romei 2008), is often quoted as an early example of this approach.

In the English language a direct equivalent for the term does not exist. The term bionics, often falsely used as an equivalent, refers to a combination of biology and electronics. It describes the replacement and/or improvement of biological (sub)systems by electronic components, as is the case, for example, in prosthetics or implants (Wanieck, 2019). According to the concept of BT presented here, the term bionics would be more likely to be assigned to biointegration or biointeraction. A more literal translation of the German term bionics into English is biomimetics or biomimicry. The term characterizes the element mimetics (ancient Greek mīmēsis) as imitation (Vincent et al., 2006; Vincent, 2009). Nature serves as the model for technology, an aspect that Benyus (2002) places in an overall social context. The revised version of VDI 6220 now distinguishes between two interpretations (Verband Deutscher Ingenieure, 2019). The correct translation of the transfer of biological solutions to technical systems to solve practical problems into the English language is biomimetics. In contrast, the term biomimicry describes the development of environmentally compatible innovations in technology and society by imitating biological phenomena. This work understands bioinspiration in the sense of the German interpretation of the term bionics and subsequently refers to the term biomimetics.

Various approaches exist for the transfer of biological phenomena to technical systems (biomimetic transfer process). Nachtigall (2010) describes three essential steps: recognition, abstraction, and implementation. ISO 18.458 provides a similar procedure (Deutsches Institut für Normung e. V., 2016). The approach presented here postulates three steps: [1] the analysis of the functions of a biological system, [2] the modelling of the biological system, and [3] the transfer and/or application of characteristics and/or functions of the biological system without using it. Wanieck distinguishes between a solution-oriented and a problem-oriented approach (Wanieck, 2019). If an accidental discovery of a biological phenomenon in basic research provides the stimulus for a technical development, this is referred to as a solution-oriented approach. This approach is frequently described as bottom-up or biology push. Wanieck names Velcro fasteners and the self-cleaning effect of the lotus plant as examples of this type of biomimetic transfer process (Wanieck, 2019). The approach of transfer to the technical system is similar to the methodology described by Nachtigall (2010). By observing nature, it attempts to generate an understanding of cause-and-effect relationships. With the help of abstraction, those functional principles are detected which are favorable to the technical solution. In a last step the implementation into the technical system is carried out.

If the starting point for development is a technical problem, the problem-oriented approach is applied. Alternative designations are top-down or technology pull. Here, the procedure described above is supplemented by a few steps. First, it is necessary to analyze, describe, and abstract the technical problem. Based on the core functions of the technical system identified in this process, an analysis of biology is conducted. Hereby, related functional principles are sought in its knowledge pool and adapted into the technical system. According to Fayemi et al. (2017) and Wanieck (2019), the procedure of the problem-oriented approach comprises seven steps. [1] First, the problem analysis is carried out in the form of a decomposition of functions, description of the target/actual situation, and definition of target functions. [2] By defining the degree of abstraction and the solution-neutral description of the problem, abstraction follows. [3] In a third step, the problem is transferred to biology by formulating 'questions to biology' in the 'language of biology'. [4] Then, biology and its sub-disciplines must be searched for suitable role models. Taking into account relevant boundary conditions and functionalities, the research is done with the help of search engines and databases. The selection of suitable role models requires the adequate description of structures and functions as well as the comparison of biology and technology. [5] The physical-chemical principles are described and a functional modelling conducted to achieve the necessary abstraction of biology. The prerequisite for this is that the biological system itself is not part of the solution. [6] The translation into technology then requires the examination of feasibility and practicability before [7] the application in the technical system is ensured by suitable tests and the production of the final product. Examples for the technology pull approach are fuel savings in vehicles, braking distance-optimized tires and self-sharpening knives.

Figure 5.2 illustrates the difference between solution- and problem-oriented approach of the biomimetic transfer process according to Fayemi et al. (2017), Speck et al. (2017), and Wanieck (2019).

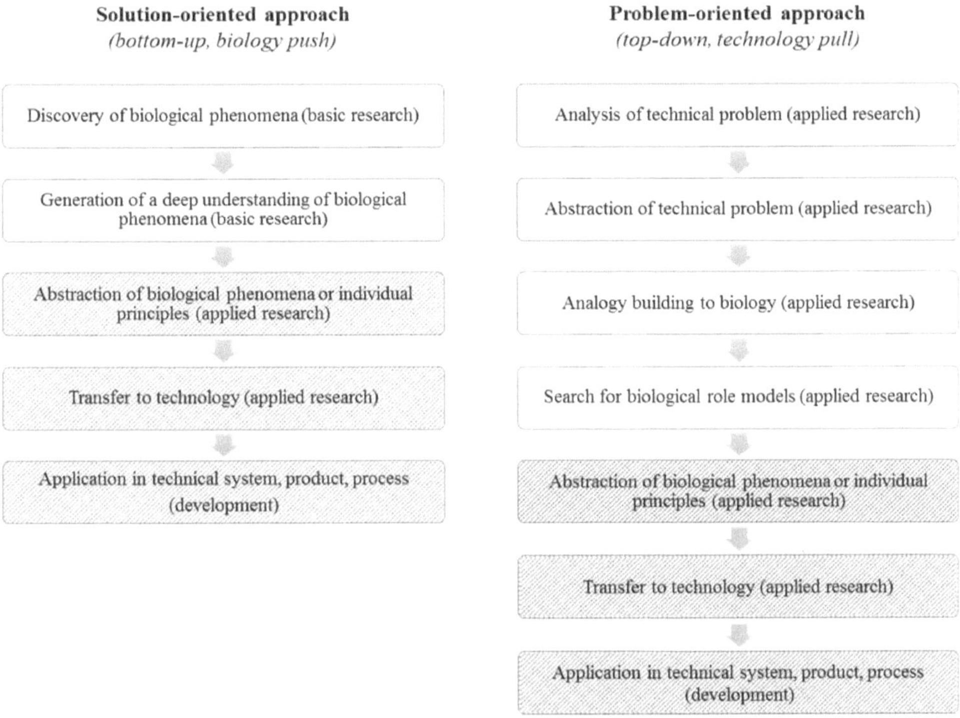

Figure 5.2 Solution- and problem-oriented approach of biomimetic transfer process according to Fayemi et al. (2017), Speck et al. (2017), and Wanieck (2019)[1)]

According to Wanieck (2019), numerous fields for biomimetic applications exist. These include architecture, fluid dynamics and aerodynamics, surfaces, robotics, materials and structures as well as lightweight construction and design (Wanieck, 2019). In order to develop innovative solutions in these fields, a comprehensive knowledge of biological phenomena and functional principles is necessary. While a systematic summary of biological principles and process descriptions for the solution-oriented approach has not yet been published in literature, a number of methods and tools that support the biomimetic transfer process exist in the context of the problem-oriented approach (Fayemi et al., 2017; Speck et al., 2017; Wanieck, 2019). However, biomimetic development can use existing process models as guidelines. For instance, Speck et al. (2017) present a decision tree for the design of sustainable, bioinspired systems based on a dataset of 70 bottom-up and top-down developments (Figure 5.3). Yet, a different understanding of bioinspiration is assumed here than the one underlying this work. As soon as an organism is applied itself, the term bioinspiration is no

1 The gray shaded steps visualize the standard building blocks of the biomimetic transfer process which have to be completed in both approaches.

longer used here, but rather biointegration. Accordingly, the graph represents an example that has still to be developed for the present interpretation of bioinspiration.

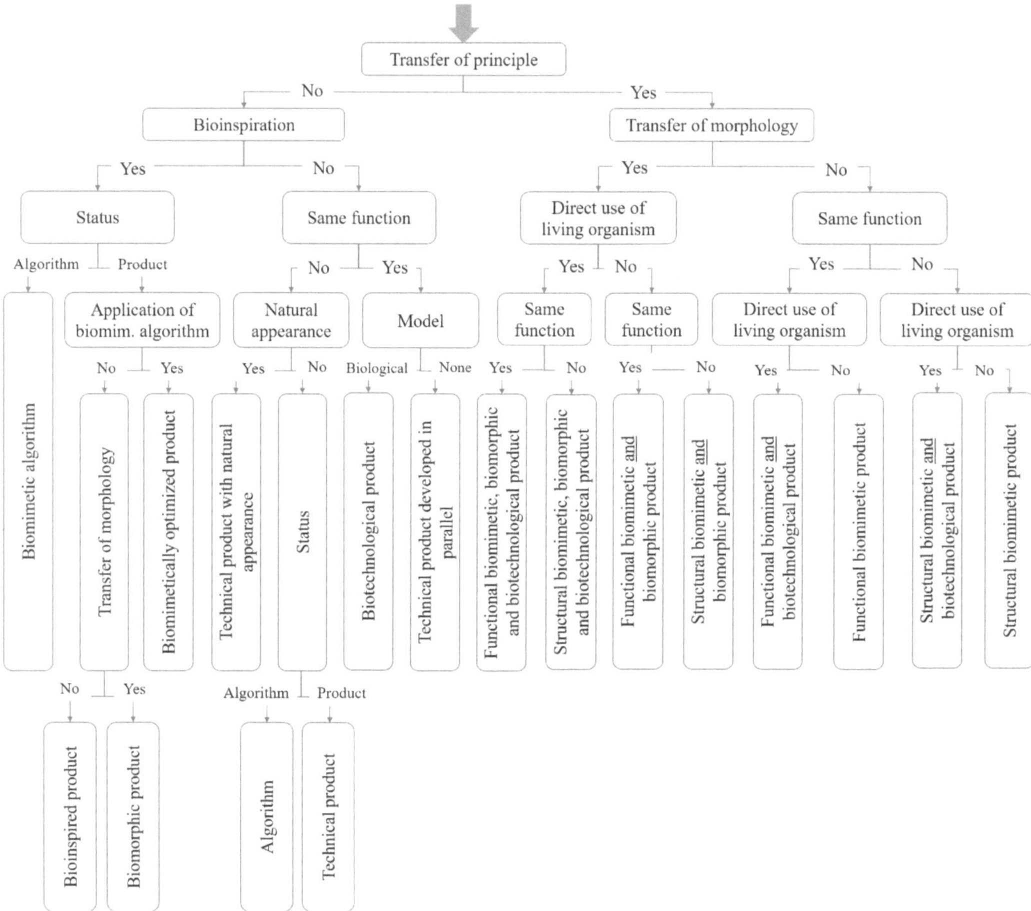

Figure 5.3 Decision tree for the design of sustainable products at the interface of bioinspiration and biointegration according to Speck et al. (2017)

Nachtigall, in turn, presents extensive reference material on the biomimetic transfer process as well as illustrations of biological examples (Nachtigall, 2010; Nachtigall and Wisser, 2015). The cataloguing of biological principles that serve as stimuli for biomimetic design is of eminent importance for bioinspiration (Nachtigall, 2005). The classification is based on elementary design principles and distinguishes 16 categories and 58 subcategories, which are assigned to specific biological mechanisms. Figure 5.4 summarizes the classification of biological principles as the basis of biomimetic designs according to Nachtigall (Nachtigall, 2005).

Connections and anchorages

amongst others

- *Connecting parts of an animal, between animal and animal, between parts of organisms etc.*
- *Vacuum devices as single suction cups, double suction cups, quadruple suction cups etc.*
- *Adhesion as simple adhesion surfaces and composite adhesion structures*

Joints and levers

amongst others

- *Joints as hinge and ball joints*
- *Levers as single and double sided levers*
- *Link chains as open and closed link chains, parallel guides and six link chains*

Lubrication and lubricants

amongst others

- *Joint lubricants using the example of hyaluronic acids*
- *Functional construction and time dependence of the load on the biomaterial 'cartilage'*

Bonds and adhesives

amongst others

- *Bonding in nature based on principles and types of bonding*
- *Adhesives in nature on the basis of suitability and special features*

Material arrangements/ lightweight construction

amongst others

- *Minimal constructions via stiffeners, radially symmetrical 'buildings' and formations*
- *Row-optimized material arrangement according to the structure of honeycombs*
- *Multi-component constructions as sandwich construction, fine construction and fibre direction*

Drive and movement mechanisms

amongst others

- *Motion drives as direct and indirect drives*
- *Jumping mechanisms as catapult and click mechanisms*
- *Running mechanisms in the form of legs and soles*

Hydraulics and pneumatics

amongst others

- *Tire concept as single and multiple tires and convertible tires*
- *Hydraulic movement mechanisms in plants and animals*
- *Hygroscopic and surface effects as hygroscopic mechanisms and surface effects*

Pumps and liquid delivery

amongst others

- *Pumps as saliva pumps and hearts*
- *Vessels using the example of arteries and veins and plant conduction vessels*

Flow control

amongst others

- *Flow-favorable body shapes in fish, dolphins, birds and water beetles*
- *Thrust and lift generators as rudder, wing and flapping wing*
- *Passive flight movement as glider and parachute*
- *Flow influence of dolphin skin and owl feather*

Burst and explosion mechanisms

amongst others

- *In the plant kingdom with pill thrower and spray cucumber, balsam fruits and fern sporangia*
- *In the animal kingdom with puncture capsules, bomb beetles and gun crabs*

Shredding and cutting equipment

amongst others

- *Chewing apparatus using the example of chewing pine and chewing stomachs*
- *Cutting devices as cutting gravers and saw blades*
- *scraping and rasping devices for snail radula, lamprey and blenny*

Net and trap constructions

amongst others

- *webs using the example of spiders and insect larvae webs*
- *Catch mechanisms as lassos and snap trap nets*

Defense and armour

amongst others

- *armour as shells, ballistic armour, variable armour and shield*
- *shells in plants and animals*

Penetrators and lancing devices

amongst others

- *Laying drills for plant wasps and grasshoppers*
- *Stinging bristles for bugs and aphids*
- *Poison injections as poison claws and teeth*

Snap and grip constructions

amongst others

- *Extreme spreads of egg snake and nightjar*
- *Snap turtle, frogfish and snap jaw ant snap*
- *Claws as predator legs, claw feet and as abdominal appendages*
- *jaws as dentures, worm mouth openings and*

Buildings and dwellings

amongst others

- *Quivers for caddisflies and bag carriers*
- *Floor constructions for wasps and hornets*
- *Grave and line constructions for solitary bees and as line nests*
- *Packs as honeycomb packs, Fibonacci spirals and split-fruit packs*

Figure 5.4 Systematization of biological principles as the basis of biomimetic constructions according to Nachtigall (2005)

Generally, evolution strategy represents a powerful tool in biomimetics to solve decision problems. According to Rechenberg (1973), evolutionary strategy is based on the four parameters of natural evolution (mutation, recombination, isolation, and selection). Whenever a mutation leads to better reproductive success, the respective trait carrier become parents for a new generation, otherwise the previous one is used. Evolution strategy considers the decision-making process as an analogy. According to this approach, the solution to any decision problem must be designed in a way that allows frequent small changes to be made in a direction that has proved successful in the search process. An evolutionary strategic approach is particularly useful where theories or calculation methods do not (yet) exist.

Nachtigall (2010) also emphasizes the elementary connection of the biomimetic transfer process with essential methods of the operational construction process. These include various creativity techniques as well as the theory of inventive problem solving (TRIZ) (Gadd and Delbrück, 2016). In the context of the broader meaning of biomimicry, Ceschin and Gaziulusoy (2020) elaborate a close connection to sustainable design (see Chapter 6).

5.3 Basic approaches and technologies of biointegration

The instruments of biointegration rely on knowledge of bioinspiration and supplement it with fundamental skills in biotechnology. In contrast to the previous development mode, here it is not primarily a matter of transferring findings from life sciences into a purely technical system, but to integrate biology as a productive element. Since this work cannot provide a complete overview of biotechnological knowledge, this section outlines the core areas of biotechnology in order to discuss basic biotechnology-related approaches and technologies with great transformative potential.

5.3.1 Classification of biotechnology

According to the OECD (2001), biotechnology is the application of science and technology to living organisms, parts of them, their products or models of them, for the purpose of modifying living or non-living matter to increase knowledge, produce goods, and provide services. Schüler (2016) specifies the definition in a pragmatic explanation: 'whenever biological systems or their parts are used to produce products in the form of a technical process, it is called biotechnology'. Fundamental in this context is the concept of the cell as a productive unit, i. e., arctium lappa, a factory. This, in turn, is the basis of microorganisms (including plants, bacteria, fungi) and complex organisms (especially animals and humans). Figure 5.5 illustrates the analogy of a cell to a production system according to Schüler (2016). The cell components pores, endoplasmic reticulum, polysomes/ribosomes, vesicles, mitochondria, and the cell nucleus fulfil related functions such as factory gates, factory roads, production, storage tanks, power plants, and management in conventional production facilities. These 'production functions' of a cell enable fermentation, i. e., all chemical transformations triggered by enzymes and microorganisms, with or without oxygen supply.

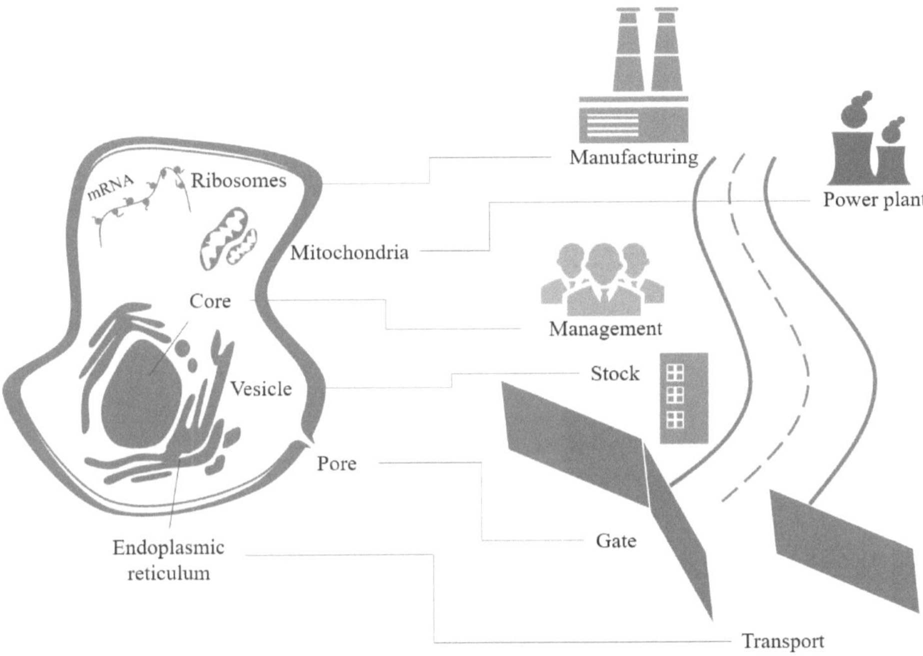

Figure 5.5 Analogy of cell and production system according to Schüler (2016)

Key areas of biotechnology have been interpreted in several ways from different perspectives. While the list-based classification of biotechnology summarizes seven different processes and methods, the sectors and markets that apply them result in the so-called color theory of biotechnology. The list-based definition divides biotechnology into individual techniques. These include DNA/RNA, proteins and other molecules, cell and tissue culture/tissue engineering, bioprocess engineering, gene and RNA vectors and nanobiotechnology. The subarea DNA/RNA includes technologies such as genomics, gene probes, DNA/RNA sequencing/synthesis/amplification and gene expression profiles. The sequencing, synthesis, and modification of proteins and peptides is as much a part of proteins and molecules as proteomics, protein isolation and purification and the elucidation of protein signaling pathways. The field of cell and tissue culture (also tissue engineering) includes instruments such as cell fusion, vaccine, and immune stimulants and embryo cultivation. Bioprocess engineering, on the other hand, includes all fermentations in bioreactors as well as bioprocessing, biological leaching, biological pulp production, biological desulphurization, microbial and plant environmental remediation and biological filtration. Methods of gene therapy and viral vectors are included in the field of gene and RNA vectors. The modelling of complex biological processes (including system biology) can be found in the field of bioinformatics as well as the creation of databases of genomes and protein sequences. Application of nano- and microsystems technology instruments for research into biological systems is included in the field of nanobiotechnology.

Among others, Renneberg and Süßbier (2005) as well as Schüler (2016) provide comprehensive explanations of the instruments of the list-based definition of biotechnology. The color theory of biotechnology, in contrast, distinguishes between its use in agriculture (green biotechnology), in medicine and pharmaceuticals (red biotechnology), in industry (white biotechnology), in waste management (grey biotechnology), in environmental protection (brown biotechnology), and for the use of marine resources (blue biotechnology). Basically, any instrument of the OECD's list-based definition can be applied in any color segment of biotechnology. Figure 5.6 illustrates the core areas of biotechnology from the point of view of methods and users, following the example of Schüler (2016) and OECD (2001).

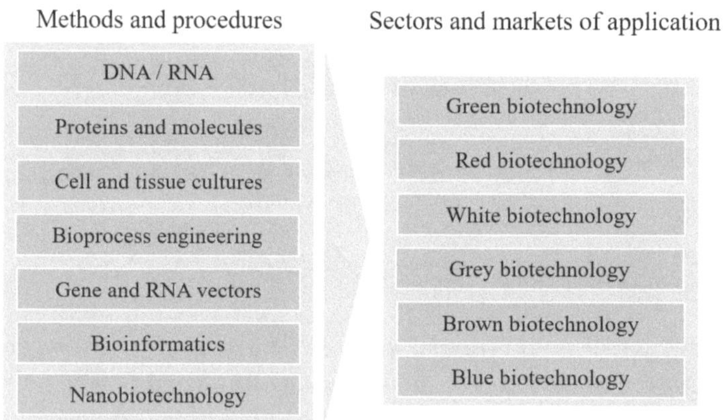

Figure 5.6 Key areas of biotechnology in terms of methods and practical application according to Schüler (2016) and OECD (2001)

The field of biotechnology comprises a multitude of methods and technologies. However, certain concepts are more important than others for the vision underlying this work, as they represent basic enabling technologies. In order to provide the reader with a deeper understanding, the following sections outline basics of bioreactors, biorefineries, biofuel cells and bioelectrolysis, additive manufacturing and bioprinting, gene editing, and others, all being closely linked to the concept of biointegration. The subsequent selection of the technologies presented here is based on Bauernhansl et al. (2019) and Miehe et al. (2020a). However, the selection should in no case be regarded as being complete, but rather as a basis for a better comprehension of the intended mindset.

5.3.2 Bioreactors

A bioreactor is an artificial technical environment that cultivates microorganisms, cells or small plants by applying biological processes, such as bioconversion or biocatalysis. A synonym for a bioreactor is fermenter (Decker and Reski, 2008). Current applications include the use in sewage treatment (to bind dissolved compounds of microorganisms of the formed biomass in the activated sludge process in order to be fermented in the digestion tower to sewage gas), in biogas plants (where the biomass used is decomposed via hydrolysis, acidogenesis, acetogenesis and methanogenesis to biogas and fermentation residue), in breweries and wineries (where the microorganism yeast converts sugar from the mash or grape juice into alcohol and carbon dioxide), and in the pharmaceutical industry (e. g., for the production of insulin) (Renneberg and Süßbier, 2005).

Compared to a chemical reactor, bioreactors differ in several points. While a chemical reactor basically always reacts identically, depending solely on composition and temperature, a bioreactor may be highly unpredictable due to differing metabolic pathways, yields, productivity, etc. of the microorganisms (Benz, 2011). Essential operating parameters of bioreactors that need to be monitored for robust functioning are nutrient supply (macronutrient elements, such as carbon, nitrogen and phosphorus, micronutrient elements, such as vitamins or amino acids, and energy-providing compounds, such as fructose, glucose), temperature (by heating and cooling circuits), oxygen content (e. g., by regulation of gas flow rate, stirrer speed, stirrer tool geometry, gas mixture composition), pH value (pumps, caustic soda, phosphoric or hydrochloric acid), and homogenization (by stirrer).

Figure 5.7 illustrates the generic construction of a bioreactor according to Singh et al. (2014). However, it has to be noted that the actual bioreactor design mostly has to be adapted to specific parameter (see below).

Bioreactors can be classified by a variety of differentiation parameter. First, discontinuous processes (filling and operation in one charge, after removal the process must be restarted, also: batch process), continuous processes (the reactor is continuously fed with starting materials and corresponding quantities of the reaction mixture are removed) and semi-continuous process (process periods of up to 90 days) can be differed (Renneberg and Süßbier, 2005; Singh et al., 2014). Table 5.2 summarizes advantages and disadvantages of discontinuous, continuous and semi-continuous bioreactor operation according to Singh et al. (2014).

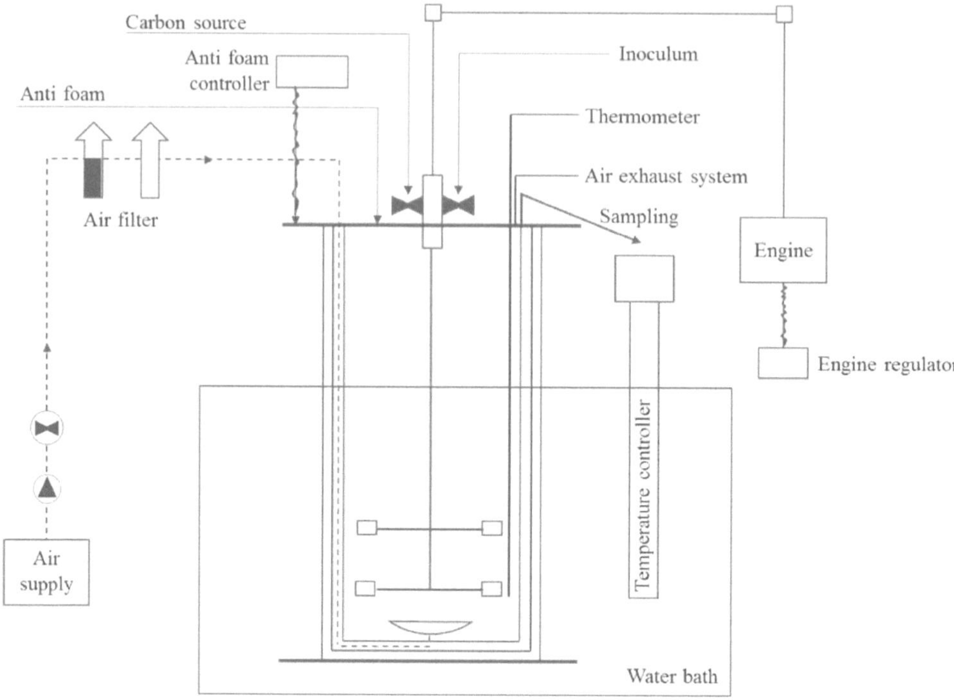

Figure 5.7 Generic construction of a bioreactor according to Singh et al. (2014)

Table 5.2 Advantages and disadvantages of discontinuous, continuous and semi-continuous bioreactors according to Singh et al. (2014)

	Advantages	Disadvantages
Discontinuous (batch) production 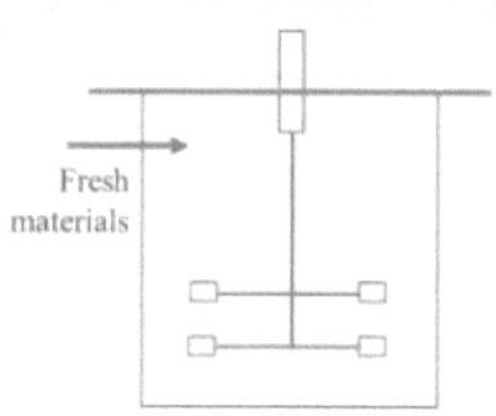 *Process time: hours to days*	▪ Simple equipment ▪ Multi-product flexibility ▪ Suitable for small production volumes	▪ Downtime for loading and cleaning ▪ Reaction conditions change with time

	Advantages	Disadvantages
Semi-continuous (fed-batch) production 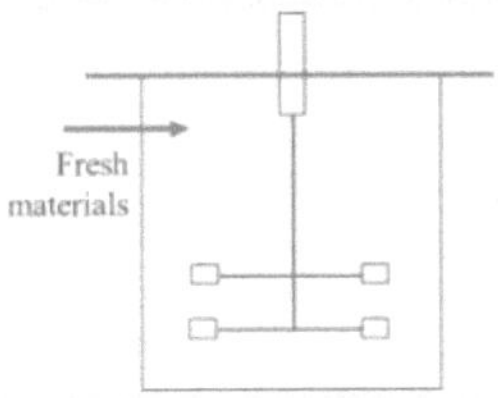 *Process time: up to ~90 days*	▪ Control of environmental conditions, e. g., substrate concentration (inhibition), induction of product formation ▪ Flexible for selecting optimal conditions ▪ Most frequently used in biotechnological processes and in chemical industry	▪ Requires feeding strategy, e. g., to keep constant temperature or substrate concentration
Continuous production 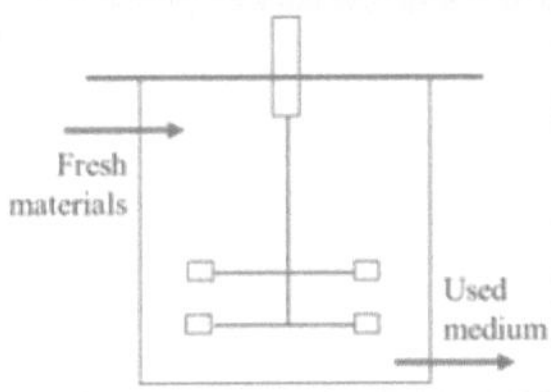	▪ High productivity and product quality due to constant conditions ▪ Good for kinetic studies	▪ Requires flow control, longevity of catalyst necessary, and stability of organisms

Another distinctive feature is the type of reaction, leading to a differentiation of aerobic and anaerobic fermenter (Benz, 2011). In anaerobic fermenters, microorganisms grow in the near absence of oxygen, while microorganisms in aerobic fermentation consume oxygen in order to grow and produce. Examples of aerobic processes include the production of antibiotics, amino acids, and enzymes, while most fuel and some organic acid productions are anaerobic processes. Other differentiation parameters are stirring technology and construction. Stirring technology fulfils the purpose of ensuring a homogeneous distribution of the three phases present in bioreactors: solid (biomass), liquid (nutrient medium) and gaseous (e. g., air, oxygen, carbon dioxide, nitrogen). Three options are available: [1] external pump circuit, which circulates the liquid through a pump, [2] movable mechanical components (stirrers) or [3] blowing the gas phase into the liquid phase. Provided that the reactor is equipped with guide tubes, propeller loop reactors (energy is introduced by a stirring element which moves axially downwards), jet loop reactors (free jet reactor with a guide tube), and mammoth loop reactors (airlift reactor or bubble column reactor with a guide tube) can be differed. The type of reactor construction leads to a division between stirred tank reactor, fixed bed reactor, trickle stream reactor, photobioreactor, tubular reactor, and membrane bioreactor (Renneberg and Süßbier, 2005).

In any case, a few preliminary steps are required, before full size bioreactor start of production (SOP). While the amount of steps depends on the complexity and scale of the process, a common approach consists of four steps (Benz, 2011). First, microorganisms are cultivated in a lab in shaker flasks under the supervision of a microbiologist. Second, experiments are carried out with a mini-fermenter, the seed tank, to test the process with an increased number of microorganisms. Once, the results prove a robust and stable process, it is transferred to a larger fermenter, the inoculum tank, and tested again with a larger number of microorganisms, before again being transferred to the final production fermenter. Shaker flasks sizes range from 100 to 1000 ml (Singh et al., 2014). Laboratory reactors have a volume of < 50 l, experimental reactors between 50 and 5000 l and operating reactors between > 50 to 1,500,000 l (Renneberg and Süßbier, 2005).

Depending on the organisms, technical and organizational operating parameters vary greatly. It is therefore often necessary to design a new bioreactor for each application. Although bioreactors exist that can be used for different applications by controlling different parameters in a wider range, there is no such thing as a flexible system. This is one of the many issues that BT can address. Further literature on bioreactor design and operation are provided by Storhas (2000), Renneberg and Süßbier (2005), Antranikian (2006), Benz (2011), and Singh et al. (2014), among others.

5.3.3 Biorefineries

The term refinery describes a technical process for cleaning, upgrading, separating or concentrating raw materials, food, and technical products. The concept of biorefineries (BR) refers to those refineries that convert biomass into various intermediate and end products (e.g., chemicals, materials, bioenergy). In contrast to crude oil, biomass contains naturally synthesized structures that are predominantly macromolecular or highly functionalized (fats, carbohydrates, lignin, proteins) (Kamm, 2008; Cherubini, 2010). Since these chemicals are the basis for a large number of existing and conceivable products, they are referred to as platform chemicals. Platform chemicals used today can be produced by biological and chemical conversation (synthesis). The synthesis building blocks can later be transformed into a variety of high-quality chemicals and materials. The basic process of a biorefinery thus attempts to isolate certain high-quality natural building blocks, the precursors, from the biomass. In doing so, the synthesis precursors of nature are used either to replace costly, artificial production processes or to produce complex, non-artificially producible compounds. These are further processed into platform chemicals, which serve as synthesis building blocks for a variety of higher-value compounds. Figure 5.8 illustrates the basic process of a biorefinery according to Kamm (2008).

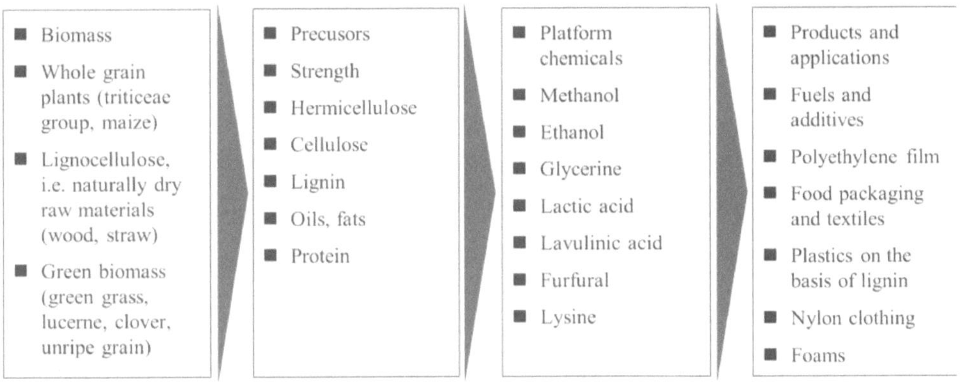

Figure 5.8 Basic process of a biorefinery according to Kamm (2008)

While the refinery concept was predominantly associated with the refinement of crude oil for technical applications, the use of biomass as feedstock is expected to reduce pollution and other environmental hazard products (Cherubini and Jungmeier, 2010). A generally accepted classification of biorefinery concepts is not yet available. The simplest distinction is based on the type of product. This can be either energetic or material, which results in two main classes (Cherubini et al., 2009): [1] energy-driven biorefinery systems (i. e., biomass is primarily used for production of secondary energy carriers such as biofuels, power and/or heat) and [2] material-driven biorefinery systems (i. e., biomass is primarily used to generate biobased products such as biomaterials, lubricants, chemicals, food, feed, etc.). Other distinctive parameters are origin of feedstock and basic processes. The feedstock can originate from the different economic sectors, i. e., agriculture (e. g., crops and crop residues), forestry (e. g., wood, short-rotation poplar, logging residues), industry (e. g., process residues and wastes), aquaculture (e. g., algae, seaweed) or households (e. g., organic residues) (Figure 5.9).

In turn, four basic process can be differed in biorefinery technology to convert biomass feedstock into marketable products (Cherubini et al., 2009):

1. *mechanical/physical*, i. e., all processes that do not change the chemical structure of biomass components, but may reduce size or separate feedstock components (e. g., pressing, milling, separation, distillation)
2. *chemical*, i. e., all processes inducing a chemical change in the substrate (e. g., hydrolysis, hydrogenation, transesterification, oxidation, pulping)
3. *biochemical*, i. e., all processes inducing a chemical change in the substance at mild conditions (lower temperature and pressure) using microorganisms or enzymes (e. g., aerobic and anaerobic fermentation, anaerobic digestion, enzymatic conversion)
4. *thermochemical*, i. e., all processes expose the feedstock to extreme conditions (e. g., pyrolysis, hydrothermal upgrading, gasification, combustion)

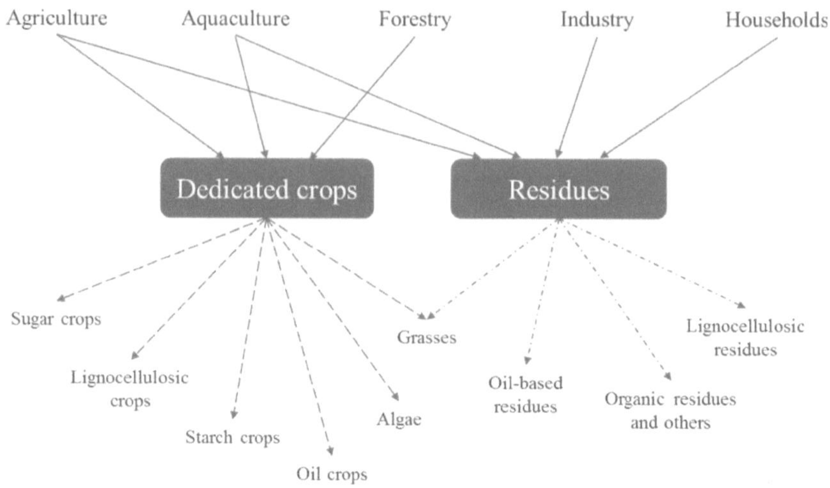

Figure 5.9 Types of feedstock for biorefineries according to Cherubini et al. (2009)

In order to develop a basic understanding of the technology, a further classification of the applied process schemes is necessary. This work adopts the conceptual thoughts of Kamm and Kamm (2004) and Kamm (2008), that differ between three to four basic biorefinery types depending on the raw material used. Alternative classifications are presented by van Ree and Annevelink (2007) and Cherubini et al. (2009), among others. In case of cereals or maize as raw material, the system is referred to as a whole crop biorefinery (WCBR). In contrast, green biorefinery (GBR) use nature-wet biomasses (e. g., green grass, lucerne, clover, immature cereal), while the raw material used in lignocellulose feedstock biorefineries (LFBR) is nature-dry such as cellulosic biomass and waste. Not least, the two platform biorefinery (TPBR) represents a combination of sugar and syngas platform. The following paragraphs outline the basics of concepts of LFBR, WCBR, and GBR according to Kamm and Kamm (2004) and Kamm (2008).

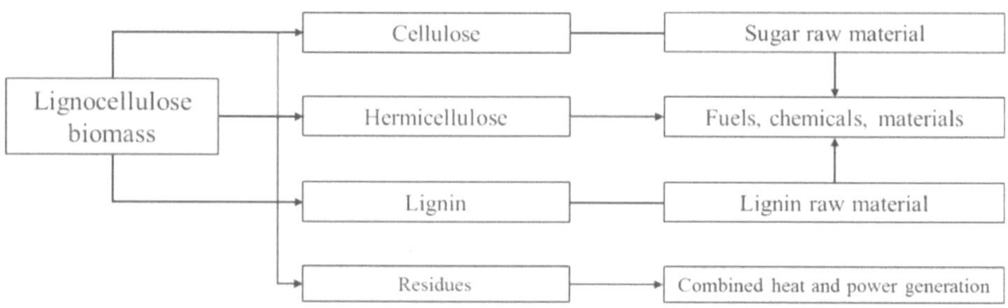

Figure 5.10 Basic principle of LFBR biorefinery according to Kamm and Kamm (2004) and Kamm (2008)

According to Kamm and Kamm (2004), LFBRs are likely to have the highest success, as the raw material situation is promising (straw, reed, grass, wood, paper-waste, etc.) and conversion products have an outstanding position on both fossil and biobased markets. Lignocellulose materials consist of three primary chemical fractions or precursors: hemicellulose/polyoses (a sugar-polymer of pentoses predominantly), cellulose (a glucose-polymer) and lignin (a phenol polymer). Figure 5.10 illustrates the basic LFBR principle according to Kamm and Kamm (2004) and Kamm (2008).

In WCBRs raw materials (cereals and maize) are first separated mechanically into corn and straw. As straw is the feedstock of LFBR, further processing is possible in such a system. Another possibility is the separation into lignin, cellulose, hemicellulose, and their further conversion in different product lines in LFBR. Also, straw is the basic material for the production of syngas via pyrolysis technologies, which again is the basic material for fuel and methanol synthesis. In turn, corn may either be converted into starch or directly used after grinding to meal. Four directions for further processing are possible: [1] break-up, [2] plasticization, [3] chemical modification, or [4] biotechnological conversion via glucose. Figure 5.11 illustrates the basic WCBR principle according to Kamm and Kamm (2004) and Kamm (2008).

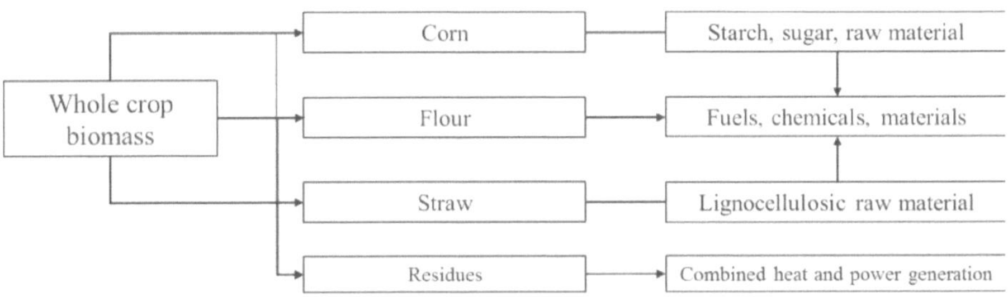

Figure 5.11 Basic principle of WCBR biorefinery according to Kamm and Kamm (2004) and Kamm (2008)

GBR represent a natural chemical and food factory that processes raw materials such as grass from cultivation of permanent grass land, closure fields, nature preserves or green crops, and immature cereals from extensive land cultivation. In a primary refinery step, careful wet fractionation technology is applied in order to isolate the content-substances in their natural form. Thereby, green crop goods (or humid organic waste goods) are separated into nutrient-rich green juice and fiber-rich press cake. While green juice contains proteins, free amino acids, organic acids, dyes, enzymes, hormones, other organic substances, and minerals, press cake includes cellulose, starch, pigments, crude drugs, valuable dyes, and other organics. Products that are generated of green juice are lactic acid and corresponding derivatives, amino acids, ethanol, and proteins, while product from press cake can be used for production of green feed pellets, as raw material for production of chemicals (e. g., levulinic acid), as well as for conversion to syngas and hydrocarbons (synthetic biofuels). Residues of conversion are applicable for combined biogas-based heat and electricity generation. Figure 5.12 illustrates the basic GBR principle according to Kamm and Kamm (2004) and Kamm (2008).

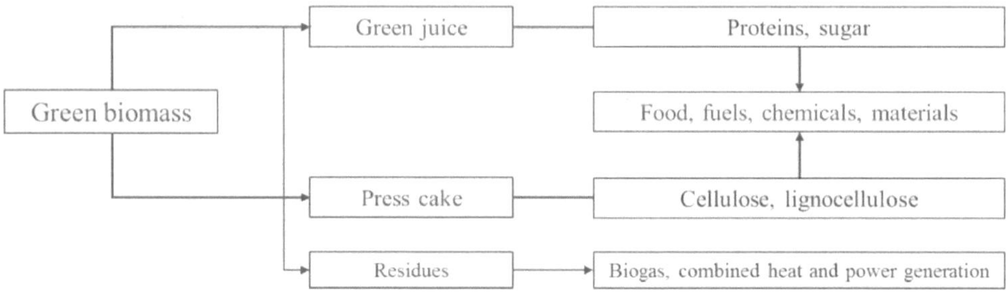

Figure 5.12 Basic principle of GBR biorefinery according to Kamm and Kamm (2004) and Kamm (2008)

5.3.4 Biofuel cell and bioelectrolysis technology

Although fuel cell and electrolysis technology are not identical systems, they are related in design, process engineering and application. Fuel cells convert chemical reaction energy of a continuously supplied fuel and an oxidant into electrical energy (Dicks and Rand, 2018; Santoro et al., 2017). As such, they represent a powertrain technology. Electrolysers, in turn, are commonly used as energy storage devices (power-to-chemistry) by using electricity to force a redox reaction in a chemical process (Carmo et al., 2013). Simultaneously, the technology can be used for the production of metals or chemical substances, which can then be further processed into valuable goods. As such, it is technologically closely related to the refinery principle described above (e. g., electrolytic refining for the purification of metals). Prominent examples of electrolysis are the extraction of hydrogen, aluminum, and chlorine. Both systems consist of electrodes that are separated from each other by an electrolyte (ion conductor) (Dicks and Rand, 2018). The electrolyte may consist of a semi-permeable membrane which is only transmissive for one type of ion, e. g., protons. The electrode plates (also bipolar plates) typically consist of metal or carbon and are coated with a catalyst (e. g., platinum, palladium). The electrolyte can be liquid (e. g., lyes or acids, alkali carbonate melts) or solid (e. g., ceramics or membranes). In the case of a fuel cell, energy is produced in a reaction of oxygen and fuel, either hydrogen or organic compounds such as methane or methanol. Both reaction components are continuously supplied by the electrodes. Electrolysers have a similar structure. In this case, however, a direct current is applied to the electrolyte through the two electrodes, whereupon reaction products are formed at the electrodes from the substances contained in the electrolyte (Carmo et al., 2013). The voltage source causes a lack of electrons in the anode and an excess of electrons in the cathode. The mixture between the cathode and anode contains positively and negatively charged ions as electrolytes. Positively charged ions (cations) or electroneutral substances absorb electrons at the cathode and are thus reduced. An opposite process occurs at the anode. Here, electrons are released into the electrode, oxidizing substances such as anions. The quantity of electrons transferred at the anode is equal to that transferred at the cathode. The transport of the substances to the electrodes follows convective mass transfer and migration. In order to illustrate the difference between the two technologies, Figure 5.13 outlines the basic concept of water based fuel cell and electrolyser according to Lamy (2016) and Dicks and Rand (2018). While electrolysis separates water (H_2O) into hydrogen (H) and oxygen (O) by passing an electric current through it ($2H_2O \rightarrow 2H_2 + O_2$), fuel cells apply the inverse process by recombining hydrogen and oxygen to water, inducing a current ($2H_2 + O_2 \rightarrow 2H_2O$) (Dicks and Rand, 2018).

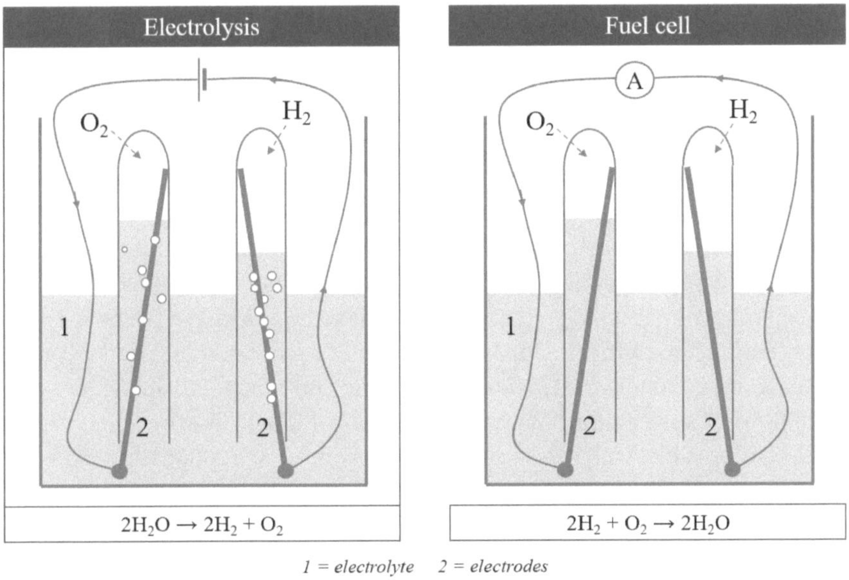

Figure 5.13 Basic concept of water based fuel cell and electrolyser according to Lamy (2016) and Dicks and Rand (2018)

Fuel cell and electrolysis types are classified according to the electrolytes, anodes, and cathodes used. The main electrolysis types are alkaline electrolysis (AEL), polymer electrolyte membrane electrolysis (PEMEL), and high temperature electrolysis (HTEL) (Carmo et al., 2013). Prominent fuel cell types are alkaline fuel cell (AFC), polymer electrolyte membrane fuel cell (PEMFC), direct methanol fuel cell (DMFC), phosphoric acid fuel cell (PAFC), molten carbonate fuel cell (MCFC), and solid oxide fuel cell (SOFC) (Dicks and Rand, 2018).

In the context of biointegration, both technological principles set the basis for bio-based designs, each existing in microbial and enzymatic form (Santoro et al., 2017; Kumar et al., 2018; Minteer et al., 2007). Since literature in the field of biofuel cells (BFC) is far more extensive, BFCs will be primarily used below to explain the differing approaches, while bridging the gap to bioelectrolysis at selected, appropriate points. In general, BFC simply represent a sub-category of fuel cells that allow a direct conversation of chemical energy of organics into electrical energy. Their primary distinctive factor is the employed biocatalysts, resulting in microbial and enzymatic fuel cells (MFC and EFC) (Minteer et al., 2007). While EFCs utilize selective enzymes to perform redox reactions that produce current, MFCs apply electroactive microbes to degrade organics and produce electricity (Santoro et al., 2017). In the case of MFC, electrochemical cells are attached to the anode and/or the cathode, while containing a membrane to separate the compartments (oxidation and reduction) (Allen and Bennetto, 1993; Pant et al., 2010; Pham et al., 2006). During oxidation, the electrons produced are transferred directly to an electrode or a redox mediator, causing the electron flux to move to the cathode. Ionic movement inside the cell maintains the charge balance of the system, usually across an ionic membrane (Lu et al., 2015). MFCs often utilize organic electron donors (e.g., sulfur compounds or hydrogen) (Pant et al., 2010). ECFs also rely on the same basic as conventional fuel cells, only differing by utilized catalysts and accepted fuels. In contrast to catalyst metals used in conventional fuel cells (e.g., platinum, nickel), ECFs apply enzymes derived from living cells. Enzymes possess several advantages compared to chemical catalysts, including higher transformation efficiency and activity under mild conditions, biocompatibility, and particularly higher specific selectivity. Due to the last two features, EFCs do not require a separation membrane, which may prove to be a decisive factor for miniaturization, potentially enabling the incorporation of ECFs in wearables and implantable devices (Calabrese Barton et al., 2004; Bandodkar and Wang, 2016). Yet, enzyme life time and pollution pose major challenges. Figure 5.14 illustrates the principles of microbial fuel cells and electrolysers according to Pham et al. (2006), Santoro et al. (2017), and Kumar et al. (2018). Further basic literature on microbial and enzymatic fuel cells and electrolysers is manifold. Extensive basic literature and reviews are provided by Allen and Bennetto (1993), Santoro et al. (2017), Dicks and Rand (2018) and Kumar et al. (2018), among others.

According to Santoro et al. (2017), MFCs are superior to EFCs regarding the bioelectrochemical oxidation of fuels as they eliminate the need for enzyme isolation and still allow several enzymatic reactions to take place under conditions close to their natural environment. Yet, they have a slower reaction time due to the more complex chemical pathways. Both systems are conceivable in different target applications over different scales, but generally differ from each other. While EFCs are outlined as a basic technology for miniaturization applications due to their compact, miniaturized and flexible bioelectrochemical design (Leech et al., 2012), MFCs are primarily suited for large-scale applications for wastewater treatment (Rozendal et al., 2008;

Sleutels et al., 2012) or on a smaller scale for portable applications, e.g., energetically autonomous robots (Melhuish et al., 2006).

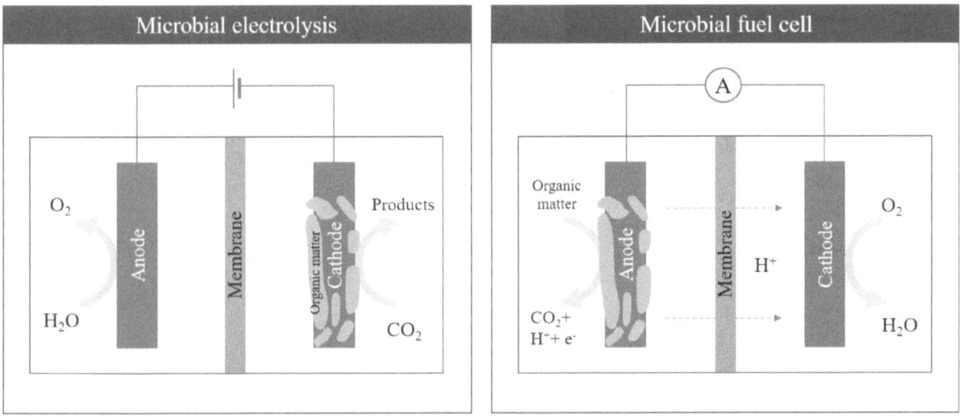

Figure 5.14 Principles of microbial fuel cells and electrolysers according to Pham et al. (2006), Santoro et al. (2017), and Kumar et al. (2018)

5.3.5 Biosensors

As a combination of biological component and physicochemical detector, a biosensor is an analytical device for detecting chemical substances (Dincer et al., 2019). A multitude of applications of biosensors already exists today, covering disease monitoring, drug discovery, hazardous chemicals, contaminants, pollutants in water and soil, disease-causing microorganisms and markers, indicators of diseases in human blood fluids (blood, urine, saliva, sweat), detection of drugs and toxicants in food, detection of vitamins and toxins, or as DNA-based molecular diagnostic for clinical assessment, etc. (Bhalla et al., 2016; Parkhey and Mohan, 2019).

For a given analyte (e.g., glucose), the sensor is usually composed of four elements (Bhalla et al., 2016). The bioreceptor is a biological recognition element (e.g., enzyme, antibodies, DNA, receptors or whole cells/cell networks). Once the bioreceptor and the analyte interact, a process of signal generation, i.e., biorecognition, is induced (in form of light, heat, pH, mass change, etc.) that is converted to a measurable signal by a transducer. A transducer is basically an energy converter, i.e., any device that converts energy from one form into another. Within a biosensor the process of energy conversation is referred to as signalization. According to Bhalla et al. (2016), the majority of transducers either generate optical or electrical signals. Usually, signalization is proportional to the quantity of bioreceptor-analyte interactions. In addition to recognition and signalization, biosensors require signal processing and display. Every biosensor thus consists of a complex electronic circuit that converts and amplifies signals from analogue into digital form. Thereupon, a display unit, consisting of hardware and software, generates output signals in form of numeric, graphic, tabular or image representation adapted to the end user interpretation system, e. g., computer display, direct printing. Figure 5.15 illustrates the setup of a biosensor according to Bhalla et al. (2016), and Parkhey and Mohan (2019).

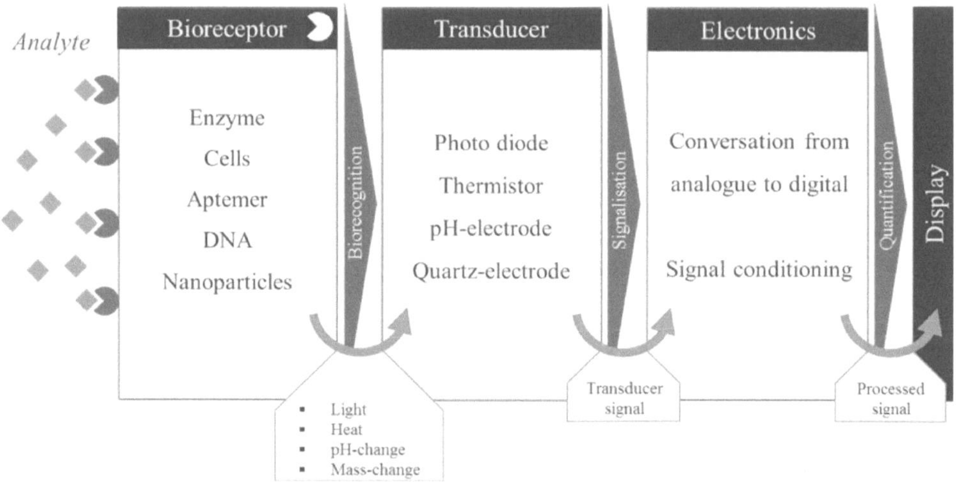

Figure 5.15 Setup of a biosensor according to Bhalla et al. (2016), and Parkhey and Mohan (2019)

A common approach to distinguish biosensor types is the applied transducer (Parkhey and Mohan, 2019). Common transducer types are based on either optical, electronic, electrochemical, piezoelectric, pyroelectric or gravimetric mechanisms. At present, electrochemical and optical biosensors are the most common. A biosensor based on an electrochemical transducer uses fluctuations in potential, current, capacity or conductivity for the testing of the analyte. Electrochemical biosensors may thus be further distinguished in amperometric, potentiometric, conductometric and impedimetric devices. In turn, optical biosensors measure fluorescence, bioluminescence, colorimetric. Other types of biosensors, which detect changes in resonating frequency of piezoelectric material depending on the adsorption or desorption of molecules from its surface, are referred to as piezoelectric or gravimetric devices. Here, a current production is measured based on resonating frequency results. Pyroelectric biosensors also measure current production, although here resulting from temperature change. If a biosensor is to analyze the variation in microbial respiration of water toxicity, baroxymeter are applied. Not least, field effect transistor (FET)-based biosensors measure the change in conductance of FET as a result of biological reaction. Parkhey and Mohan (2019) summarize selected examples of biosensors and their application based on recent literature (Table 5.3).

Table 5.3 Selected examples of biosensors and their application according to Parkhey and Mohan (2019)

Transducer type	Biosensor class	Features	Applications
Electro-chemical	Amperometric	Presence of an analyte results in redox reaction that brings a change in current between working and reference electrode which is measured	Detection and quantification of glucose, cholesterol, amino acids, urea, and alcohols
	Potentiometric	Measures potential or charge accumulation between an ion selective electrode and a reference electrode	Determination of sugars, urea, neurotransmitters, carbon dioxide, pesticides, etc.
	Conductometric	Measures change in the electrical conductivity of the medium upon entry of any analyte that changes the ionic species concentration	Heavy metals, chemicals, protein markers
	Impedance	Measures the resistivity or conductivity changes in the medium due to biorecognition event	Endocrine disrupting hormones, PCBs (polychlorinated biphenyls), milk toxins
Optical	Colorimetric	Measures the change in color or optical density of the test sample upon a chemical reaction	Food and water borne pathogen
	Bioluminescent	Based on the principle of light emission by viable bacteria in response to any chemical, biological, or physical changes in the analyte	Heavy metal contamination, food and environmental monitoring
	Fluorescence	Fluorescent-labelled biomolecules are immobilized which emit fluorescence upon reaction with an analyte.	BOD measurement, iron or water availability in plants or microbial habitat, cell populations
	Piezoelectric	Change in resonating frequency of piezoelectric material due to desorption or absorption of analyte leading to current generation	Immunosensing, nucleic acid sensing, cellular studies

Table 5.3 Selected examples of biosensors and their application according to Parkhey and Mohan (2019) (*continued*)

Transducer type	Biosensor class	Features	Applications
Optical	Pyroelectric	Measurement of change in current induced by a material upon a temperature difference	Diagnostics
	Baroxymeter	Monitoring respiration of a bacterial culture by pressure measurements and using respiration inhibition as a toxicity alert	Toxicity measurement in wastewater
	FET-based biosensor	Measurement of change in conductance of FET	Clinical investigation, IV blood pH recording

5.3.6 Additive manufacturing and bioprinting

The classification of AM (also generative manufacturing, rapid prototyping[2)]) in the context of BT is not trivial. Currently, literature often refers to bioprinting, additive biomanufacturing and/or additive manufacturing of biomaterials. However, a clear definition of these terms does not exist. Therefore, this section attempts to provide the reader with the necessary delimitation to better understand the diverse applications of additive manufacturing technology in the context of BT.

The key term here is biomaterial. Originating primarily from health industry and science, the term does not describe a substance that must necessarily consist of biological matter, but a substance that has been engineered to interact with biological systems, i.e., for therapeutic or diagnostic purposes. This is often achieved by bioimitation. In the context of biointegration, however, the aim is to describe the processing of substances that consist at least partially of biological matter. In the following, these are referred to as biobased material. This results in three possible fields of application for additive manufacturing, which can be distinguished according to basic technology and materials used (Figure 5.16):

1. *Bioprinting*, referring the flexible shaping of cell tissue, i.e., all techniques that combine cells, growth factors, and biomaterials to generate materials and products that imitate natural tissue characteristics (Murphy and Atala, 2014; Mandrycky et al., 2016; Zhang et al., 2017). These are further divided in three- and four-dimensional technologies. While 3D bioprinting generates tissue in three spatial dimen-

2 In the context of production engineering, the term is used as complementary to ablative manufacturing.

sions, 4D bioprinting adds time, referring to the possibility of product transformation by using programmable matter that reacts to parameter of the environment (humidity, temperature, etc.) and changes its form accordingly (Tibbits, 2014).

2. *Additive manufacturing of biobased or biohybrid*[3] *materials* with conventional generative technologies, i. e., all production techniques that apply material layer by layer to produce three-dimensional objects (workpieces) (Campoli et al., 2013; Zadpoor, 2017).

3. *Additive biomanufacturing* as a collective term for all biobased printing technologies that do not rely on conventional additive manufacturing technologies or generate materials and products that imitate natural tissue characteristics. An example would be an additive manufacturing based on enzymatic process technology (Zhang et al., 2014), which theoretically allows the creation of a natural cohesion at very low temperatures (< 40 °C). Compared to the high-energy thermal processes of conventional additive manufacturing, such a process would allow enormous energy savings while simultaneously processing biodegradable materials. This is primarily a field that is conceivable, but has not yet been technically implemented. Hence, this work refrains from further elaboration.

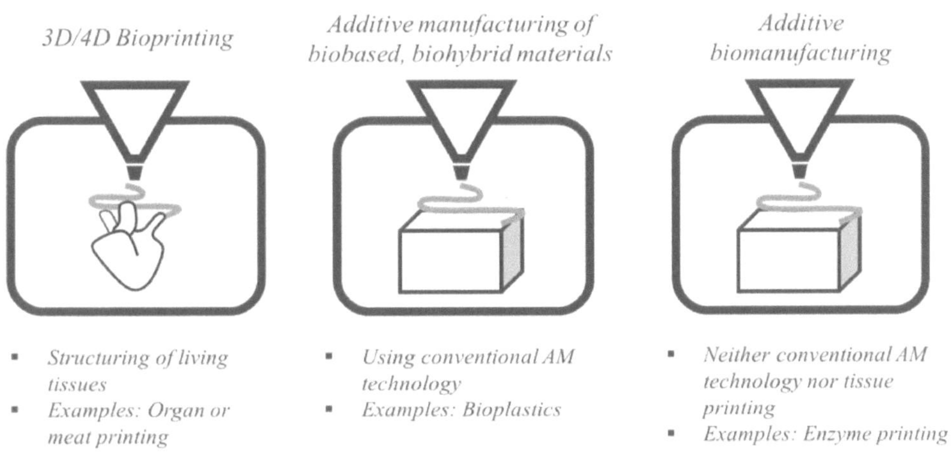

Figure 5.16 Fields of AM in the context of biointegrated systems [own illustration]

The German VDI directive 3405 lists stereolithography, laser-sintering, selective laser melting, electron beam melting, fused layer modelling, multi-jet modelling, poly-jet modelling, binder jetting, layer laminated manufacturing, digital light processing, thermal transfer sintering as conventional AM processes that are currently established in industry (Verein Deutscher Ingenieure, 2014). However, the list could be con-

---

[3] Materials, partially consisting of animate and inanimate matter, e. g., electrospinning polymer-microbe fibers (Liu et al., 2009)

siderably expanded to include other techniques that are still in their infancy. Basically, all existing AM processes are suitable for the additive processing of non-cellular biomaterial (e. g., biopolymer), depending on form of basic material and processability. Kruth et al. (1998) present an overview of common industrial AM technologies and their basic principles (Table 5.4). Since the present work cannot provide a detailed description of the individual process technologies, the reader is referred to existing basic literature, including Gebhart (2013), Gibson et al. (2014), Zadpoor and Malda (2017) and Bose et al. (2018).

Table 5.4 Overview of common industrial AM technologies and their basic principles according to Kruth et al. (1998)

Supply	Process	Layout	Layer creation technique	Phase change during layer solidification	Materials
Liquid	Stereo litho-graphy	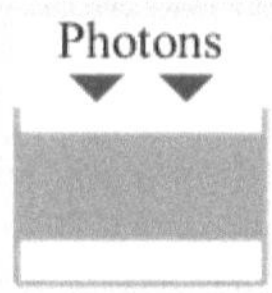 Liquid resins in a vet	Liquid layer deposition	Photo-polym-erization	Acrylates, epoxies, glass, ceramics, metal
	Fused deposition modelling	 Material melted in nozzle	Continuous extrusion and depo-sition	Solidification by cooling	Polymers, ABS, wax, filled poly-mers
	Inkjet printing	 Droplets of molten material	Drop-on-demand deposition	Solidification by cooling	Polymers, wax
Powder	Three-dimen-sional printing	 Powder in a bed	Layer of powder + drop-on-demand binder printing	No phase change	Ceramics, metals, and polymers with binder

Supply	Process	Layout	Layer creation technique	Phase change during layer solidification	Materials
	Selective laser sintering	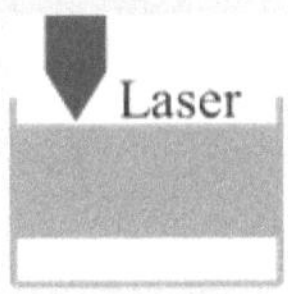 Powder in a bed	Layer of powder	Laser sintering/melting, resolidification by cooling	Polymers, metals with and without binder, sand, ceramics
Solid	Laminated object manufacturing	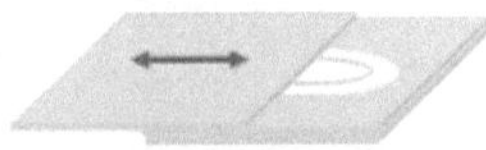 Feeding, cutting, and binding of sheets	Deposition of sheet material	No phase change	Paper, polymer, composites, ceramics, metals
Gas	Selective laser vapor deposition	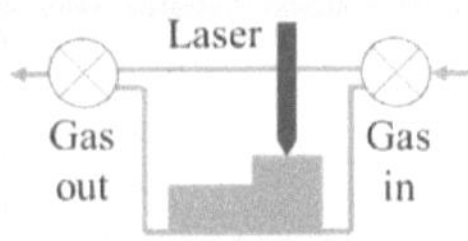	Condensation of gas	Chemical reaction	Metals, ceramics

Bioprinting is essentially based on tissue engineering techniques. This involves the production of regular structures, i. e., bioarrays, or tissues from previously cultivated individual cells (Seol et al., 2014). The technology is expected to enable the production of entire organs or food in the future. Examples of future applications of bioprinting include medicine (specific organs), food industry (artificial meat), and synthetic biology (artificial life forms). Generally, the process of bioprinting can be divided into three steps: preprocessing, processing (actual printing), and postprocessing (Mironov et al., 2006). Preprocessing covers the design of tissue or organ by using imaging and computer-aided design methods. Processing describes the actual printing in the bioprinter device, while postprocessing refers to the necessary downstream application of tissue remodeling and maturation in specially designed bioreactors. According to Dababneh and Ozbolat (2014), laser-, extrusion-, and inkjet-based bioprinting technologies are distinguished, each relying on appropriate and recommended bioinks. Each of the technologies is used to stimulate self-assembly and self-organizing capabilities of cells. Figure 5.17 illustrates a classification of existing bioprinting technologies according to Dababneh and Ozbolat (2014).

Inkjet-based bioprinting applies ink that contains living cells in the form of droplets, consisting of up to 30,000 cells, by a noncontact nozzle (Boland et al., 2006). The technology can be grouped in thermal or piezoelectric drop-on-demand methods. Thermal inkjet methods consist of a small number of nozzles and a heating element within an ink chamber. The droplet is generated by a short current pulse applied to the heat element that results in an increase of ink temperature forming a bubble that forces the ink out of the nozzle (Le, 1998). Piezoelectric inkjet bioprinters, in turn, apply piezocrystals located at the back of a chamber. The crystals are set into vibration by an electric charge, forcing a small quantity of ink to drop out of the nozzle (Stachowiak et al., 2009). Today, inkjet-based printing is a well-understood process capable of printing viable cells and biomaterials, while being comparatively cheap and can work under mild conditions (Derby, 2008). Plus, as a form of noncontact technique, inkjet-based bioprinting decreases the chance of contamination. Figure 5.18 illustrates the design of the two common approaches to inkjet-based bioprinting according to Dababneh and Ozbolat (2014).

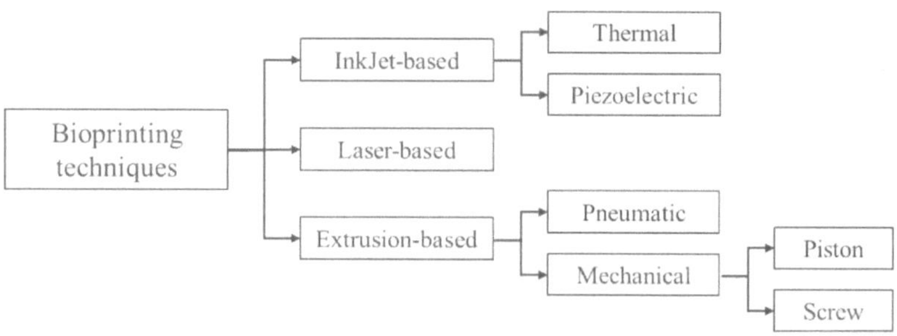

Figure 5.17 Classification of existing bioprinting technologies according to Dababneh and Ozbolat (2014)

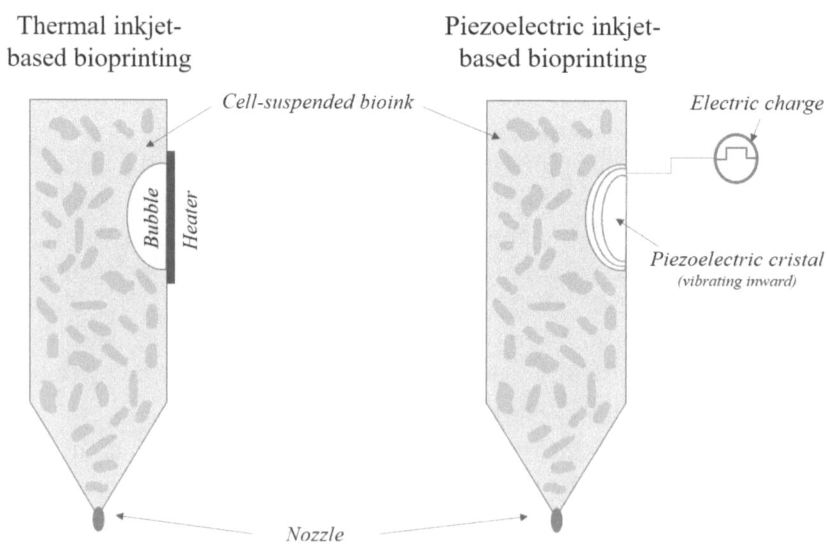

Figure 5.18 Design of inkjet-based bioprinting according to Dababneh and Ozbolat (2014)

Laser-based bioprinting is mostly associated with laser-based direct writing (LDW) (Wang et al., 2009). Here, an individual cell is guided from a donor to a collector slide by a laser pulse. This creates a bubble that again results in shock waves, forcing the cells to transfer toward the collector substrate. Common designs of LDW are laser-induced forward transfer (LIFT) and matrix-assisted pulsed laser evaporation direct writing (MAPLE DW). Both approaches are basically similar. The only difference is that MAPLE DW utilizes a lower powered pulsed laser compared to LIFT. Generally, these techniques allow a precise deposition of cells in comparatively small 3D structures. Especially MAPLE DW has proven to be effective for the processing of cellular constructs that include various cell types, e. g., fibroblasts, myoblasts, neural stem cells, etc. (Schiele et al., 2009). Unlike inkjet and extrusion-based bioprinting, LDW enables the use of high-viscosity bioink due to its nozzle-free nature. However, current limitations of the technology include a potential damaging of cells due to heat generated from laser energy, a generally lower cell viability compared to inkjet-based techniques as well as prolonged fabrication time (Ozbolat and Yu, 2013). Figure 5.19 illustrates the design of laser-based bioprinting according to Dababneh and Ozbolat (2014).

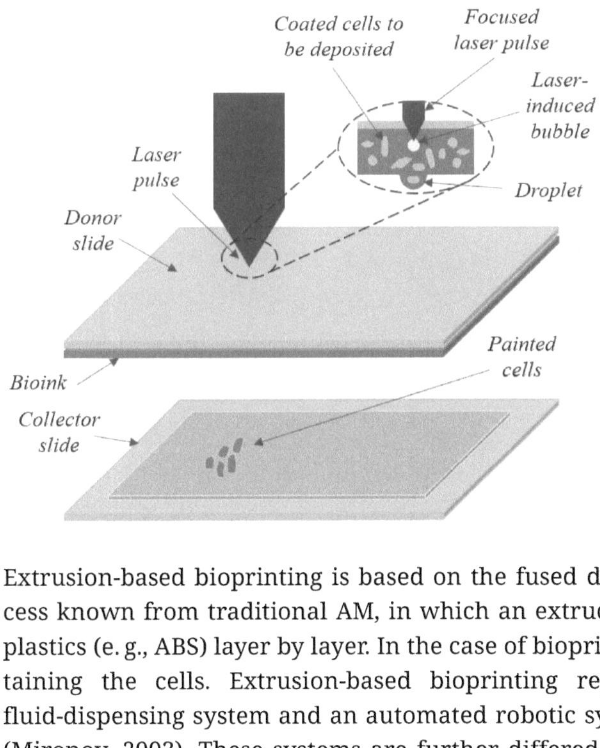

Figure 5.19 Design of laser-based bioprinting according to Dababneh and Ozbolat (2014)

Extrusion-based bioprinting is based on the fused deposition modelling (FDM) process known from traditional AM, in which an extruder creates forms from thermoplastics (e. g., ABS) layer by layer. In the case of bioprinting, a hydrogel is applied containing the cells. Extrusion-based bioprinting represents a combination of a fluid-dispensing system and an automated robotic system for extrusion and writing (Mironov, 2003). These systems are further differed by the type of fluid-dispensing system that is either based on pneumatic or mechanical principles. Mechanical systems are again divided in piston and screw-driven approaches. Generally, piston-driven deposition enables more direct control of bioink flow, while screw-driven systems provide a more spatial controllability, resulting in a better dispensing of bioinks with high viscosities. However, screw design is not trivial, as it may result in large pressure along the nozzle, which might harm the cells. Pneumatically driven systems, in turn, are beneficial with regard to the possibility of adjusting pressure and valve gating time in order to apply various types and viscosities of bioinks. Here, the bioink is dispensed by a computer-aided deposited system and crosslinked by light, chemical or thermal transitions (Ozbolat and Yu, 2013). This approach enables precise deposition of cells encapsulated in cylindrical filaments of a desired 3D custom-shaped structures. Extrusion-based bioprinting is viewed today as the most convenient technique for the manufacturing of 3D porous cell-laden structures, as it provides comparatively good structural integrity and enables the incorporation of CAD software (Yan et al., 2005). Figure 5.20 illustrates the design of extrusion-based bioprinting according to Dababneh and Ozbolat (2014).

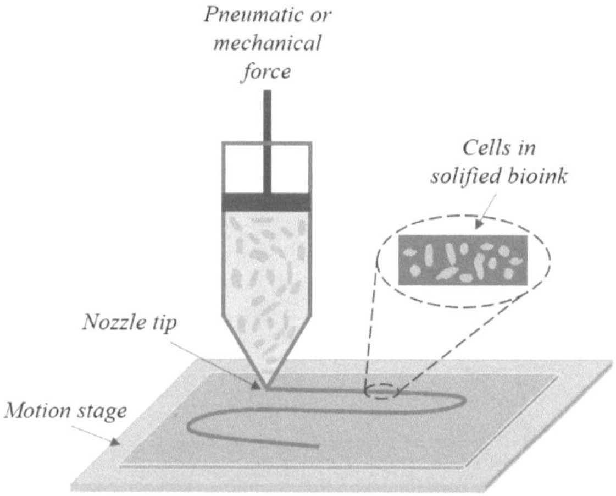

Figure 5.20 Design of extrusion-based bioprinting according to Dababneh and Ozbolat (2014)

Today, bioprinting technology is mainly used for drug testing, whilst it is not possible to print complete organs. Several key challenges and limitations remain to be solved, including repeatability, cell viability, practicality, and biocompatibility of bioprinting processes, accuracy, motion speed, automation capability, user-friendliness of bioprinters, as well as cell density, cytotoxicity, bioprintability of bioinks (Dababneh and Ozbolat, 2014). If the existing barriers are resolved, such organs would have a major advantage over current donor organs, as they can be tailored to the intended body. Similarly, it would allow for on-site production, which could also be used for in-house food production. Technologically, the printing of meat is already possible today, as the example of the company Modern Meadow proves. A comprehensive introduction of this technology, supplemented by meat replacement products, could prevent the mass farming of animals in the future.

5.3.7 Genome editing and CRISPR/Cas9 technology

The modification of genetic material is by no means a prerequisite for biointegration, yet it is capable of accelerating the process by creating tailored organisms for specific task. Before the basic methodological procedures are explained in more detail, a brief excursus on genetic engineering as a sub-area of biotechnology needs to be conducted. Genetic engineering[4)] is easily the most far-reaching tool of biotechnology. Although it is currently only explored in rudimentary form, it represents the basis for a large number of novel products, functions, and production processes (Schüler, 2016). Its initial concept is based on the theory of heredity (genetics), in which specific hereditary factors are passed on, reduced or strengthened from one generation to the next by systematic crossing. According to Schüler (2016), further milestones on the way to the development of genetic engineering were the creation of the term gene (as a carrier of hereditary characteristics), the localization of chromosomes and the development of the term nucleotide, characterizing the recurring structure of hereditary characteristics from phosphate, sugar and base, at the beginning of the 20th century. On this basis, the term deoxyribonucleic acid (DNA) emerged as the correct chemical name for the genetic material. Around the middle of the 20th century, the findings of genetics were already being used extensively in industry, for example to achieve a higher yield in the production of microorganisms. The instrumental basis was undirected mutagenesis (genetic modification). Since the mutations occurred randomly, stem selection had to follow. In the early 1970s it became possible to actively intervene in the genetic material by introducing genetic material of any origin into a cell and recombining it with its base DNA. Although this represents a natural process, which also occurs in mammals, e. g., during the fertilization of an ovule, it does not usually take place across species boundaries. This is only possible because in every living organism DNA basic patterns are identical in construction. DNA engineering enables the recombination of genetic material. Basically, it can be used to incorporate genetic material of any origin into the DNA of another living being, i. e., recombine it (e. g., into a bacterium). In order to understand the basic principles of genetic engineering, knowledge of some fundamental concepts of molecular biology is necessary. The following paragraphs outline the most significant terms according to Schüler (2016).

- *DNA* consists of four different building blocks, so-called nucleobases or nitrogenous bases: adenine (A), cytosine (C), guanine (G) and thymine (T). They can be compared to different letters from which syllables, words and sentences with a specific meaning can be formed. Their sequence varies and, as genetic informa-

[4] In public discourse, biotechnology is often falsely reduced to genetic engineering. For a clearer differentiation of the generic term biotechnology from the individual instrument of recombination of genetic material, the term 'DNA recombination engineering' (DNA engineering) would appear more appropriate.

tion, provides the blueprint for a wide variety of proteins, which in turn are the functional and structural carriers in cells.

- *Amino acids* are organic molecules consisting of four key elements: carbon (C), hydrogen (H), oxygen (O), and nitrogen (N).
- *Proteins* are biomolecules consisting of one or more chains of amino acids.
- *Codon* refers to a sequence of three base building blocks in the mRNA (messenger RNA), also called triplet, which codes for a specific amino acid. The complementary triplet in the tRNA (transfer RNA) is called anticodon. The corresponding amino acid is bound to the matching tRNA. Codons/triplets are also referred to as genetic material.
- The *genetic code* is a type of general guideline or translation key that is employed by living cells to convert information encoded in genetics into the amino acid sequence of the proteins. It codes for 22 different amino acid building blocks and is essentially the same for all living organisms with a few exceptions.
- *Translation* is carried out by the ribosome, which combines the amino acids in a sequence specified by mRNA, using tRNA molecules to transport amino acids. The mRNA reads three nucleotides ('DNA base letters') simultaneously. Each serves as a code for one amino acid. These in turn are the building blocks for the proteins. In the ribosomes the tRNA recognizes the codes in the mRNA and combines a particular amino acid bound to it into an amino acid chain (protein). In this way, a translation of the genetic information (DNA) via mRNA and tRNA into proteins occurs.
- *DNA ligase* is an enzyme that catalyzes the linking of two DNA strings.
- *DNA polymerase* is an enzyme that catalyzes the polymerization (sequencing) of nucleotides, the basic building blocks of nucleic acids. Its aim is replication (amplification) of DNA.
- *DNA sequencing* describes the determination of the sequence of individual bases (A, C, G or T), e. g., in a gene.
- *Plasmid* is bacterial genetic material that occurs in ring form in addition to chromosomes.
- *Restriction enzymes* (also restriction endonucleases) are bacterial enzymes that can cut DNA at specific points. Naturally, bacteria use them for phage defense. The nucleases recognize DNA segments of four to eight base pairs and function within these like molecular scissors.
- *Reverse transcriptase* (also RNA-dependent DNA polymerase) is an enzyme that produces complementary DNA from single-stranded RNA. The usual process direction of transcription.

- *Transcription* describes a process in which RNA polymerases read DNA and form messenger RNA (mRNA), leaving the nucleus and providing the blueprint for proteins.
- *Chromosomes* are condensed DNA spirals. Every human cell nucleus contains 23.

Figure 5.21 illustrates the basic concept (l) and the fundamental principle (r) of DNA engineering. The latter is exemplified here using a bacterium into which the foreign DNA of a human (gene donor) is incorporated. The DNA is initially found in the bacterium in ring form (plasmid). Using restriction enzymes, the DNA is cut. In a further step, the gene removed from a human cell is connected to the plasmid of the bacterium through DNA ligase. This creates a recombinant DNA molecule, which is then incorporated into the bacterium, creating a human protein, which could be used, e. g., for the production of personalized therapeutics.

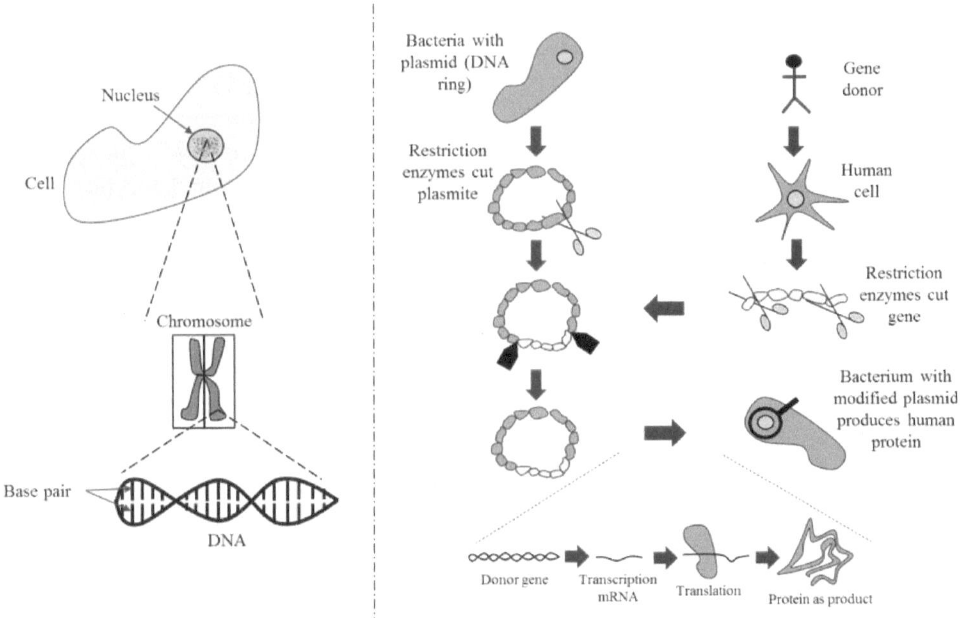

Figure 5.21 Basic concept (l) and the fundamental principle (r) of DNA engineering according to Schüler (2016)

A prominent example that was created in this way was Dolly the sheep, the first cloned mammal. The genome editing method used here was somatic cell nuclear transfer (SCNT), which is primarily based on cells grown in in-vitro culture. Homologous recombination (HR), the most prominent method for manipulation of the donor genome in this form of gene editing, is however a highly inaccurate approach. A correct incorporation of a DNA sequence is comparable with the search for a needle in the haystack. Compared to conventional homologous recombination-based gene targeting such as SCNT, recently emerged genome editors (GE, also DNA nucleases)[5] can increase the targeting rate around 10,000- to 100,000-fold. Targeted genome editing tools and applications have experienced a remarkable progress, especially in the past decade. According to Peterson (2017), the most relevant modern programmable DNA nucleases are zinc finger nucleases (ZFNs), transcription activator-like endonucleases (TALENs), and the clustered regularly interspaced short palindromic repeats/Cas9 system (CRISPR/Cas9). GE applications in organisms are manifold. Among others, the methods enable a better understanding of complex physiological systems (e. g., reproduction), the creation of transgenic animals, animal models for human diseases, specific cell lines and plants as well as the treatment of human genetic diseases. While ZFNs and TALENs utilize proteins in order to detect and bind a targeted DNA sequence, which results in a difficult and expensive synthesis of DNA nucleases, CRISPR/Cas9 employs RNA to facilitate specific binding of its DNA nuclease. The method, first published in 2012 (Jinek et al., 2012), is basically a simplified version of the antiviral defense system CRISPR-Cas9, naturally used by bacteria. It refers to a group of DNA sequences occurring in genomes of prokaryotic organisms, e. g., bacteria and archaea. These sequences are derived from DNA fragments of bacteriophages that had previously infected the bacterium or archaea. According to Barangou (2015), prokaryotes use them to detect and destroy DNA of similar bacteriophages in later infections. Closely linked to this mechanism is the CRISPR-associated protein 9 (Cas9 enzyme), which employs CRISPR sequences as a guide to recognize and split specific DNA strings that are complementary to the CRISPR sequence. The CRISPR/Cas system enables prokaryotic cells to achieve resistance to foreign genetic elements e. g., plasmids and phages (Barrangou et al., 2007). Figure 5.22 illustrates the mechanism according to Horvath and Barrangou (2010).

By incorporating the Cas9 nuclease, being bound to a synthetic guide RNA (gRNA), into a cell, the genome of the cell can be cut at a desired location, allowing existing genes to be removed and/or new genes to be added in vivo (Gasiunas et al., 2012; Hendel et al., 2015). Compared to all other GEs, CRISPR/Cas9 is thus much easier to design, while simultaneously enabling multiplex genomic modifications, resulting in considerable lower cost (Mahfouz et al., 2014; Pennisi, 2013) (Petersen, 2017). Figure 5.23 illustrates the different classes of GEs according to Petersen (2017).

[5] Basically, a new form of molecular scissors that cut DNA very precisely. Such molecules may be used to edit the genome of animals and are thus called genome editors (GEs).

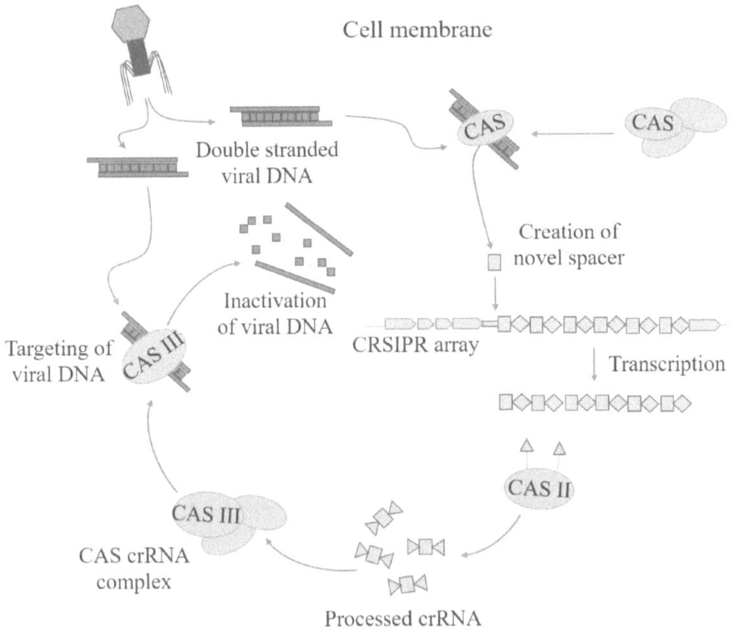

Figure 5.22 CRISPR/Cas mechanism according to Horvath and Barrangou (2010)

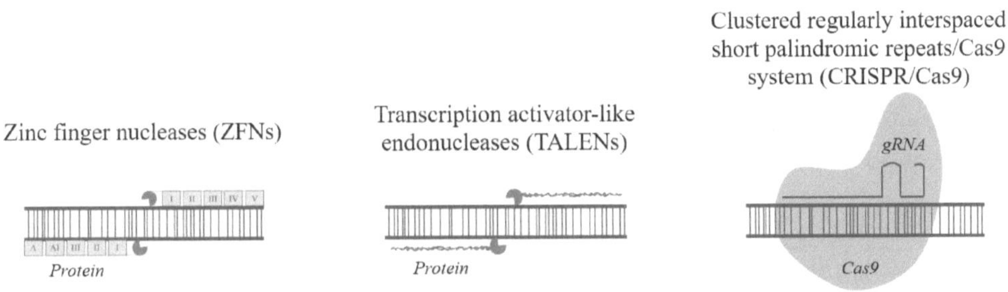

Figure 5.23 Classes of genome editors according to Petersen (2017)

All GEs are used to generate double-strand breaks (DSB) in a targeted DNA sequence that require a repair. The primary repair mechanism is non-homologous end joining, a fail-safe process that reassembles the end of the DSB, often causing small insertions or deletions (indels) and subsequent gene destruction (Petersen, 2017). Repair can be conducted in two ways: by non-homologous end connection (NHEJ) or by homology-directed repair (HDR). In most cases (up to 90%) DSBs are repaired by NHEJ, acting in all phases of the cell cycle, where the two ends of the fraction are merged and ligated. As a consequence of endonuclease activity at the interface, this process is subject to errors and often leads to the introduction of small insertions and/or deletions (indels) at the repair site. Homology dependent repair (HDR) is a robust process using a homologous template to repair the DSB. The provision of a template, mainly a single-stranded oligonucleotide (ssODN) or double-stranded plasmid DNA, allows specific modifications of the genome. Figure 5.24 illustrates the two repair mechanisms according to Petersen (2017).

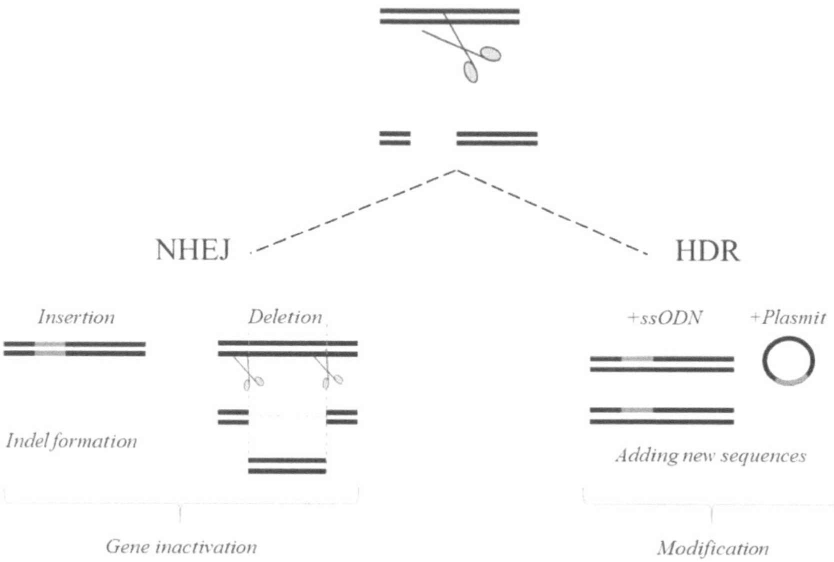

Figure 5.24 DSB repair mechanisms according to Petersen (2017)

Since its introduction, publications on a successful application of CRISPR/Cas9 as well as ZFN and TALEN have increased significantly. Prominent examples include disease resistant animals, alterations of milk composition, the production of allergen-reduced or allergen-free animal-derived products, or large animal models for cardiovascular diseases (Petersen, 2017). Yet, the technology is still in its infancy and will be rigorously further developed and tested in years and decades to come. Due to the sheer endless range of conceivable applications, it is only logical to assume that the range of possible consequences is also immense. From the perspective of sustainability, it is therefore essential to carefully assess both advantages and disadvantages of each application. As this work cannot comprehensively introduce the techniques outlined above, readers are referred to further literature, including Jinek et al. (2012), Ran et al. (2013), Petersen (2017), and Appasani (2018).

5.3.8 Evaluating technology readiness and attractiveness

Bauernhansl et al. (2019) and Miehe et al. (2020a) present a list of potential key technologies that primarily cover the bioinspiration and biointegration examples outlined above as well as a few others. According to Miehe et al. (2020a), the basic insight of the

BIOTRAIN project was, however, that not a single enabling technology but the interdisciplinary cooperation between disciplines will set the path for the BT.

- *Biosensors and neurorobotics*: Biosensors depict a basic ingredient of an interface of the biological, technical and informational systems. They are composed of a biological recognition element (e. g., enzyme, antibodies, DNA, receptors or whole cells/cell networks) and a physical sensor (transducer), which are in direct contact. Major applications of these sensors are man-machine interfaces, which are among others increasingly relevant in prosthetics, virtual reality and future manufacturing environments. In this context, neurorobotics may significantly contribute to an improvement of the current state of the art by linking human nerve cells with technical components.
- *Additive manufacturing of biobased, bio-hybrid materials and electronic components*: Especially from an ecological point of view, compostable electronic components are of highest relevance for the BT. Initial research results show success in the development of such components, in which, e. g., semiconductors made of plant extracts or insulators of gelatin can be printed on biodegradable circuit boards (e. g., compostable plastic) with additive processes. Latest research reveals progress in the required technologies that enable additive manufacturing with often highly variable and inhomogeneous biobased and biohybrid materials (Guvendiren et al., 2016).
- *Photobioreactors, switchable organisms, and biorobots*: Photobioreactors are technologies that cultivate microorganisms outside their natural and inside an artificial technical environment (Pulz, 2001). By using solar energy and CO_2 the process of photosynthesis is stimulated in order to create biomass and energy from various types of organisms, including plants, macro and microalgae and cyanobacteria, among others. Due to the controlled provision of a habitat that offers the optimal living conditions for the respective organism, photobioreactors enable significantly higher growth rates and purities than the natural environment. As such, the technology may decisively contribute to an increased functionality of products. Therefore, further development, especially with regard to efficiency criteria is required (Demirbas and Demirbas, 2010; Posten and Walter, 2012). Additionally, phototropism can also be used for the development of so-called switchable organisms that react due to light sensitivity (Wang et al., 2017). Beyond that, programmable and controllable microorganisms are viewed as highly relevant for the BT, which may pave the path for biorobots. For instance, Kriegman et al. (2020) present a concept of controllable microorganisms that are cultivated from stem cells and may serve to deliver drugs directly to the location of effect within the body. Yet, even more extensive applications of biorobot technology are to be expected in the course of BT. In the context of the present work, the term will thus be used for all controllable, purely biological organisms.

- *Biobased energy carriers*: Biobased energy carriers, e. g., formic acid for hydrogen binding, could contribute to making energy supply more flexible. Intensified research on related technologies, including the integration of efficient microbiological, enzymatically catalyzed material transformations, is regarded as increasingly relevant. The same applies for enzymatic processing technology (Miyazaki et al., 2008). Some of the innumerable naturally occurring enzymes are already of industrial relevance, e. g., in the food industry.
- *Bioleaching* is a microbial recycling technology that may significantly contribute to the closing of value chain loops (Zhu et al., 2003; Bosecker, 1997; Olson et al., 2003; Pathak et al., 2009). Existing technologies in this field, however, massively lack in efficiency and reliability.
- *Natural materials*: Some organic natural products will experience a renaissance through new manufacturing technologies such as additive manufacturing, fiber composite technology and compounding. Among others, these include lignin and chitin, whose natural occurrence is enormous and will continue to increase (Wegst and Ashby, 2004). Another current field of research with potentially massive impact is biodegradable plastics whose diffusion is still insufficient (Lenz and Marchessault, 2005). A combination of these two components, for example in the form of natural fiber-reinforced compounds with bioplastics, also has an expanded potential to find widespread applications in a timely manner.
- *Bioinspired information technologies and man-machine-interfaces*: Information technologies such as distributed ledger, pattern and emotion recognition software, predictive analytics and learning algorithms (artificial intelligence) already represent focus areas of today's production research (da Silva et al., 2017), while simultaneously bearing great potential for the BT. This likewise applies to the coupling of man and machine (Wander and Rao, 2014). Examples of these technologies in the context of the BT include adaptive exoskeletons that can physically relieve body movement in work processes (Dahmen et al., 2018). These technologies are already close to dissemination.
- *Biorefineries and adaptive bioreactors*: Biorefinery technologies are about to be transferred to industrial and large scale applications and may be used to provide biobased or platform chemicals (e. g., ethylene, propene) or likewise for a more efficient generation of biobased energy sources (e. g., methane, hydrogen) (Cherubini, 2010). Both the plant-technical know-how of these refineries and the process development for the processing of the products still have to be raised to industry level in order to achieve sufficient process stability and far-reaching application possibilities.
- *Microbial fuel and electrolysis cells* use living microorganisms that process organic substances as part of their energy metabolism to produce electricity or

hydrogen. In conjunction with other bio (electro) chemical developments (e. g., bioelectrochemical power to gas, biological methanation) these technologies can, in addition to their contribution to the flexibilization of energy supply, form a basic technology for SBMDs (Schiebahn et al., 2013). Further applied research is as well required to enable adaptivity of bioreactors or fermenters (Cusick et al., 2010). Today's bioreactors are predominantly adjusted to a single process and not easily transferable to other processes, substrates (nutrients) or organisms. Latest sensor and actuator technologies as well as learning algorithms may, however, allow flexible adaptation of the process parameters to a larger selection of organisms and substrates which could significantly contribute to a flexible production of enzymes or other fermentation products (such as personalized pharmaceuticals).

- *Interference with genetic material*: The redesign of DNA (genome editing) made possible by genetic engineering is in spite of its extremely high innovative potential, e. g., in the medical sector, associated with risks for humans and the environment, a fact that applies to synthetic biology in general (Miehe et al., 2020a). Yet, from a sole technical perspective this technology represents a key ingredient of the BT.

Bauernhansl et al. (2019) evaluate the technologies outlined above regarding their technological maturity (TRL[6]) and their attractiveness[7] for Germany. Figure 5.25 illustrates the results, revealing that most of the technologies require further basic of applied research in order to increase their technological maturity (see Chapter 7).

---

[6] Technology readiness level (TRL) represents a concept to evaluate the respective development status and applicability of technologies or systems. Originally developed by NASA to assess space technologies, the approach is used in various scientific disciplines and industrial sectors today and is standardized under ISO 16.290. The TRL approach applied here differs between nine developmental stages of a technology: observation and description of functional principle (TRL 1), description of application (TRL 2), demonstration of functionality (TRL 3), experimental set-up in laboratory (TRL 4), test setup in operational environment (TRL 5), prototype in operational environment (TRL 6), prototype in use (TRL 7), qualified system with proof of functional capability in area of application (TRL 8), qualified system with proof of successful use (TRL 9).

[7] To assess technological attractiveness, an average value was calculated from economic (including profitability, innovation, business model, markets), technical (including flexibility/mutability, integrativity, prospects/future viability, process reliability), ecological (including emissions, material resources, biodiversity, land use, biodegradability, recyclability) and social factors (job creation, qualification of employees, creation of meaning, reduction of defects, process and product hazards for people) for each technology.

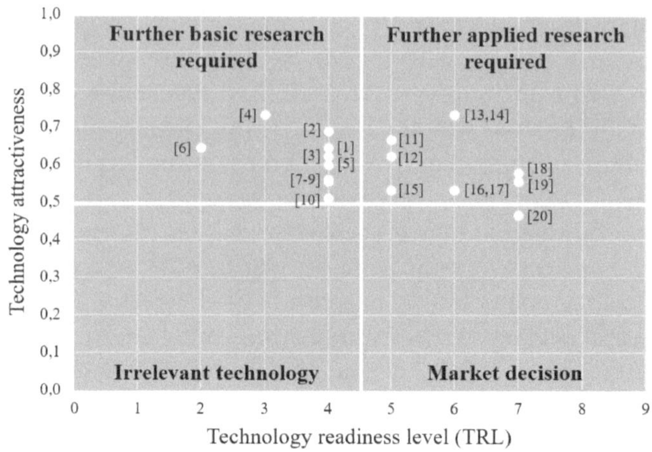

Further basic research	*Further applied research or market decision*
[1] Microbial fuel and electrolysis cells	[11] Biobased energy carriers
[2] Additive manufacturing of biomaterials (incl. 3D/4D bioprinting)	[12] Biobased platform chemicals as substitue for fossil fuel
[3] Photobioreactors	[13] Adaptive bioreactors and biorefineries
[4] Compostable electronics	[14] Predictive analytics
[5] Biosensors and actuators	[15] Pattern and emotion recognition software and hardware
[6] Programmable and controllable microorganisms	[16] Brain machine interfaces
[7] Neurorobotics	[17] Artificial intelligence
[8] Embedded functional microorganisms in materials or textiles	[18] Use of natural substances and materials
[9] Low-energy, enzymatic process technology	[19] Biodegradable polymers
[10] Bioleaching	[20] Genome Editing (incl. CRISPR/Cas)

Figure 5.25 Maturity and attractiveness of bioinspired and biointegrated technologies for Germany according to Bauernhansl et al. (2019)

5.4 Basic approaches and technologies of biointeraction

Biointeraction is the instrumental basis of biointelligence. Its basic model is the biology-technology-interface (BTI) that provides a system-theoretical template for future bidirectional, real-time capable information and control bridges between technical and biological systems. In addition to basic engineering and life science know-how, biointeraction primarily requires knowledge of basic approaches and technologies of digitization, yet with regard to biology-technology-systems. This section thus first introduces basics of intelligent manufacturing (I4.0) and AI, before outlining the BTI concept and its most promising transition technologies (soft sensors and digital twins).

5.4.1 The basics of intelligent manufacturing (Industry 4.0)

The term Industry 4.0 (I4.0) introduced by Kagermann et al. (2011) refers to the intention to initiate a fourth industrial revolution. After the first three revolutions in the manufacturing sector were primarily associated with mechanization, electrification and electronics/IT (see Chapter 2), the I4.0 concept is essentially based on the vision of comprehensive networking and decentralized decision-making (Bauernhansl et al., 2014). In English language literature smart or intelligent manufacturing are often used as synonyms. Despite fundamental criticism of the concept (e. g., proclamation of an industrial revolution before it took place, no change in the technological basis compared to the third industrial revolution), I4.0 has gained considerable leverage in research and industry in recent years. Since then, many authors have outlined the vision of I4.0.

5.4.1.1 Fundamental vision

According to Hermann et al. (2016) and Gronau et al. (2016), I4.0 is an organizational design concept that comprises four elements: [1] the networking and communication of machines, plants, and people via internet of things, [2] the expansion of information systems through sensor data to create a virtual image of products, machines, factories, and value chains, [3] the support of people through aggregated, visualized and understandable processing of information using technical assistance systems, and [4] the ability of people and machines to decide independently and decentrally through cyber-physical systems (CPS). In turn, Stock and Seliger (2016) distinguish three dimensions of I4.0: [1] horizontal integration of entire production networks (i. e., the cross-company and intra-company intelligent networking and digitization of manufacturing modules along a product life cycle and between value chains of adjacent product life cycles), [2] end-to-end engineering across entire product life cycles (i. e., intelligent cross-linking and digitalization along the entire product life cycle from raw material extraction to production, use, and end of life, and [3] vertical integration of networked manufacturing systems via CPS (i. e., the intelligent cross-linking and digitalization across various aggregation and hierarchy levels of manufacturing module). Qin et al. (2016) outline the I4.0 concept with regard to four essential elements of the industrial value chain:

- *Smart factories*: integrative systems that not only connect all production factors (sensors, actuators, machines, robots, conveyor belts, etc.) and automatically exchange information, but also consciously evaluate orders, distribute them to machines, control the production process and manage the factory system. At the same time, operational processes (product design, production planning, etc.) are simulated in a modular fashion and closely linked.

- *Smart business*: a complex communication network between different elements of the value chain (companies, factories, suppliers, logistics, resources, customers etc.) allows each section to be optimized in real time according to the requirements and status of the connected sections in the network. This enables new cooperative forms of production that provide better allocation of limited shared resources and maximum output. At the same time, costs and pollution, consumption of raw materials, etc. will be reduced.
- *Smart products*: a new type of product enabled by embedded sensors to provide customers with functional guidance and feedback from applications to the manufacturing system. Thanks to these elements, many functions such as measuring the state of products or users, tracking the product, and analyzing the results depending on the information could be added to the device. In addition, the product can be embedded with complete manufacturing information protocols to help product designers optimize design.
- *Prosumer versus consumer*: Customers receive new possibilities by intervening in the product design. New purchase methods are provided that allow customers to modify products during the order process and at any time during production based on their ideas, without incurring any costs. During use, intelligent products allow the customer not only to know the production information of the product, but also to receive instructions for use depending on his own behavior.

Stock and Seliger (2016) present a similar categorization covering aspects such as smart design, smart machines, smart monitoring, smart control, and smart scheduling. Figure 5.26 illustrates the I4.0 concept qualitatively according to Stock and Seliger (2016).

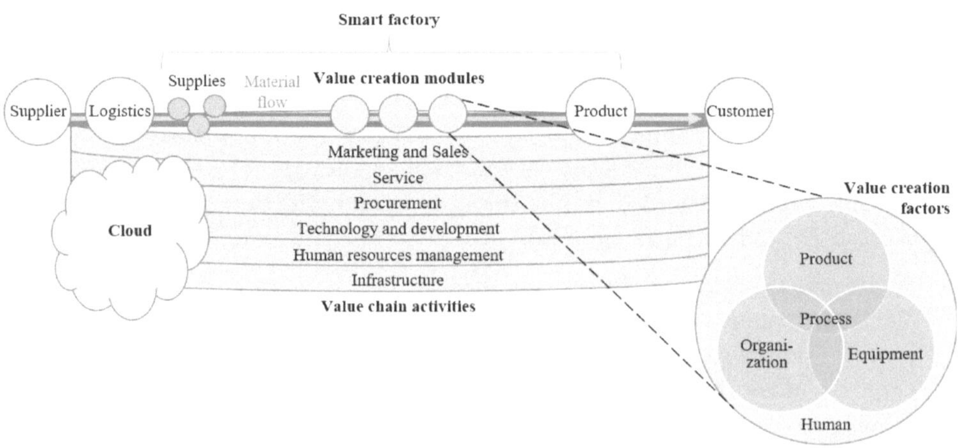

Figure 5.26 Qualitative illustration of I4.0 concept according to Stock and Seliger (2016)

Until today, the I4.0 concept has primarily entered the bulk goods industry, while establishing itself far less rapidly in other sectors, e. g., chemicals, pharmaceuticals, biotech.

5.4.1.2 Pillars, technologies and methods

To date, a number of authors have presented approaches that summarize essential pillars, technologies and/or methods of I4.0 (Bauernhansl et al., 2014; Lee et al., 2015; Qin et al., 2016; Seif et al., 2019; Stock and Seliger, 2016; Uhlemann et al., 2017a; Vaidya et al., 2018; Wollschläger et al., 2017; Zhong et al., 2017). While the categorization differs to some extent, cloud computing (CC), internet of things (IoT), cyber-physical-systems (CPS), big data analytics (BDA), and artificial intelligence (AI) are considered as consensus of elementary technologies and methods of I4.0. In the following, the individual approaches are concisely summarized in order to enable the reader to obtain a more profound understanding of the topic (all except AI, as outlined separately in Section 5.4.2). The approaches sketched here are commonly characterized by a further increase in the integration of information and communications technology (ICT) in manufacturing companies in order to enable various organizational subsystems to store, transmit, and manipulate data or information (Hashim, 2007).

1. Cloud Computing (CC)

CC describes the provision of computing services (data storage, computing power, etc.) via the Internet (Armbrust et al., 2010; Xu, 2012). CC is particularly interesting for companies because of its scalability, since the services are available in almost any size, regardless of the time (Zhang et al., 2010; Zhong et al., 2017). Cloud-based IT-platforms serve as backbones for information technical linking and communication of manifold elements of a manufacturing environment (Landherr et al., 2016). Zhang et al. (2010) present an architectural model that describes the business and operational integration of cloud computing by differing four layers: hardware/datacenter, infrastructure, platform, and application. The hardware layer, typically implemented in data centers, is responsible for managing physical factors of a cloud (incl. servers, switches, power and cooling systems). The infrastructure layer is responsible for virtualization, i. e., creating a pool of storage and computing resources. The essential feature of this layer is dynamic resource assignment via virtualization of physical entities. The platform layer depicts the operating systems and application framework. Its purpose is to reduce the burden of deploying applications directly. Finally, the application layer contains the actual automatic-scaling cloud applications. Figure 5.27 illustrates the concept according to Zhang et al. (2010).

Mell and Grance (2011) recommend five characteristics of an ideal cloud: resource pooling, rapid elasticity, on-demand self-service, broad network access, and measured service. Despite the benefits of CC, resource allocation, load balancing, scalability and availability, cyber security, compatibility and interoperability of different clouds remain critical issues of the concept (Chaves et al., 2011; Sharkh et al., 2013; Randles et al., 2010; Moreno-Vozmediano et al., 2012; Zhong et al., 2017). According to Zhong et al. (2017), typical applications of cloud computing in current research, teaching and industrial enterprises are hybrid and/or multilayer information fusion, identifying global sensitivities of input factors under uncertainty, increasing service-orientation, providing opportunities for students to test, learn, experiment, and innovate, forensic methods for efficient file extraction, solving railway container station reselection problem, etc.

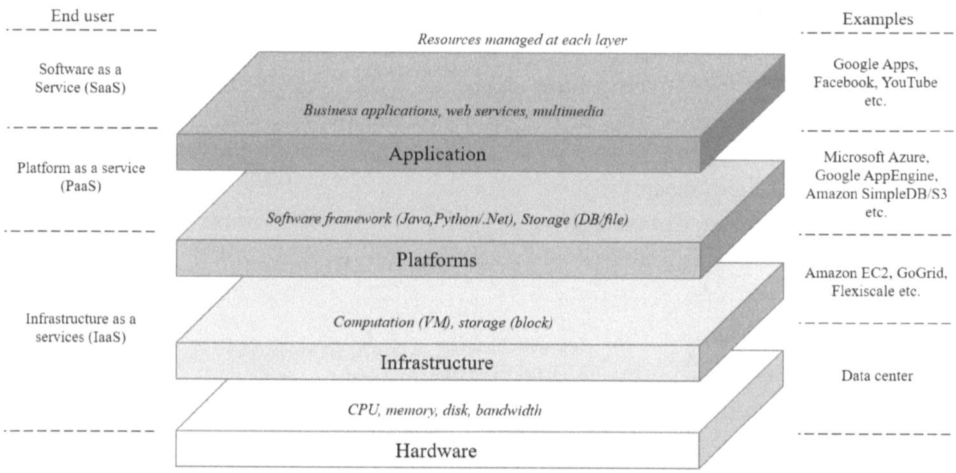

Figure 5.27 Architectural layer model of operational cloud computing according to Zhang et al. (2010)

2. Internet of Things (IoT)

The term IoT refers to a global infrastructure that enables physical and virtual objects to interconnect and cooperate through information and communication technologies (Xia et al., 2012). Essentially, it is a worldwide network of interconnected and uniformly addressed objects that communicate via standard protocols (Hozdić, 2015). An alternative term is internet of everything (IoE), consisting of internet of service (IoS), internet of manufacturing services (IoMs), internet of people (IoP), an embedded system and integration of information and communication technology (Neugebauer et al., 2016).

The three main characteristics of IoT are context, omnipresence, and optimization. An object is expected to be easily integrated into any existing environment, capable of immediate reaction, has omnipresence information about the location and physical or atmospheric conditions and provides suggestions for optimization. Compared to today's objects, which primarily have a single connection to a network of human operators, these are fully developed human-machine and machine-machine interfaces (Vaidya et al., 2018).

IoT is understood as the basis for designing an intelligent, agile and networked value chain by integrating physical objects, human factors, intelligent machines, intelligent sensors, production processes, and production lines across organizational boundaries (Schumacher et al., 2016). The key factors here are software and data for the intelligent planning and control of machines and factories (Vaidya et al., 2018). For instance, tracking and tracing of goods will become faster, more precise and safer, as will storage in warehouses (Dutra and Silva, 2016). Radio-frequency identification (RFID) is commonly viewed as a key technology for embedding physical objects in an IoT environment (Zhong et al., 2013). Examples of RFID applications include the identification of objects in warehouses, production shop floors, logistics companies, retailers, and disposal/recycle stages (Wang et al., 2010). Once identified, objects possess intelligent recognition capabilities allowing them to connect and interact with each other by specific forms of interconnectivity, which may generate huge amounts of data (big data) (Zhong et al., 2017). The interconnectivity between intelligent objects has to be predefined. For this purpose, the concept of digital shadow (DS) and/or digital twin (DT) was designed (see Section 5.4.5).

According to Zhong et al. (2017), recent examples of IoT applications are suspicious event detection in neighborhood watch, sharing data between application developers, improving the quality of health care, easing the interpretation of sensor data in machine-to-machine measurement, as well as improving energy management and energy data integration at production level.

3. Big Data Analytics (BDA)

The term big data refers to large data sets that appear to be too complex, fast-moving or weakly structured to be evaluated by conventional methods of data processing (Barton and Court, 2012). They originate from a variety of sources, such as sensors, video/audio recordings, internet, transactional applications, social media feeds, etc. (Zhong et al., 2017). The collection and comprehensive evaluation of big data is expected to become a standard feature of future production systems, supporting real-time decision-making (Rüßmann et al., 2015). BDA, based on historical data, is employed to both learn from past problems as well as to forecast emerging issues in various production processes in order to initiate targeted countermeasures. The main approaches to BDA originate from AI (see Section 5.4.2). Today, a common field of BDA approach is customer relationship management (CRM). The mining of customer data is used to advance engagement and satisfaction (Wamba et al., 2015). For instance, automotive OEMs redesign cars according to BDA of the history of customer orders and user feedback (Agarwal and Weill, 2012). Yet, BDA is gradually evolving in the manufacturing sector. While the analysis of large data sets from processes and machines can realize an increase in productivity and competitiveness in general (Lee et al., 2014), BDA proves to be beneficial in biopharmaceutical production, where a vast variety of variables have to be monitored to guarantee the accuracy and quality (Zhong et al., 2017). Big data processing enables a manufacturer to identify critical parameters (Brown et al., 2011), an approach that is also of decisive relevance for the field of predictive maintenance. Here, large amounts of data are recorded at the machine, processed in predictive models to forecast possible failures and, in the best case, to predict and plan the workpiece change (Lee et al., 2014). According to Zhong et al. (2017), other examples of BDA are identifying search patterns and recommended searches of web service providers, identifying strategic segments of customers for promotion of retailers, improving vendor-managed inventory, protecting IT infrastructure from cyber-attacks, remote monitoring for heavy-duty equipment vehicle, and utilizing process excellence and six sigma in automotive manufacturing.

4. Cyber-Physical-Systems (CPS)

The term CPS refers to the combination of information technical components (e. g., computer, software) with mechanical and electronic parts that communicate using a data infrastructure such as IoT (Lee, 2008). With the increasing connectivity and adoption of standard communication protocols associated with I4.0, organizations increasingly need to protect critical industrial systems from cyber security threats (Rüßmann et al., 2015). Unlike conventional embedded systems, a CPS contains networked interactions that are engineered with physical input and output alongside their respective control algorithms and computational capacities (Zhong et al., 2017). Accordingly, CPS require a large number of sensors, e. g., multiple sensor devices such as touch screens, light sensors, and force sensors. This enables a strong link between the physical, service, and digital worlds that can improve the quality of the information needed to plan, optimize, and operate manufacturing systems (Landherr et al., 2016). According to Lee et al. (2015), a CPS consists of two main functional components: [1] advanced connectivity that guarantees real-time data gathering from physical world and information feedback from cyber space and [2] intelligent data management, consisting of analytics and sufficient computational capability. Lee et al. (2015) present an approach to deploy CPS in a five level (connection, conversation, cyber, cognition, configuration) sequential workflow manner (5Cs). First, the connection level describes the accurate and reliable data acquisition from machines and their components as an initial step for the development of a CPS. Thereby, two factors have to be considered precisely: [1] handling of different data types and [2] selecting proper sensors (type and specification). The second stage pictures data-to-information conversion. Therefore, a multitude of different algorithms already exists and continues to be further developed. Third, the cyber stage depicts the central information hub. Here, all relevant information from every connected machine is aggregated in order to form the machine network. To cope with the massive amount of data, specific analytics have to be applied (see big data, AI). The next stage is cognition, referring to a thorough knowledge of the monitored system that supports its user to take correct decisions by constantly acquiring novel knowledge. For this stage, appropriate information graphics are required. Finally, a supervisory control of the CPS that enables a thorough self-configuration and self-adaption of machines is referred to as configuration. In this stage a CPS acts as a resilient control system that independently applies corrective and preventive measures, which have been identified in cognition. Figure 5.28 illustrates the approach according to Lee et al. (2015).

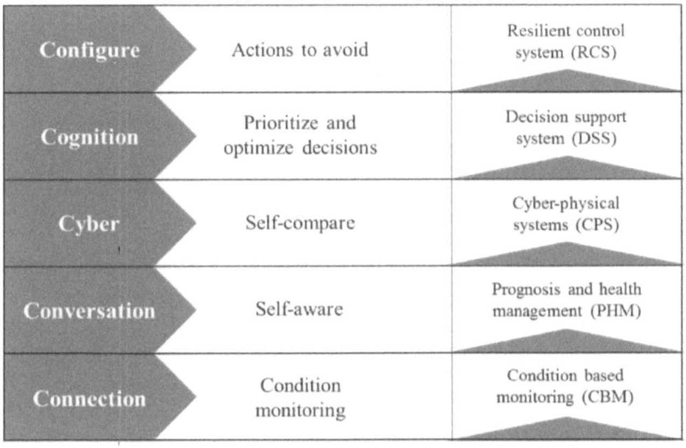

Figure 5.28 Five level approach for deployment of CPS according to Lee et al. (2015)

Today, approaches to CPS development have been reported from different industrial sectors. For instance, CPS designs of sensor-based communication-enabled autonomous systems as devices for the supervision of pipeline condition (Ali et al., 2015). Zhong et al. (2017) provide additional examples including the improvement of hydraulic actuator models in civil structures, the simulation of emergency handling problems, the identification of optimal sensing and robot scheduling policies, the easing of system implementation in decentralized or cloud environments, the development of self-adjusting and self-adopting subsystems, and the provision of sufficient information to perform medical testing. Yet, the heterogeneity and complexity of CPS still poses several challenges to production engineers, especially with regard to designing high-confidence, secure, and certifiable systems and control methodologies (Derler et al., 2011).

While the approaches listed above represent the core of I4.0, additional technologies, methods, and/or resulting features of advanced production systems are mentioned depending on the author. These mainly include:

- *Augmented reality (AR)* describes computer-aided extensions of reality perception (Azuma et al., 2001). While it basically encompasses responding to all human sensory modalities, the concept is primarily understood as the visual representation of information. Here, images or videos are supplemented with computer-generated additional information or virtual objects using fade-in/overlay. In industrial environments the technologies is applied to provide workers with real-time information to enhance decision-making and work procedures in a variety of fields, e. g., assisting the selection of parts in a warehouse or sending repair instructions via mobile devices (Vaidya et al., 2018).

- *Simulations* in facility operation are increasingly used in order to exploit real-time data to map the physical world in a virtual model. These can include machines, products, and people (Rüßmann et al., 2015). Here, digital models, shadows, and twins are distinguished (see Section 5.4.5) (Grieves, 2014). In addition to the virtual representation of entities, these can, among others, incorporate 2D and 3D simulations, cycle times, energy consumption or ergonomic aspects of a production system, e. g., to reduce downtimes and production losses during the start-up phase (Simons et al., 2017).
- *Autonomous robots*: due to increased computing power, advanced algorithms and the close information technical linking with all sort of production devices in the context of I4.0, robots are becoming increasingly autonomous, flexible, and cooperative (Vaidya et al., 2018). In a medium term they will be able to interact with one another barrier-free and work safely alongside humans, while being able to learn from them (Rüßmann et al., 2015). Autonomous robots can complete a given task precisely within a given time. As such, they will not only further autonomize production methods, but will be able to work in environments where human workers are restricted.

5.4.1.3 Industry 4.0 and sustainability

In addition to the technical renewal, various authors emphasize the potential of I4.0 for sustainability (Keeso, 2014; Beier et al., 2018). According to Stock and Seliger (2016), I4.0 optimization potentials with regard to environmental sustainability exist in nearly every field of industrial production, including product, processes, organization, human, business models as well as entire value creation networks. Possibilities for product related sustainability are primarily seen in an enhanced design of closed-loop life cycles according to the cradle-to-cradle principle, while process-related options cover process design with regard to resource efficiency or the implementation of new technologies such as internally cooled tools (Stock and Seliger, 2016). Then, decentralized factory organization focusing on an efficient allocation of products, materials, energy, and water via CPS supported by new technologies such as smart logistics, smart grids, self-sufficient supply etc. are considered to bear leverage. Rising training efficiency of workers via ICT such as virtual reality or gamification as well as fostering creativity by establishing novel forms of work organization and design are expected to boost intrinsic employee motivation. Other, yet roughly sketched ideas are the life cycle integration of new processes, such as retrofitting, via sensor and actuator systems as well as lifetime extension of products and manufacturing equipment. From a value network perspective, the increased cross-linking of value chain elements is expected to allow a more efficient coordination of substance and energy flows throughout a product's life cycle, enabling closed-loop manufacturing systems and industrial symbiosis. Not least, the use of smart data is expected to enable novel business models based on new services and functionalities. According to Bocken (2014), these can be designed in a sustainable matter that solely creates positive effects by simultaneously reducing negative impacts.

5.4.2 Basic methods of artificial intelligence (AI)

AI[8] is currently regarded as the basis for a large number of innovations in various application areas (including speech recognition and processing, image and pattern recognition, handling and automation, predictive maintenance, human-machine interaction, and autonomous robotics). Knowledge of its basic methods and application fields is likewise fundamental for biointelligence, as AI may provide a vehicle to ensure overall system intelligence.

AI can be divided into several sub-areas that emerged over the course of the historical development of the field (Boersch et al., 2007; Görz et al., 2013; Eberl, 2016; Ertel, 2016). Since it is not possible to separate the segments in some cases, the classification of AI systems is explained below by a short excursus of the genesis of the discipline. The origin of the AI, which today is essentially a field of ICT, is the logical conclusion that Gödel was able to implement in a machine for the first time in the 1930s. The discipline experienced its first real momentum with the development of neural networks. After single-layer neural networks were developed for the first time in the 1950s and machines were enabled to play simple games (e.g., checkers) against humans, systems were established from the 1970s onwards that allowed machine knowledge processing with the help of manually created knowledge bases (expert systems). With the development of learning systems in the 1980s, the sub-area of machine learning (ML) emerged, which is based on so-called learning algorithms. ML systems (MLS) are able to recognize patterns in large amounts of data and classify them. A prominent example of such a system is the spam filter in email inboxes, classifying almost any amount of incoming email based on its content and removing it according to the user's priority. After initially being considered a niche product, the importance of learning systems increased in the 2010s due to the rapidly rising amount of data available as a result of digitalization and the quickly growing capabilities of data processing systems. In this context, cognitive systems (CS) were developed for the first time, which represent a combination of knowledge-based methods and powerful learning processes. CS are usually based on complex neural networks,

8 For a definition of AI, see Section 4.2.1.3.

which, compared to simple neural networks, consist of significantly more neurons distributed and interconnected in a large number of deeply nested layers. The learning algorithms of these networks are also called Deep Learning (DL). Figure 5.29 illustrates the historically grown differentiation of the AI sub-areas according to Bitkom (2017b). DL is understood as a sub-area of ML, which in turn is a sub-area of AI.

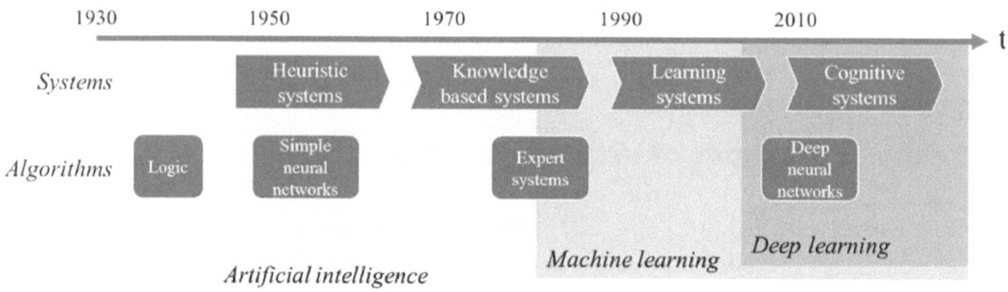

Figure 5.29 Formation of AI and its sub-areas according to Bitkom (2017b)

Another possibility of classifying AI is the differentiation based on its main algorithms. Specific algorithms are usually used for each desired function, which in turn offer different application options. Ertel (2016) distinguishes six functions of AIs, each of which is based on the systematic application of statistical methods for data mining: logic, search/problem solving, reasoning under uncertainty, combining, learning, and neural networks. Figure 5.30 summarizes essential algorithms for each desired functionality of AIs based on Ertel (2016).

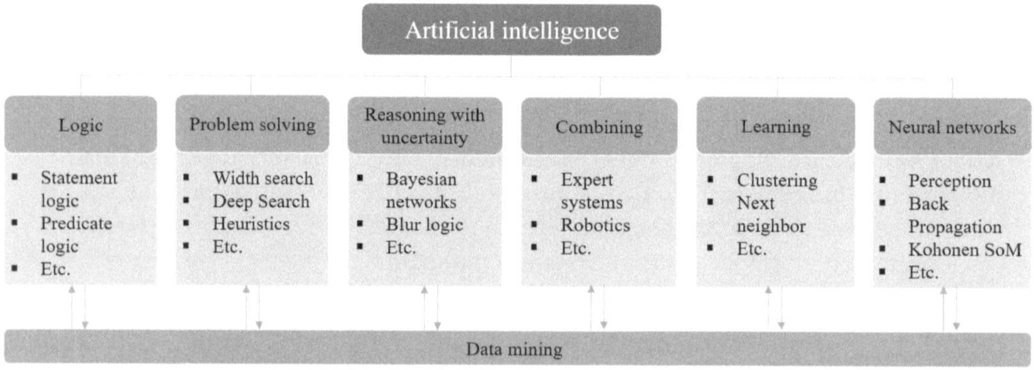

Figure 5.30 Essential algorithms depending on the desired functionality of AI according to Ertel (2016)

The following sections summarize fundamentals of non-learning AIs, learning AIs, and neural networks.

5.4.2.1 Non-learning AI

In mathematics, the classical propositional logic represents the origin of all forms of logic. Its object of investigation are statements and their logical connections. The propositional logic has two characteristic features (Beckermann, 2003): [1] a statement can be exclusively true or false (bivalence principle). The rejection of a statement is called negation.[9] [2] A compound statement is defined beyond doubt by the truth content of its partial statements (principle of extensionality). It is difficult to verify elementary statements, i. e., statements that describe the relations between a multitude of objects.[10] Here, predicate logic is applied by introducing all quantifiers and existential quantifiers (∀ and ∃) (Ertel, 2016). While propositional logic divides compound statements into elementary statements and then examines their truth content on the basis of the bivalence and extensionality principle, predicate logic focuses on examining the inner structure of elementary statements (Barwise and Etchemendy, 2005, 2006). Central elements of this form of logic are predicate and quantifier. The former is a sequence of words with blanks.[11] By inserting the subject, a statement becomes either true or false.[12] Logical reasoning, however, often proves to be ineffective[13], especially in the case of apparently contradictory statements. The AI subfield of reasoning under uncertainty focuses on these very statements and tries to map a connection with the help of probabilities.[14] In addition to Bayesian networks, fuzzy logic is often used here. The basis of this theory are so-called fuzzy sets, i. e., sets that are defined by the degree to which the elements belong (Zadeh, 1965; Goguen, 1969; Kamp and Partee, 1995). In this way, colloquial language components of statements can be mapped.[15] Further literature on the basics of logical reasoning can be found in Ertel (2016) and Kacprzyk et al. (2015).

According to Ertel (2016), the function of problem solving represents a further field of application of AI, which is significantly influenced by so-called inference systems. These algorithms are basically stepwise ramifications of search trees, which allow different search strategies depending on the width and/or depth of junctions. Often heuristics are used, i. e., analytical procedures for the development of probable statements under incomplete information. Typical for this is the development of presumable conclusions, which allow statements about a system despite limited

---

9 A statement A of this kind would be '9 times 9 equals 81'. The negation of this statement ¬ A would be '9 times 9 does not equal 81'. The propositional logic says that A and ¬ A can never be true at the same time. I. e., if A is true, ¬ A is false, and vice versa.

10 A statement of this kind would be 'All birds can fly'.

11 A sequence of words like that would be '… is a mammal'.

12 The statement becomes correct by the addition 'a lion', i. e., 'a lion is a mammal'. However, it would be wrong to say 'a closet', i. e., 'a closet is a mammal'.

13 The statements 'all birds can fly' and 'penguins are birds' are contradictory.

14 A correct statement would be 'almost all birds can fly'.

15 Fuzzy statement elements are, e. g., 'very', 'strong' or 'somewhat'.

knowledge (Gigerenzer, 1999).[16] The quality of a heuristic is determined by comparing the concluded statements with the optimal solution. In the context of AI, heuristics are increasingly used for statistical evaluations of random samples with large amounts of data (big data) (Pillkahn, 2012).

5.4.2.2 Learning AI

ML links neural networks with learning methods. Algorithms are explicitly programmed for a single case in order to learn from a certain amount of data (Bitkom e. V., 2017a). For example, this can be the recognition of patterns or a desired behavior. ML distinguishes three subareas: [1] supervised learning, [2] unsupervised learning, and [3] reinforced learning. Deep learning is often mentioned here, although it simply describes an architecture of a neural network that is composed of a large number of hidden layers. Furthermore, online and offline learning systems should not be confused. The former show an adaptive behavior during operation by continuously adjusting to new data sets. The latter do not adapt to new data sets (Auer, 2017; Skansi, 2018).

Supervised learning describes ML algorithms that link input values with corresponding output values (labels) in the training data set (Mainzer, 2016; da Silva et al., 2017). The input values are fed into the neuronal network in the form of data sets and the correlation of the output with the correct label of the data set is examined. If input and output are congruent, weightings between the neurons of different layers in the network are strengthened or weakened. Elementary fields of application of supervised learning are classification and regression. Supervised learning further differentiates between eager and lazy learning. The former requires most of the computing power during the training period, while the latter requires more computing power for classification and regression. Common methods of lazy learning are locally weighted regression and the k-nearest-neighbor method. Eager learning, instead, is associated with Bayesian networks, support vector machines, induction of decision trees, and radial basis functions (Görz et al., 2013; Ertel, 2016).

Unsupervised learning refers to ML algorithms that aim to recognize and learn structures, patterns, and correlations within an input data set. In contrast to supervised learning, no prior knowledge of the output data set is necessary (da Silva et al., 2017). Labelled data sets and feedback are thus not applied. However, the algorithm can determine the amount of output clusters ex ante. Basic algorithms of unsupervised learning include self-organizing Kohonen maps and example nearest and farthest neighbor algorithms (Ertel, 2016; Mainzer, 2016). The primary fields of application are association and clustering.

The term reinforced learning describes the acquisition of skills through multiplication (Mainzer, 2016). In this sense, reinforced learning represents an intersection be-

[16] An example of a heuristic is the trial-and-error method.

tween supervised and unsupervised learning. Although a specific goal is set, the algorithm is free to choose the way to achieve it. Although no labelled data sets are necessary, the algorithm receives feedback in each sub-section on how well or poorly the approximation to the goal is achieved. The algorithm therefore less learns how to output a correct data set, but rather a suitable procedure for a specific problem. Stochastic and probabilistic models form the basis for this type of algorithm. Basic methods of reinforced learning are value iteration, Markov decision model, Q-Learning and relational reinforced learning (Ertel, 2016).

5.4.2.3 Neural networks

Neural networks are an algorithm architecture that is based on the structure and function of human nerve cells[17] (da Silva et al., 2017; Beck et al., 2018). A neural network thus represents a bioinspired system based on artificial nerve cells (so-called neurons). Similar to the electrical impulses of dendrites, the artificial neuron receives several input signals ($x_1, ..., x_n$) from upstream neurons. By weighting ($W_1, ..., W_n$) the input signals, the 'synaptic' binding strength of the artificial neuron is simulated. A linear aggregation function (Σ) determines the activation potential ($u = \Sigma w_i \cdot x_i - \theta$), which results from the sum of the weighted input signals. If the threshold value ($-\theta$), also called bias, is exceeded, the neuron is activated and sends an output signal ($y = g(u)$) to the downstream neurons. The activation function ($g(.)$) limits the output values of the neuron to a defined range. A mathematical construct of this kind is also called a perceptron. Figure 5.31 illustrates the mathematical representation of a human nerve cell in an algorithmic structure according to da Silva et al. (2017) and Beck et al. (2018).

The architecture of a neural network comprises an input layer, consisting of neurons that pick up signals from the environment and transmit them in a normalized manner to several hidden layers. Within the hidden layers, patterns and structures are processed and the results are reflected by an output layer. The way the layers are con-

[17] Excursus: A human nerve cell consists of three main parts (da Silva et al., 2017); Beck et al., 2018): [1] the cell body (soma), [2] the dendrites, and [3] the axon. Dendrites are long, thin extensions that constantly receive electrochemical signals from an upstream nerve cell and transmit them to the cell body. The soma has a negative potential in its initial state. Until a certain activation potential is reached, all impulses converge here. When the potential is reached by the sum of the individual impulses, the cell body itself initiates an electrical impulse by the axon. This process is called depolarization. The axon also transmits the impulse to the downstream nerve cells via electrochemical processes. The connections between the axon ends and the dendrites of the subsequent nerve cell are called synapses. However, no physical contact takes place between them. The potential transfer occurs via so-called neurotransmitters over the synaptic distance between the nerve cells. The amplification or weakening of the synaptic connections in the brain is currently considered the basis of the human learning process. It must be added in relative terms that the understanding of the functioning of the human brain is constantly evolving with the findings in the neurosciences. For example, it is now assumed that dendrites not only activate signals, but also calculate signal strengths (Takahashi et al., 2016). With the so-called dendritic calcium action potential, the dendrites thus provide independent computing power. A transfer of this knowledge into the architecture of artificial neural networks is still pending.

nected can vary greatly. The most common approach is (single or multiple-layer) feedforward architecture, which is well proven for pattern recognition, classification and function approximation. The number of neurons and hidden layers depends on the quantity and quality of the available data and the complexity of the problem. Learning algorithms of single-layer feedforward networks are based on the Hebb rule, whereas multi-layer networks rely on backpropagation. As one of the basic rules of machine learning, the Hebb rule states that two repeatedly connected neurons i and j induce a change in weight w_{ij}. The change in weight (Δw_{ij}) is derived from $\Delta w_{ij} = \eta \cdot x_i \cdot x_j$. η is the learning rate, which determines the size of the learning steps (Ertel, 2016). The backpropagation, in turn, is an iterative algorithm that adjusts the weights in the entire network according to the gradient of an error function in the running process. For each output neuron, the deviation of the output value from the original value of a data set is quantified. The weighting is successively adjusted in the respective upstream layer. In this way, the algorithm iteratively approaches a local optimum (Skansi, 2018).

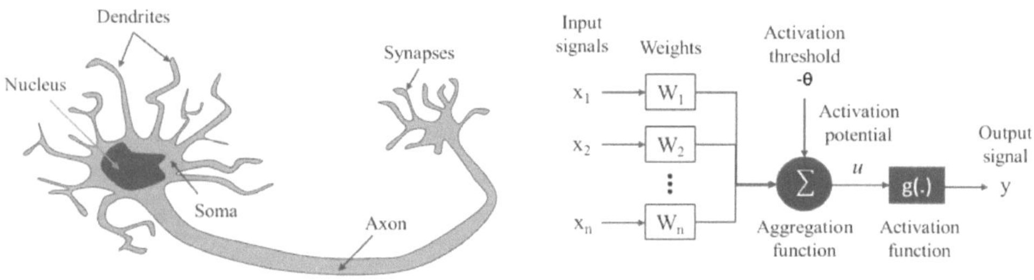

Figure 5.31 Mathematical representation of a human nerve cell in algorithmic structure according to da Silva et al. (2017) and Beck et al. (2018)

A further structural possibility is the recurrent or feedback architecture, using certain output signals as feedback input signals for neurons of upstream layers (Ertel, 2016). A feedback architecture enables dynamic information processing. Basic networks of recurrent architectures are Hopfield networks and feedback perceptrons. A prominent example of a Hopfield network is character recognition (Ertel, 2016; Skansi, 2018). Here, a pixel is represented by a neuron in the input layer and all neurons are interconnected. The weighting between the neurons can be either −1 or 1, i. e., black or white. All weights are assigned the value zero at the beginning of the learning process. In the course of the learning process, the network converges to a stable state by adjusting the weights of the neurons so that the deviation of the entire network (the so-called energy) is reduced to a local minimum.

A third option for arranging neurons in an artificial neural network is the mesh structure (Kacprzyk and Pedrycz, 2015; da Silva et al., 2017). Mesh architectures are usually the basis for self-organizing Kohonen networks, often used in the context of unsupervised learning for data classification, pattern recognition, and system optimization.

One example is the architecture of a single-layer, two-dimensional Kohonen network consisting of five by five neurons. In a learning process the Euclidean distances of the input signals (x_1, ..., x_n) to the weightings of the neurons are quantified. If a neuron has the smallest distance, it 'wins' the competition. Its weighting is increased or the distance reduced accordingly, while the ranking of the surrounding neurons is adjusted according to the Gaussian distribution.

Figure 5.32 compares the structural design of a two-layer neural network of feedforward architecture (left), a recurrent neural network of feedback architecture (middle), and a single-layer, two-dimensional Kohonen network (right) according to da Silva (2017). Further literature on these topics is provided by Kacprzyk and Pedrycz (2015) and da Silva (2017).

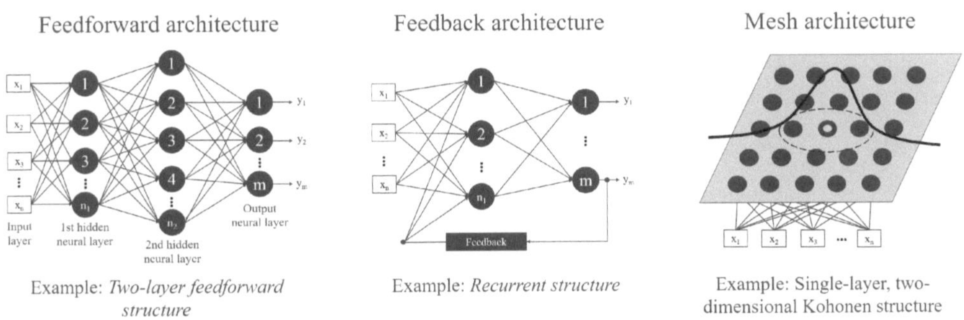

Figure 5.32 Comparison of feedforward, feedback, and mesh architecture of a neural network according to da Silva (2017)

5.4.2.4 Case study: object detection using deep neural networks

In order to gain an idea of how a neural network operates, the exemplary decoding of a pattern process provides a useful tool. A prominent example for the application of neural networks is object recognition. Analogous to character recognition, each pixel is represented by a neuron in the input layer. A frequently used architecture is the feedforward network (Figure 5.33). Three hidden layers are processed in the order illustrated here as an example. In the first layer, simple patterns are recognized, which are assembled to more complex formations in the following layer. The development of an overall image is done in the third stage by assembling the partial structures in order to enable a classification in the output by comparing the similarity with analyzed image elements.

Modern algorithm architectures for object recognition mostly rely on convolutional neural networks (CNN) (Goodfellow et al., 2016). As combinations of different network architectures, these typically consist of a random number of folding and pooling layers. In the former, also known as convolutional layer net, the input data set typically exists as a two- or three-dimensional matrix, which serves as a basis for calculating the activity of each neuron by discrete convolution. Using the activation function, an output is generated. Between two convolutional layers, pooling is used to discard redundant information. Several types of pooling layers exist, although one of the most common is the max pooling layer, which selects the maximum value from each pool of four values. After a certain number of repetitions, a fully-connected layer closes the network. Figure 5.34 illustrates the procedure of a CNN according to Skansi (2018).

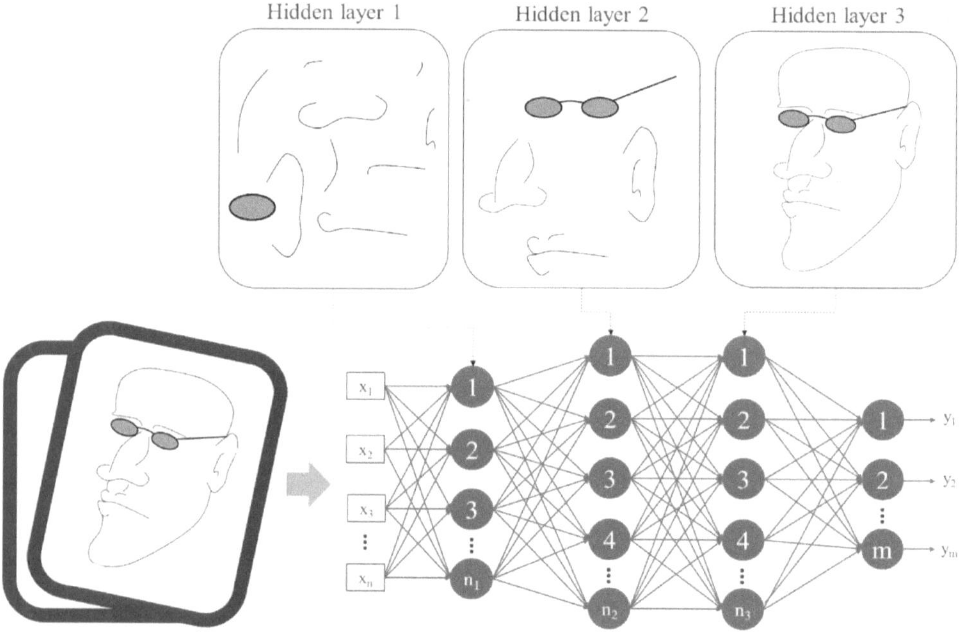

Figure 5.33 Illustration of a pattern process of image recognition using a deep neural network according to Zweig (2016)

In this case, the input data set is a color image with 10 × 10 pixels. In the convolutional layer, the input image is transformed into a smaller but deeper object. For this purpose, a receptive field[18] scans the pixel areas of the image and stores the filtered and averaged values for each color value red, green, and blue (RGB) in an 8 × 8 matrix. The word 'smaller' here describes the number of pixels of the resulting image, the term 'lower' the number of resulting images (channel). With the help of a pooling layer complexity is then reduced. The further the image is processed by the network, the smaller the images become and the more channels are created. Ultimately, all information can be reduced to one vector. A simple logistic regression is used to identify the relevant parts for the classification problem. The output value is compared with the target and in case of deviations it is propagated back through the network via feedback. Skansi (2018) provides instructions for the setup and programming of CNN.

[18] Often referred to as the kernel.

The analysis of biological systems using neural networks and imaging techniques (e. g., fluorescence of cell reactions) will be essential for the development of BIS. Consequently, a transfer of the approaches summarized here will continue to occupy the scientific community for years to come.

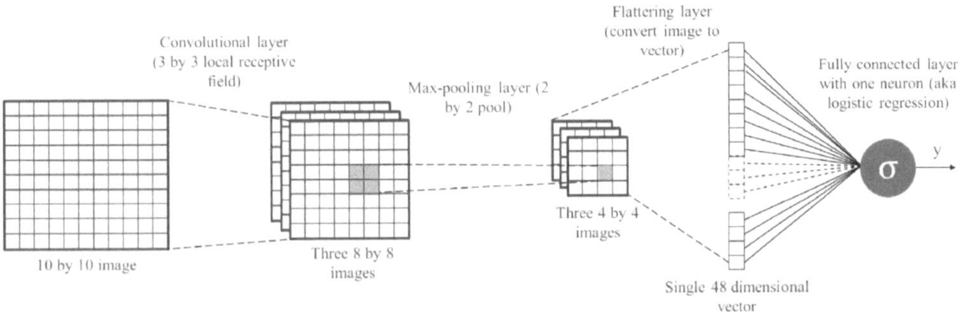

Figure 5.34 Illustration of a CNN sample process according to Skansi (2018)

5.4.3 The BTI framework model

A BIS in an explicit sense is initially based on the combination of a biological and a technological system. In order to realize an intelligent, i. e., autonomous behavior with regard to a production task, an information-technical networking is necessary. Here, the aim is communication between the biological and the technical system in order to enable autonomous control behavior. While the binary code[19] is still primarily used for processing digital information today, there is no standard for interpreting biological communication as a sign language or communication pattern. Basic considerations are made by biosemiotics. Fundamental research on this topic was conducted in the course of the collaborative research center (SFB) 563 (bioorganic function systems) of the German Research Foundation (DFG). Around the early 2000s, fundamental findings were generated at the interface between living biological systems (cells or cell models) and inorganic solid substrates (metals, semiconductors, insulators), which are already being used in various industries and contribute to the development of completely new products. In addition to the biofunctionalization of solid surfaces, this includes in particular the detailed spectroscopic characterization of the interaction between biological systems and their substrates as well as the experimental realization of bioelectronic hybrid systems, i. e., living cells that communicate with 'intelligent' substrates.

[19] Whereas one bit in today's data processing systems can adopt the value 0 or 1 (binary code), a new type of coding is necessary in the course of quantum computer technology, as the qubits found here simultaneously exhibit the values 0 and 1 (manipulable two-state quantum system). This induces the development of completely new algorithms.

Miehe et al. (2019a) transfer the generated findings into the system-theoretical basic model of a BTI, which is intended to serve as a fundamental pattern of biointelligence and consequently the development of sustainable innovations in different fields of application. In order to achieve an interaction of technical components with cellular systems, the type and level of interaction must first be differentiated. The former can basically be different physical principles (electrical, chemical, mechanical, optical). The interaction of the technical system can be carried out either with whole cells, organelles, single biomolecules or whole organisms (including humans). The exchange of information at the interface between biological and technical systems can be divided into sensory and actuator components. Actuators can be used to control the system from the technical to the biological side or vice versa. The sensor system comprises (physical, optical or spectrometric) measuring methods for the acquisition of state variables of biological or technical systems.

Well-known examples for the communication of biological cells with technical systems via electrical signals are neuron chips or multi-electrode arrays, which connect neuronal signals with electrical circuits (Wander and Rao, 2014; Paninski and Cunningham, 2018). Existing systems of this kind are already being used today for both in vitro and in vivo applications.[20] Due to their much higher sensitivity, sensor applications with neuron cells as data receiving components have a great potential for future applications, as they are expected to allow completely new functions, e. g., smelling or tasting (Agabi, 2017; Dolai et al., 2018).

Another example is the coupling of technical and biological systems via optical signals (optogenetics). Here, cells are specially equipped with photoreceptors so that cellular reactions can be controlled and read using light of different wavelengths (Pastrana, 2011; Williams and Deisseroth, 2013; Berndt and Deisseroth, 2015). Table 5.5 summarizes isolated examples of BTIs on different levels and types of interaction according to Miehe et al. (2019a).

While these examples are primarily applications that regulate biological systems using physical principles, BIS rely on real-time, bi-directional exchange of information. An example of a fully regulated system of this kind would be a bioreactor equipped with integrated soft sensors that allows continuous status monitoring and uses actuators to independently induce status changes in the bioprocess (Sommeregger et al., 2017). However, in order to ensure a robust functionality of such bioproduction systems, some research gaps still need to be closed (see Section 5.4.4).

[20] In vitro refers to organic processes that take place outside a living organism. Here the term is used to characterize that a measurement or experiment is performed in a controlled environment outside a living organism (i. e., in a test tube or Petri dish). In vivo, on the other hand, describes organic processes that take place inside a living organism.

Table 5.5 Examples of BTIs at different levels and types of interaction according to Miehe et al. (2019a)

Type of interaction	Level of interaction	
	Organism	*Cell*
Electrical	Human brain interfaces	Neuron chips
Chemical	Bacterial cells in the hydrogel	Protein receptors
Optical	Fluorescence and emotion recognition	Optogenetics, photoreceptors
Mechanical	Exoskeleton feedback	Atomic force microscopy

Prior to the introduction of the BTI framework, some preliminary considerations are necessary. Although generic models of the interaction between systems and their environment have been described by various authors in the life sciences, sociology, and control engineering (Vester, 1993, 2011; Luhmann, 1996; Bateson, 2002; Cameron et al., 2014; Gilles, 2004; Eifert et al., 2020), the close coupling of biological and technical systems under the assumption of autopoiesis, i. e., the process of self-creation and self-preservation of a complete system, has so far received little attention. In general systems theory, biological, technical, social, and mental systems are characterized by the fact that they produce and reproduce themselves materially (biologically), follow algorithms (technically), work sensibly (socially) and think (mentally) (Luhmann, 1996). To connect living and artificial systems in order to achieve an intelligent system behavior, an intermediate function is needed. Biological systems are autopoietic, self-regulating, adaptable and variable by origin – they follow the principle of 'actio et reactio' to maintain the biological balance. In contrast, technical systems are static and allopoetic. Their process regulation is developed by humans. Neither the biological nor the technical system can act by itself across logical contexts and observe its own behavior in contrast to its environment. Intelligent interface behavior, taking into account the linking of different contexts and logics, has so far only been observed in humans. Human adaptivity is a characteristic that can be interpreted as self-regulating network interaction at the interface of different systemic logics, namely biological, mental, and social systems, each of which functions in relation to its environment (Luhmann, 1996). The main task in system design across biological and technical systems is therefore to provide corrective variables to the technical system to support its development towards self-regulation (adaptability) and at the same time to create appropriate conditions for the biological system. For this purpose, it is necessary to be able to recognize both the external and the internal state of the highly variable biological system. The regulation of the entire system is equally crucial. Current approaches to the regulation of biological processes are usually non-linear model prediction controls, which are mainly used in process engineering to determine optimal control values. However, these models are not adaptive (Ashoori et al., 2009; Craven et al., 2014; del Rio-Chanona et al., 2016). Yet, this function is required, for example, to advance bioreactor systems by cell-based automation or to link individual technical solutions (bioprinters, biosensors and -actuators, self-learning algorithms) to SBMDs. Adaptive process control is indispensable for the direct coupling of biological and technical systems, since the behavior of biological systems can so far only be irritated from the outside, but not predicted accurately.

Figure 5.35 illustrates the BTI framework model in the form of an adaptive, systems-theoretical process control model between biological and technical systems according to Miehe et al. (2019a). The biological system could be an alga that produces different types of proteins. Its production rate depends on its nutritional status and light exposure. In order to achieve the production of a certain percentage of different proteins, the corresponding supply of temperature, light, and nutrients must be triggered by an intelligent control system based on trained soft sensors. Spectroscopic sensors may be used to detect the reaction of the algae and determine any possible nutrient deficiency. The BTI derives information from the sensor data, converts it into digital signals to control the technical system, which either provides the necessary inputs for controlling the biological systems or can use the information for parallel work processes. To map the dynamic variability of the biological system, statistical modelling, reservoir computing, and deep learning algorithms (see Section 5.4.2) may be used downstream of biochemical state recognition (Ferreira et al., 2013; Shang et al., 2014; Längkvist et al., 2014). The physical-mathematical models enable sensor simulation. Synthetic sensor data can be generated on the basis of robust and reliable state description models of biological objects. By applying surface defects to biological objects, the availability of synthetic sensor data sets could be scaled without limits. The special challenge of biological objects is their optical correlation with light for the visualization of defects. Continuous feedback loops are used to keep the production processes in an optimal state for the biological process by adapting their technical code characteristics. This is done by extending the range of biological behavior recognition and by extending the control possibilities. For the engineering sciences, the development of adaptivity in technology represents a major challenge. The adaptability of biological systems generally serves to maintain the balance between necessary flexibility and stability. For its transfer to technical systems, the soft sensor has to be extended by self-learning capabilities in a polycontextual environment driven by the logic of biology, technology, and information to define the functionality of the self-regulating balance in technical systems. Elementary components of BTIs are used as sensors, actuators, online and inline measurement systems, (AI) algorithms and network-based modelling (so-called digital twins).

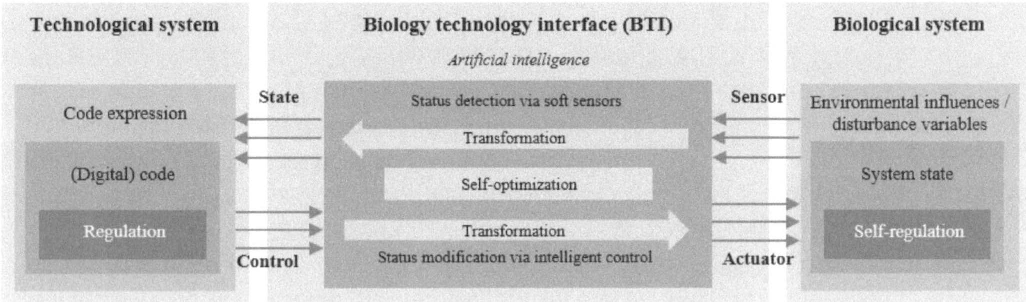

Figure 5.35 BTI framework model according to Miehe et al. (2019a)

5.4.4 Soft sensors as BTI transition technology

BTIs are intended to enable flexible structuring and regulation of the interaction between technical and complex biological systems. They form the basis for a wide range of applications, e.g., for the regulation of cell-based manufacturing processes and therapies and for intelligent orthopedics. However, the realization of BTIs is difficult today due to several fundamental challenges: first, the sensory detection of the current state of biological systems typically requires the analysis of very high-dimensional data in real time. Second, the modelling of the dynamic behavior and response characteristics of biological systems is a major challenge as they often contain a large number of internal state variables that are not directly observable. In addition, the properties of such systems change permanently through adaptation and learning. Direct modelling of such system properties is therefore often impossible. Third, the properties of biological systems are often individually specific, e.g., for the person concerned or individual differences in production batches of pharmaceuticals. BTIs thus need to be automatically adapted to such specific properties, typically on the basis of limited available data. In this context soft sensor technologies may be viewed as a vehicle of transition, as they address three fundamental aspects of BTIs. In order to provide the reader with a deeper understanding, this section outlines fundamentals of soft sensors, online and inline measurement systems, as well as AI-based state detection and control.

5.4.4.1 Basics and current application examples

The term soft sensor generally describes the combination of a software and a sensor (Fortuna et al., 2007). Other common terms are sensor fusion, inferential sensor, observer-based sensor or virtual online analyzer (Qin S. J. et al., 1997; Han and Lee, 2002; Goodwin, 2000). A soft sensor is not a single real sensor, but rather an estimator of correlating measured variables, which result from the measurement of several real in situ[21] sensors, to a control variable (Assis and Maciel Filho, 2000; Luttmann et al., 2012; Gustavsson and Mandenius, 2013; Mandenius and Gustavsson, 2015; Randek and Mandenius, 2018). In the context of bioprocess technology, the concept is directly related to online and inline measurement. A virtual sensor is defined by a model that describes the dependencies of the measured variables on the target variable. Thereby three model forms are distinguished (Hochrein and Alig, 2011). If all dependencies are known and can be represented by physical and/or chemical formulas, they are called white box models. An alternative term is rigorous modelling. If only some dependencies of the process are known, while others must be determined, the term grey box modelling is used. If the mathematical dependencies are completely unknown, it is called black box modelling. By superimposing various complex influences, the majority of biotechnical processes can only be described as approximations. This is due to the modelling and shortening, which requires simplification by making assumptions (Stachowiak, 1973). Similar to the hidden layers in neural networks, the term hidden structures is used for the unknown dependencies in grey and black box models. Besides the rigorous modelling of known dependencies, approaches of multivariate statistics (in particular analysis and regression methods) as well as learning models of AI (e. g., fuzzy logic, Kalman filter, neural networks) are becoming increasingly important. Figure 5.36 illustrates the principle of a soft sensor according to Luttmann et al. (2012) and Mandenius and Gustavsson (2015).

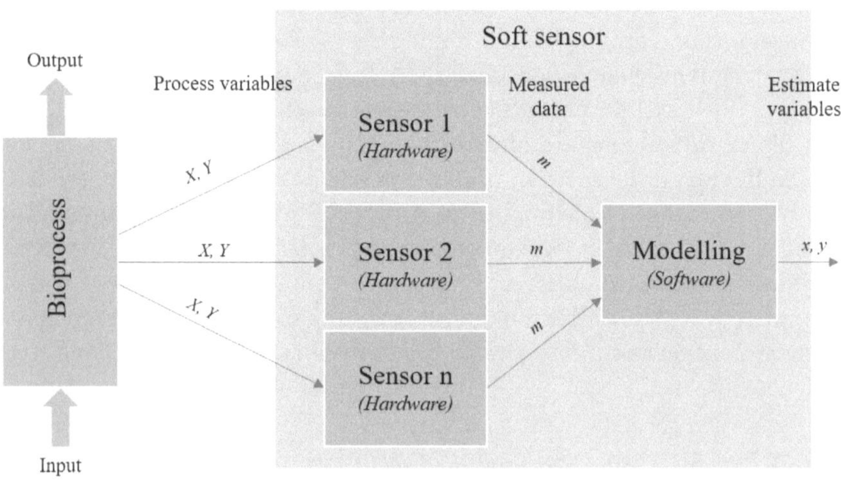

Figure 5.36 Functional principle of a soft sensor according to Luttmann et al. (2012) and Mandenius and Gustavsson (2015)

[21] In situ here indicates the positioning of the sensor directly at the location of the incident, i. e. directly in the process. In this sense, both inline and online measuring systems are in situ sensors.

While soft sensors are being used by several sectors of the manufacturing industry to improve and facilitate process flows, productivity, quality, yield, and safety (Randek and Mandenius, 2018), existing applications in biotechnology are slowly increasing due to sheer system complexity (Luttmann et al., 2012). Most known applications are based on online measurements from off-gas analyzers (Aehle et al., 2011; Luttmann et al., 2012). Very common are simple calculations of derived variables (e. g., carbon dioxide evolution rate, oxygen uptake, respiratory quotient). With more sophisticated techniques using estimators or observers, biomass concentrations, growth rates, substrate and product concentrations or substrate uptake and product formation rates are estimated (Luttmann et al., 2012). Luttmann et al. (2012) present a selection of examples of soft sensor applications (Table 5.6).

Table 5.6 Examples of soft sensor applications according to Luttmann et al. (2012)

Soft sensor application	Measurement device	Models
μ-estimation for control	Gas analyzers for O_2 and CO_2	Mass balance
μ-estimation with base titration	pH electrode, titrant volume	Mass balance
μ- and qP estimation	NIR, atline HPLC	Mass balance
Estimation by indirect measurements and correlations: OUR, CER, RQ, kLa-estimation	Gas analyzers for O_2, CO_2, dissolved O_2-probe	Mass balances, mass-transfer equation
Estimation through macroscopic balances: stoichiometric coefficient estimation	Gas analyzers for O_2, CO_2, titration, substrate conc., biomass composition (offline)	Mass balances, elemental balances, unstructured models
Extended Kalman filter for biomass, substrate and product concentrations, yield and growth rate estimation	Gas analyzers for O_2 and CO_2 as well as measurement for nitrogen uptake rate	Dynamic mass balances, unstructured models
Asymptotic observer to estimate C-source, N, biomass, product	Dissolved O_2-probe, flow $O_{2,in}$, flow $O_{2,out}$, volume, inlet fructose and nitrogen concentration (offline)	Dynamic component balances
Online detection of cell mass, viable cells, glucose, ammonia, acetate, and cell internal product	Gas analyzer for O_2 and CO_2, radio frequency impedance, 2D-fluorescence spectroscopy, FIA, HPLC	Dynamic mass balances PCA, ICA, MISO, MIMO, ANN

According to Luttmann et al. (2012), progress in the development of new and more robust online modelling analytical techniques will lead to improved soft sensors that deliver more accurate estimates of key parameters, such as growth and/or product formation rates. Bachinger and Mandenius (2000) as well as Turner et al. (2003) identify sensor arrays[22] in combination with pattern recognition systems in the gas and liquid phase (electronic nose and tongue) for the monitoring and control of bioreactors as one path to accomplish this. Among others, further applications of soft sensors are the determination of product and critical contaminants in downstream processes as well as measurement of product molecules in the production of biologics (Scheper et al., 1999; Luttmann et al., 2012). Table 5.7 summarizes a variety of applications from different industries according to Mandenius and Gustavsson (2015).

Table 5.7 Selected soft sensor applications in different industries according to Mandenius and Gustavsson (2015)

Goal	Application principle	Example of solution
Automation	Soft sensor as a part of feedback control system, estimating set-point of the controller	Set-point variable as quality variable determined by QbD (set-point derived from sensor signals Set-point
Real time release	Soft sensor provides quality parameters normally analyzed off-line	Allowing signal processing in existing process control system
Decrease use of material and energy	Soft sensor provides signals that ensure that raw materials are left unused and not converted to product	Signals from multivariate spectral data are incorporated in model used to estimate raw materials in the medium Signals from multivariate spectral data are incorporated in model used to estimate products and material-consuming by-products
Facilitate continuous processing	Soft sensor provides information for subsequent process steps on-line, enabling continuity of processing	Allowing soft sensor to be 'cloned' into another application
Prevent rejection of batches	Soft sensor provides signals that secures optimal state faster and maintained	Provided low cost, easy to discharge, in GMP/GLP

[22] The term sensor array describes a group of sensors that is deployed in order to add new dimensions to an observation, which a single sensor would not be able to measure and/or estimate. According to Albert et al. (2000), the sensors are usually grouped in a certain geometry pattern, collecting incrementally different, nonspecific data that require pattern recognition processing (via neural networks) to estimate specific variables.

Designing a sound soft sensor system that combines the appropriate analytical techniques with appropriate mathematical models and other calculations represents a challenging task (Randek and Mandenius, 2018).

5.4.4.2 Online and inline measurement systems as prerequisites

A large-scale use of soft sensors is currently primarily prevented by existing problems of online and inline analysis (Luttmann et al., 2012). Generally, bioprocess engineering distinguishes various methods of measuring the condition of a substrate.[23] Currently, the predominant methods are offline or atline measurement. Common to both approaches is that samples are taken at certain points in the biological process in order to subsequently prepare, measure, and evaluate them discontinuously. In the case of offline measurement, the examination takes place in an external laboratory. In the case of atline measurement, the laboratory is directly at the production line. Offline measurement often relies on light microscopy, requiring an extensive analysis process (preparation, resin, grinding, polishing, etching, analysis). The advantage of this approach is the high accuracy based on highly specialized equipment and expert knowledge in external laboratories, which in turn allows a high flexibility of the analysis. The disadvantage is the time required. Common methods of atline measurement are wet chemical analysis and X-ray diffraction analysis. Compared to offline analysis, a faster measurement is possible because these systems remove the sample semi-automatically and the laboratory is directly at the line. In both cases, however, continuous process control is not possible because the properties of the samples change, i. e., they cannot be reused in the active processes. For example, in flow cytometry the substrate is diluted and dyed, which means contamination of undesired substances. Figure 5.37 illustrates the difference between offline and atline measurement.

Compared to offline and atline systems, online and inline measuring systems not extract a sample from the process, as the analysis is carried out by sensors that are more or less placed in the system itself (Randek and Mandenius, 2018). The aim is to realize a continuous process control. Online systems require a branching which removes a certain amount of the substrate in a certain process step and returns it to the system after the measurement. Inline measuring systems, meanwhile, are applied directly in the actual process. Online approaches usually rely on X-ray diffraction analysis, while inline analysis is based on prompt gamma neutron activation analysis. The advantages of both systems are faster analysis results, direct connection to the process and automated procedures. The disadvantages are currently high investment and mainte-

[23] Depending on the control variable, different measurement technologies are suitable. The measurement of optical density, impedance, in situ microscopy, online flow cytometry is used to determine the condition of biomass. For the measurement of the bioactivity of a substrate, calorimetry, gas equalization techniques, in situ fluorescence monitoring, online flow cytometry as well as size determinations of particles and solutes are used. The components dissolved in the medium are determined, e. g., by flow injection analysis, chromatography, capillary electrophoresis, mass spectrometry or infra-red spectroscopy.

nance costs, greater vulnerability to failure and lower accuracy compared to offline and atline systems. Figure 5.38 illustrates the difference between online and inline analysis.

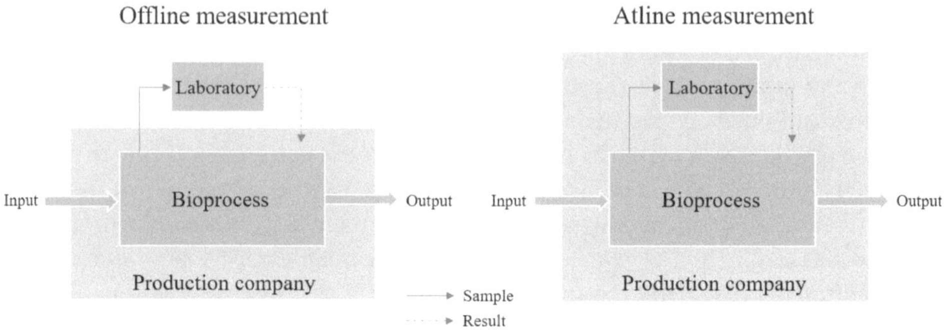

Figure 5.37 Offline vs. atline measurement of bioprocesses [own illustration]

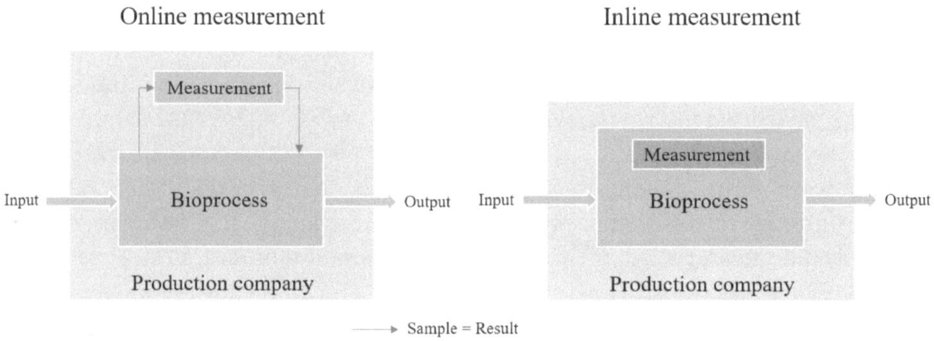

Figure 5.38 Online vs. inline measurement of bioprocesses [own illustration]

While very precise results are obtained with offline and atline analysis, online and inline measurements are often too inaccurate due to variations in dilution. Hence, the detection of alternative measurement parameters is required. One example is the fermentation process. In an offline laboratory, various concentrations of substances could be determined exactly, which would require a great amount of time (Assis and Maciel Filho, 2000). Online and inline systems therefore measure several specific parameters[24] in order to indirectly deduce a control variable (concentration). Supporting measurement instruments are near infrared (NIR) and mid-infrared (MIR) spectrometry, in situ microscopy (ISM), 2D fluorescence spectroscopy and high-performance

[24] Such parameters may be physical (stirrer flow, aeration flow, vessel weight, electrical power consumption, gas/volume/mass flow rate, etc.) or chemical parameters (protons, dissolved gas/redox pairs, pH, pCOs, etc.).

liquid chromatography (HPLC) (Scarff et al., 2006; Rhiel et al., 2002; Dietzsch et al., 2013; Randek and Mandenius, 2018). Table 5.8 summarizes selected online measurement techniques for different examples of analytes or state variables according to Luttmann et al. (2012).

Table 5.8 Selected online measurement techniques for different analytes and state variables according to Luttmann et al. (2012)

Category		Measurement device	Examples of analytes or state variables
In situ sensors/ analyzers		Physical in situ sensors/ probes	Temperature, pressure, volumetric or mass flow rates (gas & liquids), weights
Electrodes		NIR/MIR in situ probes	Media components, biomass
		Multiwavelength fluorimetry	NAD(P)H, amino acids
		Turbidimetry	Biomass
		In situ microscopy	Cell shape etc.
		Radio-frequency-impedance	Biomass, cell viability
Atline sensors/analyzers	*Liquid phase analysis*	High performance liquid chromatography	Media components (organic acids, amino acids, saccharides), recombinant proteins
		Flow injection analysis (FIA) with various detection methods	Media components, extracellular products
		Electrode-based biosensor	Enzymatic substrates (e. g., media components)
		Immuno-based biosensor	Antigens (e. g., proteins)
		Flow diffusion analyzer	Glucose, methanol
		Inline microscopy	Image analysis
		Coulter counter	Cell number, cell size distribution
		Flow cytometry	Cell viability, cell constituents (e. g., DNA, RNA, proteins, lipids, membrane characteristics, Intracellular pH, redox)

Table 5.8 Selected online measurement techniques for different analytes and state variables according to Luttmann et al. (2012) (*continued*)

Category		Measurement device	Examples of analytes or state variables
Atline sensors/analyzers	*Off-gas analysis*	Paramagnetic oxygen analyzers	O_2
		Infrared adsorption photometers	CO_2
		Gas chromatographs	O_2, volatile compounds
		Mass spectrometers	Low molecular weight compounds
		Flame ionization detectors	CH_3OH, CH_4
		Electronic noses	Microbial volatiles (ethanol, sulfides), infectants

In addition to measurement, a modelling of dependencies that are often based on non-linear relationships is required. Basic literature on state observation of non-linear systems is provided by Adamy (2012) and Khalil (2014). According to Mandenius and Gustavsson (2015), soft sensor design mostly relies on multivariate data analysis (MVDA). Usually, steady-state and dynamic models are distinguished. The former are developed from mass and component balances, mass or heat transfer laws, element balances or as purely empirical black box models, while the latter typically consist of dynamic balances in combination with kinetics (Luttmann et al., 2012). Dynamic models can be further classified into structured and unstructured designs. Structured models are used to account for changing yield coefficients, e. g., complex behavior in cultivations. Thus, these models can simulate a wide range of complex interactions and are used for example to calculate optimal nutrition profiles and other control strategies (Luttmann et al., 1994). Unstructured models are often applied for estimator-based or observer-based soft sensors due to their simplicity. Their major disadvantage compared to structured models is the use of constant flow coefficients. In most cases soft sensor models are based on artificial neural networks (Glassey et al., 1994). Common technologies for these types of models are artificial neural networks (ANN). In any case, extensive calibrations are required to adjust the parameters in MVDA and ANN models to the actual measurement situation, e. g., the analytes, the culture, and the media environment (Mandenius and Gustavsson, 2015). Yet, MVDA and ANN techniques are already utilized and are able to provide many alternatives for data prediction. Structured and unstructured mechanistic models, partial-least-square (PLS) regression model, ANN model, hybrid model, and transfer functions are used. Mechanistic models have the advantage that they can easily be linked to actual biological events and processes or recognized theoretical models in biotechnology. On the other hand, they usually have the disadvantage that they can only represent a part of reality, require parameterization and/or extensive knowledge about biological complexity. The most common modelling methods for soft sensors are PLS and ANN. They are commonly deployed in combination with electronic noses and/or tongues, where gas sensors and ion-sensitive electrode signals serve as input variables, e.g., being linked to media analyte concentrations, expressed proteins or optical spectra via modelling. While both approaches are fairly easy to set up, they require considerable effort for validation, recalibration, and/or (re)training. Combinations of modelling approaches are called hybrid models, which can be used whenever both mechanistic and regression functions need to be simulated. Yet, their understanding and validation often proves to be difficult. Mandenius and Gustavsson (2015) summarize the pros and cons of different modelling approaches outlined above based on recent literature (Table 5.9). In doing so, they note that it depends on the preferences of the user which approach he or she chooses.

Table 5.9 Advantages and disadvantages of different soft sensor modelling approaches according to Mandenius and Gustavsson (2015)

Model type	Advantages	Disadvantages
Unstructured mechanistic model	▪ Models based on first principles and may by that provide understanding	▪ Models do not mirror the full biological complexity ▪ Parameterization required ▪ Parameters must fit the design space
Structured mechanistic model	▪ Has opportunity to render increased understanding of the process	▪ Extensive biological knowledge about biological complexity required
Partial-least-square regression model	▪ Methodology fairly easy to set up	▪ Extensive validation of parameters required ▪ Non-explanatory ▪ Must be recalibrated more frequently
Artificial neural network model	▪ Easy to set up in existing software with established ANN methodology	▪ Require training for setting weight parameters ▪ Must be retrained frequently
Hybrid model	▪ Take advantage of best of mechanistic and regression models	▪ May be difficult to grasp and validate
Transfer function	▪ Straightforward, easy to implement ▪ Transparent ▪ Normally few parameters	---

5.4.4.3 Industry requirements and implementation

When developing a soft sensor, it is necessary to exactly understand the actual needs within the industrial process where it will be used (Randek and Mandenius, 2018). These needs should then guide and constrain the soft sensor design to find the configuration of analytical techniques and calculation methods that best meet these needs. Critical needs and requirements of contemporary large-scale bioproduction for online monitoring include robustness, low investment and operational costs, reliability, simplicity, precision, trainability, flexibility, durability, as well as the possibility to measure gases, substrates, impurities, main and side products (Randek and Mandenius, 2018). Depending on the process, the engineer may face completely different conditions. For instance, microbial processes work with higher cell densities, while mammalian cell processes operate with lower densities (up to 1000 times less). Mammalian cells need different, often complex nutrient supplements, while bacteria and algae have less demand for such complex supplements. Then, mammalian cells are more sensitive to nutritional and environmental changes than bacteria. Bacteria respond quickly to changes, while algae are slower (Randek and Mandenius, 2018). In order to implement a soft sensor in an industrial environment, parameterization is required. Thereby, the already realized level of automation of a process is crucial. According to Luttmann et al. (2012), the basic rule is that soft sensors that require more than a few basic calculations should be separated from the automation system and run on separate dedicated computer hardware. Communication may be realized via industrial ethernet or fieldbus. Randek and Mandenius (2018) identify six steps for implementation of a soft sensor in a biomanufacturing environment (Table 5.10).

Table 5.10 Basic steps of soft sensor implementation in biomanufacturing environments according to Randek and Mandenius (2018)

	Purpose/goal	Key actions	Decisions
1	Define objectives and targets of soft sensor setup	Clarify objectives and compare them with existing analytical methods in the company. Set analytical constraints for soft sensor (analytes, range, delay, stability, precision, automation, etc.)	Decide if objectives and targets are realistic and lead to substantial benefits in manufacturing
2	Identifying alternative soft sensor setups	Compare alternatives based on available information on performance of different techniques (hardware and software methods)	Select preferred soft sensor setup(s) based on ranking of alternatives
3	Prove feasibility of chosen soft sensor alternatives at R&D level	Test soft sensors in process R&D lab using IQ/OQ/PQ protocols, establish SOP for soft sensor	Decide whether established criteria for qualifying soft sensor are met satisfactorily

	Purpose/goal	Key actions	Decisions
4	Implement soft sensor	Install soft sensor in line and implement soft sensor algorithms in process control software	Decide upon further testing based on technical performance in the plant
5	Prove feasibility of soft sensor setup at plant	Perform feasibility testing in manufacturing process according to previous IQ/OQ/PQ protocols	Continue with GMP validation if performance requirements are met
6	Validate soft sensor setup according to regulatory requirements and GMP	Run a validation program with soft sensor	Consider redesign based on outcome, approve or disapprove

Figure 5.39 summarizes the implementation of a soft sensor in a regulatory environment of a bioprocess including inline measurement and process control with regard to productivity, yield, and economy parameter according to Randek and Mandenius (2018).

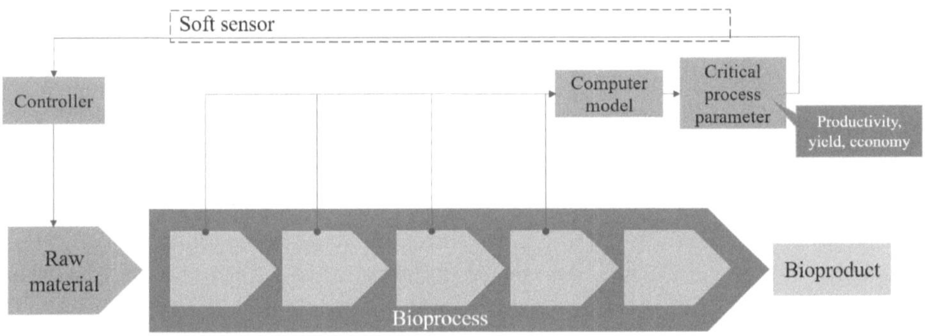

Figure 5.39 Implementation of soft sensor in a regulatory environment of a bioprocess according to Randek and Mandenius (2018)

5.4.5 Digital twin design for biology-technology-systems[25]

The concept of digital twins (DT) is increasingly gaining importance in the context of I4.0. Although the concept is currently used primarily in unit goods production, it is also emerging in biotechnology, bioprocess engineering, and medical technology. DT is also of particular importance in the context of biointelligence, in which the possibil-

[25] This section is a summary of the findings presented in Miehe et al. (2020b).

ity of exchanging information across cell boundaries is a precondition for the spontaneous and/or planned merging, cooperation, and diverging of different cells in order to fulfil a production task.

Compared to a CPS referring to an intersection of informational and technical systems (Lee, 2008; Rajhans et al., 2009; Rajkumar et al., 2010), BIS go one step further by adding a third system, the biological (Miehe et al., 2018; Miehe et al., 2019a; Miehe et al., 2020a). As discussed above, a production system can be described as biointelligent in an explicit sense, if it is based on biological principles, comprises a biological and a technical component, guarantees an interaction as self-sustaining regulating network (biology-technology interface, BTI), and shows an overall intelligent system behavior. Comprehensive data exchange and intelligent control are also essential features of CPS which are expected to be realized through the concept of a digital twin (DT), i. e., the digital representation of a physical system (Uhlemann et al., 2017a; Uhlemann et al., 2017b). The same must apply to BIS, whereas here, a digital representation for both the technical and the biological system is inevitable. Technical and biological systems, however, exhibit highly different structural features. While research and standardization have slowly agreed upon a uniform structure of technical systems' DTs, the increasing integration of biological systems represents a major challenge in terms of complexity. The concept of DTs for biology-technology-systems will continue to occupy the scientific community for years to come. Hence, this section outlines the basics of the DT concept and discusses some basic thoughts on the development of DTs for BIS.

5.4.5.1 Terminology of DTs

The concept of the DT was first mentioned by Grieves in 2003 in the context of product lifecycle management (PLM) (Grieves, 2014). The NASA adopted the term and defined it as an integrated all-physical, multi-scale, probabilistic simulation of a system applying the best available models, sensor updates etc. to reflect the lifetime of its real-life counterpart. The DT is ultra-realistic and can consider one or more important and interdependent vehicle systems (Shafto et al., 2010). While the NASA definition was primarily adopted to the aerospace sector, it utilized Grieves' approach by highlighting three major components of a DT: the real-world product, the virtual representation of it and a data flow that connects both. In the early 2000s virtual representations of products and processes were still in their infancy and production data was mainly collected manually. Around a decade later, sufficient automatically recorded data on products and processes was available. Still, there was no connection between the data sources (Grieves, 2014).

In recent years, sensing, processing, and communication technology have evolved considerably, bringing the concept of the DT one step closer to implementation. To date, the concept of a DT has been proposed for a large number of sectors, including the manufacturing industry (Negri et al., 2017). Recent scientific literature thus provides a broad range of definitions of DTs as well as various descriptions of how to implement them (Anderl et al., 2018). In this context, Kritzinger et al. (2018) present a comprehensive literature review of available DT concepts, resulting in a distinction of three stages of system integration: the digital model (DM) represents a product virtually, while not exhibiting an automated flow of information between the product and the model. The second stage, the digital shadow (DS), differs from the digital model by the incorporation of an automated upload of information from the real-world product to the virtual one. Finally, the DT enables a bi-directional information flow in real-time between both systems (Figure 5.40; Kritzinger et al., 2018).

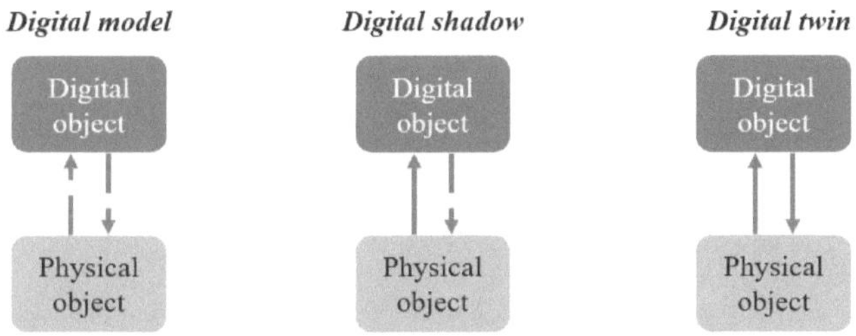

Figure 5.40 Stages of system integration by types of data flow between physical and digital object according to Kritzinger et al. (2018)

To date, however, the three terms are largely used as synonyms (Kritzinger et al., 2018). This applies in particular to the term DS and DT. From a manufacturing perspective, the DS can be understood as an accurate image of production, research and development, and adjacent processes ('as good as necessary') (Bauernhansl et al., 2016). Thus, the DS is primarily an instrument to transfer the real-world into the virtual one (Bauernhansl et al., 2018). The DT, in contrast, aims to use simulations and process models to generate an image of reality that is as accurate as possible ('as good as possible'). The DT, therefore, goes far beyond the mere simulation of a physical structure (Wagner et al., 2017; Hankel, 2017; Seif et al., 2019). As such, it contains a wide variety of information along the entire value chain that enables an autonomous production function, e. g., material composition, technical drawings, work plans, machining instructions, fasteners, parts lists, transport routes, and/or emissions. Various authors, however, agree that the development of DTs has just started (Negri et al., 2017; Anderl et al., 2018; Kritzinger et al., 2018). While the term is quite well defined, data connectivity, which is an essential prerequisite for system interoperability, is currently seen as the most difficult hurdle for implementation. Further conceptual development and real-world use cases are required in order to establish sufficient data connectivity (Anderl et al., 2018).

5.4.5.2 Asset administration shell as basic architectural model

In order to overcome the data connectivity issue, the German Platform Industrie 4.0 and the German Electrical and Electronic Manufacturers' Association (ZVEI) developed a standardized architectural model of a DT that is referred to as the asset administration shell (AAS) (Plattform Industrie 4.0, 2019d). The model aims to guide companies in implementing DTs and to ensure interoperability between devices, components and applications across different business entities. The AAS is a virtual representation of any real-world asset that contributes to an I4.0 production task, e. g., items, software or documents (Plattform Industrie 4.0, 2019d). As such, it contains all available information on properties, status, parameters, functions, and references of the asset, which are each stored in several AAS submodels. Each submodel is referring to a specific aspect, e. g., the 'positioning mode' of an axis, and may contain elements 'position [0-2]' and 'average positioning error [mm]'. The submodels and their connections within the superior model layer are organized by a generic structure specified in the AAS metamodel. Submodel features may also have external references and references to local dictionaries (Plattform Industrie 4.0, 2019a). Figure 5.41 exemplifies an XML code of a sub-model element including a reference according to the German Platform Industrie 4.0 (2019a). References may point to documents or websites that contain further explanation about the element, which is expected to help applications 'interpret' the given information semantically, enabling to apply the correct functions on the element.

```
[...]
<property>
  <idShort>>Nmax</idShort>
  <category>PARAMETER</category>
  <description lang=„EN“>maximum rotation speed</description>
  <description lang=„DT“>Maximale Drehzahl</description>
  <semanticid>
    <keys>
      <key local=„false“type=„GlobalReference“
        idType=„IRDI“>0173-1#02-baa120#007</key>
    </key>
  </semanticid>
  <valueType>double<valueType>
  <value>2000</value>
</property>
[...]
```

Figure 5.41 Exemplified XML code of sub-model element including reference according to the German Platform Industrie 4.0 (2019a)

Next to the IDs (global) and shortIDs (local) that explicitly define an AAS, asset, submodel, and property, semanticIDs are constructed for semantic references from submodels and properties. Figure 5.42 illustrates the identification, addressing and semantic referencing of an AAS according the German Platform Industrie 4.0 and the German Association for Electrical, Electronic and Information Technologies (Plattform Industrie 4.0, 2019b).

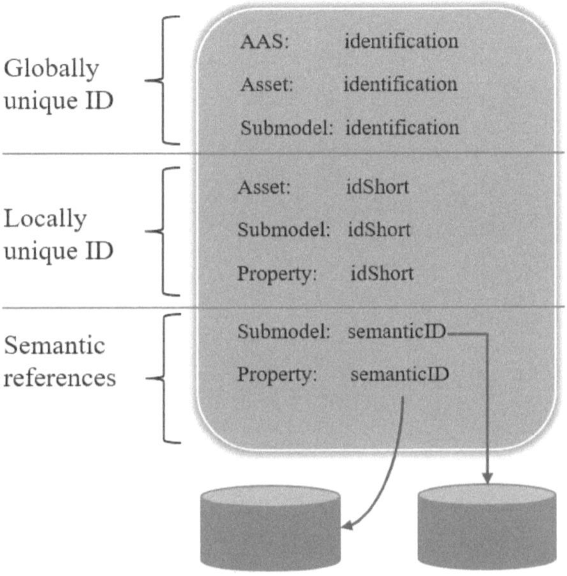

Figure 5.42 Identification, addressing, and semantic referencing of an AAS according to the German Association for Electrical, Electronic and Information Technologies (Plattform Industrie 4.0, 2019b)

As exemplified in Figure 5.43 , the combination of an asset and its AAS is referred to as an I4.0-component. It should be noted that both individual machines and entire device systems can be assigned an AAS, depending on the hierarchy level at which the system is situated. While AAS for assets of lower hierarchical levels, such as machines, or even sub-assemblies and individual parts, include specific attributes to their functionality, AAS for assets of higher hierarchical levels aggregate complex information relevant to the overall device system. The connection between the AAS and the asset is defined in a sub-model called the topology model. It connects the manufacturing device system to the AAS of the turning machine, but also to further devices such as control devices, milling machines, etc. This allows the AAS of the manufacturing device system to be directly coupled to the I4.0 system, making it the single I4.0 component and reducing the overall implementation effort. The interconnection between I4.0 components and AAS at different hierarchy levels on the one hand and their respective assets on the other hand requires seamless communication. Hence, AAS enable information transfer via various communication channels and applications, forming the linkage between locally distributed real-world assets throughout the complete life cycle. AAS thus have to ensure interoperability by a number of characteristics. The model consists of a technology-neutral unified modelling language (UML) model and is editable in extensible markup language (XML) and JavaScript object notation (JSON) formats through the AASX (secure exchange format for AAS) package explorer. On this basis, assignments to other standards such as open platform communications united architecture (OPC UA) (widely used in production and maintenance), automation markup language (AutomationML) (widely used in engineering) and resource description framework (RDF) (commonly used for ontologies) are derived and schemes for serialization are specified (Plattform Industrie 4.0, 2019d; 2019c). Together this leads to an AAS architecture model as presented by the German Platform Industrie 4.0 and the German Electrical and Electronic Manufacturers' Association (ZVEI) (Figure 5.43) (Plattform Industrie 4.0, 2019a; Plattform Industrie 4.0, 2019c; 2019d).

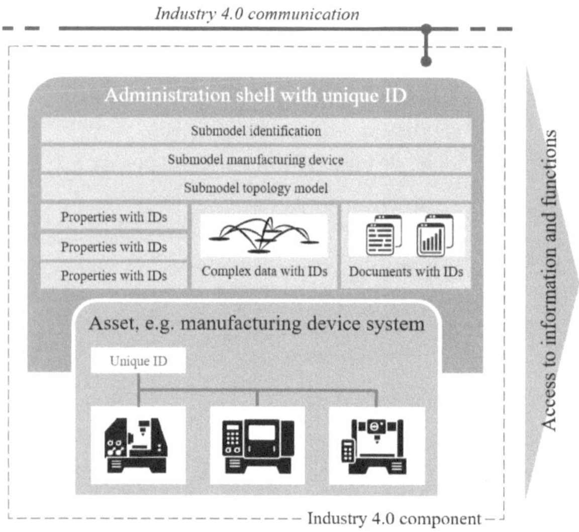

Figure 5.43 AAS architecture model [own illustration according to the German Platform Industrie 4.0 and the German Electrical and Electronic Manufacturers' Association (ZVEI) (Plattform Industrie 4.0, 2019a; 2019c; 2019d)]

In order to summarize and uniformly present data types as well as communication standards, the German Platform Industrie 4.0 and the German Association for Electrical, Electronic and Information Technologies presented the hierarchical Reference Architecture Model Industrie 4.0 (RAMI 4.0). Figure 5.44 illustrates the RAMI4.0 model according to the German Platform Industrie 4.0 and the German Association for Electrical, Electronic and Information Technologies (Plattform Industrie 4.0, 2019b).

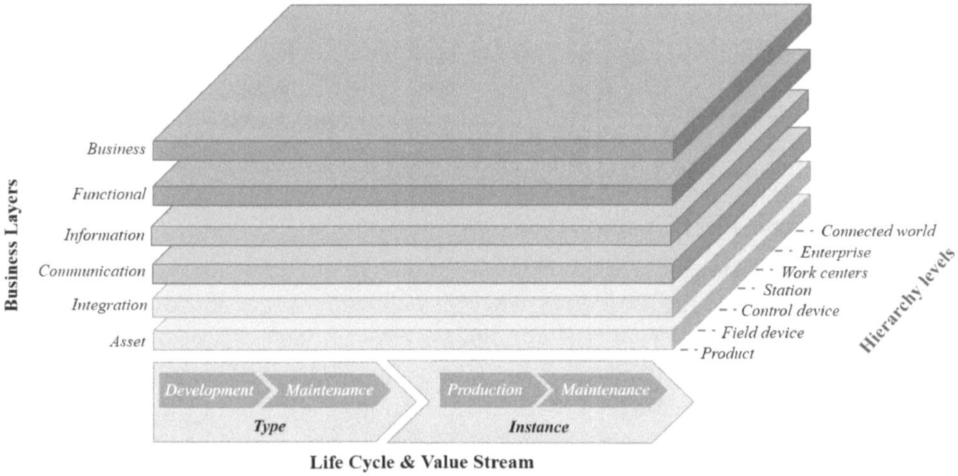

Figure 5.44 RAMI 4.0 model according to the German Platform Industrie 4.0 and the German Association for Electrical, Electronic and Information Technologies (Plattform Industrie 4.0, 2019b)

The model distinguishes between three dimensions. The y-axis summarizes the different layers of IT representation, aiming to visualize the information flow within in the IT-architecture. Assets such as machines, products or employees are connected to the IT-system via an integration layer that gathers all acquired data and makes it available for higher levels such as communication or information storage. The functional layer adds the possibility to automatically execute actions based on the incoming data. The x-axis represents various steps along the product life cycle and value streams based on IEC 62.890. When the developed product is launched into production it changes from 'type' to 'instance'. On the z-axis the hierarchical levels of the integration of ERP and control systems are derived from IEC 62.264/61.512. These levels represent the various functionalities that the assets may serve. The RAMI4.0 structural model and the I4.0 components form the core of a DT of a manufacturing system. While the considerations listed above are primarily discussed in German-speaking countries, related models are documented in international literature in English and outline a variety of similarities of the DT concepts of Germany, US, China and France (Hankel, 2017).

According to Seif et al. (2019), the introduction of AAS to real-world manufacturing facilities requires a three step approach. In a first phase, it is essential to deploy the AAS structure and integrate it into day-to-day business before actual advantages of DTs can be exploited. In this way, isolated solutions are avoided and the basis for communication capability and interoperability is ensured. In the second phase, after communication structure setup is completed, the functions can be extended to allow autonomous interaction of AAS without any central control and to enable the system to optimize on its own. The third phase comprises the deployment of the AAS via various implementations (OPC UA SDK, RESTful API, etc.) of data bases, communication protocols, and IoT integration. The final goal is to set up the communication between the real-world asset and the AAS. Use cases of AAS model deployment are documented by Kalhoff et al. (2017) as well as Tantik and Anderl (2017).

5.4.5.3 Examples in life science

The integration of biological systems into technical manufacturing systems requires an increasing focus on the life sciences. Recent publications that address DT solutions in the context of biomanufacturing and bioprocess engineering are closely linked to quality by design (QbD) and process analytical technologies (PAT) approaches (Zobel-Roos et al., 2019). Bioprocess DTs, capable of integrating atline and online bioprocess data, are used not only for virtual experiments to characterize, understand, and optimize bioproduction processes, but enable predictive monitoring and control (Nargund et al., 2019). Others describe examples for a DT in specific biomanufacturing areas, such as a concept for an aquaponic system (water circulating between fish tank and plant agriculture) (Ahmed et al., 2019). Thereby, several sensors (e. g., pH, EC, water level, temperature, water flow, humidity, light intensity) gain data which are fed into a simulation model. The validation shows that the model is capable to estimate some parameters well, while others (nitrite content, plant growth rate) could not be predicted precisely.

In healthcare, the term DT primarily relates to patient monitoring and disease treatment. DT of the coronary heart vessel apply data such as flow velocity and flow pattern, properties of the blood and the material of the vessel walls to predict the behavior of the vessel (Naplekov et al., 2018). With the help of hemodynamics and mechanical dynamics, areas of high shear stress, e. g., at junctions and valves, are identified, which helps to prevent heart failure and disease. The concept presented here, however, rather refers to a DM in classical manufacturing. Another DT concept of a human heart demonstrates a new way of detecting ischemic heart disease based on electrocardiography signals. Signals are sent from nodes to an edge computing infrastructure, which then classifies them by an artificial intelligent algorithm into non-myocardial and myocardial conditions (Martinez-Velazquez et al., 2019). The concept outlined here also represents a DS rather than a DT in the sense of I4.0. Furthermore, DT concepts for personalized medicine and virtual oncology have been discussed intensively in the community (Bruynseels et al., 2018).

As exemplified by bioprocess DT, one key challenge in the development and application of DTs with biological systems is the requirement to integrate large amounts of data (Zhang et al., 2019). Digital transformation in bioprocess industries is principally facing similar issues as classical manufacturing, such as data collection and handling, connectivity or process control algorithms. Model-based online optimization has not been widely applied to bioprocesses due to challenges in modelling complex biological behaviors, low-quality industrial measurements, and the lack of visualization techniques.

Mechanistic and statistical models are available to predict biological processes, each bearing their own advantages and disadvantages. Mechanistic approaches use knowledge about a specific system, such as a metabolic model representing the biochemical pathways of cellular metabolism, can dramatically reduce the amount of data required compared to pure data-driven approaches such as statistical modelling or machine learning. On the other hand, mechanistic models are often not capable of describing all parameters of interest and need to be complemented by data-driven approaches for factors that are not addressed mechanistically. Hybrid models integrating mechanistic and data-driven components have successfully proven to provide unprecedented predictive accuracy on the basis of limited data sets. A hybrid modelling framework has been suggested in order to exploit the benefits of both, the object-based and data-driven model, for monitoring and optimization purposes (Zhang et al., 2019; Narayanan et al., 2019).

In summary, the terms DS and DT are understood differently in biotechnology, bioprocess engineering, and medical technology than in a manufacturing context. Consequently, recent literature in this field primarily describes simulation and modelling approaches, i. e., DMs. Real-time measuring and data acquisition, a close linking of biological and informational system or even active interventions in the bioprocess have yet not been addressed sufficiently. In addition, a far more fundamental challenge is that there is no uniform concept for data structuring in a biomanufacturing context so far. Data alignment initiatives have been launched widely in the biotech industry. However, in the course of the increasing integration of biological systems in classic production environments, new approaches are required here.

5.4.5.4 Transferring the AAS architecture to biological systems

The literature review above reveals a considerable gap in application-oriented research and the current industrial state of the art, not only for data integration issues. Missing standards for data transfer and DT models severely hinder interoperability, such as the cooperation or transfer of information beyond processes and/or company boundaries. In order to enable interoperability for biointelligent manufacturing systems, the transfer of the AAS architecture, together with the different levels of AAS, submodel and data elements, offers a promising approach. According to Miehe et al. (2020b), this essentially requires a structured and standardized approach in the generation of DT models (Figure 5.45). Each biological system as well as its subsystems contain an AAS, one or more sub-models, and different data elements that have to be developed or integrated.

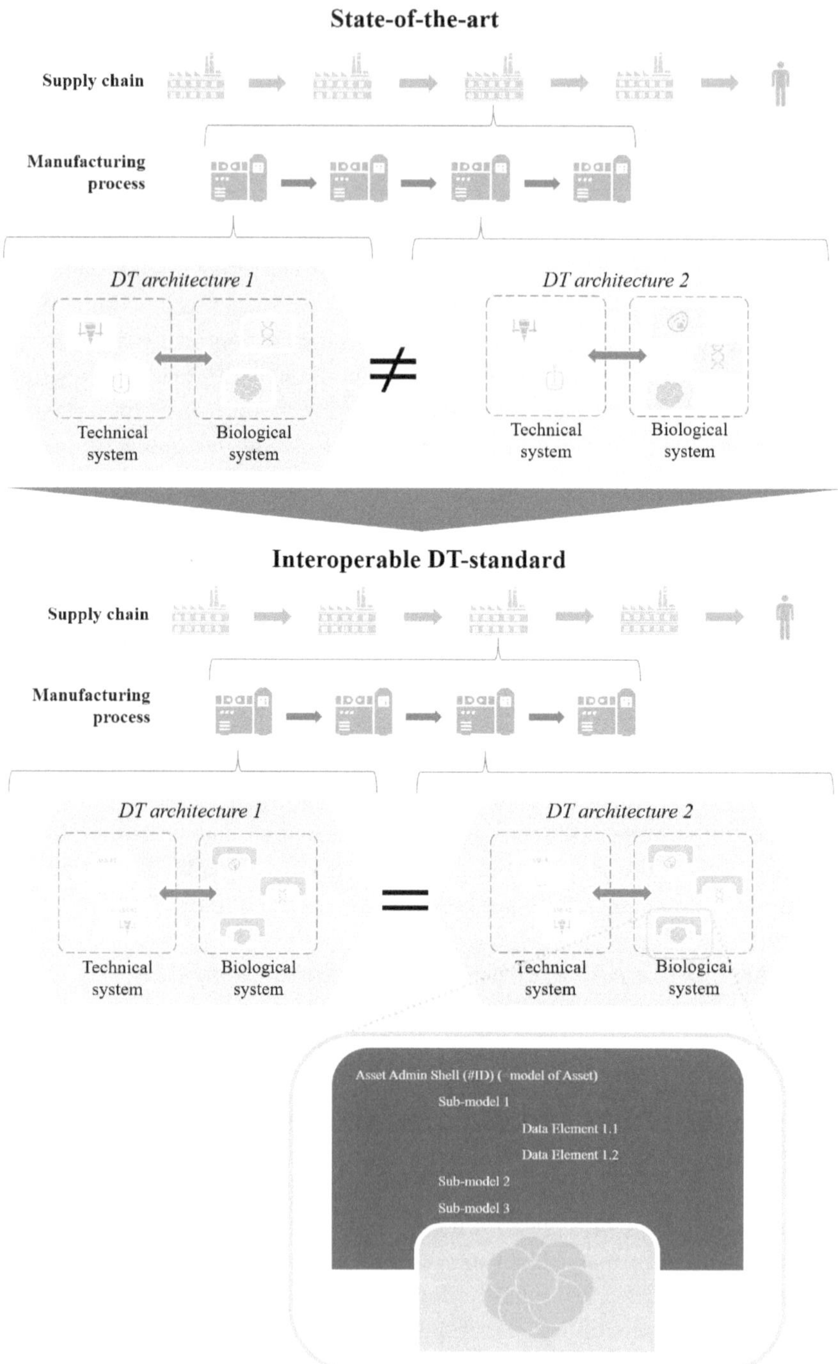

Figure 5.45 From current industry approach to future interoperable industry standard according to Miehe et al. (2020b)

To enable the transfer of the AAS architecture to biological systems as part of a production process, the information of the biological system that can or should be integrated into an AAS has to be identified. Yet, the structure of technical systems cannot be transferred directly to biological systems, since the entities are very different with regard to characteristics and organization. Technical systems are developed by humans in order to serve a specific purpose. They consist of inanimate, material-specific matter as well as a clear functional principle for the conversion, transport or storage of matter energy and/or information (Eifert et al., 2020). Generally, they appear as a spatial structure with a geometrically defined shape, they are static and allopoietic. Biological systems in contrast are autopoietic, self-regulative, adaptive, and variable (Görtz and Brümmer, 2012). They follow the principle of 'actio et reactio' to keep the biological balance. Biological systems maintain an internal state of low entropy, exhibit an ongoing energy exchange and/or metabolism. They are able to reproduce themselves, evolve, and communicate (Luhmann, 1996; Fritsche, 2015). In order to transfer the classical manufacturing DT architecture to biological systems, a classification of relevant features is necessary. According to Miehe et al. (2020b), living nature exhibits different structural layers, from single ions or elements to complex multicellular organisms (Figure 5.46). The basic unit of higher life is the cell. The components of a cell are similarly structured in all organisms. Whether plant cell, animal cell or microorganism, the most important components consist of conserved molecular structures and fulfil the same functional tasks. Lipids are involved in building cell membranes, carbohydrates form the basis for metabolism and nucleic acids are building blocks for storing genetic information. In all living organisms, the storage of the building blocks and the organization of the organism is based on the combination of just four nucleic acids, the genetic code. The next larger unit of life is the combination of several cells, ranging from a two-cell microorganism to complex living beings such as humans.

To fully describe biological behavior in ordered DT models the structure of life alone is, however, not sufficient. This also requires the mapping of functionalities and exchange relationships between different biological entities. Parameters that control of growth, proliferation, and differentiation of cells include soluble factors (e.g., nutrients, growth factors, hormones), cell architecture and polarity patterns (e.g., epithelial tissue culture, basolateral nutrient supply), cell-cell interactions (e.g., cell density, confluence, cell connections) and cell-matrix interactions (e.g., adherence and motility of cells, morphogenesis, surface discharge). Those extracellular clues regulate/interfere with intracellular responses such as metabolomic, transcriptional, translational or cellular transport events. In summary, structural elements of biological systems fulfil certain tasks/functions within larger biological processes. The structured description of all three layers is important for the development of standardized DT models (Table 5.11) (Miehe et al., 2020b).

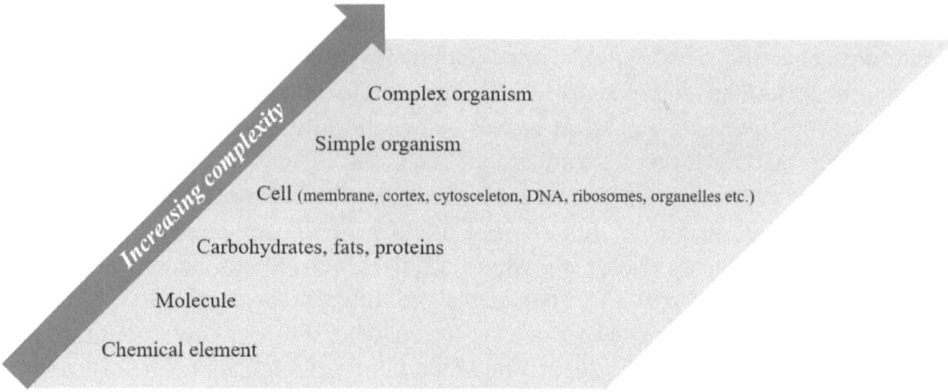

Figure 5.46 Exemplary selection of structural layers of life indicating important structural elements within cells and organisms according to Miehe et al. (2020b)

Table 5.11 Exemplary selection of structural and functional attributes of biological systems with potential relevance for a DT according to Miehe et al. (2020b)

Structural attributes	Functional attributes		
	Cell level	Organelle level	Molecule level
Molecule Cell Organelle Organism	Cell division Cellular homeostasis Differentiation Cell adherence Migration/Motility	DNA replication Transcription Cellular Transport Posttranslational modification Translation Metabolism Signal transduction	Mass transport Energy transport Conformational change Chemical modification

While future technologies will have to be able to record information and data on all these features, their current significance for the DT is determined by their application and the current state of *measurability*. If the focus is feasibility, a transfer of the RAMI 4.0 architecture represents a practical alternative compared to ASS models. While the y-axis of the RAMI 4.0 architecture (layers) is directly assignable to a biological system, it is unclear what the respective biological counterparts of the x- and z-axis levels (life cycle and value stream and hierarchical layers) are. A practical approach to the digital representation of a biological system in a DT would be the allocation of the structural and functional levels of life outlined above to the RAMI 4.0 architecture. The complexity of a biological system, however, requires an adaptation with regard to scalability. On the one hand, this is due to the fact, that the dimensions of the RAMI 4.0 architecture contain both structural and functional levels. On the other hand, biological systems have a much higher number of structural and functional levels than technical systems. Thus, the selection of the functionalities to be depicted should be based on their relevance in the production context, i. e., which functionality within a biological system actively contributes to the production output. Figure 5.47 illustrates how an example of a biological system with three process steps of protein production in a bioreactor system, including transcription, translation, and posttranslational modification, might be classified in a manufacturing context according to the RAMI 4.0 architecture.

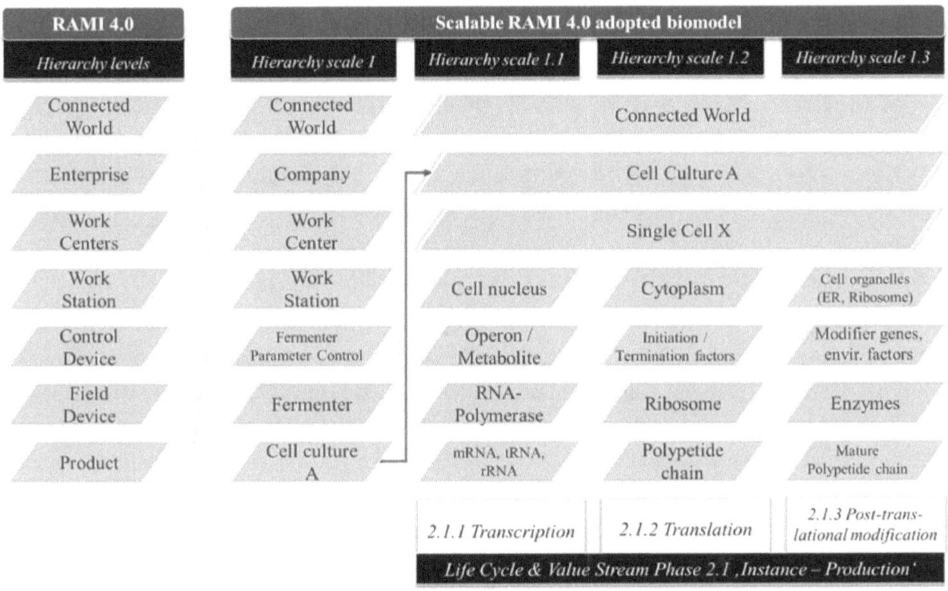

Figure 5.47 From classical RAMI 4.0 to scalable RAMI-4.0-adopted biomodel according to Miehe et al. (2020b)

In the RAMI 4.0 model the biological process steps, that represent active contributions to the production output, have to be allocated to the x-axis under the 'instance – production' phase. Depending on the granularity of the z-axis (hierarchical levels), elements of the x-axis such as development or maintenance might not be applicable, thus reducing the dimensions of the RAMI 4.0 model. Naturally, the hierarchical levels of the organization are structured similarly irrespective of the type of production industry. While the y-axis layers can thus be adopted as-is, it is different with the z-axis layers. The bioproduction process examined here may have several work centers with various workstations. Each workstation may consist of different bioreactors (field devices), which are regulated with a superordinate bioreactor control unit. As the lowest hierarchical level (z-axis), the product itself could represent a cell culture (A). An alternative approach would be to view cell culture not as the lowest unit of the z-axis but as an independent production system. In this case, cell culture A would not be assigned to the product but to the company layer of the RAMI 4.0 architecture. A further elaboration of this understanding would ultimately lead to the recognition of a single cell (X) as a work center. A production step of the cell for protein production includes gene transcription which takes place in the cell nucleus (workstation 1). In eukaryotes, gene expression is controlled by promoter regions and accessory proteins (control device). The RNA polymerase (field device 1) converts the cell's DNA into mRNA (product). This is transported from the cell nucleus into the cytoplasm (workstation 2), where the next processing step, translation, takes place. The mRNA is translated into a polypeptide chain (product 2) at the ribosome (field device 2). In various cell organelles (workstation 3), a mature peptide chain is produced from the polypeptide chain in a third process, the posttranslational modification, with the aid of enzymes and additives (field device 3). This theoretical view illustrates that, in principle, the transfer of biological processes to manufacturing-derived DT models is possible. Whether data capture on all hierarchy levels is feasible and makes sense in an industrial setting, has to be evaluated for each single use case.

5.4.5.5 Integration of biological data into the DT-system/AAS

Irrespective of the levels at which information is collected, it is necessary to define the format and communication channels in which the data are processed. For this purpose, standards must be selected or, if necessary, developed, whereby the RAMI 4.0 architecture can serve as a model. As described above, the RAMI 4.0 model uses three standards: OPC UA, AutomationML, and RDF. In bioproduction and bio-laboratories, however, the data is currently still available in many different partly proprietary (i. e., manufacturer-dependent) data formats, and is only exportable into standard/open data formats such as csv or xml partially. While a trend towards standardization of data structure and format has taken place in recent years, its realization is still in its infancy. Current approaches to standardization are the analytical information markup language (AnIML) standard, the allotrope foundation standard and the standardization in lab automation (SiLA) standard. If an attempt is made to translate these into the RAMI 4.0 model, there are strong parallels to the standards used there (Table 5.12).

SiLA provides a standardized description for the communication and data exchange between instruments and laboratory software. OPC UA was specially designed for this purpose in the field of mechanical engineering. While SiLA as a communication protocol allows the control of devices and software systems via interfaces, AnIML offers a standardized XML-based data format for the description of chemical and biological process and analysis data. AnIML can be regarded as a counterpart to AutomationML, a standardized XML-based data format for storing plant engineering data. It allows the storage of data, such as raw data, analysis results, sample and instrument data, methods, processes, manufacturers and measurement techniques in one format. By 2018, approximately 20 measurement technologies, such as balance scales, pH meters, bioreactors, cell counters, plate readers, bio-analyzers, etc., were available (for others see Table 5.13) from more than 25 manufacturers (Agilent, Sciex, etc.). The interaction of SiLA and AnIML allows an increased interoperability of data and is therefore very well suited for embedding in the DT approach.

Table 5.12 Overview of parallels between standardization approaches in biology and manufacturing based on RAMI 4.0 according to Miehe et al. (2020b)

Manufacturing: RAMI 4.0 Standards	Description	Biology: Standardization approaches
OPC UA	Standardized description of the communication/exchange of data	SiLA, OPC UA
AutomationML	Standardized XML-based data format for data storage	AnIML
RDF	Standardized description of metadata with the aim of semantic linking of data	Allotrope

Another approach is provided by the Allotrope foundation. In addition to experimentally generated process and analysis data, the Allotrop data format (ADF) stores context-dependent metadata and arbitrary additional files to map experiments holistically and is adaptable to any vendor, platform or laboratory technology. ADF consists of three main components: the data description, which describes the semantic relationships of the data, the data cube, a universal data container, and the data package, a virtual file system. The semantic relationships between data and metadata are based on allotropic taxonomies and ontologies. Furthermore, Allotrope defines data models, a procedure for the standardized use of ontologies by means of data structure schemes. When comparing Allotrope with RDF, similarities are seen as both focus on the semantic connection of data to make it interpretable for further processing. Both Allotrope and RDF can integrate different data formats, including XML. However, Allotrope appears to be more comprehensive than RDF.

Table 5.13 Supported measurement technology for AnIML according to Schäfer (2018)

Analytical chemistry	Biology	Others
HPLC, GC, UV/Vis, Infrared, FTIR, Raman, mass spectrometry, NMR	Bioreactors, micro-bioreactors, micro-plate-reader, cell counter, bioanalyses, purification chromatography	Balanced scales, pH-Meter, imaging, pipetting/liquid, handling, density

5.4.5.6 Basic designs of digital representations of biology-technology-systems

A uniform understanding of a basic architecture of the digital representation is decisive for the further advancement of biointelligent manufacturing. In terms of DT architecture, two aspects have to be distinguished: the standardization of the AAS architecture of the real-world asset (data infrastructure) and the information flow of the entire system. In any case, data is collected on the technical and biological system by sensors or (partly in the case of the biological system) soft sensors, which in combination with data storage and modelling represents the gateway to the information system. The decisive difference between DM, DS and DT is now the information flow between the real-world asset and the digital representation, from which three cases can be derived as illustrated in Figure 5.48 according to Miehe et al. (2020b):

1. The description of a real-world system takes place exclusively via physical models (e. g., functions) or CAD/FEM/etc. models without an information flow between physical and informational system, the digital representation should be referred to as a DM of a biointegrated system.
2. The description of a real-world system takes place via physical models (e. g., functions) or CAD/FEM/etc. models, while one or more sensors send data from the real-world system to the model, the digital representation should be referred to as a DS of a biointegrated system.
3. The description of a real-world system takes place via physical models (e. g., functions) or CAD/FEM/etc. models, while one or more sensors send data from the real-world system to the model. In addition, a feedback loop between the model and the real-world entity enables an active, autonomous intervention in the process (e. g., to implement a real-time control of parameters of entire processes). Only this kind of digital representation should be referred to as a DT. As such, the DT represents the information technical instrument of a BTI, while the active autonomous intervention enables the system to become biointelligent.

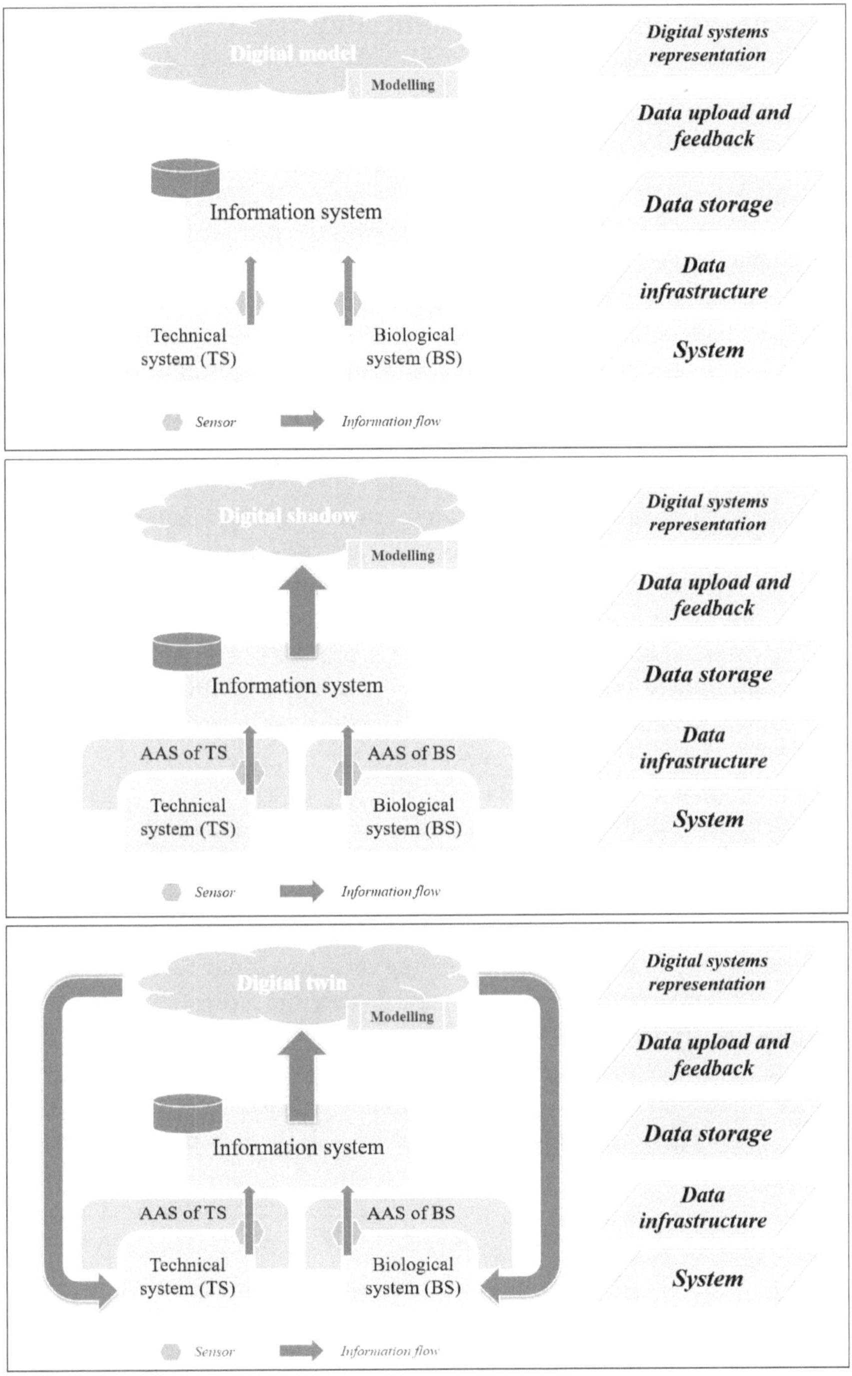

Figure 5.48 Definition of digital representation types by means of information flow and data infrastructure according to Miehe et al. (2020b)

5.4.6 Determining the degree of biointelligence

A BIS in an explicit sense is based on biological principles and requires a biological and a technical component and/or active principle as well as their interaction in the form of an autonomous regulation (biology-technology interface). The BIS must be capable of intelligent action. From the diversity of interpretations of the basic concepts outlined in Chapter 4, it is clear that no single biointelligence state exists. Rather, a system can achieve a certain degree of biointelligence. This requires the fulfilment of a number of criteria that result from the understanding of the basic concepts discussed above. The model of the system or its functional principle should exist in biology and be transferred into the system as an abstraction (bioinspiration).

The presence of a biological component can be verified by six characteristics: [1] an inner state of low entropy, [2] a continuous energy exchange, [3] the presence of a metabolism, [4] reproduction, [5] evolution, and [6] communication. Elementary criteria of the technical component are [1] the presence of inanimate material-concrete matter, [2] the generation of a specific benefit (benefit orientation), [3] an explicit functional principle for the conversion, transport or storage of matter, energy and/or information, and [4] the appearance as a spatial structure with a geometrically defined shape. The interaction in the form of the BTI provides [1] a bidirectional exchange of information between biological and technical system (in real time), [2] a control with regard to a global optimum (equilibrium orientation), and [3] a state of autopoiesis. Intelligent system behavior is expressed by the categories [1] acquisition of knowledge and memory, [2] use of stored knowledge and conclusions, [3] ability to abstract and specificize, [4] ability to plan and solve problems, [5] ability to adapt in real-time, [6] ability to learn, [7] symbol processing, [8] ability to act in incompletely described or recognized situations, [9] ability to recognize objects and patterns, and [10] ability to communicate (speech processing). The evaluation of the fulfilment of the individual categories may be carried out in the form of a binary scoring (Yes = 1; No = 0). With the different number of criteria per category, the individual components of a BIS are weighted. Yet, this cannot be justified coherently.[26)] Consequently, a standardization of the evaluation is necessary. For each category, the sum of the scoring results has to be calculated and divided by the number of criteria of the same category. The values averaged in this way are then added and divided by the number of

[26] Since other interpretations using different sub-criteria would be conceivable here as well, an equal weighting is recommended. This is also reflected in the definition in Chapter 4 by having the higher-level classes (biological role model, biological and technical component, BTI, intelligent system behavior) represent equally weighted categories.

categories (5) and multiplied by 100%. The biointelligence index (BII) is calculated as follows:

$$BII = \frac{\frac{\sum_1^n a_n}{n} + \frac{\sum_1^n b_n}{n} + \frac{\sum_1^n c_n}{n} + \frac{\sum_1^n d_n}{n} + \frac{\sum_1^n e_n}{n}}{5} \cdot 100\%$$

a Scoring result of a criterion of the category 'biological role model'

b Scoring result of a criterion of the category 'biological component'

c Scoring result of a criterion of the category 'technological component'

d Scoring result of a criterion of the category 'biology-technology-interface'

e Scoring result of a criterion of the category 'intelligence'

n Number of criteria per category

Figure 5.49 summarizes the categorization and scoring in the form of a morphological box.

Figure 5.49 illustrates the fact that the calculation of BII is based on aspects of bioinspiration and biointegration, which are elementary prerequisites for a BIS. Yet, a BIS is essentially differentiated from a biointegrated system by the BTI and its intelligent system behavior. Thus, only when individual criteria of these two categories are fulfilled one can actually speak of a 'biointelligence level'. In addition to the BII, the threshold value transmission (TVT) is introduced, calculated from the standardized scoring results of the categories biology-technology-interface (d) and intelligence (e).

$$TVT = \left(\frac{\sum_1^n d_n}{n} + \frac{\sum_1^n e_n}{n} \right) \cdot 100\%$$

The exemplary evaluation of a random system in Figure 5.49 results in a BII of 62%, with an excess of the threshold value of about 8%. The prerequisites are therefore sufficiently fulfilled to speak of a BIS. If a value of 0% is calculated for the TVT, it is not a BIS, but a biointegrated or bioinspired system

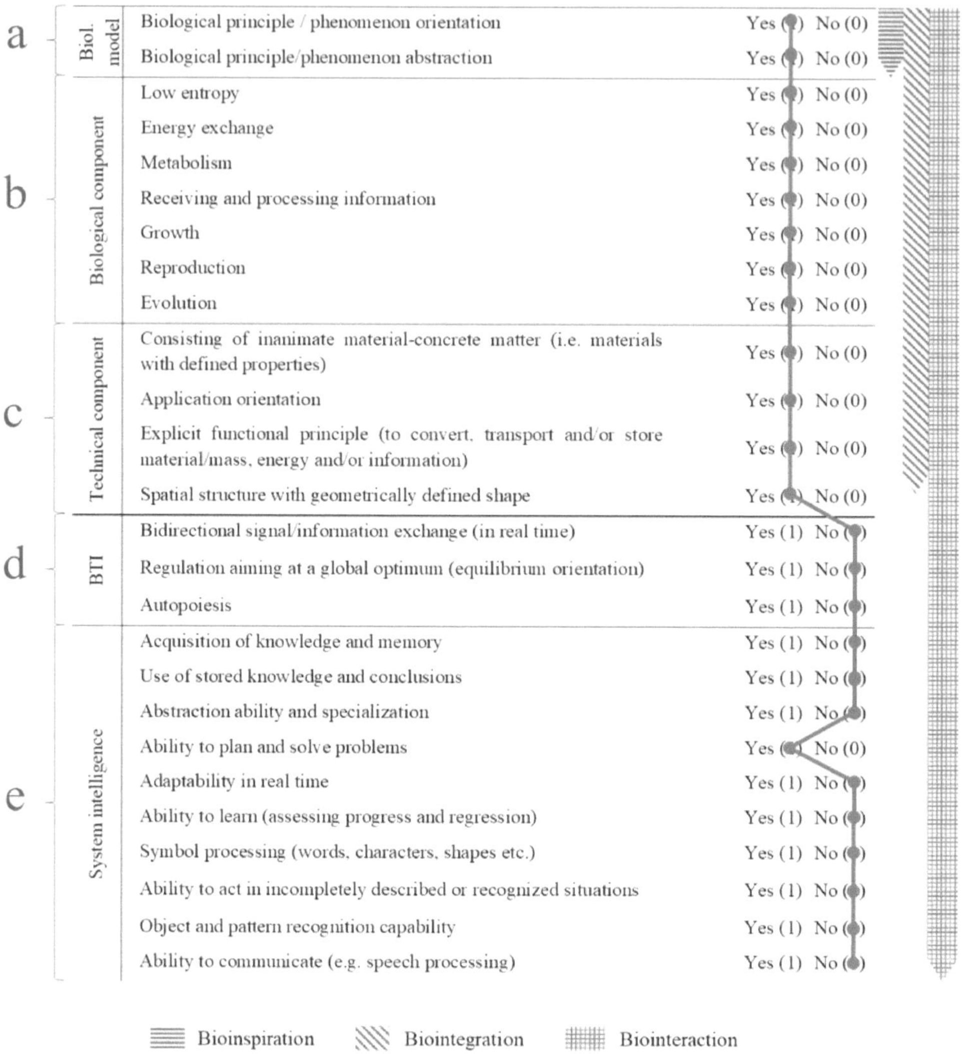

Figure 5.49 Categorization and scoring as foundation for assessing the level of biointelligence of a given system [own illustration]

5.5 Chapter review and conclusions

The development of BIS in an explicit sense is decisive for the creation of a new, sustainable technological basis. Yet, the concept does not constitute an entirely new field, it rather aims to (further) develop individual basic technologies by adding methods from various fields of research. Methodological principles for the development of BIS in an explicit sense are thus based on the three development modes of BT. Thereby, it is necessary that each subsequent level continues to build upon the knowledge of the previous one. For bioinspiration, basic engineering and biological knowledge as well as profound experience of the biomimetic transfer process are of particular importance. The most important difference in the context of the biomimetic transfer process is top-down and bottom-up approach, which both have to be applied in a suitable way in manufacturing companies. The instrumental foundation of biointegration is the knowledge of basic biotechnology approaches, yet in addition to the prior know-how of bioinspiration. The technology examples presented here and selected from the pool of biotechnology (including bioprinting, biorefineries, bioreactors, biosensors, and CRISPR/Cas9) only represent a small excerpt, yet shall illustrate the potential of combination, even within conventional production sectors (e. g., machinery and plant engineering). Based on the previous development modes, basic ICT know-how is fundamental for biointegration.

Similar to the concept of intelligence and life, no single explicit biointelligence state exists for a system. Rather, a system can be evaluated in terms of its biointelligence degree. Its classification has to be based on common criteria for grading its main components, i. e., presence of biological principles, biological and technical component and/or active principle, interaction in form of autonomous regulation (biology-technology interface) as well as overall system intelligence.

Lessons learned

- Biointelligence aims to (further) develop and/or recombine individual basic technologies by adding methods from various fields of research. In fact, it is supposed to stimulate thinking in certain structures.
- The creation of BIS in an explicit sense is based on the development modes of BT: bioinspiration, biointegration, biointeraction.
- While bioinspiration primarily relies on basic biological and engineering know-how combined with the application of the biomimetic transfer process (either top-down or bottom-up), biointegration largely represents the application of biotechnology to production systems.
- Advanced biointegrated technologies that already exist today providing a basis for future recombination, decentralization and personalization include: biosensors, bioprinting, additive biomanufacturing, additive manufacturing of biomaterials, biorefineries, bioreactors, as well as microbial and enzymatic fuel cells. The most far-reaching approach, however, is the CRISPR/Cas9 system, which will enable several technological leaps in existing systems as well as completely new combinations in the future.
- While numerous examples and methods of bioinspired and biointegrated systems exist, BIS in an explicit sense result from their further development and/or recombination via information technology. Here especially I4.0 approaches (including CPS, AI, BDA, DT) are essential.

- However, compared to CPS, BIS represents a further level of complexity. This becomes quite clear in the example of DTs. The concept, which has primarily been discussed in conventional manufacturing and is well advanced in standardization, requires extensive transitional adjustments for its transfer to biological systems.
- BIS in an explicit sense require an adaptive, systems-theoretical process control model between both the biological and the technical system. Therefore, the generic BTI approach was developed in order to enable flexible structuring and regulation on various levels of biology-technology interaction. As such, they form the basis for a wide range of applications, (e. g., regulation of cell-based manufacturing processes, intelligent orthopedics).
- As BTIs need to be automatically adapted to specific properties, typically on the basis of limited available data, soft sensor technologies may be viewed as a vehicle of transition.

Follow-up questions

- What design approaches are required for the management of BIS?
- How can the holistic benefit criterion of BIS be ensured?
- Which research and development fields currently exist?

6 A framework for the management of biointelligent systems

After the previous section presented basic technologies and approaches for the development of BIS in an explicit sense, this section discusses issues of the management of BIS.[1] The core objective of the management of BIS is to ensure compliance with the normative, holistic benefit criterion (see Chapter 4). Therefore, a basic framework is introduced, followed by an in-depth examination of methods to facilitate systems-oriented life cycle thinking in cellular units and the discussion of design approaches for future value creation systems.

6.1 Conceptual structure

Before introducing the conceptual design of the BIS management framework, a few general thoughts are necessary. Various approaches to management frameworks exist in literature. Among the most prominent are the St. Gallen Management Model and the Stuttgart Enterprise Model (Westkämper, 2006; Rüegg-Stürm and Grand, 2015). However, these primarily assume a single company rather than a production network of decentralized production units, which include not only companies but also individuals, households, non-specialized producers, and spontaneously emerging production teams outside the conventional factory system. In addition to addressing these issues and regarding production as a spontaneous merging of cells (cellular thinking), a framework for the management of BIS must be able to portray the life cycle perspective (life cycle thinking) and its system-theoretical interdependencies. Moreover, it has to provide a normative framework that aligns the entirety of production with the overall capacity of the ecosystem. Simultaneously, the framework should be able to

[1] This section largely represents a summary of the findings from Miehe et al. (2021a).

map the essential basic elements of doing business within this new understanding of value creation, i. e., to address the following questions:

1. What are the basic technologies for BIS?
2. What markets and customers exist for BIS?
3. What strategies and business models (BM) exist for BIS?
4. What are biointelligent innovations and how are biointelligent products developed?
5. How can a cellular production network be organized and managed?
6. How is ensured that producers and consumers take the necessary responsibility for their actions?

Biointelligence seeks to develop systems that represent an integrative optimum of efficiency, effectiveness, and sufficiency. Hence, the management framework can only be based on the priority model of sustainability (i. e., strong/absolute sustainability). Maintaining the planetary boundaries (i. e., the natural overall system) is thus of utmost importance. Concurrently, it must be ensured that the basis of the prosperity of societies (often expressed by the human affluence = material standard of living) is maintained and distributed equally around the globe. The conceptual design presented here is thus partially based on the framework of life cycle engineering developed by Hauschild et al. (2017; 2020). Here, scope of environmental concern and scope of temporal concern are distinguished. While the former is determined by planetary boundaries, the latter is based on the perspective of the entire span of civilization. A related approach is described by Stead and Stead (2008). The production system serves the manufacturing of products, which require biotic and abiotic cellular systems that are in constant interrelations with their surrounding cells (information networking). Their configuration in form of products and productive systems is primarily determined by systems-oriented life cycle engineering. Leadership, organization, planning, and control of companies, households as well as company-household-networks are responsibilities of systems-oriented life cycle management. Technology deployment essentially relies on the available basic technologies (Chapter 5), their convergence via interdisciplinary research and development (Chapter 4) as well as their appropriate placing on markets. As consumers/households are likely to increasingly become producers of goods (e. g., via horticulture, tissue printing), producers and consumers in certain cases merge to prosumers. In case of entrepreneurial activity, a strategy and a corresponding BM are necessary. Furthermore, innovation and product development as well as production organization and management must ensure efficient and sustainable operations within the cellular network. Last but not least, it is important to seize that all participants involved along these fields of biointelligent value creation are hold accountable for their actions. Accordingly, the framework for the management of BIS (Figure 6.1) distinguishes between a core, i. e., systems-oriented life cycle thinking in cellular units as the fundamental mindset for decision-making, as well as the scope of environmental concern and the scope of management concern.

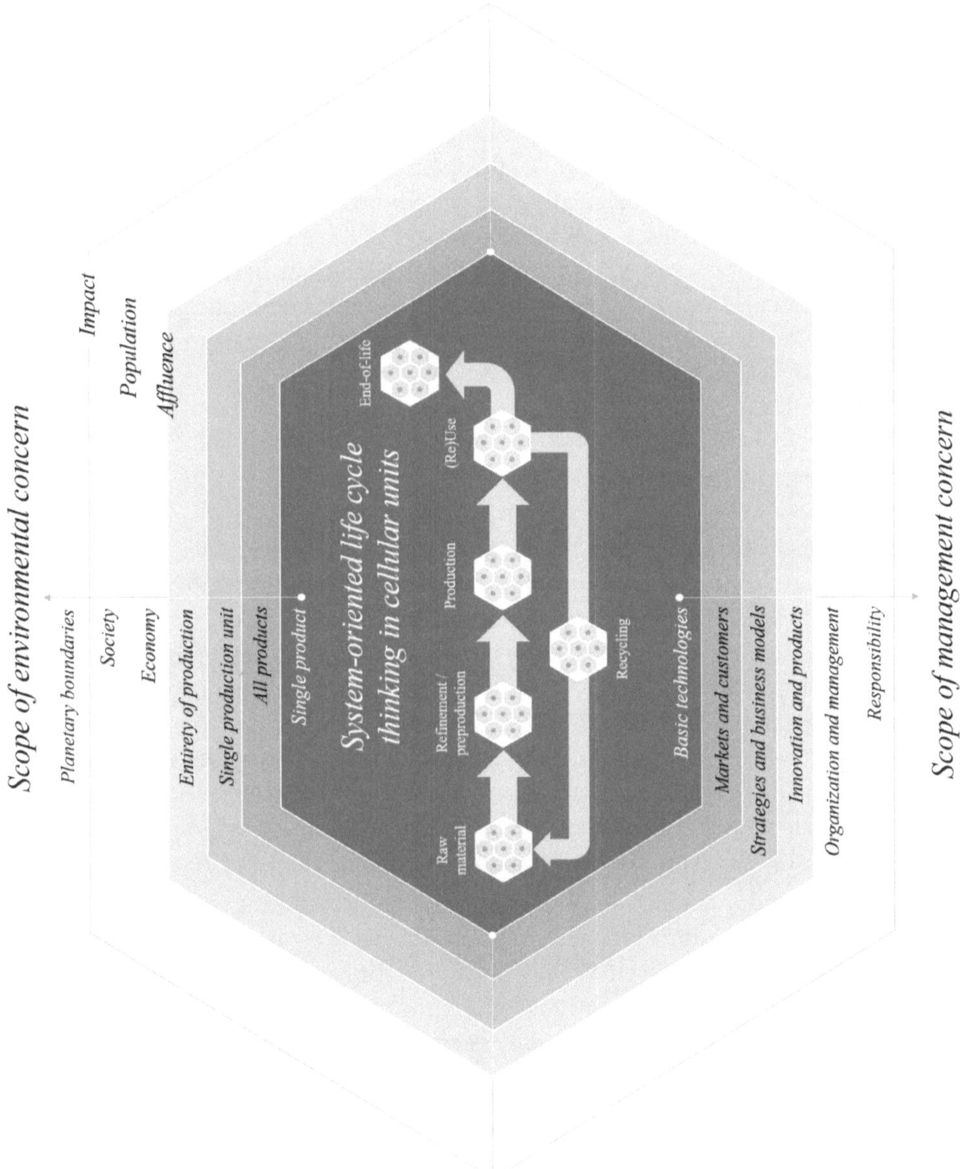

Figure 6.1 Framework for the management of BIS [own illustration based on Stead and Stead (2008) and Hauschild et al. (2017; 2020)]

The following sections introduce the individual elements of the framework by providing an overview of the basic approaches and their relationship to BIS.

6.2 Supporting the fundamental mindset

Systems-oriented life cycle thinking in cellular units represents the core of BIS management. Each of its sub-areas can be sustained by more or less established methods. Life cycle thinking may be supported by the quantification of life cycle inventory data and their impact from an ecological, social or economic perspective. Its life cycle understanding (see Section 4.4.1) and basic approaches have been transferred into several management approaches. In turn, system-thinking has become much more established in management, although not in the sense of biocybernetics, as it is intended in this work. Its primary method of raising awareness of dependencies is systems-theoretical interdependence analysis (STIA).

6.2.1 Life cycle assessment (LCA)

LCA is the most comprehensive instrument to date for determining the impact of technologies and individual actions on the ecosphere and human beings within the framework of a single analysis. The method can be applied both ex ante and ex post to assist decision-making across entire life cycles. Two options are distinguished when applying an LCA. A descriptive LCA aims to present all environmental impacts caused by a specific good or a specific action in a way that is appropriate to its cause. In contrast, a decision-oriented study aims at comparing two or more alternatives (Frischknecht, 2013). Thus, its purpose is to illustrate increases and decreases in one or more environmental impacts of different options. The following paragraphs outline the methodological basics as well as different approaches to impact assessment.

The impact of products and their manufacturing on surrounding systems is manifold. In literature, natural environment, human health, and resources are usually considered. LCA has established itself as the instrument for assessing and evaluating these impacts of products and processes throughout their life cycle. In companies, it serves to prepare and support environmentally relevant decisions (Finnveden et al., 2009; Guinée et al., 2011). Although initial approaches can be found as early as the 1880s, the resource environmental profile analysis (REPA) was regarded as a direct precursor of LCA in the 1960s (Klöpffer and Grahl, 2014; Horne et al., 2009). After case studies at Coca Cola, the US EPA (1974) as well as the Swiss Federal Office for Environmental Protection (1984) published first studies under the term LCA. According to Klöpffer and Grahl (2014), the dissemination of the method was subsequently promoted in particular by the work of the Society for Environmental Toxicology and Chemistry (SETAC). Since 2006, LCA is internationally standardized (Deutsches Institut für Normung e. V., 2009; 2006). Subject of any LCA are all elementary flows arising during production, use and disposal, i. e., all material and energy flows which are added to or removed from a system. According to ISO 14.040 and 14.044, an LCA consists of four phases (Deutsches Institut für Normung e. V., 2009, 2006): [1] definition of the goal and specification of scope, [2] life cycle inventory, [3] impact assessment, and [4] evaluation. According to ISO 14.040, the following seven principles must be followed when planning and conducting an LCA (Deutsches Institut für Normung e. V., 2009):

1. systematic consideration of entire life cycle – from raw material extraction to ultimate disposal
2. environmental orientation – disregarding economic and social aspects
3. relative approach – relation to functional unit
4. iterative approach – findings from previous phases provide the basis for subsequent phases
5. transparency – open and comprehensive depiction of information
6. holistic approach – consideration of aspects of natural environment, human health, and resources and their interaction
7. priority of scientific approach – decisions in LCA should preferably be based on scientific knowledge

The following sections describe the procedure for conducting an LCA in detail and discuss supporting instruments.

6.2.1.1 Definition of goal and scope of analysis

Results of an LCA are largely based on the definition of its objective and the choice of its scope. For example, the goal of an LCA may be the comparison of two product alternatives. According to Klöpffer and Grahl (2014), the definition of objectives includes the scope of the investigated subject as well as the reason and target group of the study, which again determines depth and accuracy of the study. The European Commission (2010c) considers LCA to be effective in three cases. Decision support at micro level focuses on product-relevant issues, each of which is exclusively within the decision-making authority of the initiator. In contrast, if an LCA serves to support decision-making at meso or macro level, the system boundaries are comparatively wide. Actions can also have consequences outside the decision-making sphere of authorized persons. Furthermore, an LCA study can fulfil a purely descriptive function. The European Commission (2010c) also distinguishes between consequential and attributional analyses. While consequential studies analyze the effects of a decision at the macro level, attributional LCAs determine all impacts of a single subject within a certain period of time.

Once the objectives have been defined, the scope of the investigation has to be determined. According to ISO 14.040, this step contains a description of the system to be investigated and its functions, a definition of a functional unit, the system boundaries, an allocation, a choice of impact categories, and a selection of evaluation to be applied (Deutsches Institut für Normung e. V., 2009). Further aspects influencing the scope of the study are the requirements for data depth and quality, the type of critical review, the form of the report and a list of assumptions and limitations.

Since systems usually have a wide range of functions, the first step is to determine the purpose to be investigated. This is expressed by the functional unit. A functional unit describes one or more characteristic properties of the system to be analyzed, which serve as a reference for the investigation. A functional unit can, for example, be the bottling of half a liter of lemonade. All material and energetic input and output quantities are subsequently related to this reference. A certain amount of output of a system is required to fulfil the previously defined function. These are summarized in a reference flow, e. g., a quantity of a cable of a special material to ensure the transmission of the previously defined amount of electricity.

In addition, the system boundaries have to be outlined in this phase. Here the goal is to determine which process modules are included in the analysis. ISO 14.040 proposes that the system boundary should be selected in a manner that guarantees exclusively elementary flows to cross the separation line (Deutsches Institut für Normung e. V., 2009). Therefore, the systems theory understanding has to be aligned with the view of ecology. According to the European Commission (2010c), this requires a distinction between foreground and background systems. The distinction is determined by the extent to which a system can be influenced. The background system thus represents those upstream and downstream processes which the investigator cannot influence. The investigated subject is embedded in the technosphere, the system of man-made technologies. The technosphere is a subsystem of the ecosphere, the animate and inanimate system earth. Figure 6.2 illustrates the understanding according to the European Commission (2010c).

Finally, requirements for data quality are defined in this phase. Characteristic features of data quality include the time-related, geographical, and technological scope of coverage as well as completeness, consistency, comparability, and uncertainty. In addition, in case of multi-product systems, suitable allocation procedures can be agreed upon in order to realize an appropriate distribution of the environmental impacts to individual products.

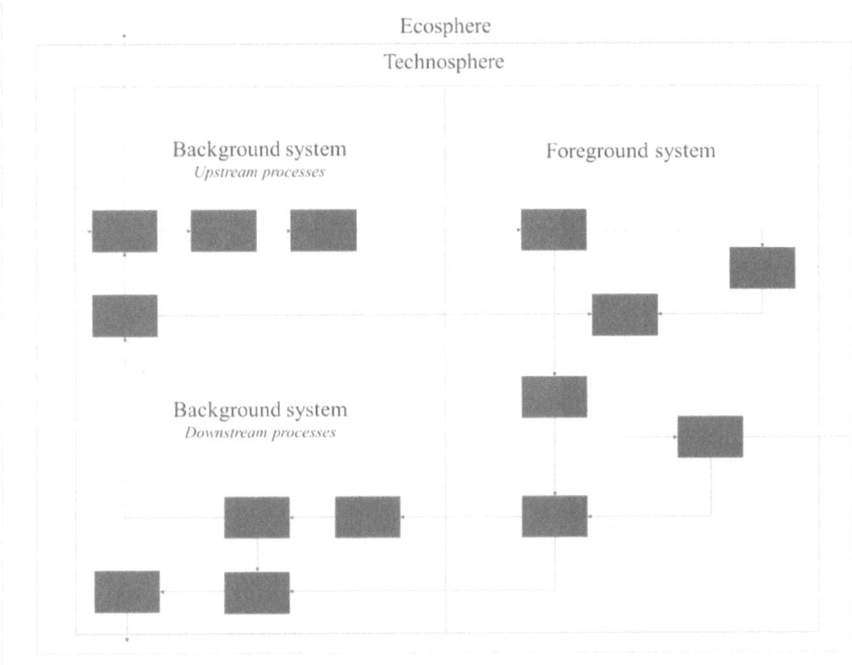

Figure 6.2 Interactions between foreground and background systems according to the European Commission (2010c)

6.2.1.2 Life cycle inventory (LCI)

Life cycle inventory analysis (LCI) is used to quantify relevant material and energy flows via data collection and/or the use of suitable calculation methods in relation to the given system (European Commission, 2010d). This requires a comparison of input and output flows. According to Klöpffer and Grahl (2014), this phase of LCA represents a simplified linear system analysis. Once objective and scope of the study have been defined, the first step is to collect data. For this purpose, a model is developed to subdivide the overall system into sub-steps (process modules). For each process module, the relevant quantitative data in terms of input and output flows are collected and aggregated to the overall system. According to ISO 14.044, the data to be collected can be divided into four classes (Deutsches Institut für Normung e. V., 2009):

1. energy, raw material, fuel and other physical inputs
2. products, by-products and waste
3. emissions to air, water and soil
4. other environmental aspects

Data collection can basically be carried out by measurement, calculation or estimation (internal) or be obtained from specialized data sources (external). Usually, the requirements for the data collection process are determined by the definition of the foreground and background systems. While for the background system, generally valid data based on a database search are often sufficient, the data of the foreground system should be based primarily on a targeted measurement, calculation or estimation by experts. Various databases exist for external data collection. These are primarily made available by national and international institutions. An example is the free ProBas database (Umweltbundesamt, 2016). Further databases are provided by industry associations (e. g., PlasticsEurope, World Steel Association) as well as private organizations in form of fee-based systems (GaBi, ecoinvent, etc.).

The next step is data calculation, serving to verify if data quality requirements are met. ISO 14.044 recommends mass and energy balances or other comparative studies, e. g., emission factors (Deutsches Institut für Normung e. V., 2006). Various instruments for calculating and visualizing material balances are available. The main calculation tools used in literature are process-oriented, economic input-output analysis, and hybrid life cycle inventories. Flow diagrams (e. g., material and energy flow diagram, Sankey diagram) serve as visualization tools. Dependencies between input and output flows or between individual process modules are usually represented by ratios or matrices (Suh and Huppes, 2005). A reference to the process module and the functional unit has to be ensured at any time. An iterative adjustment of the system boundary is possible at this point, e. g., to exclude unimportant components from the analysis. The data collected up to this point are purely quantitative in nature and do not contain any evaluation.

6.2.1.3 Life cycle impact assessment (LCIA)

According to Klöpffer and Grahl (2014), the third phase of LCA serves to determine the environmental impacts and consequences of the investigated system. The life cycle impact assessment (LCIA) is based on the data determined in LCI. ISO 14.044 divides this phase into mandatory and optional components (Deutsches Institut für Normung e. V., 2006). Binding elements include the selection of impact categories, impact indicators and characterization models, as well as the allocation of LCI results (classification) and the calculation of impact indicator values (characterization). Normalization, ordering, weighting, and analysis of data quality are optional. The following paragraphs outline the general procedure in this phase.

First, impact categories have to be identified. ISO 14.044 defines an impact category as a group representative of a specific environmental topic to which the results of LCI can be assigned (Deutsches Institut für Normung e. V., 2006). Examples of impact categories are climate change or acidification. None of the LCA standards provides a fixed list of impact indicators. The definition depends on the respective application.

For the quantitative representation of the impact categories, quantifiable indicators are required. For climate change, for example, the amplification of infrared radiation is regarded as an impact indicator. The driving force behind the increase in infrared radiation is the amount of greenhouse gases in the atmosphere and their respective contribution to climate change. Various characterization models exist for this purpose. Mostly the IPCC 100-year baseline global warming scenario is used for climate change. LCI results (here: quantity of greenhouse gases emitted), the so-called elemental flows, are aggregated to an impact indicator value (kg CO_2 equivalent) using the characterization factors from the IPCC scenario (GWP100). Figure 6.3 illustrates the LCIA procedure for the midpoint category climate change.

Term	Example climate change
Impact category	Climate change
Inventory result	Quantity of GHGs related to functional unit
Characterization models	IPCC-Scenarios 20, 100 or 500 years
Impact indicator	Amplification of infrared radiation (W/m²)
Characterization factor	Global warming potential GWP100 for each greenhouse gas (conversation factors)
Impact indicator unit	kg CO_2e / functional unit
Impact endpoint	Coral reefs, forests, harvests etc.
Environmental relevance	The amplification of infrared radiation is an example of possible impacts on the climate, which depend on the integrated atmospheric heat absorption, caused by emissions and the distribution over the duration of heat accumulation.

Chemical formula	*GWP time horizon 100 years*
CO_2	1
CH_4	25
N_2O	298
CCl_3F	4,750
CCl_2F_2	10,900
$CClF_3$	14,400
CCl_2FCClF_2	6,130
$CClF_2CClF_2$	10,000
$CClF_2CF_3$	7,370
$CBrF_3$	7,140
$CBrClF_2$	1,890
$CBrF_2CBrF_2$	1,640
CCl_4	1,400
CH_3Br	5
CH_3CCl_3	146
$CHClF_2$	1,810
$CHCl_2CF_3$	77
$CHClFCF_3$	609
CH_3CCl_2F	725
CH_3CClF_2	2,310
$CHCl_2CF_2CF_3$	122
$CHClFCF_2CClF_2$	595
CHF_3	14,800
CH_2F_2	675
CHF_2CF_3	3,500
CH_2FCF_3	1,430
...	...

Figure 6.3 LCIA procedure for the midpoint category climate change [own illustration according to ISO 14.040/44 and IPCC]

The assignment of LCI results to the corresponding impact categories is also referred to as classification. This basically enables individual LCI results to be assigned to several impact categories. According to ISO 14.044, however, characterization refers to the calculation of indicator values (Deutsches Institut für Normung e. V., 2006). In a final step, the impact categories are assigned to the impact endpoints or damage categories. The European Commission (2010b) distinguishes between three fields: human health, natural environment, and resources. Figure 6.4 summarizes the impact assessment procedure based on the European Commission (2010b).

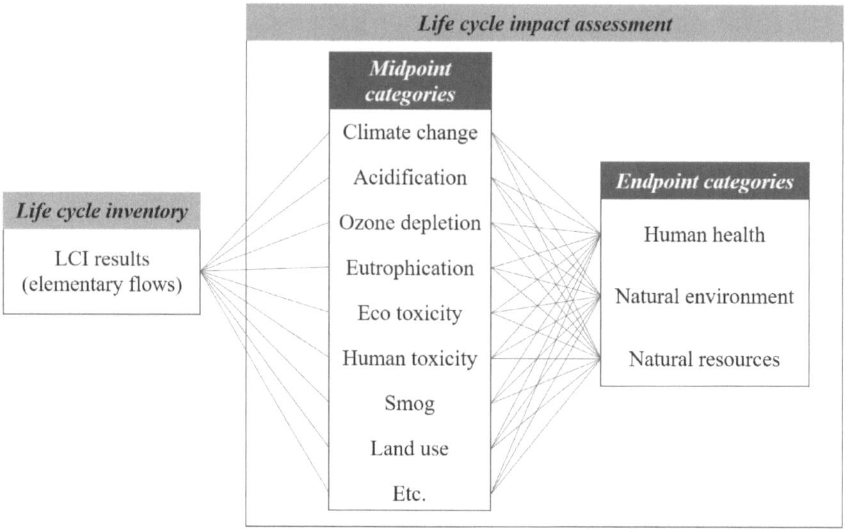

Figure 6.4 Structure of impact categories for characterization at midpoint and endpoint level according to the European Commission (2010b)

Optional steps such as normalization, ordering, weighting, and analysis of data quality can be conducted next. It remains entirely up to the user to decide whether all or only some of the optional elements are to be applied additionally. Normalization describes a summary of the impact indicator values by reference to a characteristic reference value. According to Klöpffer and Grahl (2014), regional, national or international values are primarily suitable as a reference. Miehe et al. (2016d) illustrate this in an exemplary manner referring to the turnover of a company and its industry. Although this step is useful for bundling the large amount of data from an LCA study and to support decisions in companies more effectively, it bears the risk of misleading main results. In contrast, ordering is characterized by the sorting and possible ranking of the impact categories, e. g., using a nominal scale or predefined hierarchy. Additionally, the indicator values of different impact categories can be combined and weighted differently. According to Klöpffer and Grahl (2014), weighting has a relativizing function. Since these optional steps are based on individual judgement that is often not supported by scientific evidence, results can vary substantially. Finally, an analysis of data quality can be an integral part of an LCA study. ISO 14.044 lists various possible methods for this: focal point analysis, error estimation, and sensitivity analysis (Deutsches Institut für Normung e. V., 2006).

6.2.1.4 Evaluation

The final phase of an LCA study seeks to provide a comprehensive overview of the results and to derive recommendations for action. The evaluation consists of several components. According to ISO 14.044, first those parameters should be identified which are responsible for significant impacts (Deutsches Institut für Normung e. V., 2006). Then, completeness, sensitivity, and consistency analysis have to be conducted. Conclusions, limitations, and recommendations for action should be derived last in order to meet the objective and the scope of the study.

6.2.1.5 Selected LCIA methods and tools

After outlining the basics of LCA in the previous sections, this section summarizes some basic methods of LCIA and discusses its advantages and disadvantages regarding application. Literature contains various LCIA instruments. A distinction is usually made between impact and damage-oriented methods. The former provide results for midpoint categories, while the latter deliver results for endpoint categories. Table 6.1 presents a classification of existing LCIA procedures according to Miehe (2018).

Table 6.1 Classification of selected LCIA methods according to Miehe (2018)

Impact-oriented	▪ Critical volumes ▪ Cumulative energy demand ▪ Ecological, carbon and/or water footprint ▪ CML ▪ TRACI ▪ EDIP 2003 ▪ Material intensity per service unit (MIPS) ▪ Abatement cost approach (e. g., Eco Costs 99)
Damage-oriented	▪ ReCiPe ▪ EcoIndicator 99 ▪ Impact 2002+ ▪ Method of ecological scarcity ▪ JEPIX ▪ LUCAS ▪ Damage cost approach (e. g., EPS)

Below, a selection of individual methods is roughly outlined based on Frischknecht (2013) and Klöpffer and Grahl (2014). Detailed literature concerning their application can be found there.

Two comparatively simple, widely used methods are cumulative energy demand (CED) and material intensity per service unit (MIPS). CED refers to the sum of primary energy associated with the production, use, and disposal of an economic good (product or service) (Huijbregts et al., 2010; Frischknecht et al., 2015). Thereby, net energy is converted into primary energy in relation to a functional unit using semi-standardized conversation factors. In contrast, MIPS calculates the required quantity of material in relation to a functional unit as the sum of the categories abiotic and biotic material, soil movement in agriculture and forestry, water and air (-components). Its aim is dematerialization. Generally, the method allows a weighting of each category, which is however not intended. The MIPS indicator enables easy visualization of material savings (Schmidt-Bleek, 2014b). This provides companies a comparatively simple tool to increase material efficiency.

One of the most widely used methods for LCIA is CML. This physical-impact-oriented approach is primarily based on the work of Heijungs and Guinée (Heijungs, 1992; Guinée et al., 2001; Guinée, 2004). A characteristic feature of the concept developed at the Centrum voor Milieukunde (CML) at the University of Leiden is the aggregation of life cycle inventory data within individual impact categories using equivalence factors (e. g., CO_2e, SO_2e). Its basis is a predefined assignment of LCI results to nine environmental impact categories: [1] global warming potential (GWP), [2] abiotic depletion factor (ADF), [3] stratospheric ozone depletion potential (ODP), [4] photochemical ozone creation potential (POCP), [5] aquatic ecotoxicity (ecological classification factor for aquatic ecosystems, ECA), [6] terrestrial ecotoxicity (ecological classification factor for terrestrial ecosystems, ECT), [7] human toxicity (human toxicological classification factor, HC), [8] acidification potential (AP), and [9] eutrophication (nutrification potential, NP).

A further concept is the method of ecological scarcity (MES), which seeks to quantify the environmental impact on the basis of socially accepted targets (in this case the environmental goals of a region) (Frischknecht et al., 2009). Compared to the previous approaches, this approach represents an endpoint methodology that aggregates different midpoint categories via eco-factors and environmental impact points. Figure 6.5 illustrates an application of the approach for a comparison of the environmental impact of different beer production and packaging options according to Dinkel and Kägi (2014).

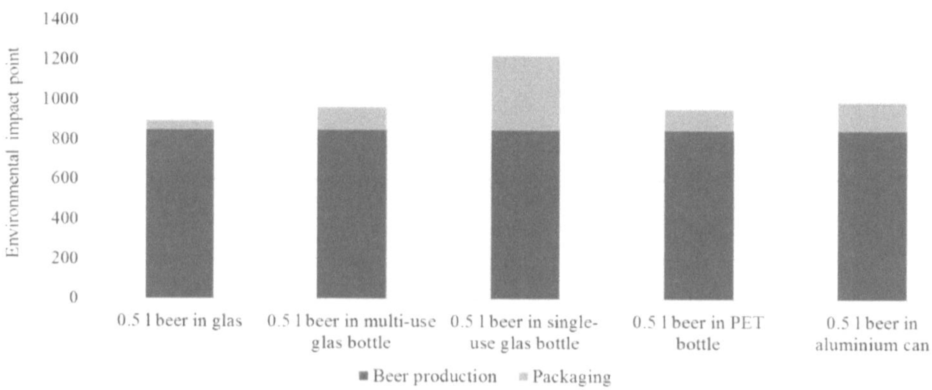

Figure 6.5 Example of MES application for comparison of environmental impact of different beer production and packaging options according to Dinkel and Kägi (2014)

In addition, the physical results of an LCIA, e.g., assessed via CML (quantity of CO_2e, SO_2e, R11e etc.), can be converted into costs. This requires a calculation of external cost factors to be used to multiply the physical results. As discussed above, two basic approaches are distinguished: abatement and damage cost approach (Adensam et al., 2002; Quinet, 2004). While the former seeks to estimate the damage caused, the latter can be understood as costs of prevention (Kesicki and Ekins, 2012; Vogt-Schilb and Hallegatte, 2014; Ward, 2014). A prominent example of the abatement cost approach in the context of LCA is the EcoCosts99 concept that originates from the work of Vogtländer et al. (2000; 2001). It is intended to serve as an extension of the equivalence factor concept with regard to the quantification of external costs. For this purpose, Vogtländer et al. (Vogtländer et al. 2000; Vogtländer et al. 2001) present cost factors for selected impact categories (including climate change, air pollution, eutrophication, acidification). Figure 6.6 illustrates an example of a marginal abatement cost calculation (MACC) for individual climate mitigation strategies of a German standard household (2.1 persons) based on the results of Miehe et al. (2016b).

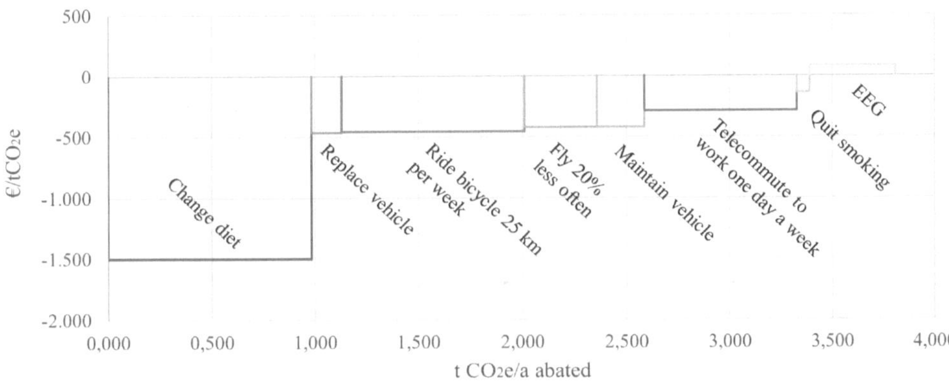

Figure 6.6 Example of a marginal abatement cost calculation for individual climate mitigation strategies of a German standard household (2.1 persons) [own calculation based on the results of Miehe et al. (2016b)]

However, the abatement cost approach is viewed as highly controversial, as it assesses measures on suspicion without reliable information on causal relationships or damage costs (Adensam et al., 2002). The concept is widely used because it requires little effort compared to damage cost-based approaches. In turn, the damage cost approach is much more complex, as it requires an extensive stakeholder query. Usually, selected stakeholders are questioned about the price they would be willing to pay for the acceptance of an environmental impact (willingness-to-accept approach) or how much they would pay additionally to eliminate a certain problem (willingness-to-pay approach). An example of the damage cost approach is documented by Andensam et al. (2002). Each of the LCIA methods has different scopes (regarding the integration of impact categories) as well as differing advantages and disadvantages. Table 6.2 summarizes the scope of the methods discussed above according to Frischknecht (2013).

Table 6.2 Scope of different LCIA methods according to Frischknecht (2013)

		CED	MIPS	CML	MES	Eco Cost 99
Resources	Energy carrier	+	+	+	+	–
	Ores and minerals	–	+	+	+	–
	Water	–	+	–	+	–
	Biotic resources	–	+	+	–	–
	Habitat/land	–	–	+	+	–
Emissions	Climate change	–	–	+	+	+
	Ozone depletion	–	–	+	+	–
	Human toxicity	–	–	+	+	(+)
	Eco toxicity	–	–	+	+	(+)
	Formation of photooxidants	–	–	+	+	(+)
	Acidification	–	–	+	+	+
	Overfertilization	–	–	+	+	(+)
	Microcontamination	–	–	–	+	–
	Smell	–	–	+	–	–
	Noice	–	–	(+)	–	–
	Radioactive emissions	–	–	+	+	–
	Accident victims	–	–	+	–	–
	Waste	–	–	–	+	–

The variation in focus and scope leads to very different application scenarios. Table 6.3 summarizes fields of application as well as pros and cons of the methods discussed above.

Table 6.3 Advantages, disadvantages and fields of application of selected LCIA methods according to Miehe (2018)

Method	Pros	Cons	Application areas
CED	▪ Simple approach ▪ Comparability of the results, as only one indicator	▪ Sole focus on energy ▪ No holistic assessment possible	▪ Energy management, monitoring, and efficiency ▪ Energy intensive industry
MIPS	▪ Simple approach ▪ Comparability of the results, as only one indicator	▪ Sole focus on material ▪ No holistic assessment possible	▪ Material efficiency ▪ Material intensive industry ▪ Valuable materials
CML	▪ Transparency ▪ Distribution ▪ Number of impact categories	▪ Complexity ▪ Comparability of results	▪ Specific environment-oriented optimization ▪ Virtually in all cases a broad investigation is required
MES	▪ Comparability of results via non-dimensional indicator ▪ Focus on socially accepted goal ▪ Number of effect categories considered	▪ Low comparability due to regional differences ▪ High implementation costs (especially if eco-factors for the region are not known) ▪ Low transparency for outsiders	▪ Aggregation of various impact categories ▪ Publications not intended to reveal substantial data
Abatement cost approach (e. g., Eco-Costs 99)	▪ Tangible indicator (€) ▪ Comparatively simple approach ▪ Comparability of results, as only one indicator	▪ Calculation depending on time (best-available-solution) ▪ No independent application, as based on other LCIA methods ▪ High uncertainty	▪ Communication towards management board ▪ Consolidation of impact categories ▪ Holistic evaluation of investment alternatives ▪ Marketing via eco-prices
Damage cost approach	▪ Tangible indicator (€) ▪ Comparability of results, as only one indicator	▪ Extensive ▪ No independent application, as based on other LCIA methods ▪ Query of sensitivities of stakeholders results in high level of uncertainty	▪ Communication towards management board ▪ Consolidation of impact categories ▪ Involvement of stakeholders ▪ Marketing via eco-prices

6.2.2 Social life cycle assessment (SLCA)

Similar to the quantification of environmental impacts over the life cycle, societal challenges can be identified via SLCA (Dreyer et al., 2006; Jørgensen et al., 2008; Jørgensen et al., 2010; Benoît et al., 2010). According to Dreyer et al. (2006) and Hunkeler (2006), SLCA is largely based on the conventional LCA, as its aim is to complement environmental with social impact categories for the same object of investigation. Hence, it relies on the same procedure, system boundaries, and functional unit. As human health is already evaluated within one or more LCIA categories, depending on the research design, SLCA focusses on human dignity and well-being. Similar to LCI and LCIA, inventory and impact of SLCA are measured by specific indicators that define the data to collect (quantitative or qualitative) (Benoît et al., 2010). After these have been defined in the context of goal and scope (step 1), a reference for the respective indicator is selected for each process step for which an input-output balance is developed in LCI. In most cases, this requires extensive stakeholder dialogue. Figure 6.7 illustrates the difference between LCI and SLCI according to Dreyer et al. (2006).

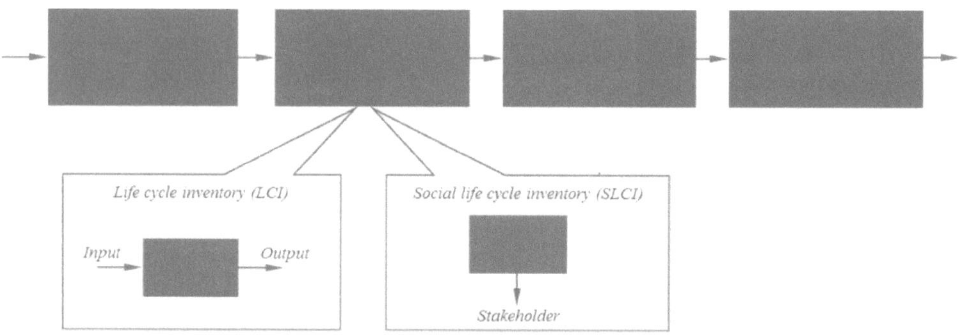

Figure 6.7 Difference between LCI and SLCI according to Dreyer et al. (2006)

The inventory indicators are related to subcategories, grouped into impact and stakeholder categories. Each impact category relates to social issues of stakeholders and decision-makers. Stakeholder classification and classification by impact categories are supplementary. Figure 6.8 illustrates the classification of stakeholder and impact categories according to Benoît et al. (2010).

Stakeholder relations of an organization are very specific and complex, making it difficult to establish a general social impact model. Yet, production impacts society in various fields. Starting from the factory and/or the production system, the impact of the company on people becomes more indirect as the level increases. Internal processes may also have an impact in the local community or society. The necessary inclusion of impacts on other relevant stakeholders requires an alternative perception of LCIA elements. According to Dreyer et al. (2006), SLCA does not require a traditional characterization. A characterization model in LCA serves the purpose to quantitatively translate the results of LCI into environmental impacts as expressed by the respective impact indicators. In SLCA the category indicators are developed to measure the social impacts directly in the company and/or the supply chain. Figure 6.9 illustrates the social life cycle impact levels according to Dreyer et al. (2006).

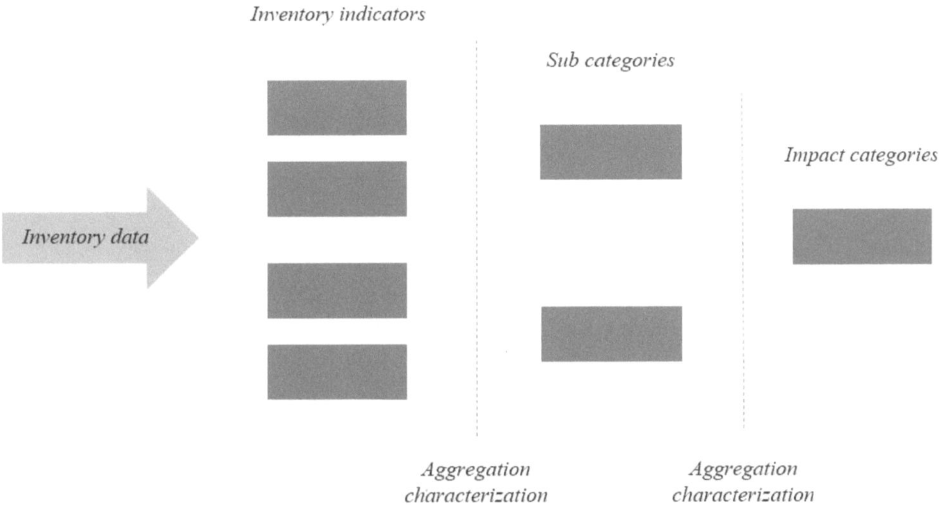

Figure 6.8 Classification of stakeholder and impact categories according to Benoît et al. (2010)

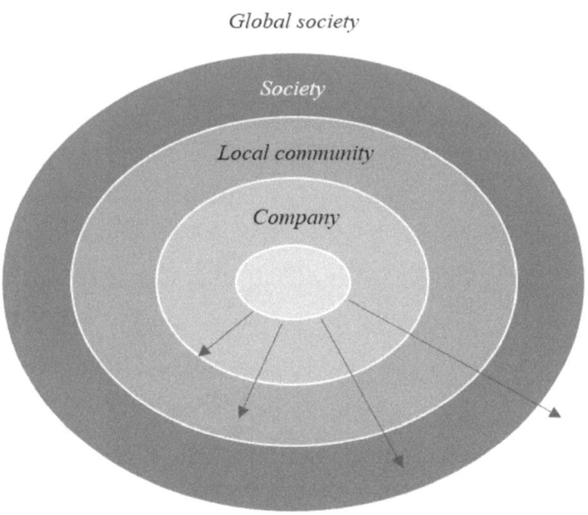

Figure 6.9 Social life cycle impact levels according to Dreyer et al. (2006)

Among the stakeholders to be addressed in each step of the life cycle are workers/employees, local community, society (national and global), consumers (covering B2B and B2C), and value chain actors (including suppliers) (Benoît et al., 2010). Examples for social issues include human rights, working conditions, cultural heritage, poverty, disease, political conflict, indigenous peoples rights etc. Table 6.4 lists a selection of social issues relevant to different stakeholder categories according to Benoît et al. (2010).

Table 6.4 Selection of social issues relevant to different stakeholder categories according to Benoît et al. (2010)

Stakeholder	Social issues
Worker	Freedom of association and collective bargaining; working hours; child labor; health and safety; social forced labor; fair salary; equal opportunities/discrimination; benefits/social security
Consumer	Health and safety; consumer privacy; transparency; feedback mechanism; end of life responsibility
Local community	Access to material resources; access to immaterial resources; respect of indigenous rights; delocalization and migration; safe and healthy living conditions; community engagement; local employment; cultural heritage; secure living conditions
Society	Public commitments to sustainability issues; prevention and mitigation of armed conflicts; technology development; contribution to economic development; corruption
Value chain actors (excl. consumers)	Fair competition; supplier relationships; promoting social responsibility; respect of intellectual property rights

A uniform set of impact categories is not available (Benoît et al., 2010). Yet, the variety of possible indicators to quantify social impact is large (Jørgensen et al., 2008). Table 6.5 summarizes a set of selected indicators according to Jørgensen et al. (2008).

Table 6.5 Selected social impact indicators according to Jørgensen et al. (2008)

Level	Social impact category	Indicators
Mid-point	*Human rights*	Non-discrimination (incl. indicators on diversity, e. g., composition of employees on all levels according to gender, age group, disabled, part-time workers, and other measures of diversity); freedom of association and collective bargaining; child labor (incl. hazardous child labor); forced and compulsory labor

Level	Social impact category	Indicators
	Labor practices and decent work conditions	Wages (incl. equal remuneration on diverse groups, regular payment, length and seasonality of work and minimum wages); benefits, including family support for basic commodities and workforce facilities; physical working conditions (incl. rates of injury and fatalities, nuisances, basal facilities, and distance to workplace); psychological and organizational working conditions, e. g., maximum work hours, harassments, vertical, two-way communication channels, health and safety committee, job satisfaction, and worker contracts; training and education of employees
	Society	Corruption (incl. incidents/press reports concerning fraud, corruption and illegal price-fixing, and violation of property rights; development support and positive actions towards society (incl. job creation, support of local suppliers, general support of developing countries, investments in research and development, infrastructure, and local community education programs); local community acceptance, e. g., complaints from society and presence of communication channels; ensuring of commitment to sustainability issues from and towards business partners
	Product responsibility	Integration of costumer health and safety concerns in product, e. g., content of contaminants/nutrients, other threats/benefits to human health (including special groups) due to product use, and complaint handling system; information about product to users, e. g., labelling, information about ingredients, origin, use, potential dangers, and side effects; marketing communications, e. g., ethical guidelines for advertisements
End-point	Mortality; morbidity; autonomy; safety, security, and tranquility; unequal opportunities; participation and influence	

Recent scientific literature reveals initial approaches to combined LCA and SLCA approaches. For instance, Foolmaun and Ramjeeawon (2013) merge LCA and SLCA in order to assess environmental and social impact of used polyethylene terephthalate (PET) bottles in Mauritius. Chang et al. (2015) use a similar approach to evaluate welding technologies in order to evaluate salary sufficiency of welders in Germany regarding environmental impact and potential health risks.

A similar approach exists within the framework of the general reporting initiative (GRI). Although the indicators used here are largely consistent, it should not be regarded as equivalent. While sustainability reporting is primarily a company-level document for marketing and financing purposes, SLCA offers a tool for detailed and targeted analysis of social issues throughout the entire life cycle of a product or production system.

6.2.3 Life cycle costing (LCC)

While LCA and SLCA basically originate from the concept of sustainability, LCC represents a pure economic analysis that emerged from the total costs of ownership approach (TCO). Today it is necessary to distinguish between two schools of thought: [1] the classic TCO/LCC approach (Figure 6.10) and [2] the integrative approach that primarily serves to add the economic perspective to LCA and SLCA.

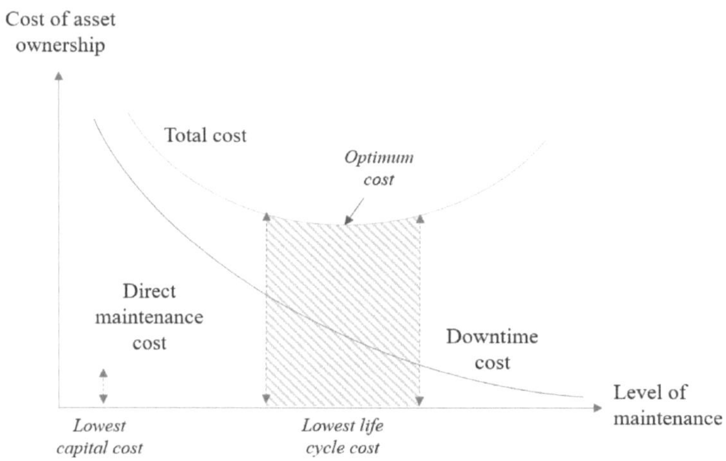

Figure 6.10 Conventional approach to LCC according to Woodward (1997)

The *conventional approach* is based on its origins. The method was initially developed in recognition of the fact that some very capital-intensive goods generate much higher costs in the use and disposal phase (e. g., through maintenance, operating expenses, significant depreciation). Woodward (1997) illustrates this via maintenance costs (Figure 6.9). While these tend to be high at a project's launch due to initial setup, they stabilize at a comparatively low level during the life cycle before rising rapidly again at its end. Optimal cost-effective operation is thus only possible if the entire life cycle is considered.

Since defense applications usually have a long service life as well as frequent maintenance and upgrades (battle value enhancements), the approach originates from military considerations in the 1930s. Thus, the American General Accounting Office included the (future) maintenance costs in its tenders for the purchase of tractors. In 1971 Directive 5000.1 ('Acquisition of Major Defence Systems') was released, establishing the application of LCC as mandatory in the military sector. Accordingly, various standards exist in Germany as well, which are intended to enable a uniform approach (e. g., VDI 2884, VDMA 34.160, DIN EN 60.300-3-3). Yet, the mere fact that different standards exist proves this argument ad absurdum. Figure 6.11 illustrates the basic life cycle model assumed here according to Woodward (1997), Asiedu and Gu (1998) as well as Korpi and Ala-Risku (2008).

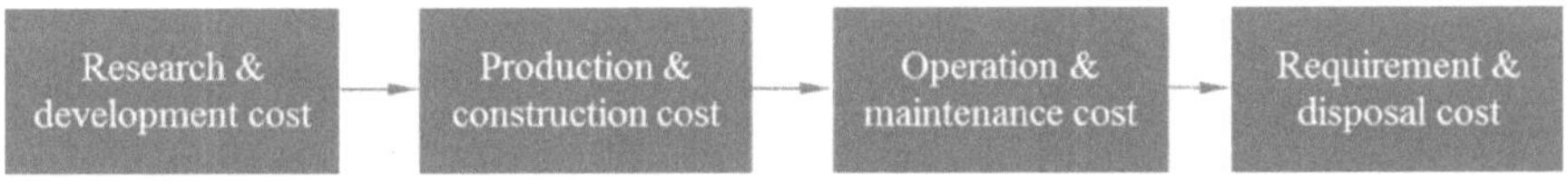

Figure 6.11 Basic life cycle model of conventional LCC according to Woodward (1997), Asiedu and Gu (1998) and Korpi and Ala-Risku (2008)

As the life cycle thinking approach here is considerably different to the one outlined above, this approach cannot easily be combined with an LCA. Yet, according to Norris (2001), a number of further differences exist between LCA and conventional LCC regarding purpose and approach exist that prevent a simple combination (Table 6.6).

Table 6.6 Differences between LCA and conventional LCC according to Norris (2001)

Aspects	LCA	LCC
Purpose	Comparison of environmental impact of alternative product and/or production systems regarding the same end-use function from broad societal perspective	Determination of cost-effectiveness of alternative investments and/or business decisions from purely economic perspective, e. g., manufacturing company or consumer
Steps considered part of life cycle	All processes causally connected to the physical life cycle of product or production system (incl. entire upstream and downstream supply chain, use and end-of-life phase)	Activities causing direct costs or benefits to the decision-maker within economic life time of investment
Flows considered	Resources, substances, materials, energy, pollutants	Monetary flows directly impacting decision-maker in terms of costs and benefits
Units for tracking flows	Mass, volume, energy, and other physical units	Monetary units
Time treatment and scope	Timing of processes, release, and consumption flows traditionally ignored; impact assessment often assessed within time window of impacts (e. g., GWP IPCC 100 year scenario)	Timing as crucial factor, present valuing (discounting) of costs and benefits; analysis relies on specific time horizon scope, any costs or benefits occurring outside that scope are ignored

Hence, a second approach was developed in recent years. The integrative approach primarily serves to add an economic and/or cost perspective to LCA and SLCA (Norris, 2001; Kicherer et al., 2007; Finkbeiner et al., 2010). In few cases it is referred to as environmental LCC (Hunkeler et al., 2008). Consequently, its procedure is based on the concept of LCA, relying on the same system boundaries, functional unit, and life cycle thinking. Yet, instead of defining specific LCIA indicators, only costs serve as measurement. An LCC of this kind is to be carried out in relation to LCA and SLCA and has to be established in the scope and objectives (step 1) accordingly at the beginning of the study. A later merging of separately performed LCCs and LCAs for the same product and/or production system is hardly possible, since different preconditions (system boundaries, functional unit, cost categories etc.) were applied and/or differing assumptions (process steps, handling of peripherals etc.) were made.

While life cycle thinking as well as related assumptions taken into account may vary considerably, the fundamental concept behind both approaches remains the same: to assess all costs along the life cycle of a product or production system. While conventional methods of investment assessment primarily consider purchase price and operating costs, LCC extends the scope beyond company boundaries. According to Norris (2001), five types of costs may be considered relevant for an LCC, depending on the purpose and design of the intended study (Table 6.7).

Table 6.7 Relevant cost types for LCC according to Norris (2001)

Cost type		Examples
1	Direct	Direct costs of capital investment, labor, raw material, and waste disposal; recurring and non-recurring costs possible; both capital and O&M costs included
2	Indirect	Indirect costs not allocated to the product or process (overhead); both recurring and non-recurring costs possible; both capital and O&M costs included
3	Contingent	Contingent costs, e. g., fines and penalties, costs of forced clean-up, personal injury liabilities as well as property damage liabilities
4	Intangible	Difficult to measure costs (incl. consumer acceptance, customer loyalty, worker wellness and morale, corporate image, community relations)
5	External	Costs of affected third parties (especially society)

6.2.4 Life cycle sustainability assessment (LSCA)

The approaches described above address ecological, social, and economic issues along the entire life cycle. Yet, an integrative assessment within the triple-bottom-line-framework of sustainability is not possible. Among others, Finkbeiner et al. (2010) outline the concept of life cycle sustainability assessment (LCSA), seeking to aggregate the results of the three separate analysis, i. e.,

$$LCSA = LCA + SLCA + LCC$$

Here a utility value (W) is calculated for each of the results of the separate analysis via normalization. By using a sustainability triangle visualization, deflections in one or the other dimension become visible. Figure 6.12 illustrates the approach according to Finkbeiner et al. (2010).

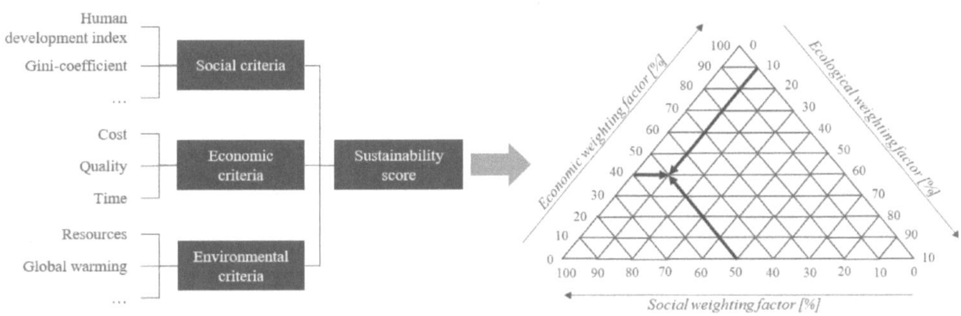

Figure 6.12 Life cycle sustainability assessment according to Finkbeiner et al. (2010)

Yet a number of issues remain unsolved for the scientific community. Among others, these include the handling of uncertainty due to low data availability and/or an overly high level of aggregation, the limited operability of results due to complexity and the handling of varying system boundaries. Reap et al. (2008) present a detailed analysis of unresolved problems in each of the four steps of LCA, which are listed here as representative for all approaches outlined above.

- Problems in *step 1 (definition of goal and scope)* arise from inclusion and exclusion decisions. In defining functional units, it is necessary to select the functions to be included as well as to decide how to quantify them. Choosing boundaries requires decisions about the inclusion and exclusion of processes. Excluding economic and social impacts from LCA imposes fundamental limits to the completeness of the instrument. In addition, the choice of alternative scenarios affects decisions in the interpretation phase. These decisions and processes reduce the accuracy of the assessment and lead to uncertainties.

- Problems in *step 2 (LCI)* primarily concern flows and transformations. An allocation problem arises from the necessity to associate flows from a multifunctional process precisely with each of its functions or the products to be evaluated. Criteria used to identify and eliminate unimportant resource and waste streams become critical when attempting to weight information costs against potentials for missing significant environmental effects. Local technical consistency becomes an issue when average or generic data are used to represent processes that deviate significantly from the norm.
- Reductions and assumptions about global homogeneity and steady states lead to the most serious errors in *step 3 (LCIA)*. By selecting environmental impact categories, the types of damage considered in a study are effectively truncated, resulting in inaccuracies. Setting arbitrary time horizons biases the results towards short or long-term effects. If spatial variation, local uniqueness and environmental dynamics are ignored, impacts of pollution concentrations are weakened, resulting in inaccurate estimates of potential environmental damage.
- Problems in *step 4 (interpretation)* primarily arise from aggregation. The aggregation of inventory or impact data into a single figure requires some form of weighting or evaluation. These, however, lead to subjectivity, which is not always handled satisfactorily by current methods of decision science. Inaccuracies in other phases as well as the inherent variability of modelled systems cause a high degree of overall uncertainty. Given the level of uncertainty, generating useful decisions thus represents a severe challenge.

Reap et al. (2008) conclude that it is necessary to further improve the concept of LCA in the long term, as otherwise it is unlikely that environmental damage can be avoided if current LCA results are used as a basis. The solution of the listed problems would improve the LCA in this respect.

6.2.5 System-theoretical interdependency analysis (STIA)

In contrast to the life-cycle-oriented approaches outlined above, the monitoring of system thinking requires a different approach. STIA aims to analyze the interactions (cause-effect relationships) between an observed system and all relevant elements/subsystems of the complex overall system it is embedded in. As such, it represents an elementary component of the development and management of BIS. The approach outlined here is based on the system dynamics (SD) approach of Forrester (1971; 1993; 1994) and the interdependency analysis outlined by Vester (1993; 2011) (see Section 4.4.1). According to Forrester (1994), the approach to SD comprises six steps. Before a system can be improved, it must first be understood. Each investigation thus starts with a detailed description of the system to be investigated (step 1). Here, it is imperative to formulate hypothesis for how the system is creating a certain problematic behavior. Thereupon, a simulation model has to be formulated that transfers the system description into level and rate equations of an SD model. While step 1 requires a general and incomplete description of the system, the writing of equations in step 2 reveals gaps and inconsistencies that have to be remedied. The simulation of the model takes place in step 3. Here, SD software may be applied. In the following stage policy alternatives for testing are developed (step 4) via intuitive insights or from experience of the analyst. Other sources are proposals by experts within the operating system or a testing of parameter changes. Each alternative has to be tested via simulation in order to determine which policy proves to be the best choice. Thereupon, a consensus for implementation has to be identified. Therefore, an education and discussion with relevant decision-makers is necessary (step 5). Not least, step 6 aims to implement the new policies. Figure 6.13 illustrates the general systems dynamics approach according to Forrester (1994).

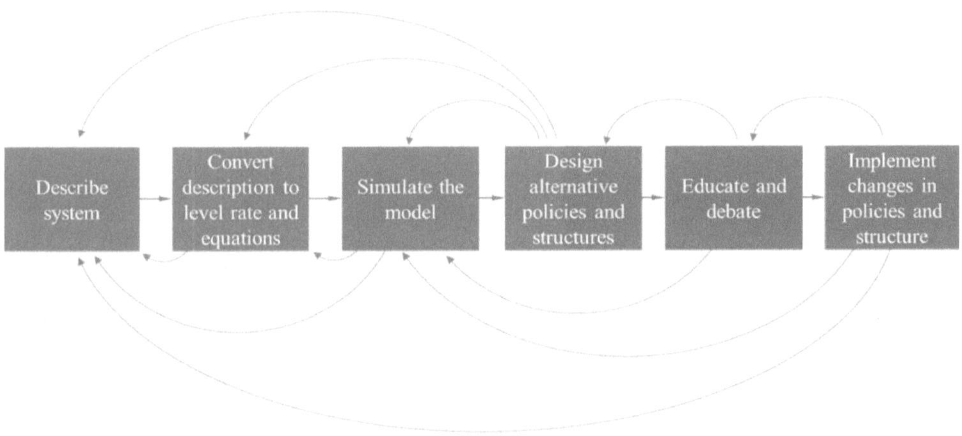

Figure 6.13 General systems dynamics approach according to Forrester (1994)

A similar approach is the sensitivity model of Vester (2011). Based on the concept of fuzzy logic, a general pattern is developed, which views the system as a patient, who has to be treated with regard to viability, self-regulation, flexibility, and controllability. This requires nine steps: [1] system description, [2] assessment of influencing variables, [3] testing of system relevance, [4] questioning of interactions, [5] determining individual role in system, [6] examining of overall networking, [7] cybernetics of individual systems, [8] prognoses and policy tests as well as [9] system evaluation and strategy. The SD approach enables a holistic analysis of complex, dynamic systems. Thereby, qualitative and quantitative methods are distinguished. The former primarily serves the identification of closed positive or negative chains of effects (reinforcing or balancing loops), which is also possible without quantitative data. The latter represents a modelling of often non-linear relationships, which is possible exclusively on the basis of quantitative factors (Roberts, 1978; Randers, 1980; Sterman, 2000). Depending on goal and scope of analysis as well as type of information and perceived reality, different supporting methods are available (Table 6.8) (Vester, 2011).

Table 6.8 Supporting methods of system dynamics approach depending on goal and scope of analysis according to Vester (2011)

		Type of information		
Perceived reality		*Classification*	*Relation*	*Relevance*
	Distinct but separate	Variables set	Influence matrix	
		Criteria matrix		Control cycle
	Complete but blurred	Impact structure		
		System description	Role distribution	
		Partial scenarios		Evaluation
	Complete and distinct		Policy tests	
		Simulation models		Strategy

While SD often requires extensive mathematical modelling, which is rarely practicable for solving real problems, simpler approaches exist that have to be regarded as a minimum standard for the development and management of BIS. Among these, the influence matrix of Vester (1993; 2011) deserves special attention, as it enables both detailed analysis via extensive mathematical modelling as well as pragmatic applications within industry environments. The approach is essentially based on three steps: [1] identification of relevant elements/subsystems of the complex overall system that are related to the observed element/subsystem, [2] (visual) comparison of each element/subsystem with each other and evaluation of mutual interaction in form of a scoring and [3] derivation of recommendations for action for decision-makers. The influence matrix is a two-dimensional matrix containing the individual factors (elements/subsystems) to be considered both horizontally and vertically. The effect by and on each factor (F) is then evaluated by assigning a numerical value. Typically, a scale from zero (no influence) to three (high degree of influence) is applied, although other scales are equally possible. The subsequent evaluation is carried out by adding the evaluations for each factor. Here, a difference is created between asset and liability totals (AT, LT). The sum of each line is the AT, the sum of each column is the LT. The former indicates the extent to which a single factor affects all other factors, the latter indicates the extent to which a single factor is affected by all other factors. The influence matrix is one of the basic principles of systems-oriented management (Wilms, 2000). As a tool for identifying the systems relations, a visualization in form of a dependency diagram can be helpful. According to the SD approach, all corresponding systems are mapped, connections between them are defined and evaluated in a similar logic. Table 6.9 illustrates the influence matrix according to Vester (1993; 2011).

Table 6.9 Blank influence matrix according to Vester (1993; 2011)

		Impact on								
		F1	F2	F3	F4	F5	F6	F7	F8	AT
Impact caused by	F1		0	0	0	0	0	0	0	0
	F2	0		0	0	0	0	0	0	0
	F3	0	0		0	0	0	0	0	0
	F4	0	0	0		0	0	0	0	0
	F5	0	0	0	0		0	0	0	0
	F6	0	0	0	0	0		0	0	0
	F7	0	0	0	0	0	0		0	0
	F8	0	0	0	0	0	0	0		0
	LT	0	0	0	0	0	0	0	0	

6.3 Rationale of design options

After introducing the conceptual framework and summarizing fundamental supporting instruments for systems-oriented life cycle thinking in cellular units, this section presents basic thoughts on design approaches for biointelligent value creation systems. Its foundation are the levels of the scope of management concern of the framework introduced in Section 6.1. This excludes the basic technologies, which were discussed in Chapter 5. Selected methodological basics for the subsections presented here are summarized in Annex 1 to Annex 5.

6.3.1 Markets and customers

The markets and customer groups that might be covered by biointelligent products and/or biointelligent production cannot be assessed conclusively at present. In addition to the prominent examples of on-site production of personalized medicine and food, a multitude of other biointelligent applications appear promising from today's perspective, including decentral energy generation, circular on-site production of consumer goods as well as biointelligent buildings (Bauernhansl et al., 2019). While this is subject to speculation at present, it is clear that the concept of biointelligence envisages a shift in the focus of value creation towards the customer. In this context, Miehe et al. (2019b) picture trends in four essential fields of human consumption:

- *Health*: from active ingredients to personalized therapy!
- *Nutrition*: from mass production to just-in-time production!
- *Consumption*: from linear to circular products!
- *Housing*: from consuming to self-sustaining life styles!

Figure 6.14 Shift of focus of value creation towards the customer [own illustration]

Consequently, a new approach is required to the concept of customer centricity, which is well-known from marketing science (Shah et al., 2006; Fader, 2020). While the classic view of this concept advocates a shift from product to customer focus, the consumer always remains the individual. In the context of biointelligence, however, it is necessary to establish society and the environment as full-worthy customer groups. This approach is expressed in framework of holistic needs satisfaction (see Section 4.4.3). In addition to a conceptual structure, the expansion of the benefit context requires the creation of suitable framework conditions, which largely depend on legislation (see Section 6.3.5). Biointelligence requires not only the personalization of products, but also the ability of the customer to act as producer. The already widely accepted concept of prosumers thus needs to be expanded as well. While the classical understanding of this approach primarily characterizes the producing activity of the customer by minor interventions (e.g., the articulation of wishes) in the production process (Kotler, 2010; Ritzer et al., 2012), biointelligence regards individuals as full-worthy producers.

6.3.2 Strategy and business models

The question which strategies and BMs are relevant for BIS cannot be answered conclusively in this work. In many cases, however, it is clear that these will be very different from traditional strategies and BMs. This is due to the fact that the perceptions of production largely differ (Table 6.10). Annex 1 provides a summary of the basics of strategic management in order to provide the reader with a brief overview of the state of the art.

Table 6.10 Comparison of traditional and biointelligent understanding of production [own interpretation]

	Traditional understanding of production	Biointelligent understanding of production
Definition	Production as a planned combination of input factors enabling a satisfaction of individual customer needs	Production as a spontaneous or planned combination and recombination of cellular units to create or dismantle a cellular structure capable of intelligent behavior enabling a satisfaction of individual, social, and environmental needs
Scope	Production (enterprises) as individual units for the purpose of economic benefit to the shareholders	Entirety of production as an organism aligned with planetary boundaries, individuals, communities, enterprises as subsystem production units
Scale	Mostly medium sized or large enterprises	Networks of small enterprises, communities, individuals, and households
Volume	Mostly centralized production units operating in series and mass production processes; partly small series production; individual production only in few cases	On-site production of single products in individual spontaneous production networks as standard model
Decision context	Free, i. e., fully independent decision-making	Normative, i. e., regulated decision-making directed to planetary boundaries/ budgets
Decision parameter	Mostly efficiency and costs	Increasingly effectivity and sufficiency driven
Scientific basis	Engineering and information science	Engineering, information, and life sciences

Since individuals, households and non-specialized companies are intended to increasingly fulfil production tasks within a biointelligent production network[2)], two perspectives have to be differentiated in the context of strategic management. Individual (corporate) strategies depict the vision, mission, and targets of a single production unit, while production network strategies result from the spontaneous merging of

[2] Examples could be the on-site production of nutrition (e.g., in vitro meat production in a fridge) or the on-site production of personalized medicine at a local hospital.

various single production units in order to fulfil a collaborative vision, mission, and targets. While individual (corporate) strategies are likely to continue to be developed according to the basic approaches of strategic management (see Annex 1), several examples of network strategies have been reported under the term network based view of strategic management (Lavie, 2006; Stokes, 2006). Generally, the network-based view is understood as an extension of the resource-based view of strategic management. The question of whether this also applies to biointelligent production remains to be examined. In any case, a transfer of the network-based view to biointelligent production will have to provide a much more granular understanding of production networks and apply the systems-oriented life cycle thinking in cellular units outlined above. Meanwhile, the perspective of value creation must be expanded, especially to include reuse and recycling options as well as new network participants (e. g., households, individuals). Accordingly, three hypotheses can be formulated that will have to be examined by future research:

- Individual strategies must match production network strategies in order to effectively produce sustainable products.
- Individual strategies have a comparatively long-term orientation, while production network strategies focus on short-term and medium-term intervals, i. e., the duration of production networks depends on the utility of cooperation. Once the network no longer offers a common benefit to the participants, it dissolves.
- The fulfilment of socially and environmentally beneficial strategies will continue to largely require incentives and regulation by the government, both for individual production units and for production networks (see Section 6.3.5).

Closely related to the concept of strategy is the configuration of the appropriate BM. Traditionally, the term describes the functional nature of a company and its basic approach to generating profits (Morris et al., 2005; Zott et al., 2011; Burkhart et al., 2011). An elementary aspect of BMs is the search for success factors. Annex 2 summarizes basics of BMs.

According to Burkhart et al. (2011), the emergence of the internet in particular has led to a significant increase in the number of BMs. This trend continues with I4.0 (Man and Strandhagen, 2017). Due to the addition of a third dimension (biology), accompanied by the normative approach described above, further options are to be expected in the context of biointelligence. Compared to traditional production, biointelligent production represents a fundamental change of framework conditions. The production networks that until now only exist in theory constitute completely new constellations of actors. Accordingly, knowledge, skills, technology, infrastructure, and governance must be handled in a fundamentally different way. This also applies to the role of governments. Before discussing potential BMs, it is thus necessary to introduce a thought model by distinguishing between an unstructured basic system, a structured production system and a semi-structured consumption system.

The basic system covers the entirety of input factors necessary for a production task including basic materials, knowledge provider/experts, technology and infrastructure provider etc. A production task has to be triggered by an impulse. This may either result from customer-pull or technology-push. This again requires a mechanism that mediates the input factors, e. g., platforms. In this model, the government, which has the mandate to provide environmental and social protection, acts as a genuine market participant that buys the services needed to preserve both the state of society and the environment. This represents a triggering moment for the emergence of production networks. After a production network has fulfilled a given task and provided the output (i. e., the product), it dissolves. The consumption system is to be understood as semi-structured, since not all utilization paths and abrasions of the product can be traced and it remains impossible to ensure that the product is used according to its original purpose. After the use of a product, another production task is triggered, in this case with the focus on reproduction. Whereas in the case of initial production the government appears as an actor in individual cases, it is responsible for creating the framework conditions for the productive utilization and/or safe disposal of all substances contained in the product. This may also be understood as a triggering moment, i. e., the commissioning by the government. Figure 6.15 illustrates this simplified model of thought that sets the basis for the understanding of biointelligent BMs.

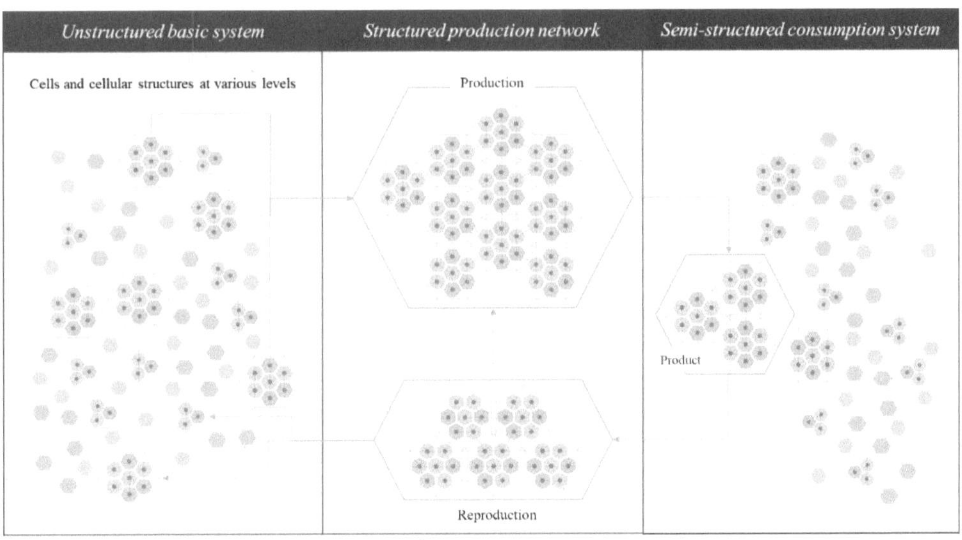

Figure 6.15 Thought model for the description of biointelligent BMs [own illustration]

Since this understanding of biointelligent production currently remains a theoretical concept, the BMs presented below are not based on observation, but on the author's basic considerations regarding the fundamental structure of a biointelligent value creation system. The models presented should therefore rather be understood as design approaches for future BMs, derived from a number of approaches that have been formulated in the scientific community (Dijkman et al., 2015; Ibarra et al., 2018; Müller, 2019). Therefore, three fundamental questions are addressed: [1] What services are likely to be needed in the marketplace? [2] Who could provide them? [3] Who could buy them? Further essential questions in collaborative BMs including the questions of which partners and what extend of customer centricity are required (Weiblen, 2015), will intentionally not be considered here, as this would require real world examples that simply do not exist.

A prerequisite for biointelligent value creation is a broader distribution of expertise. Although it can be partially bundled via internet and open access formats, experts will continue to exist whose knowledge is crucial for a production task. In a biointelligent production system, these experts represent a single cell that can offer its services to the market through established advisory and consultancy earnings models (know-how as a service). Another form of a know-how as a service model could be the provision of construction plans. As production equipment will increasingly be located on-site, this concept, which has been discussed for years in the context of additive manufacturing, will gain importance.

Likewise, human labor will continue to be necessary in the production system, although its nature is changing constantly due to the emergence of new technologies. Unlike the provision of knowledge, individuals will continue to monetize (skilled and unskilled) physical labor capabilities in a biointelligent production system. In this labor as a service model, an increase in the share of freelancers is expected in comparison to common employment models.

Then again, production requires equipment and infrastructure. Increasing decentralization may lead to a rising demand for contracting services, especially if production equipment remains costly and households and individuals emerge as producing units in networks. A BM in which the provider retains ownership and is compensated only for the use of the equipment is referred to here as equipment as a service.

Equally necessary for production is the capital to buy the input factors and carry out the production task. Individuals and companies (mostly banks) can also offer their services in a biointelligent production system via established investment and financing models. This capital as a service model, however, requires that the capital can be distributed much more rapidly to previously separate, spontaneously emerging alliances. This again requires the establishment of appropriate structures.

Apart from the provision of know-how and technology, the ability to spontaneously create production networks largely depends on the existence of mechanisms that mediate the input factors. A solution may be a platform as a service BM, in which the provider provides the environment and the participants pay for access to it. Similarly, individual companies and households may not always have the complex expertise required to operate not just a single technology, but an entire technology platform. Another possibility for a platform as a service model could be the provision of a technology platform (including relevant know-how for its operation). In this case, even non-experts could easily benefit from a more complex technology.

The future importance of product recovery and recycling can be financed through a product as a service model. Similar to the equipment as a service approach, the producer retains ownership of the consumer good and the client pays for it via common leasing models.

Companies or individuals with the necessary infrastructure, technology, and know-how may naturally choose to employ the rather standard, more integrated (re)production as a service model in a biointelligent system. Initial approaches and frameworks for BM design in the context of sustainability and circular economy are presented by Abdelkafi and Täuscher (2016), Schaltegger et al. (2016), and Lüdeke-Freund et al. (2019) (see Annex 2).

Both production and reproduction, especially in spontaneously emerging production networks, require an efficient allocation. This is the task of management. In the context of biointelligence, management competencies are likely to experience increased attention. Providing ad hoc management for the spontaneously merged production networks in form of a management as a service approach thus appears as a promising model.

Ultimately, it is an elementary component of biointelligent production to view nature as a scarce rather than a free good. This provides governments with the opportunity to provide portions of ecosystem services in the sense of an environment as a service approach. In this context, the active commissioning of a production network to solve an environmental problem is equally conceivable. The range of earnings models includes environmental prizing and the entire spectrum of existing regulatory instructions (see Section 6.3.5). Table 6.11 summarizes design approaches for potential BMs in the context of biointelligent production.

Table 6.11 Design approaches for potential BMs in biointelligent value creation system [own portrayal][3)]

Business model	Value proposition	Provider	Buyer	Earnings model
Know-how as a service	Providing know-how, design plans, and skills	All market participants	All market participants	Common advisory and consultancy models, pay-per-unit
Labor as a service	Providing necessary manpower	Individuals	All other market participants	Pay-per-hour, common employment models

[3] A possible approach to the further development of biointelligent business models could be based on the work of Weiblen (2015), who presents fundamental considerations on the topic of collaborative business models.

Table 6.11 Design approaches for potential BMs in biointelligent value creation system [own portrayal] (*continued*)

Business model	Value proposition	Provider	Buyer	Earnings model
Equipment as a service	Providing equipment and infrastructure for production task	All market participants	All market participants	Common contracting models
Capital as a service	Providing required capital for production task and/or spontaneous emergence of production networks	Individuals, companies	All market participants	Common investment and financing models
Platform as a service	Providing platforms, either technological or as mechanism for input factor allocation	All market participants	All market participants	Participant fee, pay-per-use
Product as a service	Providing products that have to be returned after use	All market participants	All market participants	Common leasing models
(Re)Production as a service	Manufacturing of goods	All market participants	All market participants	Common approaches of production related earnings models
Management as a service	Providing ad hoc management for the spontaneously merged production networks	Individuals, companies	Networks	Share of profit, pay-per-hour etc.
Environment as a service	Providing portions of the scarce good of ecosystem services	Government	All other market participants	Rigorous prizing[4)] and common approaches of regulation (see Section 6.3.5)

[4] Environmental prizing is typically a common instrument of regulatory economics. However, this understanding is based on the assumption that nature is not a scarce but a free good. In the context of biointelligence, this understanding is rejected.

6.3.3 Innovation and product development

Complementary to Schumpeter's interpretation of an innovation as 'creative destruction' (Schumpeter, 2005; 2007) (see Section 4.2.3), a biointelligent innovation requires a positive contribution to ensuring the viability of the complex natural overall system. In addition to the recombination of production factors (creation) and the replacement of old structures (destruction), which primarily describe actions at the micro level (e.g., a technical development in an explicit sense of biointelligence), biointelligent innovation must ensure that no adverse effects occur within the framework of its complex structure with the diverse elements/subsystems of the complex natural overall system (macro level). Hence, sustainability represents an integral aspect of biointelligent innovations. According to Fichter et al. (2005), sustainable innovation strategies can be divided into six paths of origin in business practice:

1. sustainability as the primary goal of an innovation (i.e., a shortcoming is identified and eliminated at an early stage)
2. sustainability as an integral corporate goal (i.e., sustainability as a strategic success factor, which is a normative variable in innovation management)
3. sustainability as an accidental discovery in an existing development process (appropriate awareness of participants required)
4. sustainability as a corrective in an existing innovation process (e.g., due to public criticism)
5. sustainability as a subsequent attribute of the product for the purpose of marketing
6. sustainability as an invisible hand (sustainability potential is identified, for example, by monitoring external actors, but was not intended)

In contrast to this descriptive approach of current business practice, the normative approach of biointelligence requires to prioritize sustainability in any case. Figure 6.16 illustrates the development paths of sustainable innovations according to Fichter et al. (2005) as well as the classification of a biointelligent innovation.

A biointelligent innovation always pursues the same main objective (increasing and securing the viability of the complex natural overall system). To meet this goal, the identification of trends and the development of ideas must always be complemented by system-theoretical interdependencies analyses (Vester, 2011).

Innovation management provides the basis for product development (see Annex 3 and Annex 4). Biointelligent products represent cluster of biotic and abiotic cells connected via information technical networking that have a proven benefit to the individual, societal, and environmental benefit system (see Chapter 4). While classical product development generally follows a strict process in manufacturing companies (see Annex 4), biointelligent products will increasingly be produced by non-expert

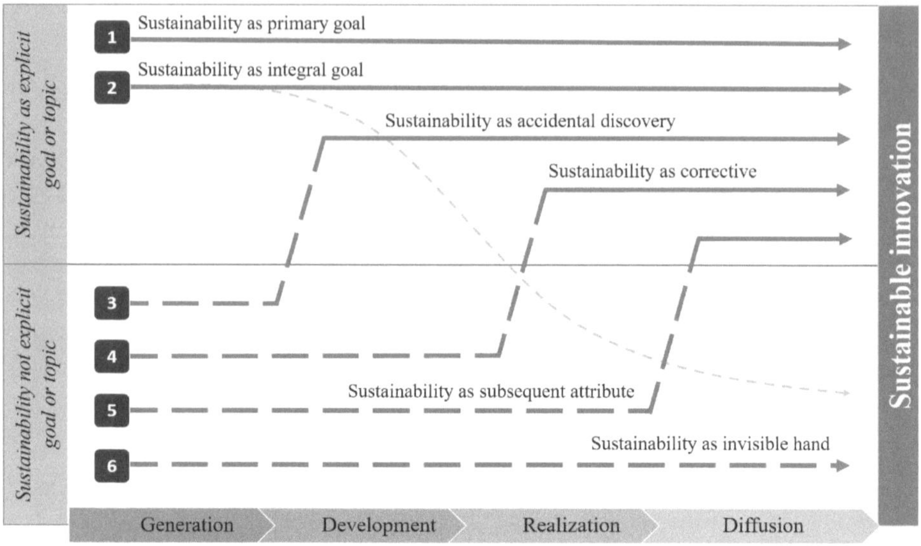

Figure 6.16 Paths of sustainable innovations according to Fichter et al. (2005) and classification of biointelligent innovation

individuals or small groups. In order to meet the requirements of a biointelligent innovation, however, this requires the extension of scope of classical product development. Hence, other factors are becoming increasingly important in addition to the actual product, including services as well as the interaction with or satisfaction of communities and societies. A variety of concepts, primarily in scientific sustainability related literature, exist that aim to extend the scope of product development, e. g., design for environment (DFE), design for recycling (DFR), design for disassembly (DFD), design for sustainability (DFS) or life cycle design (LCD). At the same time, the discipline is gradually evolving. Ceschin and Gaziulusoy (2020) show that the field of sustainable design has broadened in the past couple of years, expanding its scope from improving single product elements to entire product life cycles. In recent years, research has been increasingly focusing on socio-technical system innovations, which in turn requires the selection of different system boundaries, methods, and optimization strategies as well as the incorporation of other stakeholder requirements. The development of the scientific field can be understood as a hint that even companies that manufacture individual products in the future will increasingly have to work and design in systemic alliances. Entirely independent product design that follows the philosophy of unconstrained growth has to be replaced by a holistic approach (Vester, 2011). Individual entities can only implement what does not harm the overall organism, i. e., the ecosphere and sociosphere. Figure 6.17 illustrates the evolution of sustainable design approaches according to Ceschin and Gaziulusoy (2020) and defines the scope of 'design for biointelligence (DFB)'.

Although a biointelligent product specifies the material basis, it can only represent an intermediate or finished product (see Chapter 4). DFB therefore starts with the development of products and includes the product-service system, the spatio-social system and the socio-technical system. It combines approaches from green design, biomimicry, the cradle-to-cradle principle, eco design, design for sustainable behavior, product service system design, design for social innovation, systemic design, and design for sustainable transformation. Extensive works on the methodological basis of these existing approaches are provided by Bhamra and Lofhouse (2007), Ramani et al. (2010) and Ceschin and Gaziulusoy (2020). At the same time, it must be clear that the methodological basis must continue to be supplemented in order to optimally reproduce the holistic benefit criterion. Annex 3 and Annex 4 summarize basic supporting instruments for classical innovation and product development processes in manufacturing companies with regard to sustainability issues to provide the reader with a better understanding of the state of the art.

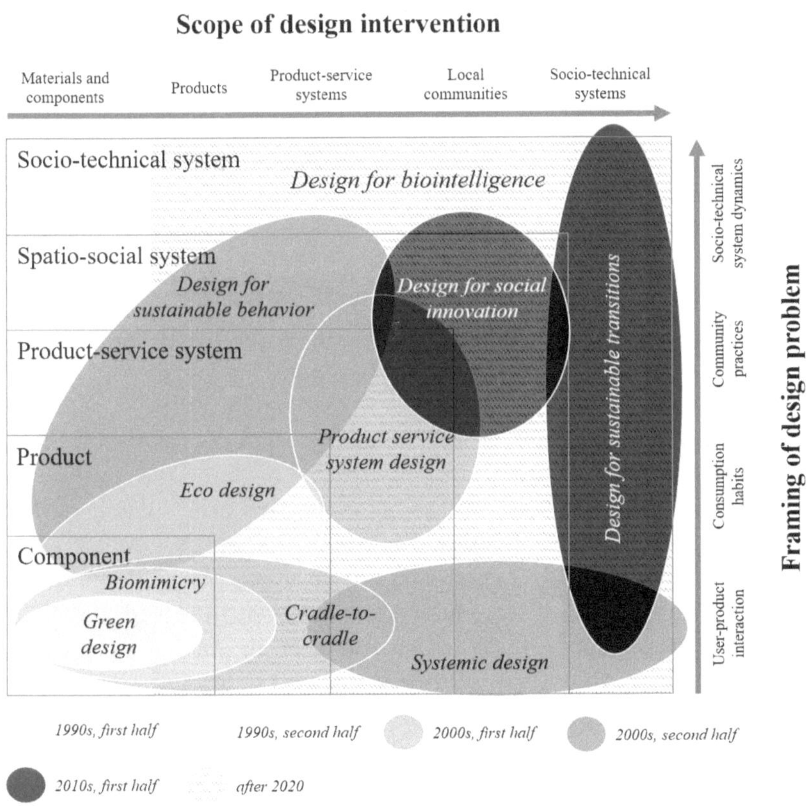

Figure 6.17 Evolution of sustainable design approaches according to Ceschin and Gaziulusoy (2020) and scope of 'design for biointelligence'

6.3.4 Organization and management

For organization and management, certain challenges arise from the outlined approach to biointelligent value creation, which can only be partially addressed by the existing toolbox of production science and practice. The desired state of a biointelligent production network that spontaneously joins forces as a result of a triggering moment is the equal, voluntary collaboration of cellular units that are initially completely detached from each other. Compared to the mechanistic view of the classical and neoclassical periods, this represents a fundamental change. Even compared to todays 'digital age', the transition is immense. Figure 6.18 illustrates the difference between a biointelligent production organization and existing approaches from the classical, neoclassical, and digital era.

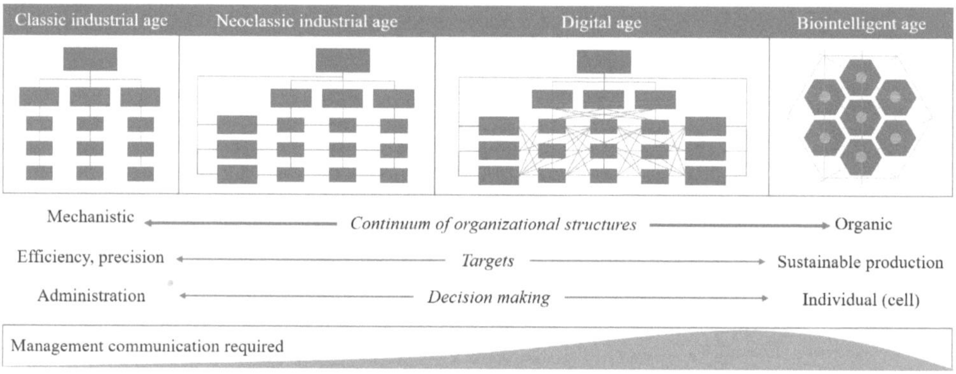

Figure 6.18 Biointelligent organizational structure in comparison to existing industrial structures [own illustration based on Westkämper (2006)]

Basic approaches to a systems-oriented cellular understanding of organizations and factories have been formulated in literature since the 1960s, but have rarely been transferred into practice. The most prominent examples are Beer's viable system model (VSM) (Beer, 1967; 1972; 1995), Miller's living system theory (LST) (Miller, 1978), and Warnecke's concept of the fractal factory (Warnecke, 1992). Related approaches are documented in the field of evolutionary management (Stead and Stead, 2008). The characteristic feature of these concepts is the perception of production as an organism. Especially VSM, which is complementary to STIA (see Section 6.2), represents a fundamental approach to the organization and management of biointelligent production systems (see below). In turn, the fractal factory approach, although being derived from VSM and LST, is not easily transferable to biointelligence. This becomes clear by the understanding of a fractal. According to Warnecke (1992), a fractal is an independently acting corporate unit whose goals and performance can be clearly described. They are self-reliant, i. e., organize themselves, independently determine their internal and external relationships and the services to be provided, and formulate their own goals. Furthermore, fractals are spontaneous (i. e., they transform themselves, emerge and dissolve) and networked via high-performance information and communication systems. As such, the concept of fractals forms a basis for the cellular thinking of biointelligence. As introduced in Chapter 4, the concept of biointelligence divides reality into cellular units with each cell comprising a segment of reality. Cells occur at different levels from single molecules to entire ecosystems) and are sufficiently described by their type of matter, internal processes, interrelations, purpose for application, functional principle, and conditions to fulfil its purpose. Whereas all the characteristics of a fractal can be transferred to cells of a biointelligent production system, the scope of cellular thinking in the context of biointelligence clearly extends the approach (e. g., fractals cannot be used to describe relationships between biological and technical systems or the alignment of single production units and planetary boundaries). Moreover, biointelligence extends the system boundary beyond the factory and integrates previously excluded groups into the value creation process (individuals, households, non-specialized actors, etc.). The concept of the fractal factory formulated in the 1990s is therefore rather a visionary basis for today's I4.0 standard than for BT and sustainability.

In order to adopt VSM to biointelligent production, a short introduction is necessary. Generally, the approach serves as a reference model for describing, diagnosing and designing the management of organizations (Beer, 1967; 1972; 1995). The instrument allows to identify relevant management functions on different organizational levels and to visualize information flows between them. The core objective of the approach is to develop a viable system that is able to adapt flexibly to external and internal changes (Pfiffner, 2010). This requires the system to sense internal and external changes and utilize them in a reasonable fashion in order to independently evolve. The elementary goal is not profit maximization, but survival. Vester (2011) adopts this understanding and transfers it to the natural overall system as the core objective of any human action. Beer (1967; 1972; 1995) distinguishes between four systems. Production (system 1) consists of operational units that perform a value-adding activity. While system 1 focuses primarily on production within fixed boundaries (i. e., the factory), biointelligence assumes a less defined structure. In both cases, however, it is indispensable that each production units must be viable itself. System 2 coordinates system 1 by providing a venue for the self-organization of all systems 1. Optimization of current resource allocation is handled by system 3, supported by its subsystem (system 3*), which is responsible for auditing the information flow through various channels. Future resource allocation involving planning and analysis is handled by system 4. To date, various applications of VSM have been presented in literature, which mainly address internal business interrelationships (Espejo, 1990; Schwaninger, 2006; Leonard, 2009; Hoverstadt, 2009; Pfiffner, 2010; Vahidi et al., 2019). Simultaneously, the instrument is applied for the analysis of systemic interrelations on a macro level. A particularly valuable approach to biointelligence is presented by Schwaninger (2015). Here, VSM is used to develop a framework for sustainable renewal that is capable of including all relevant contexts, including individuals, organizations, communities, regions, and the world. Therefore, a further system is introduced. This system 5 seeks a balance between the present and the future. The system is responsible to develop supreme norms and values that govern the system. Systems 1-2-3 represent the operative, system 4 the strategic and system 5 the normative system. 3-4-5 interact in a fictive meta-system. To date, however, cellular thinking has not been applied to VSM. An appropriate application of the method in the context of this work would thus have to incorporate the key characteristics of cellular units of biointelligent systems for each unit of analysis. Figure 6.19 thus illustrates a simplified approach to VSM with respect to the cellular thinking of biointelligence based on Beer (1967; 1972; 1995) and Schwaninger (2015). Whether additional granularity in form of cellular thinking adds value to VSM remains to be demonstrated. It is clear, however, that VSM is a fundamental component of the organization and management of biointelligent production.

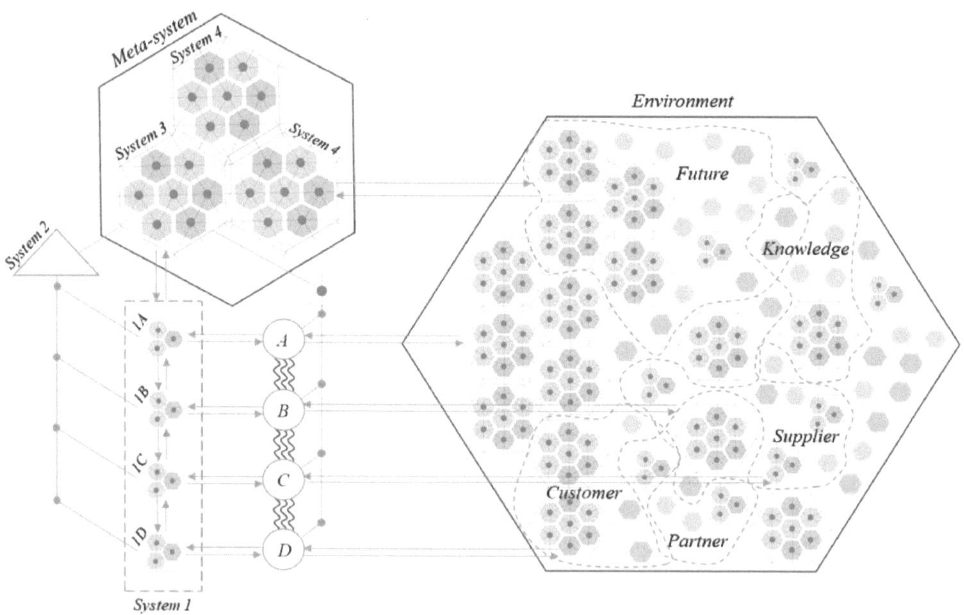

Figure 6.19 Simplified approach to VSM with respect to the cellular thinking of biointelligence [own illustration based on Beer (1967; 1972; 1995) and Schwaninger (2015)]

Apart from providing methodological support for the mapping of system interrelationships, the practical management of BIS, especially in the context of established entrepreneurial activity, requires the integration of social and environmental aspects at every level of decision-making. The above considerations provide guidance for the expansion of the scope of current management systems. In this context, Annex 5 presents extracts from the state of the art of holistic corporate management. The approaches discussed here serve the reader as additional indications for the further development of the state of the art.

6.3.5 Responsibility

It is imperative that each BIS contributes to the persistence of human life on earth (holistic benefit criterion). This requires a high degree of responsibility of management. Two basic alternatives exist to ensure this: individual voluntary action and government regulation.

Individual voluntary action is to be accomplished primarily via the framework described above and the specific application and improvement of the current method base. Therefore, the foundation must be the systemic life cycle monitoring of all activities (which already represents an enormous challenge due to the diversity of methods and the unsolved problems of life cycle analysis) and the thorough compliance with all applicable regulations over the entire life cycle of an entity. At the same time, the accounting of individual economic activities does not ensure that the entirety of production remains within planetary boundaries. One approach to linking the concept of microeconomic accounting via life cycle analysis with the planetary boundaries is the IPAT equation (Hauschild et al., 2017; Hauschild et al., 2020). The formula that was developed in the 1970s by Ehrlich and Holden (1971) is a mathematical expression for describing the impact of human activities on the environment. The impact (I) is the product of the factors population (P), affluence (A) and technology (T), i. e.,

$$I = P \cdot A \cdot T$$

A fundamental principle is this: If population and affluence increase, the same must apply to the eco-efficiency of the technology used to provide it (Hauschild et al., 2017; Hauschild et al., 2020). The eco-efficiency (EE) is calculated from the inverse of T, i. e.,

$$EE = \frac{1}{T} = \frac{P \cdot A}{I}$$

Assuming a similar normative framework as the one outlined above, Hauschild et al. (2017; 2020) differentiates between two approaches to the application of the IPAT formula in microeconomic management: top-down and bottom-up. The top-down approach is based on a specific interpretation of the IPAT formula. In this approach, product output is introduced as a frictional variable and expressed in relation to population and pollutants emitted. This results in a separate interpretation of affluence as the ratio of product output and population, and of technology as the ratio of emitted pollutants and product output. The basis for this is the microeconomic accounting via the life cycle analysis. Hauschild et al. (2017; 2020) describe three innovation paths towards sustainability: [1] incremental innovations and redesign are targets for the further development of technologies, [2] function and system innovations are determined by the ratio of product output and population (which is most likely to be changed by a transformation of life style), and [3] genuine sustainable innovations are directly related to population growth. Figure 6.20 illustrates the top-down and bottom-up approach of IPAT applications for microeconomic management according to Hauschild et al. (2017; 2020).

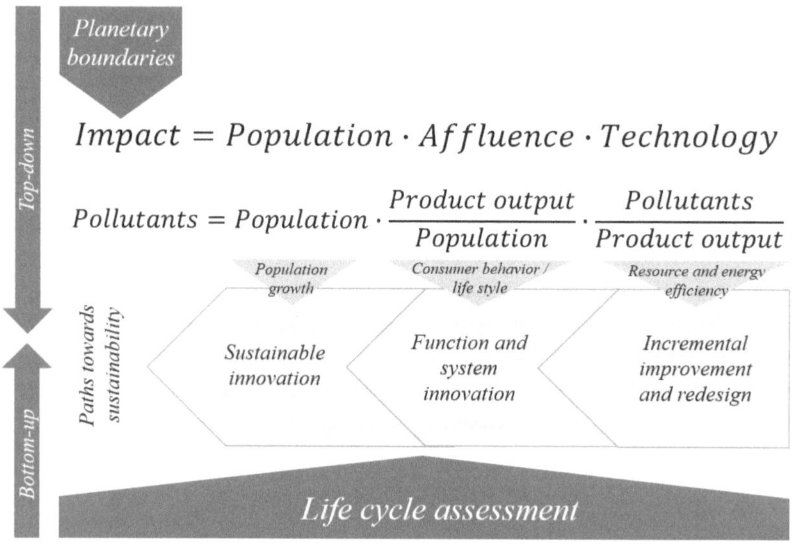

Figure 6.20 Top-down vs. bottom-up approach of IPAT applications for microeconomic management according to Hauschild et al. (2017; 2020)

Although the approach provides a generally useful conceptual structure, the link to the planetary boundaries remains vague. By simply assuming that the uncoordinated micro-level endeavors (e. g., by applying methods of life cycle design in product development) result in a global optimum, it fuels the notion of solving the environmental problem via microeconomic technical solutions. A thorough optimization of a product along its entire life cycle represents an improvement for this very product, but not for the overall system. As outlined in Chapter 3, the optimization of a system within a too narrow framework always leads to a shift of problems into precisely those areas that are beyond consideration. This again emphasizes the importance of the system-theoretical embedding of every life cycle in the context of the overall system (Figure 6.21).

While the science-based targets initiative represents an attempt to address the issue of translating global into business goals (Science Based Targets Initiative, 2020), a credible implementation in the implicit sense of biointelligence would require a more comprehensive procedure. Ensuring production within planetary boundaries clearly represents a regulatory concept, which largely implies a rejection of purely autonomous decision-making as well as uncontrolled technology development and application at the micro level (Miehe et al., 2021b).

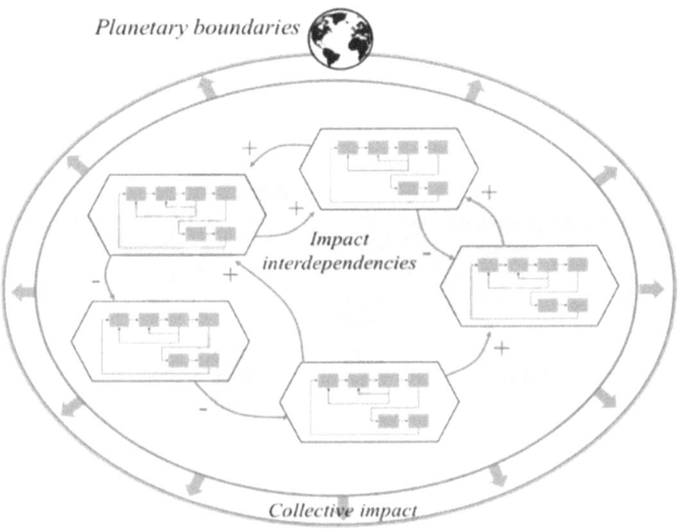

Figure 6.21 Systemic embedding of entirety of product life cycles in the context of the natural overall system according to Miehe et al. (2021b)

If it is to be ensured that industrial production remains within the framework of planetary boundaries, the general understanding of production must change (see Chapter 4). Biointelligent production in an implicit sense is not a single economic act, but a self-contained organism whose natural basis (the planetary boundaries) must not be exceeded. This understanding of production presumes the existence of a globally acting controlling unit that acts as a regulatory authority (normative approach). It would have to monitor and control essential aspects of production (e. g., input factor type and provision, scope of factor combination, volume of goods in circulation). An authority of this nature would be indispensable to assess the current status of planetary stability at regular intervals, to calculate a safe-operating space for a certain period according to standardized procedures, to break down individual units (planetary budgets) into single economic units (companies, households), and to ensure that these are adhered to by means of monitoring and enforcement (Miehe et al., 2021b).

Assuming this (under current conditions) highly unlikely scenario, the responsibility issue is returned to the political level. In economic theory, government regulations are applied due to market failures. The concept of market failure and regulation has a long tradition in economics (especially in the context of regulatory economics). Basic literature is provided by Stigler (1978), Laffont and Tirole (1993) and Cebula and Clark (2014). Two essential concepts of regulation are commonly differed positive and normative theories. While the former analyze how regulation occurs, the latter provide supporting instruments for the regulator. Both approaches are relevant to determine appropriate regulations in the context of biointelligence. This especially applies to environmental market regulations, whose basic concepts hardly changed in years. According to Tietenberg (1990) and Ren et al. (2018), these are:

- *Command-and-control environmental regulations*, e.g., certain emission trading systems that establish a highest allowable concentration of emissions in a certain category. Further examples of this approach are discharge standards and certain restrictions on production technologies or material use.
- *Market-based environmental regulations*, e.g., certain forms of emission trading, i.e., giving out certificates in form of emission budgets that may be traded in case of over-performance. Other examples of this category are various forms of emissions charges (e.g., environmental taxes), i.e., establishing a reporting system and imposing of tariffs on certain amounts of emissions. This approach is suitable to raise revenue for the funding of measures to improve or mitigate a certain environmental impact. Generally, two types of emission charges have to be distinguished. Efficiency charges seeks to achieve a certain target by forcing the polluter to compensate completely for the damage caused. A cost-effective charge in turn aims to achieve a predefined standard at the lowest possible cost. Subsidies represent another example of this category aiming to incentivize a certain pro-environmental behavior.
- *Voluntary environmental regulations*: Examples of this category are environmental standards, labels, certification, and information disclosure. Other examples include voluntary external cost accounting and corporate social responsibility.

However, besides environmental protection biointelligent production requires a variety of further considerations for governance. These include labor market, workplace design and intellectual property rights in decentralized, cellular production networks, as well as the anticipation of technology consequences and the corresponding regulation of research and basic technologies corresponding to societal will (Miehe et al., 2020a). Among others, major factors for establishing a uniform regulatory basis for biointelligent production are:

- education for sustainability, expertise and interdisciplinary collaboration
- rigorous measures against monopolies in all markets
- incentives for on-site production and regional supply chains
- challenge current intellectual property legislations with regard to open access formats
- transfer of production competencies and know-how to individuals, households and non-specialized actors
- strengthening of collaboration within and between communities and regions
- establishing remanufacturing as an elementary component of biointelligent production through regulation and incentivization
- commitment to the creation of a globally active regulatory authority and the concept of an economy based on the exchange of knowledge instead of physical products

As this work cannot provide conclusive solutions to these matters, they provide a basis for further interdisciplinary research. Simultaneously, it is imperative to seek completely new ways of environmental protection that might involve a voluntary technological standard for a self-regulation of the entirety of production via global interlinking of technologies. Miehe et al. (2021b) present basic considerations on how this might be possible using the DT concept introduced in Chapter 5. In this context, the DT has to be understood as a vehicle of an IT-supported transformation towards a sustainable production by acting as an enabling technology for an improved execution of relevant strategies, thought patterns, methods, and decisions. It is therefore essentially a tool that can support individual decisions or independently decide in terms of sustainability within a constant sequence of operations. But how exactly can this be accomplished? Therefore, Miehe et al. (2021b) define three superior categories and ten major demands for the role of DTs with regard to the implicit understanding of biointelligence that will have to be addressed by future research. While becoming relevant and/or technically feasible at different points in time, they remain equally crucial.

1. *Requirements regarding data interoperability*
   a) Ensuring interoperability of relevant data, documents, models etc. via harmonized formats, design patterns, and interfaces along the entire life cycle of an entity (including connections to existing databases such as GaBi, EcoInvent, iPoint, CDX, BOMcheck, etc.) within a standardized, adaptable and modular structure.
   b) Definition of appropriate access rights in order to provide the required data to the relevant entity.

2. *Requirements regarding the data itself*
   a) Extensive compilation of all relevant data on substances, materials, and energy (an/or metabolic pathways) used as well as emissions and waste generated along the entire life cycle of an entity.
   b) Transparency on chemical composition and physical design principles for the purpose of environmental compliance as well as disassembly, e. g., via technical documents, dismantling instructions and the like.
   c) Constant checking of standards and databases and linking them to material and substance composition model.
3. *Requirements regarding specific submodels*
   a) Submodels and simulations relevant to environmental impact assessment (e. g., characterization models, life cycle impact models, Monte Carlo simulation etc.) within the framework of existing impact categories.
   b) Predictive models for the development of scenarios for the secondary use of all substances used at different points in the life cycle (exploitation options).
   c) Predictive sensitivity models to anticipate and incorporate previously unknown relationships and impacts with and on people and the environment.
   d) Collective ecosystem impact submodel of the 'organism production' to conclusively map impact interrelations between all subsystems of the overall system as well as the collective impact to PBs according to a standardized procedure model.
   e) Active control of individual units and/or entirety of production via AI-based independent, interlinked ecosystem control submodel(s)

Figure 6.22 summarizes the role of DTs in the context of the implicit understanding of biointelligence as well as essential features in the sequence of growing contribution to sustainable production within planetary boundaries and the current scope of research. Further elaborations on the topic are documented by Miehe et al. (2021b).

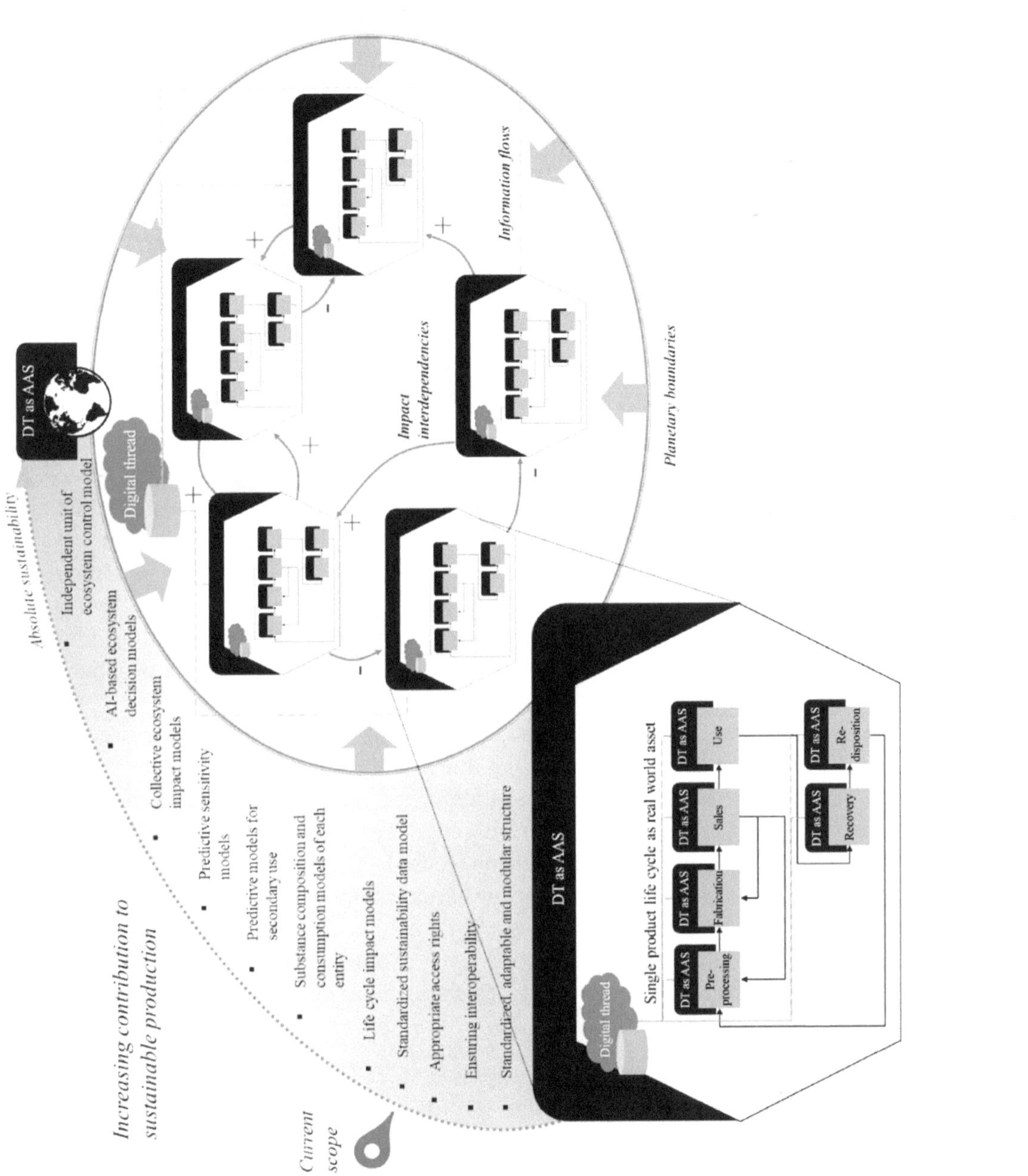

Figure 6.22 Role and essential features of DTs in the context of normative systems-thinking (biointelligent) approach of production according to Miehe et al. (2021b)

6.4 Chapter review and conclusions

The management of BIS requires a framework, a fundamental mindset, and basic approaches for the design of future value creation systems. Compared to present-day concepts of production, biointelligent production envisages a radical change. The act of production is understood as the spontaneous or planned combination and recombination of biotic and abiotic cellular units to create or dismantle a cellular structure capable of intelligent behavior. Instead of large-scale centralized factories, production takes place on-site in networks of small enterprises, communities, individuals, and households. The decisive factor for productive success is not only the fulfilment of customer needs, but also the satisfaction of social and environmental demands. The latter requires a holistic view of production as an organism that is aligned with planetary boundaries.

This section thus presented initial thoughts on an appropriate management framework for BIS. Its conceptual design is based on recent advances in the fields of LCE and business management. Two perspectives are distinguished (the scope of environmental concern and the scope of management concern), each of which is subdivided into different sequential levels. The core of the management framework is the fundamental mindset, which is systems-oriented life cycle thinking in cellular units (introduced in Chapter 4). It represents the basis for any decision in the production context and can be supported by a number of methods, each of which being subject to further research in the context of biointelligence. While life cycle thinking is predominantly supported by LCA, LCC, SLCA, and LCSA, systems thinking mostly relies on systems dynamics and its STIA. The supporting methods of the fundamental mindset also provide the main platform for the first framework dimension (scope of environmental concern). Further basic considerations included in this section focused on design options for the sublevels of the second framework dimension (scope of management concern). These included the discussion of established approaches and their fit to the concept of biointelligence in the fields of markets and customers, strategies and BMs, innovation and product development, organization and management, as well as responsibility.

Lessons learned

- The management of BIS serves primarily to ensure the holistic benefit criterion of the intended use.
- Crucial to the management of BIS are a conceptual framework, a fundamental mindset, and basic approaches to the design of future value creation systems.
- The conceptual framework presented here distinguishes between two essential concerns for future decision-makers: environmental and managerial.
- The fundamental mindset for the management of BIS is systems-oriented life cycle thinking in cellular units which can be supported by LCA, LCC, SLCA, LCSA, and STIA.

- Various approaches exist in literature to support decision-making in essential field of managerial concern, albeit partially requiring expansion to ensure appropriate application. Among principal models of thought and approaches that require an extension and/or further scientific development are:
  - the concepts of prosumer and customer centricity
  - the concept of strategy (especially when differentiating between individual and network strategies with a biointelligent production system)
  - the concept of design for biointelligence (DFB)
  - the conceptual framework for BM innovations within a biointelligent production system
  - the incorporation of cellular thinking in the concept of the viable system model and its application to biointelligent production concepts
  - the design of a regulatory framework to promote a responsible biointelligent production

7 Biointelligence in future R&D&I

After outlining basic approaches for the development and management of BIS in the previous sections and frequently addressing open questions for the scientific community, this section summarizes a selection of unresolved issues for research, development, and innovation (R&D&I) that will occupy the scientific community for years and decades to come. This section primarily relies on the findings of Miehe et al. (2018; 2020a) and Bauernhansl et al. (2019), yet is structurally adapted to this work and specifically expanded based on own research. According to Miehe et al. (2020a), the following sections provide a synthesis of existing trends and gaps in research and development in ten sub-categories. Based on the technology management perspective of this work, four domains are distinguished: [1] research and development issues that are fundamental for all other fields and/or for the comprehensive implementation of the here presented patterns of thought, [2] research and development issues that primarily address challenges in single enterprises (intraorganizational), [3] research and development issues that require at least two partners from different disciplines (interorganizational), and [4] research and development issues that demand for socio-political actions. Figure 7.1 illustrates the framework according to Miehe et al. (2020a).

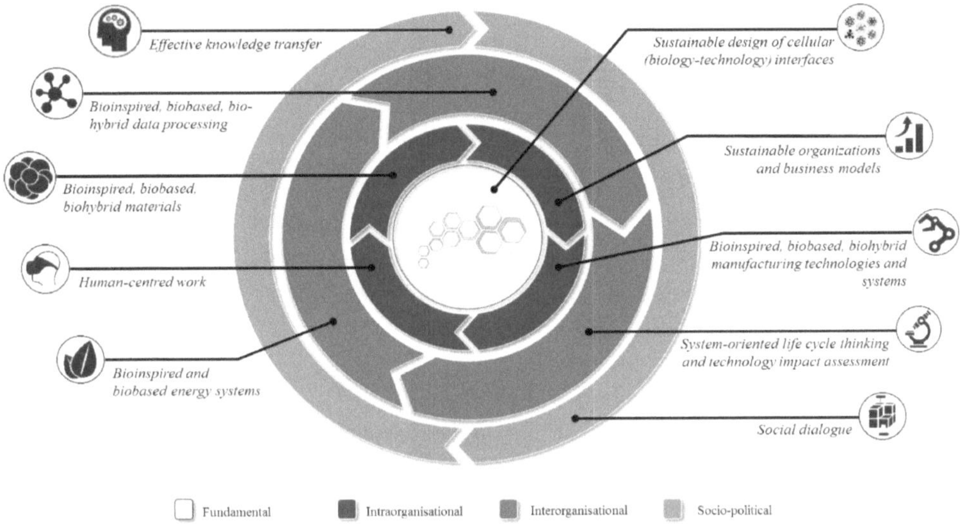

Figure 7.1 Field of action framework of BT of industrial value creation according to Miehe et al. (2018; 2020a)

7.1 Sustainable design of cellular (biology-technology) interfaces

The deliberations presented here on BTIs as a foundation for BIS in an explicit sense (see Section 5.4) represent general considerations on the coupling of biological and technical system for the purpose of production. However, this is still in its very infancy. Hence, it is necessary to further elaborate the concept and to significantly expand its scope.

Among others, further elaboration is required regarding the handling of complexity. In this context, Gilles (2004) identifies a range of similarities between the construction of cellular biological systems and the construction of complex technical processes. In order to exploit biological processes in technical systems in a way that possibly enables autopoiesis, modularity, hierarchical structuring of regulations, redundancy and diversity are essential design features. Cellular functional units are characterized by the fact that they contain a multitude of molecular biological components which are interlinked in a complex way. Therefore, highly efficient regulation processes are required that act upon sensory information and enable a targeted mode of operation as well as its maintenance despite disturbing influences. The complex molecular structures of biological systems also require powerful signal transductions and regulations in order to understand their behavior and utilize their function. Holistic behavioral mechanisms of biological systems can only be truly understood if they are subjected to a systems-theoretical analysis on the basis of molecular biological models (Gilles, 2004). While current examples of BTIs generally are developed in order to control biological systems via technology, a biointelligent manufacturing system requires a bidirectional real time information exchange between the systems (Miehe et al., 2020a). Although basic scientific research in this field lead to individual solutions of fully and/or semi-controlled systems (e.g., biointerfacing of diamond thin films with proteins or local biosensing using silicon-based photonic nanostructures, bioreactors equipped with integrated online sensors allowing constant control) (Deutsche Forschungsgemeinschaft, 2007; Sommeregger et al., 2017), fundamental issues remain unanswered. Especially the fields of synthetic biology and systems biology offer great potential for integrating biological molecules and components in technical systems in a targeted manner, to reduce their complexity and to make them manageable through a selection of functional components (semisynthetic systems) (Cameron et al., 2014). Future research topics to be addressed in interdisciplinary formats under the guidance of systems biology and biocybernetics include:

- development of system-theoretical concepts for the structuring of different biological (cell, organism, etc.) and technical systems as well as their interfaces/interaction mechanisms
- development of system-theoretical methods and tools to decompose different biological systems into suitable functional units (e.g., physiological function, genetic function, regulation, signal transduction)
- identification of reaction kinetic and thermodynamic key parameters for the description of the holistic behavior of different biological systems
- modelling of different functional units of different biological systems using learning algorithms

The further development of BTIs is also closely linked to the advancement of biosensors and soft sensors. Although the basic principle of biosensors is not entirely new, recent successes in biotechnology (e.g., CRISPR/Cas9) offers great potential for the development of novel approaches that are to be systematically investigated. Due to the complexity of biological signals, also the use of machine learning needs to be explored. Soft sensors provide a second option for the extensive acquisition of alternative technical measurement variables as input for a rigorous implementation of BTIs. They are designed to enable quantitative, automatable system monitoring and control. Due to their indirect approach to system analysis by using neural networks with complex substructures, natural disadvantages (flexibility, transferability) arise compared to direct measurement, which have to be overcome. For instance, if the initially specified parameters are not sufficient for a research object, or if existing measured variables are to be transferred to new products or processes, the neural network has to be retrained. In addition to transferability of neural networks and systematic transfer from atline/offline to online/inline measurement systems in the context of industrial production processes, the following issues require further scientific engagement (Luttmann et al., 2012; Mandenius and Gustavsson, 2015; Randek and Mandenius, 2018; Miehe et al., 2020a):

- expansion of multivariate data acquisition and data analysis to advance the signal processing and data communication across different and parallel systems as well as the identification of critical process parameters
- extension, validation, and implementation of bioanalytical methods, which are capable of real-time and on-site monitoring, especially those compatible with online and inline analytical instruments (including spectrometric and optical methods). In this context, miniaturization and microfluidic approaches deserve special attention.
- development of continuous systematic knowledge management methods and databases for the identification of process weaknesses and implementation and monitoring of process improvement
- promotion of soft sensors in other industrial applications, not only bioprocesses
- alignment of soft sensors to the goal of simplifying production (non-expert systems). A use of soft sensors should not lead to a further increase in complexity and/or decrease of controllability.
- stimulation of knowledge and information transfer on soft sensors with the aim of raising awareness among young researchers, researchers and industry
- standardization of implementation methods for soft sensors
- development of convenient, user-friendly interfaces to facilitate the adaptation and integration of soft sensors into existing process software environments

An expansion of BTI scope is crucial especially with regard to the underlying systems thinking in cellular units, which requires that each individual entity (production, treatment, product, ecosystem impact etc.) is to be regarded as a spontaneous merging, interacting, and instance disperse of different cells. BTIs can thus occur on a wide variety of levels, not only in the context of technical operations. This results in widely differing research topics, which can only be solved satisfactorily if several sciences collaborate on an interdisciplinary basis. Figure 7.2 exemplifies a selection of different entities that are interpreted as cellular units that develop BTIs once they merge. Each merger results in a set of research topics involving different scientific disciplines, which must cooperate in order to solve it.

While the individual research topics listed here as examples are beyond the focus of a BTI, their respective research design requires the examination of the following fundamental issues for BTIs of various types:

- design of classification systems of cells at multiple levels for diverse applications
- investigation and description of cell aggregation types as well as their spontaneous merging and diverging

- investigation and description of the exchange relationships of different cell clusters as well as their BTIs
- development of BIS designs in the explicit and implicit sense

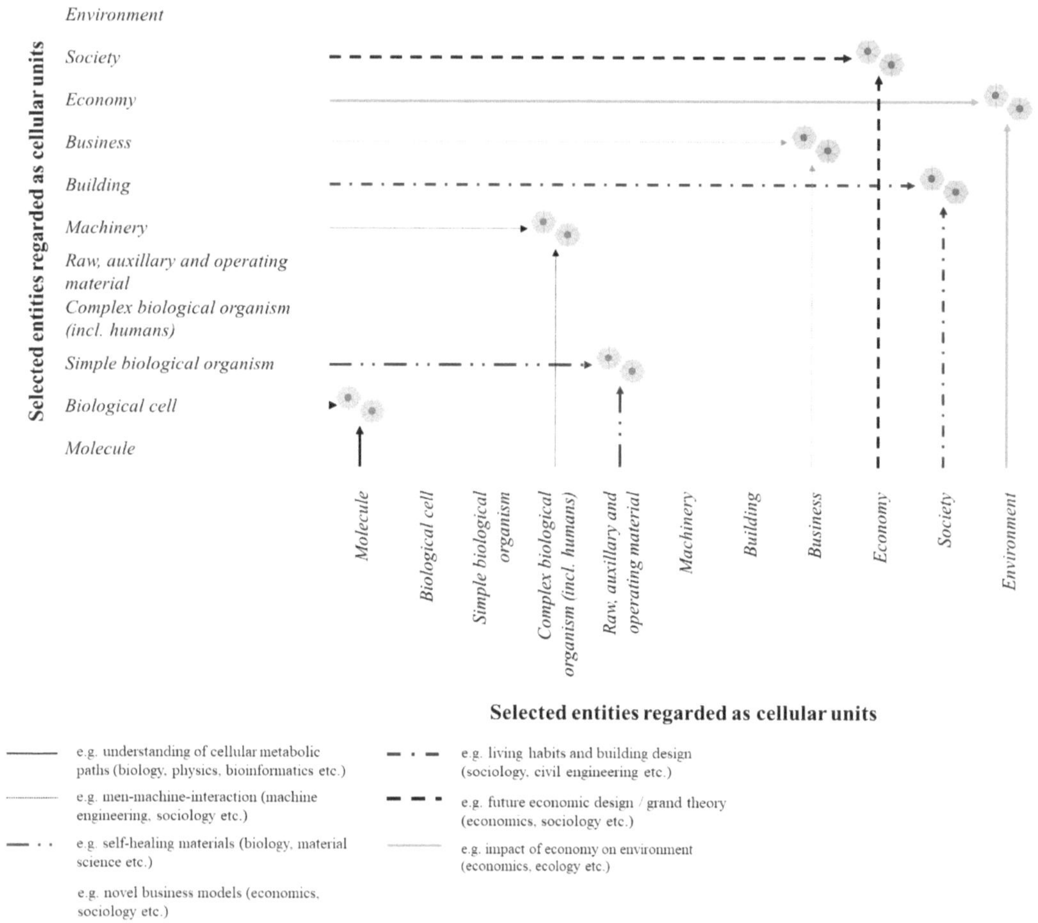

Figure 7.2 Selection of cellular entities at different levels, creating BTIs once they merge [own illustration]

7.2 Bioinspired, biobased, biohybrid materials

In the development, fabrication, and utilization of new materials two main directions can be distinguished: [1] the necessary substitution of the fossil-based resource base with renewable, biobased, biodegradable resources and [2] the development of functional, intelligent materials for adding valuable characteristics to B2B and B2C goods.

The latter also includes the further development of bioinspired materials such as different biomaterials in the health industry.

A large number of new raw materials based on natural precursors or biomass are conceivable, a fact that is primarily addressed in the bioeconomy. Here, the main aim is to replace conventional chemical synthesis paths with biological ones, e. g., in order to increasingly process lignin, chitin, cellulose, etc. Changing the raw material base is essentially a task of bioprocess engineering. However, the fabrication and utilization of these new materials in an industrial environment require an integrative approach, which is only possible through the convergence of the disciplines involved (see Chapter 4). Besides bioprocess engineering, the development, scaling, and optimization of relevant machines and equipment which enable novel synthesis paths (e. g., biorefineries, bioreactors, bioelectrolysis) is the core business of the machinery and equipment industry (see Section 5.3). In addition, biointelligent production demands an appropriate handling of material flows over their entire life cycle and possible subsequent use (re-fabrication). Hence, the exploitation of possibilities of subsequent use (re-fabrication) of each material in the loop is an essential issue for future research. Besides the chemical industry, especially sectors with high material intensity (e. g., construction) represent interesting platforms for the utilization of biogenic and recycled materials. Thus, future research and development is required for the further exploration and novel applications of biogenic and recycled materials, the precise forecasting and targeted activation of the degradation process of biobased materials as well as their adaptation to usage cycles.

In addition to the increased use of biobased materials, their functionalization also bears great potential. The high functionality of organic molecules is already being used for self-organized structures such as DNA origami (Khoo et al., 2015; Praetorius et al., 2017). Intelligent composite materials, such as shape memory alloys and polymers, self-evolving structures or tissue engineering, combined with molecular control possibilities form the basis for nano-scale bio-based and functional structures (Khoo et al., 2015; Gao et al., 2016). Research in this field is regarded as the basis of disruptive developments in bioelectronics and biocomputing as well as sensor technologies. At present, applications are still lagging behind in their possibilities. Here, further research and development activities are needed, which should be supported accordingly. 3D printing processes using biological material promise new approaches in the area of tissue engineering or 3D bioprinting (see Section 7.3). Another field of future research is smart materials, a term commonly referring to materials that can change their properties in response to external stimuli, such as temperature, pH, pressure or humidity. Such materials are already widely used as shape memory materials, chromogenic materials or piezoelectric materials (Khoo et al., 2015). However, biological materials or surfaces can be selectively accessed and activated through a wider range of stimuli. Self-X functions, such as self-induced reshaping or self-healing, could be realized. Then again, the integration of living biological organisms into material networks has predominantly been researched in biomedicine. However, the potential of organisms as actuators, sensors, and controllers for advanced material functions is far from exhausted. Ultimately, this can be used to develop 'living materials', i. e., materials that have functionalities such as learning. As a step towards this biological functionalization, which is already comprehensively addressed in research at the molecular level, must continue to add more complex functions, including the integration of living organisms. Great opportunities are seen in the emerging new gene editors (see Section 5.3) that may set the basis for an efficient and targeted realization of complex functionalities in biomolecules.

7.3 Bioinspired, biobased, biohybrid manufacturing technologies and systems

The development of new manufacturing technologies and processes bears manifold potentials. Therefore, two perspectives have to be differed: [1] time (short, medium, long term) and [2] BT development mode (bioinspiration, biointegration, biointeraction). From the very beginning, it is crucial to focus on a technology-open, interdisciplinary development with regard to decentralization, process coupling, simplification, and self-x capability. Provided the framework conditions are appropriate (see Section 7.9 and Section 7.10), sustainable biointelligent production becomes possible in the long term. While on the short- and medium-term research is likely to focus on bioinspiration of existing manufacturing systems as well as semi-interdisciplinary advancements of biointegrational aspects (e. g., biotechnology), research will lead to increasing convergence in the long term, which in turn promotes integrative production concepts, facilitating the technology-oriented demand economy. Figure 7.3 illustrates the development qualitatively in terms of the scope of production networks required for an individual baskets of goods in both economic forms.

Research for rather short-term improvements of existing manufacturing systems is required in the field of bioinspiration in order to create stable, quickly customizable and self-configurable production systems as well as organizational forms. In this context, optimization solutions based on bioinspired algorithms will increasingly replace classical methods (Dalavi et al., 2016). Basically, the goal is to significantly increase the resilience in production systems and processes, for example by self-adaptation, i. e., the ability of systems to adapt independently to changing requirements and conditions of an environment. Prerequisites for this are robust sensors, actuators, and computer-aided techniques, which allow to continuously recognize and monitor relevant states of processes and production systems as well as the workpieces to be manufactured. As the amount of information access is likely to increase, efficient data processing and interoperability is a key technology (see Section 7.6). On the short-term, research activities will thus not substitute, but rather complement current developments (e. g., CPS, DTs, see Section 7.6) in order to increase productivity and product quality and to create new possibilities of process transparency and planning. Analogies from biological systems can also be applied to production planning (Ueda et al., 2001; Schatten and Žugaj, 2011). One example are single-cell organisms that are able to form 'organization teams' instantly when exploitable opportunities are identified. Similar to unicellular organisms, organizational units will have to change their form agile and adaptively by adjusting their internal relationships, teams, and members.

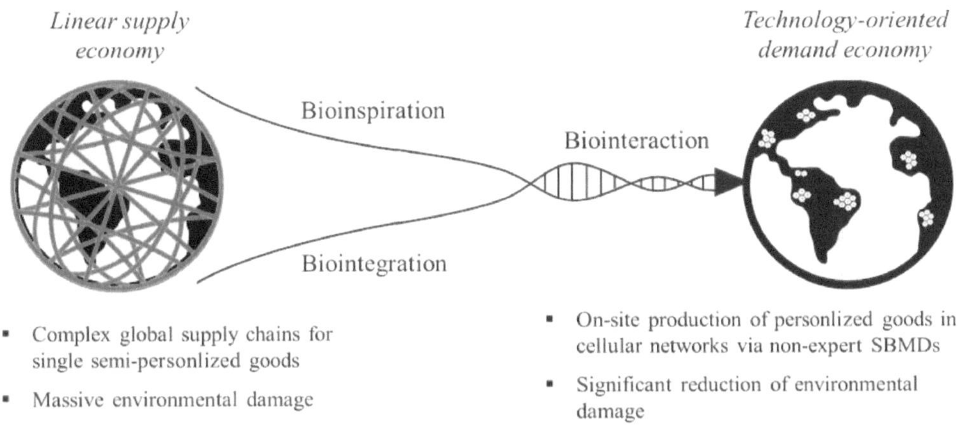

Figure 7.3 From a linear supply economy to a technology-oriented demand economy [own illustration]

From a medium- and long-term perspective, biointegrational and biointeractional aspects will increasingly dominate research. Initially, this is closely related to the technologies discussed in Chapter 5 (additive manufacturing/bioprinting, bioreactors, biorefineries, biofuel cells, bioelectrolysis, biosensors, etc.). Bioprinting and additive manufacturing represent the most promising approach for on-site manufacturing of consumer goods, food, medical devices as well as entire organs. While 3D bioprinting is already used in several industrial applications, 4D bioprinting potentially enables the processing of functional, smart materials (Chia and Wu, 2015; Jang et al., 2018). Yet, these technologies still face a variety of challenges. Among others, essential future research topics for bioprinting are (Dababneh and Ozbolat, 2014; Seol et al., 2014):

- enhancement of repeatability, resolution, cell viability, practicality, and biocompatibility of the bioprinting processes, e. g., standardization of scalable fabrication methods for the robotic delivery of cells
- advancement of mechanical and chemical properties, bioprintability, cell density and viability, solidificability and solidification speed, affordability and abundancy,

as well as long-term cell functionality of bioinks, e. g., improving nozzle and functional bioink with high cell density or developing enabling technologies for multi-bioink multiscale hybrid bioprinting processes

- increase of compactness, resolution, accuracy, high-degree-of-freedom, motion capability and motion speed, full-automation capability, user-friendliness, sterilibility, affordability, and versatility of bioprinting devices

The use of 3D/4D bioprinting beyond biomedical applications depicts massive potential to produce new, complex products and machinery. Hence, it is of great importance to further expand research these fields (Chia and Wu, 2015; Jang et al., 2018). In addition, further research is required in the field of additive manufacturing materials especially with regard to processability of novel biobased or biohybrid materials, while additive biomanufacturing might enable the emergence of completely new forms of printing technology and materials, e. g., enzyme printing.

One prerequisite for the development of fabrication technologies for biobased and biohybrid materials is the substitution of the resource base from fossil to biobased. This strongly depends on the concepts of biorefineries, bioreactors, and bioelectrolysis. Biorefineries face several challenges regarding industrial use. These primarily include the necessary increase of basic substances (cellulose, starch, sugar, and oil) based on biogenic raw materials as well as the further introduction and sample construction of biorefinery demonstration plants (Kamm and Kamm, 2004; Kamm, 2008). In addition, further research and development is necessary in order to increase the understanding of biomass as a raw material, improve sustainable systems design for developing, harvesting, and processing biomass resources, as well as advancing efficiency and performance of conversion and distribution processes. Not least, an appropriate regulatory design as well as a market is necessary in order to boost the technology, especially with regard to established industries and disciplines such as chemistry (Kamm and Kamm, 2004; Kamm, 2008; Cherubini, 2010). In the context of the provision of biobased raw materials, bioelectrolysis technology may occupy an important spot as well, yet rather as a key technology for flexible decentralized production networks (so-called hubs). Here it can be applied, e. g., for cascading production processes and the coupling of energy and material flows, i. e., simultaneously storing energy and providing basic chemicals and materials (see Section 7.7). Especially with regard to the on-site production of raw materials and biopharmaceuticals, bioreactors are of crucial importance. In the context of biointelligence, essential research required here is closely linked to the BTI concept. Consequently, the focus of research has to be miniaturization, flexibilization, simplification and intelligent control. Among others, this requires rigorous sensory data acquisition using soft sensors, extensive bioanalytics (including the development of digital twins), as well as the incorporation of switchable organisms (as actuators).

By increasingly applying information and biotechnological advancements (e. g., biosensors, soft sensors, digital twins, genome edited organisms, etc.) to production systems, technological possibilities for simple, on-site, coupled, autonomous production concepts of individualized goods will evolve (Miehe et al., 2020a). In the long term, research will thus increasingly focus on the thorough incorporation of various technologies and methods (convergence), which eventually enables SBMDs. For this purpose, a wide variety of existing individual technologies (additive manufacturing, bioprinting, biosynthesis, bioreactors, etc.) are to be purposefully combined in collaborative projects. Figure 7.4 exemplifies an SBMD as a miniaturized integrated concept of an intelligent bioreactor, a bioprinter and a real-time bidirectional information exchange via BTI and DT according to Miehe et al. (2020a).

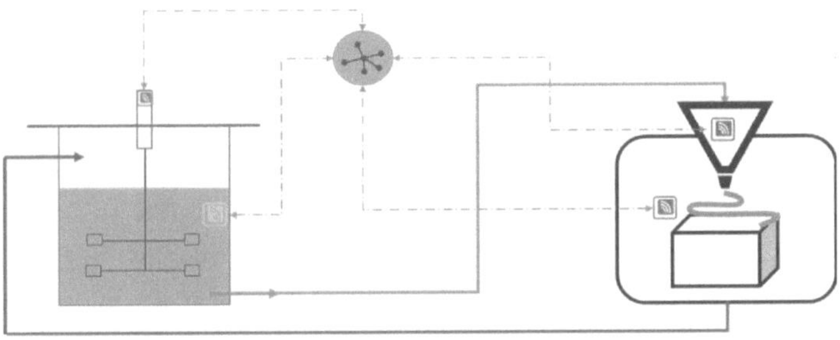

Figure 7.4 Example of SBMD according to Miehe et al. (2020a)

Robotics bears great potential as well. Among others, this includes the (further) development of controllable microorganisms, e. g., for the 'robotic' transport of substances in fluid systems, as well as the improvement of soft sensor technology through the application of new, bioinspired, biobased, functional materials. In the long term, robots might become available that are completely energy self-sufficient through enzymatic electrolysis and independently adapt their source code to adapt to changing environments. Similar applications, potentials, and research topics exist for wearables and exoskeletons.

7.4 Human-centered work

A transformation towards a sustainable economy can only succeed if people continue to occupy a significant, meaningful position. Similar to machinery and equipment, work contents, places, and forms in modern manufacturing environments are increasingly affected by the deployment of information and communication technology. Since technology and men increasingly interact at the workplace (human-machine interface), a constant need for further training measures exists as a task for employers, employees and authorities. In the short to medium term, employees are expected to become an active part of a more and more transformable factory. The redesign of work place layout is thus of great interest for future research in order to improve both ergonomics and productivity. Already today, research is increasingly focusing on man-machine interaction, leading to various new approaches in fields such as adaptive, anthropometrically and dermatologically compatible portable assistance systems (e. g., exoskeletons and wearables). Whilst the existing anthropometric wearables represent heavy, slow individual technologies, future biocompatible human-machine interfaces will enable rapid adaptability and easy portability in addition to person-specific intelligent assistance systems. Therefore, considerable research is required for the acquisition and interpretation of human biomarkers, especially for robust, non-invasive measurements of muscle and mental activities (Munneke et al., 2018). Furthermore, the development of technically supported emotions and facial expressions is of central importance in future production environments. This applies both to the optimization of the workplace with regard to the individual movement behavior of employees and to the real-time evaluation of work movements (Reinhart et al., 2017). As a result, new possibilities for the individual regulation of break times and the optimization of motions in terms of ergonomics and process speed are emerging. Another field of research and development that is related to the appropriate handling of biomanufactured (semi-finished) products. As employees are likely to increasingly face such challenges, the advancement of suitable safety and qualification measures is essential. In the long term, biointelligent production will enable increased self-sufficiency. Employees will gradually become producers and consumers (i. e., prosumers) at once. Working and living are moving closer together again. Long-term research issues involve the compatibility and design of workplaces in on-site cellular units.

7.5 Sustainable organization and business models

Biointelligent production is understood as the spontaneous or planned combination and recombination of biotic and abiotic cellular units to create or dismantle a cellular structure capable of intelligent behavior. Instead of large-scale centralized factories, production takes place on-site in networks of small enterprises, communities, individuals, and households. The decisive factor for productive success is not only the fulfilment of customer needs, but also the satisfaction of social and environmental demands. In this context, a biointelligent production enterprise represents a goal-oriented, participatory socio-technical system of individual cellular units. Sustainable organizational design is based on a fundamental insight: an organization shall never exist solely for its own profit, but for the integrated satisfaction of individual, social, and ecological needs within the framework of a complex network of dependencies and responsibilities. At the microeconomic level, an increasing demand exists to consider sustainability criteria in decision-making at all levels and stages of the organizational value chain (Miehe et al., 2016a). Future research in this field includes the advancement of integrative management systems, which increasingly incorporate aspects such as society and environment into decision-making. This requires the further refinement of existing management systems (strategy, innovation, lean, quality, environment, energy, etc.) regarding the considerations outlined in Chapter 6 as well as the advancement, simplification, and standardization of supporting measures from various disciplines, including economics, ecology, sociology, industrial engineering, as well as information and communication technology. Elementary fields of research include improving the concepts of prosumer and customer centricity, systematically analyzing the interdependencies of individual and network strategies within a biointelligent production system, supplementing the DFB approach, expanding the conceptual framework for BM innovations, incorporating and testing of cellular thinking in VSM, as well as developing voluntary and regulatory approaches to promote a responsible biointelligent production. Annex 1 and Annex 5 provide excerpts of the state of the art as starting points for scientific advancement.

Given the preconditions, an enterprise will always need to be worthwhile for the owner. Otherwise, it will not exist. Decisive for this are BM (Bocken et al., 2014). While BMs are currently being increasingly adapted to digitization, a shift towards sustainability is to be expected in the context of biointelligence (Bauernhansl et al., 2019; Miehe et al., 2020a). This applies in particular to the vision of a technology-based demand economy, leading to a largely decentralized production (see Chapter 6). Consequently, a demand arises not only for new BMs, but also for their ratings and development strategies. For future BM ratings, it is crucial to consider not only the economic value, but also social and ecological factors. Models to be developed could be applied by the legislator, e. g., for subsidization or taxation. Fields of research and development the further methodical development of BMs taking into account economic, social and ecological aspects (e. g., Canvas), the development of novel approaches for the integration of consumers in value-added processes, and the promotion of interdisciplinary projects in BM development and interdisciplinary education (Joyce and Paquin, 2016). In addition, it is essential to provide sufficient financial resources for founders, to promote the connection of inventors and investors on tailored platforms, and to reduce bureaucratic burdens.

7.6 Bioinspired, biobased, biohybrid data processing

Data processing is a prerequisite for biointelligence. As such, it is closely linked to the BTI concept. The ultimate goal of biointelligent production has to be complete interconnectivity and interoperability – not only between technical systems and/or companies, but across all conceivable cellular entities (i. e., from molecular to ecosystem level). This requires the stringent further development of cross-linking of production technology and economy, which was initiated under I4.0, as well as the rigorous integration of biological systems. The latter is associated with an immense demand for knowledge generation about biological systems led by systems biology and bioinformatics. Research issues arising from the current state of I4.0 thus depict coherent structures of precondition and reinforcing factors of biointelligence. Thereby, main areas of research are connectivity, interoperability, intelligent data processing, and autonomous regulation (Vaidya et al., 2018; Qin et al., 2016), each of which bearing a wide range of potential research areas in the context of biointelligence. A decisive approach is the concept of DT, which is currently in an initial development phase (Uhlemann et al., 2017b; Uhlemann et al., 2017a; Váncza and Monostori, 2017).

In the context of biointelligence, DTs are expected to adopt the function of a genotype to its respective component state, the phenotype. This 'manufacturing gene pool' is likely to expand in existing manufacturing environments with regard to the development of CPS, using bioinspired algorithm cycles, such as swarm intelligence and ant algorithms (Werfel et al., 2014). Meanwhile, as discussed in Section 5.4, the concept needs to be extended to different biological systems and processes, which poses a number of challenges. These include dealing with complexity, data collection, as well as reviewing and adapting existing DT design patterns. Other future research in this field range from bioinspired software (e.g., algorithms and communication mechanisms) (Dalchau et al., 2018; Kar, 2016; Navlakha and Bar-Joseph, 2011) to far-reaching approaches of data storage and processing using biological structures (hardware) (Bonnet et al., 2013; Lv et al., 2018).

Information technologies are already the most important drivers in biological research and development today (May, 2017). The complex diversity and enormous amounts of data required for understanding relationships in complex systems are challenges that can only be met by modern ICT-tools. Although great advances in molecular biology have provided a profound understanding of how information from the genome is translated by chemical reactions, a comprehensive understanding of biological phenomena at the next level remains largely open. Information processing is a crucial part of every biological system, appropriate theoretical models for the description of these information flows are however missing (Nurse, 2008). The increasing sensorization of production processes and products creates new technological systems whose development and stability are to be controlled.

A bioinspired system design applies fundamental systemic principles of nature to design and control such equilibria (Kitano, 2004). Dynamic problems, changing constraints, faulty or incomplete data sets, and insufficient computing power are new challenges in the development of data-driven production technologies that have to be designed in accordance with large ICT networks. Smart bioinspired algorithms can provide solutions for effective and decentralized communication and control mechanisms in such technical systems. This field of research is currently dominated by the development of AI. Whereas recent research has focused on a few algorithms whose origins date back to the 1970s to the 1990s (e. g., neural networks, genetic algorithms, particle swarm, ant colonies), latest algorithms which may provide superior solutions are still under-utilized in industrial applications (Dressler and Akan, 2010).

Due to the large amounts of data and complex links in BIS, further fields of research and development are data storage, data structures, and data security. On the medium and long term, bioinspired or biobased hardware such as molecule-based systems are possible. These systems are currently in their infancy, yet theoretically capable of extremely high computing power and storage capacity (Bonnet et al., 2013; Lv et al., 2018; Church et al., 2012; Daniel et al., 2013). Especially DNA computing and programmable cell networks might lead to a variety of innovations. Politics and industry are thus increasingly forced to deal with a great variety of design issues in the field of data processing, including the further expansion of the ICT infrastructure as a prerequisite for comprehensive information technology networking in industries, the further development of data protection, transparency, and processing standards and/or regulations to be completed faster and better in line with the progressing developments as well as the appropriate addressing of ethical issues such as responsibility of algorithm-based decisions.

7.7 Bioinspired, biobased and biointelligent energy systems

Energy is the basis of any form of production as well as a foundation for a wide range of consumer applications. The transformation towards a sustainable energy system can only succeed by considering a multitude of possible future technologies, their interaction and consequences in an integrative manner, especially with regard to a much closer coupling of energy generation and manufacturing. This requires a way of thinking whose dissemination is one of the core objectives of this work. In the context of BT, this is extensively demonstrated by Full et al. (2019), who provide a case study on the potentials of bioinspiration and biointegration in the energy system. Accordingly, the following paragraphs are primarily based on this paper, only adding additional thoughts regarding the transfer towards biointelligence.

Generally, sustainable energy supply can only be realized if the amount of energy provided by the sun is exploited with optimal efficiency during conversion, distribution, storage, and use. This requires the least possible loss of exergy in each step, as well as being robust and economical. Hence, today's predominantly linear energy value chain (generation, distribution, storage, use) has to develop into a more complex renewable, biobased energy value creation system that is similar to the natural one (bioinspiration) (Figure 7.5). Full et al. (2019) describe the necessary steps as follows: first, the energy provided by the sun has to be converted into a technically usable form, i. e., primary conversation into electricity, solar heat and/or biomass/biomolecules etc. Thereupon, several sub-process technologies exist that are capable of optimizing its provision at the location of use by 'keeping it in the loop'. Energy can be transferred to sub-processes either directly, i. e., without intermediate process steps, or indirectly. Sub-process technologies cover further conversion stages, storage, distribution, and utilization processes (e. g., manufacturing processes). For instance, waste heat from an utilization process (e. g., high temperature industrial production process) can be applied to further processes, e. g., via remote heat pipes (transport) at a lower temperature level for further use (e. g., climate control of buildings) before it dissipates. Sub-processes do not follow a rigid order, yet should follow the principle of least resistance, which is determined by entropy production or exergy loss. Exergy loss is to be minimized in any stage of the energy value chain, while entropy production in the form of waste heat should in any case be used as far as possible until it dissipates in the last step. This approach requires a paradigm shift in energy process design. Processes can no longer be seen exclusively as producers or consumers, but must be seen as both. Simultaneously, it is necessary to consider the link to the utilization system (e. g., production) when designing the energy system. In a biointelligent production system, among others, resulting from the convergence of energy, process and production technology, all necessary steps (energy and raw material extraction, storage, and refinement as well as production) ideally occur in a closed loop structure at a single location. Figure 7.5 illustrates the design focus of a future energy system and the vision of an integrated biointelligent production cell according to Full et al. (2019).

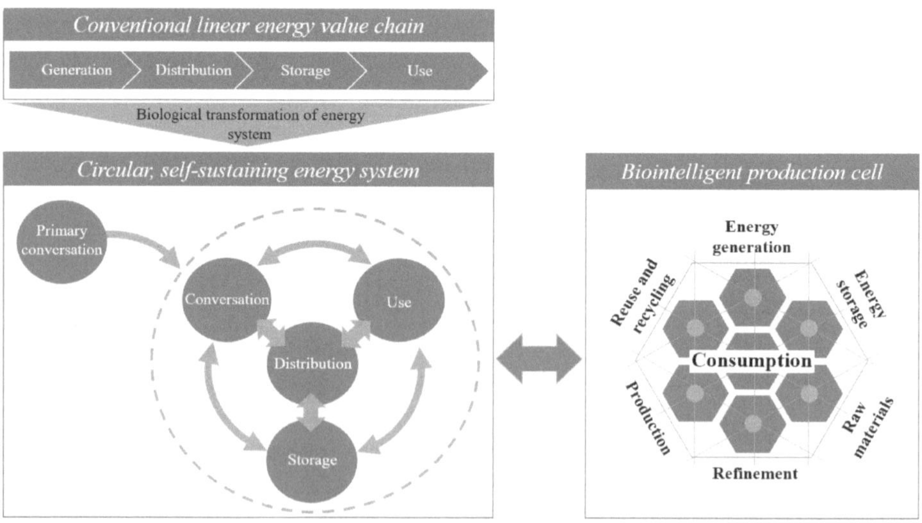

Figure 7.5 Conventional and biointelligent energy value creation system (l) according to Full et al. (2019) and integrated biointelligent production cell (r) [own illustration]

In the context of energy system design, bioinspiration and biointegration are advantageous and in many cases interrelated (Full et al., 2019). This requires additional thought: carbohydrates and fats represent the basic sources of energy in plants, generated by photosynthesis from solar energy, exhibiting the main reason for the flexibility of natural energy systems. In nature, chemically bound energy represents the predominant way of storage. Thereby, the energy is bound in energy-rich molecules via anabolic processes for storage (and material use at the same time). In order to release the energy, the molecules are split again. This principle already exists today in a technical form known as 'power-to-chemical'. Basic technologies of this field that mimic the evolution-proven principle of cold combustion of hydrogen are electrolysis and fuel cells (da Silva Veras et al., 2017). Besides significant cost advantages compared to predominantly direct electrification in the heat and transport, these technologies enable a direct production of basic chemicals (e.g., methanol, Fischer-Tropsch fuels, higher order alcohols, etc.). The linking of material and energy flows is viewed as another basic principle of nature that bears great potential in the context of biointelligence, when being transferred into future energy and manufacturing systems. Although coupled energy flows (heat, electricity, and chemical energy sources) already partially exist in industry known under the term 'sector coupling', the simultaneous production of basic chemicals and materials is usually strictly separated from one another. Besides fuel cells and electrolysis, biorefineries represent another path. These technologies produce food and feed, as well as basic and fine chemicals as products from biogenic material, while at the same time being energy self-sufficient units that receive the required energy directly from the biomaterial being processed. The substances produced here (e.g., bioplastics), can be energetically recycled after use. However, fuels or basic chemicals can also be produced directly as products via photosynthesis from biogenic material. Biorefineries may significantly contribute to a comprehensive utilization of biobased raw materials and residues as well as a regenerative supply of energy via liquid or gaseous organic energy carriers (i.e., e-fuels, biofuels, or synfuels) (Chew et al., 2017; Mohan et al., 2016). The rapid progress in ICT represents another important basis for successful bioinspiration, e.g., by significantly increasing the flexibility and efficiency of power systems. For instance, organisms always had to deal with rapid, unforeseen changes and react to them immediately. Their internal control mechanisms are adapted to frequent changes in their environment by learning from experience, a principle that is likewise important for future electricity grids. Bioinspired algorithms (e.g., artificial neural networks, genetic algorithms, and swarm intelligence) may thus represent and interesting asset for smart grid implementation (Goebel et al., 2014), enabling a quick response to differences between availability and use profiles. Such algorithms will also play a major role in achieving self-regulation of different sub-processes. Instead of switching off power generation plants due to low demand, storage represents a much more balanced approach, which could be increasingly based on scalable power-to-chemical systems. Here, further biomimetic research is required to transfer of evolutionary proven structures and surfaces into in the field of energy supply and storage (e. g., principles for electrochemical energy storage in batteries or natural generation and storage of hydrogen as an energy source). In addition, future research in the field of bioinspiration of energy systems includes the further development of bioinspired algorithms, intelligent bioinspired sector coupling, efficient power-to-chemical processes and the bionic optimization of structures and surfaces (Full et al., 2019).

Simultaneously, biotechnological production of substances has revolutionized numerous processes in the chemical, food, detergent, agricultural and pharmaceutical industries. Often biotechnological systems have both economic and environmental advantages compared to chemical processes. This especially applies to bioprocesses operating under comparatively mild conditions such as microorganisms (e. g., bacteria, microalgae) that accomplish complex chemical synthesis at room temperature and normal pressure. Chemical processes require much higher temperatures and high pressure, mostly resulting in higher exergy loss. Already today, several examples of biointegration exist in the field of energy generation and storage. The concepts of microbial and enzymatic electrolysis that produce chemical energy carriers via biological processes or organisms are just one example. Others are microbial methanization, biogas plants, and biorefineries. Thus, intensified research on related technologies, including the integration of efficient microbiological, enzymatically catalyzed material transformations, becomes increasingly relevant. The same applies for enzymatic processing technology (Miyazaki et al., 2008). Theoretically, a wide range of organic energy carriers and products can be produced by organisms like bacteria or algae including hydrocarbons, hydrogen as well as complex basic chemicals (e. g., as a basis of biopolymers) (Trchounian, 2015; Burk and van Dien, 2016; Chen et al., 2016; Santillan-Jimenez et al., 2016; Jozala et al., 2016). Microorganisms capable of efficiently and flexibly transferring and providing chemical energy sources for industrial use from non-usable forms of energy (potentially boosted by modern GE methods) are most likely to have the greatest significance in future coupled energy and manufacturing systems. Example of these in current research include specific cyanobacteria producing large amounts of hydrogen as well as purple bacteria generating hydrogen from wastewater as a by-product of bacterial photosynthesis or electrolysis (Bandyopadhyay et al., 2010; Vasiliadou et al., 2018). Another promising path for further research is artificial photosynthesis, which aims to develop processes that produce energy from solar radiation and CO_2 (Bonke et al., 2015; Kim et al., 2015; Zhou et al., 2010). As photosynthesis represents a very complex process, several challenges have to be addressed by future research.

Besides the integrative design of coupled energy and production systems, information technology networking, sensory equipment and intelligent control, being closely linked to the concepts of BTI and DT, sufficiently outlined in Section 5.4, represent the most important aspects of research in the context of biointeraction.

7.8 Systems-oriented life cycle thinking and technology impact assessment

Systems-oriented life cycle thinking depicts the key to biointelligence. Yet, its implementation is difficult since it involves assessing technological impacts ex ante as well as ex post for entities for which no reference state and/or models exist. This results in two primary fields of research: bridging the gap between existing challenges of the individual methods (especially LCA, SLCA, LCC, LCSA) as well as adapting them to the new challenges of potentially completely autonomous systems, partly based on genetically engineered basic substances. Simultaneously, a cross-sectional field of research to other topics described here exists in the development, validation and integration of supporting systems (especially AI, DT, BTI) (Falcone et al., 2015). General unsolved issues of LCA, SLCA, LCC, and LCSA are outlined in detail in Section 6.1. These primarily include the handling of uncertainty due to low data availability and/or an overly high level of aggregation, the limited operability of results due to complexity, and the handling of varying system boundaries. Generating useful decisions based on these methods thus still represents a severe challenge. According to Reap et al. (2008), it is unlikely that environmental damage can be avoided if current LCA results are used as a basis. It is thus of utmost importance to further develop the method, increase standardization and impact assessment as well as massively expand databases. An essential research path of the further methodological development in the context of biointelligence are predictive approaches, which are closely related to the overall handling of uncertainty. Decision-makers in politics, research and industries require valid supporting measures of economical, ecological and social impact of future technologies (Cucurachi et al., 2018; Klöpffer and Grahl, 2014). If no reference is available that could provide a model of the behavior of an entity, it is hardly possible to use historical data. This is precisely why supporting systems are necessary. Due to the expanded understanding of biological systems at all levels of cells to be generated with BTIs, system-theoretical models capable of predicting the effects in different categories could be developed in the medium and long term. In addition, the extensive networking and communication between cells enables the retrieval of relevant information in real time. Yet, to reach this point it needs several decades of development work. In the sense of systems-oriented life cycle thinking, further regulation of microeconomic organizations is likewise required (see Section 7.5). For instance, standardized DfE/LCD approaches could be promoted, i. e., to establish sustainability as a key strategic element in the development process. This however requires an extensive social dialogue and knowledge transfer (see Section 7.9 and Section 7.10).

7.9 Effective knowledge transfer

Knowledge transfer is an essential aspect of the interdisciplinary scientific field of biointelligence, requiring tailored approaches in three crucial fields of action:

- *Creation of common scientific basis and terminology*: The novelty of the subject as well as its vision of a convergence of different fields of science implies that insights have to be transferred from one discipline to another. In addition to suitable platforms, research designs and moderation, this requires first and foremost a coherent terminology, an aspect to which this work is intended to contribute. However, this cannot be achieved comprehensively within the framework of this work, but requires the collaboration of different scientists from a wide range of disciplines in diverse research projects over a long period of time.
- *Preparation of a new enlightenment*: Knowledge transfer is an elementary building block of a necessary new enlightenment (Weizsäcker and Wijkman, 2017). As discussed in Chapter 3, it is urgently recommended to educate the population as well as decision-makers in science, politics, and economy about misinterpreted approaches (e. g., of national economics) in the context of sustainable value creation in order to create the basis for a reorientation through social dialogue.
- *Creation of transparency about technological alternatives*: The example of genome editing demonstrates that an assessment of consequences is currently virtually impossible. However, GE represents both an integral component and a prerequisite for the technical feasibility of several sustainable solutions outlined here. Due to existing reservations among the general public, it is necessary to ensure the greatest possible degree of transparency regarding advantages and disadvantages of technological alternatives for the population in order to address them within a social dialogue.

Effective knowledge transfer thus demands a positive vision, a social momentum via public presence (media), opportunities for participation (discussion) as well as a targeted knowledge transfer (school, training, media) (Miehe et al., 2020a). Therefore, a political strategy is needed in order to create appropriate early education in schools as well as appropriate study programs for generalists and specialists. A fundamental transformation of curricula and pedagogical concepts into more project-related work and interdisciplinary subject networks appear promising in this context.

7.10 Social dialogue

Biointelligence is both a path towards a new technological basis (explicit understanding) as well as a mind-set of sustainable design of an economy within planetary boundaries (implicit understanding). As such, it may significantly contribute to science-based innovations, simultaneously resulting in economic prosperity, social welfare, and environmental protection (Sinemus and Egelhofer, 2007). Scientific progress is, however, strongly dependent on social perception and predominant ideologies (Kranzberg, 1986; Harari, 2014). As discussed in Chapter 2, the prevailing ideologies are primarily based on intentionally or unintentionally misinterpreted theories. A re-orientation of economy, technologically driven by scientific approaches based on the concept of biointelligence, is only possible if the population and decision-makers in science, politics, and industry engage in a comprehensive second enlightenment. This needs to be addressed within a social dialogue. However, a mere discussion will not suffice, as some 40 years of sustainability research have confirmed. Instead, social dialogue has to be targeted towards a consensus on the design of the economy and thus the categorical interpretation of the sustainability concept in the foreseeable future. Research and development in this area particularly challenges economics (together with social and environmental science) to develop a new sustainable economic theory that effectively accompanies and supports the transformation by providing a suitable framework. Although ecological economics and industrial ecology offer some promising approaches, the ultimate goal must be the development of a holistic theory at the economic-ecological interface that is capable of attracting a majority within the framework of a social discourse. Failing this, neither BT nor biointelligence will be achieved in the sense presented here. Provided a clear, generally accepted objective of a future economy, further knowledge-based, transparent social discourse is required, addressing the risks and benefits of technology alternatives (Sinemus and Egelhofer, 2007). Current examples of discussion topics that involve a variety of social stakeholders are the application limits of AI and the handling of generated data in a progressive digitization. The discourse is now less concerned with the technical feasibility, but is much more influenced by ethical issues such as the decision-making of AI systems or their permissible autonomy degrees (Hengstler et al., 2016). Potential fields of research and development, led by social and environmental science, are the assessment of impacts of biologically engineered technologies on humans, nature, and society, the definition of the role of scientists and managers as communicators in the social discourse as well as the establishment of secure dialogue formats for open, transdisciplinary discourses (e. g., e-governance).

7.11 Chapter review and conclusions

This section summarized a selection of unresolved issues for research, development, and innovation in the context of biointelligence, providing a synthesis of existing trends and gaps in research and development in ten sub-categories. First, the sustainable design of cellular (biology-technology) interfaces requires both a further development and application of soft sensor technology as well as a significant expansion of its current scope. As the spontaneous merging of cellular units can basically occur at any level between molecules and the ecosystem, a wide range of BTIs is possible that is not yet adequately addressed in the scientific community. Second, the development, fabrication, and utilization of new materials may be distinguished in two directions that require a wide variety of further research: [1] the necessary substitution of the fossil-based resource base with renewable, biobased, biodegradable resources and [2] the development of functional, intelligent materials for adding valuable characteristics to B2B and B2C goods. The latter also includes the further development of bioinspired materials such as different biomaterials in the health industry. Then, the field of bioinspired, biobased, biohybrid manufacturing technologies and systems covers technology-open, interdisciplinary developments with regard to decentralization, process coupling, simplification and self-x capability. While on the short- and medium-term research is likely to focus on bioinspiration of existing manufacturing systems as well as semi-interdisciplinary advancements of biointegrational aspects (e. g., biotechnology), in the long term this will lead to increasing convergence, which in turn promotes integrative production concepts, facilitating the technology-oriented demand economy. Apart from the (further) development of biotechnological concepts (e. g., biorefineries, bioprinting, bioreactors, biosensors) and their integration into existing value chains, decentralization, process coupling, simplification and self-x capability are increasingly important. Simultaneously, a transformation towards a sustainable economy can only succeed if people continue to occupy a significant, meaningful position. Similar to machinery and equipment, work contents, places, and forms in modern manufacturing environments are increasingly affected by the deployment of ICT. The redesign of workplace layout as well as the development of advanced wearables and exoskeletons are thus of great interest for future research in order to improve both ergonomics and productivity.

In the long term, biointelligent production will enable increased self-sufficiency. Employees will gradually become producers and consumers (i. e., prosumers) at once. Working and living are moving closer together again. Long-term research issues involve the compatibility and design of workplaces in on-site cellular units. Another crucial field of research is sustainable organizational design, including the development of a basic theory of sustainable organizations, integrative management systems, and supportive measures. Yet, real-work organizational behavior strongly depends on existing framework conditions. An enterprise will always need to be worthwhile for the owner. Otherwise, it will not exist. Consequently, a demand arises not only for new BMs, but also for their ratings and development strategies that will increasingly have to cover sustainability issues. A further field of research is data processing, which is closely linked to the BTI concept. The ultimate goal of biointelligent production has to be complete interconnectivity and interoperability not only between technical systems and/or companies, but across all conceivable cellular entities (i. e., from molecular to ecosystem level). This requires the stringent further development of cross-linking of production technology and economy, which was initiated under I4.0, as well as the rigorous integration of biological systems. Apart from the further development of the DT concept of biology-technology-systems, research in this field ranges from bioinspired software to approaches of data storage and processing using biological structures. Then again, bioinspiration, biointegration, and biointeraction are valuable strategies for realizing a sustainable energy supply that optimally exploits the amount of energy provided by the sun. Research in this field range from bioinspired concepts such as the coupling of material and energy flows to approaches of artificial photosynthesis. Additional research is required for the rigorous implementation of systems-oriented life cycle thinking, involving the ex ante and ex post assessment of technological impacts for entities for which no reference state and/or models exist. Fundamental for the progress of biointelligence is also an effective knowledge transfer focusing on the creation of a common scientific basis and terminology, the preparation of a new enlightenment as well as the creation of transparency about technological alternatives. As scientific progress is strongly dependent on social perception and predominant ideologies, a social dialogue has to be established in order to engage the population and decision-makers in science, politics and industry in a comprehensive second enlightenment.

Lessons learned

- Biointelligence requires a multitude of future research in a wide range of basic and applied sciences.
- The fundamental principle is always the interdisciplinary orientation and the holistic benefit criterion.
- Future research demands may be divided into ten fields:
  - Sustainable design of cellular (biology-technology) interfaces
  - Bioinspired, biobased, biohybrid materials
  - Bioinspired, biobased, biohybrid manufacturing technologies and systems
  - Human-centered work
  - Sustainable organizations and business models
  - Bioinspired, biobased, biohybrid data processing
  - Bioinspired, biobased, and biointelligent energy systems
  - Systems-oriented life cycle thinking and technology impact assessment
  - Effective knowledge transfer
  - Social dialogue

8 Conclusion

8.1 Summary

Industrial production is both a source of prosperity and one of the main causes of multiple social distortions and environmental damage. Simultaneously, a rethinking process has emerged that expresses the vision of a resilient and adaptable industrial production. Although the necessity of reorienting current production technology and economics already became apparent in the 1970s, a far-reaching redesign according to the concept of sustainability did not yet occur. The newly emerged concept of biointelligence is intended to enable this very reorientation. Yet, due to its recent emergence, the concept has not been sufficiently established in the scientific community, which hinders both its urgently required further academic advancement as well as a comprehensive dissemination and awareness of future decision-makers in higher education. In this context, the goal of this work was to constitute biointelligence in science from the perspective of a sustainability-oriented technology management. Therefore, previous theoretical deliberations should be summarized and further developed in order to stimulate scientific discourse and progress. As a result, a methodological-normative conceptual framework should be developed from a production science perspective, providing a systematic basis for sustainability-oriented technology management along the entire biointelligent value chain. The reader is invited to explore existing patterns of thought, behavior, and technological design and to identify potentials for their modification towards a sustainable production.

Currently prevailing patterns of thought and behavior of decision-makers that are responsible for the contemporary economic situation are deeply rooted in the history of mankind and the development of industrial production. In order to understand them, this work first discussed the most influential developments in the history of mankind by examining major social and technological upheavals in both the pre-industrial and the industrial age (Chapter 2). Decisive stages for today's socio-economic order were: the formation of intersubjective realities (cognitive revolution), the development of writing systems, the emergence of systematic modes of production in agriculture (Neolithic revolution), the invention of modern printing, the emergence of the guiding ideas of humanism and rationalism (scientific revolution), the systematization of industrial production processes (industrial revolution), the methodological theory revolution in economics and politics, and the emergence of ICT and the internet. Generally, the formation and change of societies, ideals, scientific theories, and technology is closely interrelated. Western societies in particular have incorporated two ideals into their constitutions, which still influence theory and practice of production today, although being partly contradictory: freedom and equality. Since the first industrial revolution, the diffusion of the technological basis of a society represents a history of continuous socio-technical paradigm shifts, known today as Kondratieff cycles. Although humanity has always affected its environment, only industrialization initiated a process that permanently shifted the limits of human influence on the ecosystem. The rise in prosperity resulting from the increase in productivity led to an almost exponential worldwide population growth, resulting in a consumption of goods that massively exceeds the level of natural regeneration capacity. The majority of economic theories that still dominate today's production, were formulated under completely different basic conditions, i. e., in an empty world framework. In contrast, future-oriented theories and concepts must be adapted to the new framework conditions of a full world.

The framework for a fundamental reorientation was created in the late 1980s with the sustainability concept (Chapter 3). However, the lack of a uniform interpretation of the concept massively impedes its operationalization. To date, a number of elementary questions that would require a social consensus remain unanswered. This includes the trade-off between ecosystem preservation and its productive capacity as well as the compatibility of various macroeconomic guiding principles (i. e., circular economy, demand economy, bioeconomy). Yet, some implications for manufacturing sciences at macro and micro level can be derived. Of substantial importance is the development and pursuit of a positive vision of sustainable value creation that is based on seven elementary design principles, which are derived from already formulated approaches of circular economy, bioeconomy and demand economy: substitution of fossil fuels by renewable raw materials and energy sources, calibration of the prevailing liberal economic logic by aspects of a socio-ecological economic logic (e. g., stronger regulation of markets, dismantling of the principle of division of labor, increase of self-sufficiency, regionalization of supply chains), minimization of energy and material use for the production of goods, orientation towards natural principles, phenomena and metabolisms, complete transparency over and recycling of every substance, material and assembly used. From a sole production technology perspective, innovative solutions are required in areas of need that are highly critical from an environmental point of view (housing, energy, nutrition, health, consumption, mobility).

For the very purpose of a realistic positive vision of sustainable value creation, the concept of biointelligence was developed (Chapter 4). Its goal towards technology development is the conception and application of so-called smart biomanufacturing devices, which enable the manufacturing and re-fabrication of personalized goods on site, based on local resources. In order to achieve this, two interpretations of the term are required: explicit and implicit. While the explicit understanding interprets the term as the systematic information-technological networking of biological and technical systems, the implicit interpretation covers the responsible handling of biological resources. From a sustainability perspective, the concept thus implies a profound technological renewal through a rigorous combination of information-driven intelligence with the development and establishment of new biological, medical or therapeutic processes, systems and devices, under the condition of responsibility for the persistence of human life on earth (holistic benefit criterion). The understanding, development and management of BIS demands a specific mind-set. This systems-oriented life cycle thinking in cellular units implies a separation of reality in cellular segments, ranging from single molecules to entire ecosystems. Each cell is sufficiently described by six factors: type of matter, internal processes, relations with surrounding cells and overall system, as well as purpose, functional principle and conditions for and to application. In addition, the mind-set requires the assumption that each cellular unit is directly and/or indirectly interconnected with each other cellular unit, resulting in a circular logic of consequences. The impact of a system on its surrounding systems and the overall system can only be thoroughly described over its entire life ('cradle to grave'). Consequently, the concept implies a balanced generation of benefits for individual, society and ecosystem via function orientation, not technology and/or product focus. Meanwhile, the comparison to related scientific disciplines proves that biointelligence represents an independent scientific discipline which, although relying on existing disciplines, generates new knowledge through a strict interdisciplinary application orientation and a holistic benefit criterion. Accordingly, not single enabling technologies or thought patterns of individual existing scientific disciplines are decisive for biointelligence, but rather systematic interdisciplinary cooperation, leading to an increasing convergence of the sciences and culminating in a sustainable technical renewal in the long term. However, it remains clear that the vision outlined here can only become reality if it is accompanied by a sustainable transformation of society that adapts to the demands of the full world.

The development of BIS in an explicit sense is decisive for the creation of a new, sustainable technological basis (Chapter 5). Its methodological principles are based on the three development modes of BT. Thereby, it is necessary that each subsequent level continues to build upon the knowledge of the previous one. For bioinspiration, basic engineering and biological knowledge as well as profound experience of the biomimetic transfer process are of particular importance. The most important difference in the context of the biomimetic transfer process is top-down and bottom-up approach, which both have to be applied in a suitable way in manufacturing companies. The instrumental foundation of biointegration is the knowledge of basic biotechnology approaches (including bioprinting, biorefineries, bioreactors, biosensors, and CRISPR/Cas9), yet in addition to the prior know-how of bioinspiration. Based on the previous development modes, basic ICT know-how is fundamental for biointegration. Similar to the concept of intelligence and life, no single explicit biointelligence state exists for a system. Rather, a system can be evaluated in terms of its biointelligence degree. Its classification has to be based on common criteria for grading its main components, i. e., presence of biological principles, biological and technical component and/or active principle, interaction in form of autonomous regulation (biology-technology interface) as well as overall system intelligence.

In turn, the management of BIS (Chapter 6) serves primarily to ensure the holistic benefit criterion of the intended use. This requires a framework, a fundamental mind-set and basic approaches for the design of future value creation systems. Compared to present-day concepts of production, biointelligent production envisages a radical change. The act of production is understood as the spontaneous or planned combination and recombination of biotic and abiotic cellular units to create or dismantle a cellular structure capable of intelligent behavior. Instead of large scale centralized factories, production takes place on-site in networks of small enterprises, communities, individuals, and households. The decisive factor for productive success is not only the fulfilment of customer needs, but also the satisfaction of social and environmental demands. The latter requires a holistic view of production as an organism that is aligned with planetary boundaries. The management framework presented here distinguishes two perspectives (the scope of environmental concern and the scope of management concern), each of which is subdivided into different sequential levels. The core of the management framework is the fundamental mind-set, which is the already introduced systems-oriented life cycle thinking in cellular units. It represents the basis for any decision in the production context and can be supported by a number of methods, each of which being subject to further research in the context of biointelligence. The most significant areas of future research in the field of BIS management include: improving the concepts of prosumer and customer centricity, systematically analyzing the interdependencies of individual and network strategies within a biointelligent production system, supplementing the DFB approach, expanding the conceptual framework for BM innovations, incorporating and testing of cellular thinking in VSM, as well as developing voluntary and regulatory approaches to promote a responsible biointelligent production.

Biointelligence as a discipline, however, demands far more extensive future research in a wide range of basic and applied sciences (Chapter 7). Thereby, the elementary principle is always the interdisciplinary orientation and the holistic benefit criterion.

This work provided a synthesis of existing trends and gaps in research and development in ten sub-categories and outlined their potentials in the context of biointelligence. Fundamental to the concept of biointelligence is future research in the field of sustainable design of cellular (biology-technology) interfaces at different levels ranging from molecules to ecosystems. Simultaneously, the development of bioinspired, biobased, biohybrid materials, manufacturing technologies and systems will increasingly have to focus on decentralization, process coupling, simplification, and self-x capability. As a transformation towards a sustainable economy can only succeed if people continue to occupy a significant, meaningful position, novel approaches for the redesign of workplace layouts as well as the development of advanced wearables and exoskeletons are crucial. In the long term, biointelligent production will enable increased self-sufficiency. Employees will gradually become producers and consumers (i. e., prosumers) at once. Working and living are moving closer together again. Long-term research issues involve the compatibility and design of workplaces in on-site cellular units. Another crucial field of research is sustainable organizational design, including the development of a basic theory of sustainable organizations, integrative management systems and supportive measures as well as appropriate sustainability-oriented business models and their ratings. As the ultimate goal of biointelligent production is complete interconnectivity and interoperability not only between technical systems and/or companies, but across all conceivable cellular entities (i. e., from molecular to ecosystem level), data processing represents another crucial field of research, among others, covering the further development of the DT concept of biology-technology-systems, bioinspired software as well as approaches of data storage and processing using biological structures. Concurrently, bioinspiration, biointegration and biointeraction are valuable strategies for realizing a sustainable energy supply that optimally exploits the amount of energy provided by the sun. Research in this field range from bioinspired concepts such as the coupling of material and energy flows to approaches of artificial photosynthesis. Additional research is required for the rigorous implementation of systems-oriented life cycle thinking, involving ex-ante and ex-post assessment of technological impacts for entities for which no reference state and/or models exist. Fundamental for the progress of biointelligence is also an effective knowledge transfer focusing on the creation of a common scientific basis and terminology, the preparation of a new enlightenment as well as the creation of transparency about technological alternatives. As scientific progress is strongly dependent on social perception and predominant ideologies, a social dialogue has to be established in order to engage the population and decision-makers in science, politics and industry in a comprehensive second enlightenment.

8.2 Critical review and outlook

This work provided a technology management summary of basic reflections on sustainable production according to the concept of biointelligence. The main objective was to provide the reader with a conceptual basis in the form of a mind-set, selected methods and approaches for the design of future value creating systems as well as research directions in order to actively participate in the renewal of the socio-technical basis of an economy. In this context, it is important to emphasize that even in the distant future, not all production will be biointelligent. Whether the concept will represent a vision for a market niche or will establish itself as a standard model in the long term remains a matter of debate.

In any case, the development of a technology, which under current economic conditions primarily occurs on a microeconomic level, can only be sustainable if the holistic benefit criterion is strictly adhered from the very beginning. However, it has to be clear that the notion of a purely technological-oriented renewal in terms of sustainability is an illusion. Rather, science and technology development are always embedded into the context of social values and ideologies, which in turn determine the economic framework conditions. If these are not addressed via modern enlightenment and intensive social dialogue, the exaggerated emphasis on liberal ideologies in economics will continue, which is likely to prevent an appropriate technological renewal in the sense of the equity-oriented approach of sustainability. As a result, economic development within the framework of natural planetary boundaries would be unlikely. Instead of a preservation of the current ecosystem, mankind would depend on a maximization of its productive capacity, especially by using modern technologies (e. g., geoengineering).

Yet, even if these requirements were met, this work is not free of criticism. Reasons for criticism mainly result from the perspective adopted by the work and its resulting scope for portraying the state of knowledge. From a sustainability-oriented technology management perspective, it is imperative to develop an understanding of the origins and the current state of the sustainability problem as well as to acquire insights into as many areas as possible that are relevant for the design of viable production technology and economics. This requires a broad portrayal of topics from various fields, which might be regarded as partially limited by experts of the respective fields. In addition, integrative innovations that demonstrate the intended technological convergence could only be presented in extracts. Yet, the reader of this work is not intended to develop an independent technical solution in a special field, but rather to acquire a way of thinking that is decisive for a sustainability-oriented design of production systems. However, a conversion into actual systems requires existing prior knowledge and/or further training in specific areas.

Future research and development in the context of production technology and economics increasingly requires interdisciplinary approaches that are aligned with the framework of planetary boundaries. Whether these borders shall be natural and/or artificially extended demands a global social consensus and a corresponding political framework. This is fundamentally based on a general societal dialogue and modern enlightenment, whose core objective involves the weighting of three basic ideals: freedom, equality and responsibility.

9 References

Abdelkafi, N. and Täuscher, K. (2016), 'Business models for sustainability from a system dynamics perspective', *Organization & Environment*, Vol. 29 No. 1, pp. 74–96.

Abramson, S. D., Alexe, G., Hammer, P. L. and Kohn, J. (2005), 'A computational approach to predicting cell growth on polymeric biomaterials', *Journal of Biomedical Materials Research Part A*, Vol. 73 No. 1, pp. 116–124.

Ackerman, P. L., Beier, M. E. and Boyle, M. O. (2005), 'Working memory and intelligence: The same or different constructs?', *Psychological Bulletin*, Vol. 131, pp. 30–60.

Adamy, J. (2012), *Nichtlineare Systeme und Regelungen*, 2nd ed., Springer Vieweg, Berlin.

Adensam, H., Bruck, M. and Geissler, S. e.a. (2002), *Externe Kosten im Hochbau, Externe Kosten*, Wien, available at: *http://www.bmwfw.gv.at/Tourismus/HistorischeBauten/Documents/Externe%20Kosten/ExterneKosten1.pdf* (accessed 1 March 2017).

Adorno, T. W. (1992), *Prismen – Kulturkritik und Gesellschaft*, 4th ed., Suhrkamp, Berlin.

Aehle, M., Kuprijanov, A., Schaepe, S., Simutis, R. and Lübbert, A. (2011), 'Simplified off-gas analyses in animal cell cultures for process monitoring and control purposes', *Biotechnology Letters*, Vol. 33 No. 11, pp. 2103–2110.

Agabi, O. E. (2017), *Cell culture, transport and investigation: U. S. Patent Application No. 15/211,557.*

Agarwal, R. and Weill, P. (2012), 'The benefits of combining data with empathy', *MIT Sloan Management Review*, Vol. 54 No. 1, p. 35.

Ahmed, A., Zulfiqar, S., Ghandar, A., Chen, Y., Hanai, M. and Theodoropoulos, G. (2019), 'Digital Twin Technology for Aquaponics: Towards Optimizing Food Production with Dynamic Data Driven Application Systems', *Methods and Applications for Modeling and Simulation of Complex Systems*, Vol. 1094, pp. 3–14.

Alaerts, L., Augustinus, M. and van Acker, K. (2018), 'Impact of bio-based plastics on current recycling of plastics', *Sustainability*, Vol. 10 No. 5, p. 1487.

Albert, K. J., Lewis, N. S., Schauer, C. L., Sotzing, G. A., Stitzel, S. E., Vaid, T. P. and Walt, D. R. (2000), 'Cross-reactive chemical sensor arrays', *Chemical Reviews*, Vol. 100 No. 7, pp. 2595–2626.

Algedri, J. and Frieling, E. (2015), *Human-FMEA: Menschliche Handlungsfehler erkennen und vermeiden*, Hanser, München.

Ali, S., Qaisar, S. B., Saeed, H., Khan, M. F., Naeem, M. and Anpalagan, A. (2015), 'Network challenges for cyber physical systems with tiny wireless devices: A case study on reliable pipeline condition monitoring', *Sensors*, Vol. 15 No. 4, pp. 7172–7205.

Allen, R. M. and Bennetto, H. P. (1993), 'Microbial fuel-cells', *Applied Biochemistry and Biotechnology*, Vol. 39 No. 1, pp. 27–40.

Alon, U. (2006), *An Introduction to Systems Biology: Design Principles of Biological Circuits*, Taylor & Francis, London.

Altshuller, G. (1984), *Creativity as an Exact Science*, Gordon & Breach, Amsterdam.

Altshuller, G. (1999), *The Innovation Algorithm: TRIZ, systematic innovation, and technical creativity*, Technical Innovation Center, Worcester.

Amadi, P., Ogunka-Nnoka, C. and Bene, A. (2018), 'Biotransformation of plantain pseudostem fibres using local enzyme sources; analysis of their potential as commercial poultry feed', *Biocatalysis and Biotransformation*, Vol. 36, pp. 1–9.

Anderl, R., Haag, S., Schützer, K. and Zancul, E. (2018), 'Digital twin technology – An approach for Industrie 4.0 vertical and horizontal lifecycle integration', *it – Information Technology*, Vol. 60 No. 3, pp. 125–132.

Andersen, M. S. (2007), 'An introductory note on the environmental economics of the circular economy', *Sustainability science*, Vol. 2 No. 1, pp. 133–140.

Antranikian, G. (2006), *Angewandte Mikrobiologie*, Springer, Berlin.

Appasani, K. (2018), *Genome editing and engineering: from TALENs, ZFNs and CRISPRs to Molecular surgery*, Cambridge University Press, Cambridge.

Aratani, N., Kim, D. and Osuka, A. (2009), 'Discrete cyclic porphyrin arrays as artificial light-harvesting antenna', *Accounts of Chemical Research*, Vol. 42 No. 12, pp. 1922–1934.

Arechederra, R. L., Treu, B. L. and Minteer, S. D. (2007), 'Development of glycerol/O2 biofuel cell', *Journal of Power Sources*, Vol. 173 No. 1, pp. 156–161.

Armbrust, M., Fox, A., Griffith, R., Joseph, A. D., Katz, R. and Konwinski, A. e.a. (2010), 'A view of cloud computing', *Communications of the ACM*, Vol. 53 No. 4, pp. 50–58.

Arndt, H.-K. (1997), *Betriebliche Umweltinformationssysteme: Gestaltung und Implementierung eines BUIS-Kernsystems*, Gabler, Wiesbaden.

Ashauer, R., Hintermeister, A., O'Connor, I., Elumelu, M. and et al. (2012), 'Significance of Xenobiotic Metabolism for Bioaccumulation Kinetics of Organic Chemicals in Gammarus pulex', *Environmental Science & Technology*, Vol. 46 No. 6, pp. 3498–3508.

Ashoori, A., Moshiri, B., Khaki-Sedigh, A. and Bakhtiari, M. R. (2009), 'Optimal control of a nonlinear fed-batch fermentation process using model predictive approach', *Journal of Process Control*, Vol. 19 No. 7, pp. 1162–1173.

Ashton, K. (2009), 'That "internet of things" thing', *RFID journal*, Vol. 22 No. 7, pp. 97–114.

Asiedu, Y. and Gu, P. (1998), 'Product life cycle cost analysis: state of the art review', *International Journal of Production Research*, Vol. 36 No. 4, pp. 883–908.

Aspromourgos, T. (1995), *On the Origins of Classical Economics: Distribution and Value from William Petty to Adam Smith*, Routledge, Abingdon, United Kingdom.

Assaf, T., Stefanini, C. and Dario, P. (2013), 'Autonomous underwater biorobots: A wireless system for power transfer', *IEEE Robotics & Automation Magazine*, Vol. 20 No. 3, pp. 26–32.

Assis, A. J. de and Maciel Filho, R. (2000), 'Soft sensors development for on-line bioreactor state estimation', *Computers & Chemical Engineering*, Vol. 24 No. 2-7, pp. 1099–1103.

Auer, P. (2017), 'Online Learning', in Sammut, C. and Webb, G. I. (Eds.), *Encyclopedia of Machine Learning and Data Mining*, Springer, Boston, pp. 929–937.

Autumn, K., Liang, Y. A., Hsieh, S. T., Zesch, W., Chan, W. P. and Kenny, T. W. et al. (2000), 'Adhesive force of a single gecko foot-hair', *Nature*, Vol. 405 No. 6787.

Ayres, R. U. and Ayres, L. W. (Eds.) (2002), *A Handbook of Industrial Ecology*, Edward Elgar Publishing, Inc., Northampton, Massachusetts, USA.

Azevedo, F., Vale, Z. A., Oliveira, P. M. and Khodr, H. M. (2010), 'A long-term risk management tool for electricity markets using swarm intelligence', *Electric Power Systems Research*, Vol. 80 No. 4, pp. 380–389.

Azuma, R., Baillot, Y., Behringer, R., Feiner, S., Julier, S. and MacIntyre, B. (2001), 'Recent advances in augmented reality', *IEEE Computer Graphics and Applications*, Vol. 21 No. 6, pp. 34–47.

Babensee, J. E. (2008), 'Interaction of dendritic cells with biomaterials', *Seminars in Immunology*, Vol. 20 No. 2, pp. 101–108.

Babu, R. P., O'Connor, K. and Seeram, R. (2013), 'Current progress on bio-based polymers and their future trends', *Progress in Biomaterials*, Vol. 2 No. 1, p. 8.

Bach, R. (2018), *Rousseau und die Physiokraten: Politische Ideengeschichte im begrifflichen Wandel zwischen Aufklärung und Revolution*, Böhlau, Köln.

Bachinger, T. and Mandenius, C. F. (2000), 'Searching for process information in the aroma of cell cultures', *Trends in Biotechnology*, Vol. 18 No. 12, pp. 494–500.

Bailon, P. and Won, C. Y. (2009), 'PEG-modified biopharmaceuticals', *Expert Opinion on Drug Delivery*, Vol. 6 No. 1, 1-16.

Bain, R. (1937), 'Technology and State Government', *American Sociological Review*, Vol. 2 No. 6, pp. 860–874.

Balint, R., Cassidy, N. J. and Cartmell, S. H. (2014), 'Conductive polymers: Towards a smart biomaterial for tissue engineering', *Acta Biomaterialia*, Vol. 10 No. 6, pp. 2341–2353.

Balzer, V., Schrodi, T. and Wiendahl, H.-H. (2020), 'Strategien und Struktur produzierender Unternehmen', in Bauernhansl, T. (Ed.), *Fabrikbetriebslehre 1: Management in der Produktion*, Springer Vieweg, Berlin, pp. 35–66.

Bandodkar, A. J. and Wang, J. (2016), 'Wearable biofuel cells: a review', *Electroanalysis*, Vol. 28 No. 6, pp. 1188–1200.

Bandyopadhyay, A., Stöckel, J., Min, H., Sherman, L. A. and Pakrasi, H. B. (2010), 'High rates of photobiological H 2 production by a cyanobacterium under aerobic conditions', *Nature Communications*, Vol. 1 No. 1, pp. 1–7.

Banham, R. (2003), *Das Ford Jahrhundert*, Tehabi.

Bar-Cohen, Y. and Breazeal, C. (2003), 'Biologically inspired intelligent robots. In Smart Structures and Materials 2003: Electroactive Polymer Actuators and Devices', *International Society for Optics and Photonics*, No. 5051, pp. 14–20.

Barrangou, R. (2015), 'The roles of CRISPR–Cas systems in adaptive immunity and beyond', *Current Opinion in Immunology*, Vol. 32, pp. 36–41.

Barrangou, R., Fremaux, C., Deveau, H., Richards, M., Boyaval, P. and Moineau, S. e.a. (2007), 'CRISPR provides acquired resistance against viruses in prokaryotes', *Science*, Vol. 315 No. 5819, pp. 1709–1712.

Barthlott, W. and Neinhuis, C. (1997), 'Purity of the sacred lotus, or escape from contamination in biological surfaces', *Planta*, Vol. 202 No. 1, pp. 1–8.

Barton, D. and Court, D. (2012), 'Making advanced analytics work for you', *Harvard Business Review*, Vol. 90 No. 10, pp. 78–83.

Barwise, J. and Etchemendy, J. (2005), *Sprache, Beweis und Logik. Band 1: Aussagen- und Prädikatenlogik*, Mentis, Paderborn.

Barwise, J. and Etchemendy, J. (2006), *Beweis und Logik. Band 2: Anwendungen und Metatheorie*, Mentis, Paderborn.

Basel, N. (2010), 'Green Minds – Michael Braungart. Der Anti-Müllmann', *Financial Times Deutschland*.

Basmer, T., Kulse, P. and Birkholz, M. (2010), 'Systemarchitektur intelligenter sensorimplantate', *Biomedical Engineering/Biomedizinische Technik*, Vol. 55, pp. 43–46.

Baßeler, U., Heinrich, J. and Utecht, B. (2006), *Grundlagen und Probleme der Volkswirtschaft*, 18th ed., Schäffer-Poeschel, Stuttgart.

Bateson, G. (2002), *Mind and nature: A necessary unity*, Hampton Press, New York.

Bauernhansl, T., Beckett, M., Brecher, C. and Demmer, A. e.a. (2019), *Biointelligenz – Eine neue Perspektive für nachhaltige industrielle Wertschöpfung – Ergebnisse der Voruntersuchung zur Biologischen Transformation der industriellen Wertschöpfung (BIOTRAIN)*, Fraunhofer Verlag, Aachen u. a.

Bauernhansl, T., Hartleif, S. and Felix, T. (2018), 'The Digital Shadow of production – A concept for the effective and efficient information supply in dynamic industrial environments', *Procedia CIRP*, Vol. 72, 69–74.

Bauernhansl, T., Hompel, M. ten and Vogel-Heuser, B. (Eds.) (2014), *Industrie 4.0 in Produktion, Automatisierung und Logistik: Anwendung, Technologien, Migration*, Springer Vieweg, Berlin.

Bauernhansl, T., Krüger, J., Reinhart, G. and Schuh, G. (2016), 'WGP-Standpunkt Industrie 4.0', available at: *https://wgp.de/de/wgp-standpunktpapier-fuehrt-durchs-schluesselloch-zu-industrie-4-0/* (accessed 23 April 2020).

Bauernhansl, T. and Miehe, R. (2020), 'Industrielle Produktion – Historie, Treiber und Ausblick', in Bauernhansl, T. (Ed.), *Fabrikbetriebslehre 1: Management in der Produktion*, Springer Vieweg, Berlin, pp. 1–34.

Baumast, A. (2013), 'Perspektive Nachhaltigkeit. Effizienz, Konsistenz und Suffizienz als Unternehmensstrategien', in Baumast, A. and Pape, J. (Eds.), *Betriebliches Nachhaltigkeitsmanagement*, UTB GmbH, Stuttgart, pp. 360–373.

Baumgärtner, S. (2000), *Ambivalent joint production and the natural environment: An economic and thermodynamic analysis*, Physica, New York.

Beattie, B. R. (2009), *The economics of production*, McGraw-Hill, New York.

Beck, H., Anastasiadou, S. and Meyer zu Reckendorf, C. (2018), *Faszinierendes Gehirn,* 2nd ed., Springer, Berlin.

Beckermann, A. (2003), *Einführung in die Logik,* 2nd ed., De Gruyter, Berlin.

Bedau, M. A. (2010), *The nature of life. Classical and contemporary perspectives from philosophy and science*, Cambridge University Press, Cambridge.

Beer, S. (1967), *Cybernetics and Management*, John Wiley & Sons, Hoboken, New Jersey, USA.

Beer, S. (1972), *Brain of the firm: A Development in Management Cybernetics*, Herder and Herder, Munich.

Beer, S. (1995), *Diagnosing the System for Organizations,* 2nd ed., Wiley, Chichester.

Behlau, L. (2012), *Die Dimensionen der Nachhaltigkeit: Ein Überblick*, München, available at: *http://www.muenchner-wissenschaftstage.de/2012/upload/download/Behlau_Lothar_Dimensionen_der_Nachhaltigkeit.pdf* (accessed 1 March 2017).

Beier, G., Niehoff, S. and Xue, B. (2018), 'More sustainability in industry through industrial internet of things?', *Applied Sciences*, Vol. 8 No. 2, p. 219.

Bellwood, P. S. (2004), *First farmers: the origins of agricultural societies*, Wiley-Blackwell, Hoboken, New Jersey, Vereinigte Staaten.

Benner, S. A. and Sismour, M. A. (2005), 'Synthetic Biology', *Nature Reviews Genetics*, Vol. 6, pp. 533–543.

Benoît, C., Norris, G. A. and Valdivia, S. e.a. (2010), 'The guidelines for social life cycle assessment of products: just in time', *The International Journal of Life Cycle Assessment*, Vol. 15 No. 2, pp. 156–163.

Bensaude-Vincent, B. and Stengers, I. (1997), *A History of Chemistry*, Harvard University Press, Cambridge.

Bentham, J. (1823), 'An Introduction to the Principles of Morals and Legislation', available at: *http://www.earlymoderntexts.com/assets/pdfs/bentham1780.pdf* (accessed 1 March 2017).

Benyus, J. M. (2002), *Biomimicry: Innovation Inspired by Nature*, William Morrow, New York.

Benz, G. T. (2011), '(2011). Bioreactor design for chemical engineers', *Chemical Engineering Progress*, Vol. 107 No. 13, pp. 21–26.

Berekoven, L., Eckert, W. and Ellenrieder, P. (2009), *Marktforschung,* 12th ed., Gabler, Wiesbaden.

Berggren, C., Bjarnason, B. and Johansson, G. (2001), 'Capacitive biosensors', *Electroanalysis*, Vol. 13 No. 3, pp. 173–180.

Bergmiller, G. G. and McCright, P. R. (2009), 'Parallel Models for Lean and Green Operations', *Proceedings of the 2009 Industrial Engineering Research Conference*.

Berkhout, P. H.G., Muskens, J. C. and Velthuijsen, J. W. (2000), 'Defining the Rebound Effect', *Energy Policy*, Vol. 28 No. 6–7, pp. 425–432.

Berndt, A. and Deisseroth, K. (2015), 'Expanding the optogenetics toolkit', *Science*, Vol. 349 No. 6248, pp. 590–591.

Besi, M. de and McCormick, K. (2015), 'Towards a bioeconomy in Europe: National, regional and industrial strategies', *Sustainability*, Vol. 7 No. 8, pp. 10.461–10.478.

Bhalla, N., Jolly, P., Formisano, N. and Estrela, P. (2016), 'Introduction to biosensors. Essays Biochemistry', Vol. 60, pp. 1–8.

Bhamra, T. and Lofthouse, V. (2007), *Design for sustainability: a practical approach*, Gower Publishing, Ltd.

Bhatia, S. N. and Ingber, D. E. (2014), 'Microfluidic organs-on-chips', *Nature Biotechnology*, Vol. 32 No. 8, p. 760.

Biermann, W. E. (1904), *Zur Lehre von der Produktion und ihrem Zusammenhang mit der Wert-, Preis- und Einkommenslehre*, Probevorlesung gehalten am 6. Juli 1904 an der Universität Leipzig, Leipzig.

Biju, P. L., Shalij, P. R. and Prabhushankar, G. V. (2015), 'Evaluation of customer requirements and sustainability requirements through the application of fuzzy analytic hierarchy process', *Journal of Cleaner Production*, Vol. 108, pp. 808–817.

Bilo, D. (1971), 'Flugbiophysik von Kleinvögeln', *Zeitschrift für vergleichende Physiologie*, Vol. 71 No. 4, pp. 382–454.

Binder, P.-M. (2009), 'Computation: The edge of reductionism', *Nature*, Vol. 459 No. 7245, pp. 332–333.

Binet, A. (2015), *The Development of Intelligence in Children*, Palala Press.

Binswanger, H. C. (1991), *Geld und Natur: Das wirtschaftliche Wachstum im Spannungsfeld zwischen Ökonomie und Ökologie*, Weitbrecht, Stuttgart.

Binswanger, H. C. (2016), *Die Wirklichkeit als Herausforderung*, Murmann Publishers GmbH, Hamburg.

Bitkom e. V. (2017a), *Entscheidungsunterstützung mit Künstlicher Intelligenz: Wirtschaftliche Bedeutung, gesellschaftliche Herausforderungen, menschliche Verantwortung*, Kaiserslautern, available at: *https://www.bitkom.org/Bitkom/Publikationen/Entscheidungsunterstuetzung-mit-Kuenstlicher-Intelligenz-Wirtschaftliche-Bedeutung-gesellschaftliche-Herausforderungen-menschliche-Verantwortung.html* (accessed 23 April 2020).

Bitkom e. V. (2017b), 'Künstliche Intelligenz – Wirtschaftliche Bedeutung, gesellschaftliche Herausforderungen, menschliche Verantwortung', available at: *https://www.bitkom.org/Bitkom/Publikationen/Entscheidungsunterstuetzung-mit-Kuenstlicher-Intelligenz.html* (accessed 18 May 2020).

Blackmore, S. (2000), *Die Macht der Meme oder Die Evolution von Kultur und Geist,* Spektrum Verlag, Berlin.

Blumenthal, K. von (2007), 'Die Steuertheorien der Austrian Economics', Universität Hamburg, Weimar, 2007.

Bocken, N. M., Short, S. W., Rana, P. and Evans, S. (2014), 'A literature and practice review to develop sustainable business model archetypes', *Journal of Cleaner Production*, Vol. 65, pp. 42–56.

Boersch, I., Heinsohn, J. and Socher, R. (2007), *Wissensverarbeitung – Eine Einführung in die Künstliche Intelligenz,* Spektrum, Heidelberg.

Bofinger, P. (2011), *Grundzüge der Volkswirtschaftslehre: Eine Einführung in die Wissenschaft von Märkten,* 3rd ed., Pearson Studium, München.

Böhme, G. (1986), *Bildungsgeschichte des europäischen Humanismus*, Wissenschaftliche Buchgesellschaft, Darmstadt.

Boland, T., Xu, T., Damon, B. and Cui, X. (2006), 'Application of inkjet printing to tissue engineering', *Biotechnology Journal: Healthcare Nutrition Technology*, Vol. 9 No. 1, pp. 910–917.

Bonke, S. A., Wiechen, M., MacFarlane, D. R. and Spiccia, L. (2015), 'Renewable fuels from concentrated solar power: towards practical artificial photosynthesis', *Energy & Environmental Science*, Vol. 8 No. 9, pp. 2791–2796.

Bonnet, J., Yin, P., Ortiz, M. E., Subsoontorn, P. and Endy, D. (2013), 'Amplifying genetic logic gates', *Science*, Vol. 340 No. 6132, pp. 599–603.

Borgmann, A. (2006), 'Technology as a Cultural Force: For Alena and Griffin', *The Canadian Journal of Sociology*, Vol. 31 No. 3, pp. 351–360.

Borisov, S. M. and Wolfbeis, O. S. (2008), 'Optical biosensors', *Chemical Reviews*, Vol. 108 No. 2, pp. 423–461.

Bose, S., Ke, D., Sahasrabudhe, H. and Bandyopadhyay, A. (2018), 'Additive manufacturing of biomaterials', *Progress in Materials Science*, No. 93, pp. 45–111.

Bosecker, K. (1997), 'Bioleaching: metal solubilization by microorganisms', *FEMS Microbiology Reviews*, Vol. 20 No. 3-4, pp. 591–604.

Box, G. E.P., Hunter, J. S. and Hunter, W. G. (2005), *Statistics for Experimenters. Design, Innovation, and Discovery,* 2nd ed., John Wiley & Sons, Hoboken.

Brand, K.-W. and Jochum, G. (2000), *Der deutsche Diskurs zu nachhaltiger Entwicklung*, available at: *http://www.sozialforschung.org/wordpress/wp-content/uploads/2009/09/kw_brand_deutscher_nachh_diskurs.pdf* (accessed 14 July 2015).

Braungart, M. and McDonough, W. (2014), *Cradle to Cradle: Einfach intelligent produzieren*, Piper, München.

Brem, A. (2008), *The Boundaries of Innovation and Entrepreneurship – Conceptual Background and Essays on Selected Theoretical and Empirical Aspects*, Gabler, Wiesbaden.

Brian, A. W. (2009), *The Nature of Technology*, Free Press, New York.

Brown, B., Chui, M. and Manyika, J. (2011), 'Are you ready for the era of 'big data'', *McKinsey Quarterly*, Vol. 4 No. 1, pp. 24–35.

Brüggemann, H. and Bremer, P. (2012), *Grundlagen Qualitätsmanagement. Von den Werkzeugen über Methoden zum TQM*, Vieweg, Wiesbaden.

Bruynseels, K., Santoni de Sio, F. and van den Hoven, J. (2018), 'Digital twins in health care: ethical implications of an emerging engineering paradigm', *Frontiers in Genetics*, Vol. 9, p. 31.

Buchheim, G. and Sonnemann, R. (1993), *Geschichte der Technikwissenschaften*, Edition, Leipzig.

Bugge, M. M., Hansen, T. and Klitkou, A. (2016), 'What is the bioeconomy? A review of the literature', *Sustainability*, Vol. 8 No. 7, p. 691.

Buijs, B., Tercero Espinoza, L. A. and Sievers, H. (2012), 'Limits to the critical raw materials approach', *Proceedings of the ICE – Waste and Resource Management*, Vol. 165 No. 4, pp. 201–208.

Bundesamt für Umweltschutz (1984), *Ökobilanz von Packstoffen, Schriftenreihe Umweltschutz*, Bern, Schweiz.

Bundesministerium für Wirtschaft und Energie (2018), 'Fakten zum deutschen Außenhandel', available at: *www.bmwi.de/Redaktion/DE/Publikationen/Aussenwirtschaft/fakten-zum-deuschen-aussenhandel.html* (accessed 6 January 2020).

Burk, M. J. and van Dien, S. (2016), 'Biotechnology for chemical production: challenges and opportunities', *Trends in Biotechnology*, Vol. 34 No. 3, pp. 187–190.

Burkhart, T., Krumeich, J., Werth, D. and Loos, P. (2011), 'Analyzing the business model concept – a comprehensive classification of literature', *ICIS Proceedings*, pp. 1–19.

Burschel, C. J., Losen, D. and Wiendl, A. (2004), *Betriebswirtschaftslehre der Nachhaltigen Unternehmung*, Oldenbourg Wissenschaftsverlag, München.

Butler, M. (2005), 'Animal cell cultures: recent achievements and perspectives in the production of biopharmaceuticals', *Applied microbiology and biotechnology*, Vol. 68 No. 3, pp. 283–291.

Byrne, G., Dimitrov, D., Monostori, L., Teti, R., van Houten, F. and Wertheim, R. (2018), 'Biologicalisation: Biological transformation in manufacturing', *CIRP Journal of Manufacturing Science and Technology*, Vol. 21, pp. 1–32.

Calabrese Barton, S., Gallaway, J. and Atanassov, P. (2004), 'Enzymatic biofuel cells for implantable and microscale devices', *Chemical Reviews*, Vol. 104 No. 10, pp. 4867–4886.

Caldwell, L. K. (1987), *Biocracy: Public Policy And The Life Sciences*, Westview Press, Boulder, USA.

Cameron, D. E., Bashor C. J. and Collins J. J. (2014), 'A brief history of synthetic biology', *Nature Reviews Microbiology*, Vol. 12 No. 5, p. 381.

Campana, A. (1946), 'The Origin of the Word Humanist', *Journal of the Warburg and Courtauld Institutes*, Vol. 9, pp. 60–73.

Campbell, C. J., Liesenborghs, F., Schindler, J. and Zittel, W. (2007), *Ölwechsel!: Das Ende des Erdölzeitalters und die Weichenstellung für die Zukunft*, Deutscher Taschenbuch Verlag, München.

Campbell, N. A. and Reece, J. B. (2005), *Biologie*, 6th ed., Pearson, London.

Campoli, G., Borleffs, M. S., Yavari, S. A., Wauthle, R., Weinans, H. and Zadpoor, A. A. (2013), 'Mechanical properties of open-cell metallic biomaterials manufactured using additive manufacturing', *Materials & Design*, Vol. 49, pp. 957–965.

Caplin, J. D., Granados, N. G., James, M. R., Montazami, R. and Hashemi, N. (2015), 'Microfluidic organ-on-a-chip technology for advancement of drug development and toxicology', *Advanced Healthcare Materials*, Vol. 4 No. 10, pp. 1426–1450.

Carlowitz, H. C. von (1713), *Sylvicultura Oeconomica oder haußwirthliche Nachricht und Naturgemäßige Anweisung zur Wilden Baum-Zucht*, Johann Friedrich Braun, Leipzig.

Carmo, M., Fritz, D. L., Mergel, J. and Stolten, D. (2013), 'A comprehensive review on PEM water electrolysis', *International Journal of Hydrogen Energy*, Vol. 38 No. 12, pp. 4901–4934.

Carnau, P. (2011), *Nachhaltigkeitsethik: Normativer Gestaltungsansatz für eine global zukunftsfähige Entwicklung in Theorie und Praxis*, Hampp, Mering.

Carter, S. S.D., Costa, P. F., Vaquette, C., Ivanovski, S., Hutmacher, D. W. and Malda, J. (2017), 'Additive biomanufacturing: an advanced approach for periodontal tissue regeneration', *Annals of Biomedical Engineering*, Vol. 45 No. 1, pp. 12–22.

Cassel, G. (1921), *Theoretische Sozialökonomie*, Winter, Heidelberg.

Cebula, R. and Clark, J. (2014), 'Economic Freedom, Regulatory Quality, Taxation, and Living Standards', available at: *https://mpra.ub.uni-muenchen.de/58.108/1/MPRA_paper_58.108.pdf* (accessed 13 January 2021).

Ceschin, F. and Gaziulusoy, I. (2020), *Design for Sustainability: A Multi-level Framework from Products to Socio- technical Systems*, Routledge, New York.

Chang, Y. J., Sproesser, G. and Neugebauer, S. e.a. (2015), 'Environmental and social life cycle assessment of welding technologies', *Procedia CIRP*, Vol. 26, pp. 293–298.

Chapin, F. S., Matson, P. A. and Mooney, H. A. (2011), *Principles of Terrestrial Ecosystem Ecology*, 2nd ed., Springer, Berlin.

Chaubey, A. and Malhotra, B. (2002), 'Mediated biosensors', *Biosensors and Bioelectronics*, Vol. 17 No. 6-7, pp. 441–456.

Chaves, S. A. de, Carlos, B., Westphall, C. M. and Gerônimo, G. A. (2011), 'Customer security concerns in cloud computing', *Proceedings of the Tenth International Conference on Networks*, pp. 7–11.

Chemical Abstract Service (2015), 'CAS Assigns the 100 Millionth CAS Registry Number® to a Substance Designed to Treat Acute Myeloid Leukemia. Media Releases', available at: *https://www.cas.org/news/media-releases/100-millionth-substance*.

Chen, T., Barton, S. C., Binyamin, G., Gao, Z., Zhang, Y., Kim, H. H. and Heller, A. (2001), 'A miniature biofuel cell', *Journal of the American Chemical Society*, Vol. 123 No. 35, pp. 8630–8631.

Chen, Y. W., Lee, H. V., Juan, J. C. and Phang, S. M. (2016), 'Production of new cellulose nanomaterial from red algae marine biomass Gelidium elegans', *Carbohydrate Polymers*, Vol. 151, pp. 1210–1219.

Cheng, N. G., Gopinath, A., Wang, L., Iagnemma, K. and Hosoi, A. E. (2014), 'Thermally tunable, self-healing composites for soft robotic applications', *Macromolecular Materials and Engineering*, Vol. 299 No. 11, pp. 1279–1284.

Cherubini, F. (2010), 'The biorefinery concept: using biomass instead of oil for producing energy and chemicals', *Energy conversion and Management*, Vol. 51 No. 7, pp. 1412–1421.

Cherubini, F. and Jungmeier, G. (2010), 'LCA of a biorefinery concept producing bioethanol, bioenergy, and chemicals from switchgrass', *The International Journal of Life Cycle Assessment*, Vol. 15 No. 1, pp. 53–66.

Cherubini, F., Jungmeier, G., Wellisch, M., Willke, T., Skiadas, I., van Ree, R. and Jong, E. de (2009), 'Toward a common classification approach for biorefinery systems', *Biofuels, Bioproducts and Biorefining*, Vol. 3 No. 5, pp. 534–546.

Chew, K. W., Yap, J. Y., Show, P. L., Suan, N. H., Juan, J. C. and Ling, T. C. et al. (2017), 'Microalgae biorefinery: high value products perspectives', *Bioresource Technology*, Vol. 229, pp. 53–62.

Chia, H. N. and Wu, B. M. (2015), 'Recent advances in 3D printing of biomaterials. 9(1), 4.', *Journal of Biological Engineering*, Vol. 9 No. 1, pp. 1–14.

Chiarini, A. (2014), 'Sustainable manufacturing-greening processes using specific Lean Production tools: an empirical observation from European motorcycle component manufacturers', *Journal of Cleaner Production*, Vol. 85, pp. 226–233.

Christensen, C. M. (1997), *The Innovator's Dilemma: When New Technologies Cause Great Firms to Fail*, Harvard Business Review Press, Watertown, Massachusetts.

Christoffersson, J., van Noort, D. and Mandenius, C. F. (2017), 'Developing or-gan-on-a-chip concepts using bio-mechatronic design methodology', *Biofabrication*, Vol. 9 No. 2, p. 25.023.

Chu, C. C. (2007), 'Biodegradable polymeric biomaterials: an updated overview', *Biomaterials*, Vol. 6, pp. 1–6.

Chui, M., Evers, M., Manyika, J., Zheng, A. and Nisbet, T. (2020), *The Bio Revolution – Innovations transforming economies, societies, and our lives*, San Francisco.

Church, G. M., Gao, Y. and Kosuri, S. (2012), 'Next-generation digital information storage in DNA', *Science*, Vol. 337 No. 6102, p. 1628.

Clausewitz, C. von (2010), *Vom Kriege*, Anaconda, Köln.

Clavé, G., Garoux, L., Boulanger, C., Hesemann, P. and Grison, C. (2016), 'Ecological Recycling of a Bio-Based Catalyst for Cu Click Reaction: A New Strategy for a Greener Sustainable Catalysis', *ChemistrySelect*, Vol. 1 No. 7, pp. 1410–1416.

Coase, R. H. (1960), 'The Problem of Social Cost', *Journal of Law and Economics*, Vol. 3, pp. 1–44.

Coelli, T. J., Rao, D. S.P. and O'Donnell, C. J. et al. (2005), *An Introduction to Efficiency and Productivity Analysis*, 2nd ed., Springer, Berlin.

Coenenberg, A. G. and Salfeld, R. (2007), *Wertorientierte Unternehmensführung: Vom Strategieentwurf zur Implementierung*, 2nd ed., Schäffer-Poeschel, Stuttgart.

Comanns, P., Buchberger, G., Buchsbaum, A., Baumgartner, R., Kogler, A., Bauer, S. and Baumgartner, W. (2015), 'Directional, passive liquid transport: the Texas horned lizard as a model for a biomimetic "liquid diode"', *Journal of the Royal Society Interface*, Vol. 12 No. 109, p. 20.150.415.

Cooper, M. (2014), 'Leben jenseits der Grenzen. Die Erfindung der Bioökonomie', in Folkers, A. and Lemke, T. (Eds.), *Biopolitik: Ein Reader*, Suhrkamp, Berlin, pp. 468–526.

Cooper, R. G. (2010), *Top oder Flop in der Produktentwicklung. Erfolgsstrategien: von der Idee zum Launch*, Wiley-VCH, Weinheim.

Corning, P. A. (2002), 'The Re-Emergence of 'Emergence': A Venerable Concept in Search of a Theory', *Complexity*, Vol. 7 No. 6, pp. 18–30.

Corsten, H. (1985), *Die Produktion von Dienstleistungen: Grundzüge einer Produktionswirtschaftslehre des tertiären Sektors*, (Betriebswirtschaftliche Studien, Band 51), Erich Schmidt Verlag GmbH & Co, Berlin.

Corsten, H. (Ed.) (2007), *Produktionswirtschaft – Einführung in das industrielle Produktionsmanagement*, Oldenburg Wissenschaftsverlag GmbH, München.

Costanza, R. and Daly, H. E. (1992), 'Natural Capital and Sustainable Development', *Conservation Biology*, Vol. 6 No. 1, pp. 37–46.

Cottingham, J. (1988), *The Rationalists: History of Western Philosophy*, Oxford University Press, Oxford.

Cournot, A. A. (1924), *Untersuchung über die mathematischen Grundlagen der Theorie des Reichtums*, Gustav Fischer, Jena.

Craven, S., Whelan, J. and Glennon, B. (2014), 'Glucose concentration control of a fed-batch mammalian cell bioprocess using a nonlinear model predictive controller', *Journal of Process Control*, Vol. 24 No. 4, pp. 344–357.

Cristóbal, J., Matos, C. T., Aurambout, J. P., Manfredi, S. and Kavalov, B. (2016), 'Environmental sustainability assessment of bioeconomy value chains', *Biomass and Bioenergy*, Vol. 89, pp. 159–171.

Crittenden, V. L., Crittenden, W. F., Ferrell, L. K., Ferrell, O. C. and Pinney, C. C. (2011), 'Market-oriented sustainability: a conceptual framework and propositions', *Journal of the Academy of Marketing Science*, Vol. 39 No. 1, pp. 71–85.

Crommelin, D. J., Storm, G., Verrijk, R., Leede, L. de, Jiskoot, W. and Hennink, W. E. (2003), 'Shifting paradigms: biopharmaceuticals versus low molecular weight drugs', *International Journal of Pharmaceutics*, Vol. 266 No. 1-2, pp. 3–16.

Crowther, P. (2005), 'Design for disassembly–Themes and principles', *Environment Design Guide*, pp. 1–7.

Cruse, H., Dean, J., Kindermann, T. and Schmitz, J. (1997), 'Simulation komplexer Bewegungen mit Hilfe künstlicher neuronaler Netze', *e-Neuroforum*, Vol. 3 No. 4, pp. 133–139.

Crutzen, P. J. (2011), 'Die Geologie der Menschheit', in Crutzen, P. J., Davis, M. and Mastrandrea, M. D. et al. (Eds.), *Das Raumschiff Erde hat keinen Notausgang*, Suhrkamp, Berlin, pp. 7–10.

Cucurachi, S., van der Giesen, C. and Guinée, J. B. (2018), 'Ex-ante LCA of emerging technologies', *Procedia CIRP*, Vol. 69, pp. 463–468.

Cusick, R. D., Kiely, P. D. and Logan, B. E. (2010), 'A monetary comparison of energy recovered from microbial fuel cells and microbial electrolysis cells fed winery or domestic wastewaters', *International Journal of Hydrogen Energy*, Vol. 35 No. 17, pp. 8855–8861.

D'Anna, W. and Cascini, G. (2011), 'Supporting sustainable innovation through TRIZ system thinking', *Procedia Engineering*, Vol. 9, pp. 145–156.

da Silva, I. N., Hernane Spatti, D., Andrade Flauzino, R., Bartocci Liboni, L. H. and dos Reis Alves, S. F. (2017), *Artificial Neural Networks – A practical course*, Springer, Berlin.

da Silva Veras, T., Mozer, T. S. and da Silva César, A. (2017), 'Hydrogen: trends, production and characterization of the main process worldwide', *International Journal of Hydrogen Energy*, Vol. 42 No. 4, pp. 2018–2033.

Dababneh, A. B. and Ozbolat, I. T. (2014), 'Bioprinting technology: a current state-of-the-art review', *Journal of Manufacturing Science and Engineering,*, Vol. 136 No. 6.

Dagamseh, A. M.K., Wiegerink, R. J., Lammerink, T. S. and Krijnen, G. J. (2012), 'Towards a high-resolution flow camera using artificial hair sensor arrays for flow pattern observations', *Bioinspiration & Biomimetics*, Vol. 7 No. 4, p. 46.009.

Dahmen, C., Wöllecke, F. and Constantinescu, C. (2018), 'Challenges and possible solutions for enhancing the workplaces of the future by integrating smart and adaptive exoskeletons', *Procedia CIRP*, Vol. 67.

Dai, J. and Blackhurst, J. (2012), 'A four-phase AHP–QFD approach for supplier assessment: a sustainability perspective', *International Journal of Production Research*, Vol. 50 No. 19, pp. 5474–5490.

Daka, A. and Peer, D. (2012), 'RNAi-based nanomedicines for targeted personalized therapy', *Advanced Drug Delivery Reviews*, Vol. 64 No. 13, pp. 1508–1521.

Dalavi, A. M., Pawar, P. J. and Singh, T. P. (2016), 'Optimal sequence of hole-making operations using particle swarm optimization and modified shuffled frog leaping algorithm', *Engineering Review*, Vol. 36 No. 2, pp. 187–196.

Dalchau, N., Szép, G., Hernansaiz-Ballesteros, R., Barnes, C. P., Cardelli, L., Phillips, A. and Csikász-Nagy, A. (2018), 'Computing with biological switches and clocks', *Natural Computing*, Vol. 17 No. 4, pp. 761–779.

Daly, H. E. (1990), 'Sustainable Development: From Concept and Theory to Operational Principles', *Population and Development Review*, No. 16, pp. 25–43.

Daly, H. E. and Farley, J. (2003), *Ecological economics: principles and applications*, Island press, Washington, D. C.

D'Amato, D., Droste, N., Allen, B., Kettunen, M., Lähtinen, K. and Korhonen, J. e.a. (2017), 'Green, circular, bio economy: A comparative analysis of sustainability avenues', *Journal of Cleaner Production*, Vol. 168, pp. 716–734.

Daniel, R., Rubens, J. R., Sarpeshkar, R. and Lu, T. K. (2013), 'Synthetic analog computation in living cells', *Nature*, Vol. 497 No. 7451, pp. 619–623.

Dean, A. and Voss, D. (2017), *Design and Analysis of Experiments*, 2nd ed., Springer, New York.

Deb, K. (2009), *Multi-objective optimization using evolutionary algorithms*, John Wiley & Sons, Hoboken.

Decker, E. L. and Reski, R. (2008), 'Current achievements in the production of complex biopharmaceuticals with moss bioreactors', *Bioprocess and Biosystems Engineering*, Vol. 31 No. 1, pp. 3–9.

Del Priore, L. V. (2005), 'Effect of sham surgery on retinal function after subretinal transplantation of the artificial silicone retina', *Archives of Ophthalmology*, Vol. 123 No. 8, p. 1156.

del Rio-Chanona, E. A., Zhang, D. and Vassiliadis, V. S. (2016), 'Model-based real-time optimisation of a fed-batch cyanobacterial hydrogen production process using economic model predictive control strategy', *Chemical Engineering Science*, Vol. 142 No. 289-298.

Demirbas, A. (2008), 'Biofuels sources, biofuel policy, biofuel economy and global biofuel projections', *Energy conversion and Management*, Vol. 49 No. 8, pp. 2106–2116.

Demirbas, A. and Demirbas, M. F. (2010), *Algae Energy: Algae as a New Source of Biodiesel Green Energy and Technology*, Springer, Berlin.

Derby, B. (2008), 'Bioprinting: inkjet printing proteins and hybrid cell-containing materials and structures', *Journal of Materials Chemistry*, Vol. 18 No. 47, pp. 5717–5721.

Derler, P., Lee, E. A. and Vincentelli, A. S. (2011), 'Modeling cyber–physical systems', *Proceedings of the IEEE*, Vol. 100 No. 1, pp. 13–28.

Descartes, R. (2011), *Meditationen: Mit sämtlichen Einwänden und Erwiderungen*, Meiner, Hamburg.

Deutsche Forschungsgemeinschaft (2007), 'Projekt: Bioorganische Funktionssysteme auf Festkörpern', available at: *http://gepris.dfg.de/gepris/projekt/5.484.320* (accessed 1 May 2020).

Deutscher Bundestag (1998), *Konzept Nachhaltigkeit.: Vom Leitbild zur Umsetzung*, (Abschlußbericht der Enquete-kommission 'Schutz des Menschen und der Umwelt' des 13. Bundestages), available at: *http://dipbt.bundestag.de/doc/btd/13/112/1.311.200.pdf* (accessed 14 July 2015).

Deutsches Institut für Normung e. V. (2003), *Fertigungsverfahren – Begriffe, Einteilung* No. DIN 8580, Beuth, Berlin.

Deutsches Institut für Normung e. V. (2006), *Umweltmanagement – Ökobilanz – Anforderungen und Anleitungen (ISO 14.044:2006); Deutsche und Englische Fassung EN ISO 14.044:2006* No. DIN EN ISO 14.044:2006-10, Beuth, Berlin.

Deutsches Institut für Normung e. V. (2009), *Umweltmanagement – Ökobilanz – Grundsätze und Rahmenbedingungen (ISO 14.040:2006); Deutsche und Englische Fassung EN ISO 14.040:2006* No. DIN EN ISO 14.040:2009-11, Beuth, Berlin.

Deutsches Institut für Normung e. V. (2016), *DIN ISO 18.458:2016-08 Bionik – Terminologie, Konzepte und Methodik*, Beuth, Berlin.

Devine, P. and Devine, P. J. (2010), *Democracy and Economic Planning*, Polity Press, Cambridge, UK.

Dey, P. K. (2003), 'Analytic hierarchy process analyzes risk of operating cross-country petroleum pipelines in India', *Natural Hazards Review*, Vol. 4 No. 4, pp. 213–221.

Diaz-Elsayed, N., Jondral, A. and Greinacher, S. e.a. (2013), 'Assessment of Lean and Green strategies by simulation of manufacturing systems in discrete production environments', *CIRP Annals – Manufacturing Technology*, Vol. 62, pp. 475–478.

Dicks, A. L. and Rand, D. A. (2018), *Fuel cell systems explained*, 978-1.118.613.528, 3rd ed., Wiley, Hoboken.

Diesner, T. (2015), *Die Allgemeine Systemtheorie bei Ludwig von Bertalanffy: Eine Begründung interdisziplinärer Wissenschaftspraxis*, Logos, Berlin.

Dietzsch, C., Spadiut, O. and Herwig, C. (2013), 'On-line multiple component analysis for efficient quantitative bioprocess development', *Journal of Biotechnology*, Vol. 163 No. 4, pp. 362–370.

Dijkman, R. M., Sprenkels, B., Peeters, T. and Janssen, A. (2015), 'Business models for the Internet of Things', *International Journal of Information Management*, Vol. 35 No. 6, pp. 672–678.

Dincer, C., Bruch, R., Costa-Rama, E., Fernández-Abedul, M. T., Merkoçi, A. and Manz, A. e.a. (2019), 'Disposable sensors in diagnostics, food, and environmental monitoring', *Advanced Materials*, Vol. 31 No. 30, p. 1.806.739.

Dinkel, F. and Kägi, T. (2014), *CustomLCA – Ökobilanz Getränkeverpackungen*, Basel, available at: *www.carbotech.ch* (accessed 21 April 2020).

Dittmann, J. (2008), 'Einstellungen zur Umwelt', in Statistisches Bundesamt, Gesellschaft Sozialwissenschaftlicher Infrastruktureinrichtungen and Wissenschaftszentrum Berlin für Sozialforschung (Eds.), *Datenreport 2008: Ein Sozialbericht für die Budesrepublik Deutschland*, Bonn, pp. 356–362.

Dobbs, M. E. (2014), 'Guidelines for applying Porter's five forces framework: a set of industry analysis templates', *Competitiveness Review*, Vol. 24 No. 1, pp. 32–45.

Dolai, S., Leu, H. Y., Magda, J. and Tabib-Azar, M. (2018), 'Bio-Mimetic synthetic cell hydrogel magnetometer', *Bioinspiration & Biomimetics*.

Doran, G. T. (1981), 'There's a SMART way to write management's goals and objectives', *Management review*, Vol. 70 No. 11, pp. 35–36.

Döring, R. and Ott, K. (2001), 'Nachhaltigkeitskonzepte', *Zeitschrift für Wirtschafts- und Unternehmensethik (zfwu)*, Vol. 2 No. 3, pp. 315–339.

Dressler, F. and Akan, O. B. (2010), 'A survey on bio-inspired networking', *Computer Networks*, Vol. 54 No. 6, pp. 881–900.

Dreyer, L., Hauschild, M. and Schierbeck, J. (2006), 'A framework for social life cycle impact assessment', *The International Journal of Life Cycle Assessment*, Vol. 11 No. 2, pp. 88–97.

Drucker, P. F. (1963), 'Managing for Business Effectiveness', *Harvard Business Review*, No. 3, pp. 53–60.

D'souza, S. F. (2001), 'Microbial biosensors', *Biosensors and Bioelectronics*, Vol. 16 No. 6, pp. 337–353.

Durakovic, B. (2017), 'Design of experiments application, concepts, examples: State of the art', *Periodicals of Engineering and Natural Sciences*, Vol. 5 No. 3, pp. 421–439.

Dürre, P. (2007), 'Biobutanol: an attractive biofuel', *Biotechnology Journal: Healthcare Nutrition Technology*, Vol. 12 No. 2, pp. 1525–1534.

Dutra, D. D.S. and Silva, J. R. (2016), 'Product-service architecture (psa): toward a service engineering perspective in industry 4.0', *IFAC-PapersOnLine*, Vol. 49 No. 31, pp. 91–96.

Dyckhoff, H. (2008), *Betriebliche Produktion: Theoretische Grundlagen einer umweltorientierten Produktionswirtschaft*, 2nd ed., Springer, Berlin.

Dyckhoff, H. and Spengler, T. S. (2010), *Produktionswirtschaft – Eine Einführung*, 3rd ed., Springer-Verlag, Berlin.

Eagly, R. V. (1974), *Structure of Classical Economic Theory*, Oxford University Press, Oxford.

Eberl, U. (2016), *Smarte Maschinen: Wie Künstliche Intelligenz unser Leben verändert*, Hanser, München.

Edler, F., Soden, M. and Hankammer, R. (2015), *Fehlerbaumanalyse in Theorie und Praxis: Grundlagen und Anwendung der Methode*, Springer, Berlin.

Eggins, B. R. (1997), *Biosensors: an introduction*, Wiley, Hoboken.

Ehrhard Bahr (2008), *Was ist Aufklärung?: Thesen und Definitionen*, Reclam, Stuttgart.

Ehrlich, P. R. and Holdren, J. P. (1971), 'Impact of population growth', *Science*, Vol. 171 No. 3977, pp. 1212–1217.

Eiben, Á. E., Hinterding, R. and Michalewicz, Z. (1999), 'Parameter control in evolutionary algorithms', *IEEE Transactions on Evolutionary Computation*, Vol. 3 No. 2, pp. 124–141.

Eifert, T., Eisen, K., Maiwald, M. and Herwig, C. (2020), 'Current and future requirements to industrial analytical infrastructure – part 2: smart sensors', *Analytical and Bioanalytical Chemistry*, pp. 1–9.

Eigen, M., Gardiner, W., Schuster, P. and Winkler-Oswatitsch, R. (1981), 'The Origin of Genetic Information', *Scientific American*, Vol. 244, pp. 88–118.

Eisenstein, E. (1997), *Die Druckerpresse. Kulturrevolutionen im frühen modernen Europa*, Springer, Wien.

El-Chichakli, B., Braun, J. von, Lang, C., Barben, D. and Philp, J. (2016), 'Policy: Five cornerstones of a global bioeconomy', *Nature*, Vol. 535 No. 7611, pp. 221–223.

Eldin, A. B. (2011), 'General Introduction to Design of Experiments (DOE)', in Akyar, I. (Ed.), *Wide spectra of quality control*, IntechOpen, London, pp. 21–26.

Ellen MacArthur Foundation (2015), 'Towards a Circular Economy: Business rationale for an accelerated transition', available at: *https://www.ellenmacarthurfoundation.org/assets/downloads/TCE_Ellen-MacArthur-Foundation_9-Dec-2015.pdf* (accessed 5 May 2020).

Enders, M. and Henschel, J. (2000), 'Nebel: Wasserquelle in der Wüste', *Spektrum der Wissenschaft*, Vol. 2, pp. 38–41.

Engels, W. (1962), *Betriebswirtschaftliche Bewertungslehre im Licht der Entscheidungstheorie: Beiträge zur betriebswirtschaftlichen Forschung*, VS Verlag, Wiesbaden.

Erdmann, G. and Zweifel, P. (2008), *Energieökonomik: Theorie und Anwendung*, Springer, Berlin.

Erdmann, L., Behrendt, S. and Feil, M. (2011), *Kritische Rohstoffe für Deutschland: Identifikation aus Sicht deutscher Unternehmen wirtschaftlich bedeutsamer mineralischer Rohstoffe, deren Versorgungslage sich mittel- bis langfristig als kritisch erweisen könnte*, Berlin, available at: *https://www.kfw.de/Download-Center/Konzernthemen/Research/PDF-Dokumente-Sonderpublikationen/Kritische-Rohstoffe-LF.pdf* (accessed 5 August 2015).

Erdmann, L. and Graedel, T. E. (2011), 'Criticality of Non-Fuel Minerals: A Review of Major Approaches and Analyses', *Environmental Science & Technology*, Vol. 45 No. 18, pp. 7620–7630.

Erk, C. (2016), *Was ist ein System?: Eine Einführung in den klassischen Systembegriff*, Lit Verlag, Münster.

Erlach, K. and Sheehan, E. (2014), 'Die CO2-Wertstrom-Methode zur Steigerung von Energie- und Materialeffizienz in der Produktion', *ZWF – Energie- und Materialeffizienz*, Vol. 109 No. 9, pp. 1–4.

Erlach, K. and Westkämper, E. (2009), *Energiewertstrom: Der Weg zur energieeffizienten Fabrik*, Fraunhofer Verlag, Stuttgart.

Ertel, W. (2016), *Grundkurs Künstliche Intelligenz: Eine praxisorientierte Einführung,* 4th ed., Springer Vieweg, Wiesbaden.

Espejo, R. (1990), 'The viable system model', *Systems practice*, Vol. 3 No. 3, pp. 219–221.

Eucken, W. (2013), *Die Grundlagen der Nationalökonomie*, 6th ed., Springer, Berlin.

Europäische Kommission (2013), 'Umwelt: Neues Maßnahmenpaket für saubere Luft in Europa', available at: *http://europa.eu/rapid/press-release_IP-13-1274_de.htm* (accessed 3 October 2016).

European Chemicals Agency (2015), 'Liste der für eine Zulassung in Frage kommenden besonders besorgniserregenden Stoffe', available at: *http://echa.europa.eu/de/candidate-list-table*.

European Commission (2010a), *Critical raw materials for the EU: Report of the Ad-hoc Working Group on defining critical raw materials*, Brussels, Belgium, available at: *http://ec.europa.eu/enterprise/policies/raw-materials/critical/index_en.htm* (accessed 5 August 2015).

European Commission (Ed.) (2010b), *International Reference Life Cycle Data System (ILCD) Handbook: Framework and requirements for Life Cycle Impact Assessment models and indicators*, Publications Office of the European Union, Luxembourg.

European Commission (Ed.) (2010c), *International Reference Life Cycle Data System (ILCD) Handbook: General guide for Life Cycle Assessment – Detailed guidance*, Publications Office of the European Union, Luxembourg.

European Commission (Ed.) (2010d), *International Reference Life Cycle Data System (ILCD) Handbook: Specific guide for Life Cycle Inventory data sets*, Publications Office of the European Union, Luxembourg.

European Commission (2014), *Report on Critical Raw Materials for the EU: Report of the Ad hoc Working Group on defining critical raw materials*, Brussels, Belgium, available at: *http://ec.europa.eu/enterprise/policies/raw-materials/files/docs/crm-report-on-critical-raw-materials_en.pdf* (accessed 5 August 2015).

European Commission (2018), 'Public Consultation on EU Summertime Arrangements. Report of Results', available at: *https://eur-lex.europa.eu/legal-content/DE/TXT/?uri=CELEX:52018SC0406* (accessed 18 January 2020).

European Environment Agency (2017), 'Energy and climate change', available at: *https://www.eea.europa.eu/signals/signals-2017/articles/energy-and-climate-change* (accessed 17 May 2020).

Fader, P. (2020), *Customer centricity: Focus on the right customers for strategic advantage*, Wharton School Press, Philadelphia.

Falcone, G., Strano, A., Stillitano, T., Luca, A. I. de, Iofrida, N. and Gulisano, G. (2015), 'Integrated sustainability appraisal of wine-growing management systems through LCA and LCC methodologies', *Chemical Engineering Tansactions*, Vol. 44, pp. 223–228.

Fankhauser, U. (2014), 'Mystery. Lokal, selbstbestimmt und nachhaltig. Weltweite Bedeutung des Family Farming', available at: *https://catalogue.education21.ch/de/mystery-lokal-selbstbestimmt-und-nachhaltig* (accessed 5 May 2020).

Farley, T., Kiefer, J., Lee, P., Hoff, D. von, Trent, J. M., Colbourn, C. and Mousses, S. (2013), 'The BioIntelligence Framework: a new computational platform for biomedical knowledge computing', *Journal of the American Medical Informatics Association*, Vol. 20 No. 1, pp. 128–133.

Faulkner, W., Templeton, W. and Gullett, D. e.a. (Eds.) (2012), *Visualizing Sustainability Performance of Manufacturing Systems using Sustainable Value Stream Mapping (Sus-VSM), Proceedings of the 2012 International Conference on Industrial Engineering and Operations Management*.

Fava, J., Denison, R. and Jones, B. e.a. (1994), *A Technical Framework for Life-Cycle Assessment*, SETAC, Pensacola, USA, Brussels, Belgium.

Fayemi, P. E., Wanieck, K., Zollfrank, C., Maranzana, N. and Aoussat, A. (2017), 'Biomimetics: Process, tools and practice', *Bioinspiration & Biomimetics*, Vol. 12 No. 1, 011.002.

Felber, C. (2010), *Die Gemeinwohl-Ökonomie: Das Wirtschaftsmodell der Zukunft,* 9th ed., Deuticke, Wien.

Felderer, B. and Homburg, S. (2005), *Makroökonomik Und Neue Makroökonomik,* 9th ed., Springer, Berlin.

Ferreira, A. A., Ludermir, T. B. and Aquino, R. de (2013), 'An approach to reservoir computing design and training', *Expert Systems with Applications*, Vol. 40, pp. 4172–4182.

Fetisova, Z. G., Shibaeva, L. V. and Terol, S. (1987), 'Principles for designing optimal artificial light-harvesting molecular systems', *Proceedings of 1986 International Congress Renewable Energy Sources, Madrid, Spain*, Vol. 1, pp. 80–89.

Fichter, K., Loew, T. and Seidel, E. (1997), *Betriebliche Umweltkostenrechnung: Methoden und praxisgerechte Weiterentwicklung*, Springer, Berlin.

Fichter, K., Paech, N. and Pfriem, R. (2005), *Nachhaltige Zukunftsmärkte: Orientierungen für unternehmerische Innovationsprozesse im 21. Jahrhundert*, Metropolis, Marburg.

Field, C. B., Campbell, J. E. and Lobell, D. B. (2008), 'Biomass energy: the scale of the potential resource', *Trends in Ecology & Evolution*, Vol. 23 No. 2, pp. 65–72.

Figge, F. and Hahn, T. (2002), *Environmental Shareholder Value Matrix: Konzeption, Anwendung und Berechnung*, Universität Lüneburg, Lüneburg.

Figge, F. and Schaltegger, S. (2000), *Was ist 'stakeholder value'? Vom Schlagwort zur Messung*, Universität Lüneburg, Lüneburg.

Fink, A. and Siebe, A. (2011), *Handbuch Zukunftsmanagement*, 2nd ed., Campus, Frankfurt.

Finkbeiner, M., Schau, E. M., Lehmann, A. and Traverso, M. (2010), 'Towards life cycle sustainability assessment', *Sustainability*, Vol. 2 No. 10, pp. 3309–3322.

Finkel, D., Reynolds, C. A., McArdle, J. J. and Pedersen, N. L. (2007), 'Age changes in processing speed as a leading indicator of cognitive aging', *Psychology and Aging*, Vol. 22, pp. 558–568.

Finkenstadt, V. and Willett, J. L. (2005), 'Preparation and characterization of elec-troactive biopolymers', *Macromolecular symposia*, No. 227, pp. 367–372.

Finnveden, G., Hauschild, M. Z. and Ekvall, T. e.a. (2009), 'Recent developments in Life Cycle Assessment', *Journal of Environmental Management*, Vol. 91 No. 1, pp. 1–21.

Fischer, G. and Schrattenholzer, L. (2001), 'Global bioenergy potentials through 2050', *Biomass and Bioenergy*, Vol. 20 No. 3, pp. 151–159.

Fischer, H. (2012), *Stoff-Wechsel: Auf dem Weg zu einer solaren Chemie für das 21. Jahrhundert*, Verlag Antje Kunstmann, München.

Fitzgerald, D. P., Herrmann, J. W., Sandborn, P. A., Schmidt, L. C. and Gogoll, T. H. (2007), 'Design for environment (DfE): strategies, practices, guidelines, methods, and tools', *Environmentally conscious mechanical design*, Vol. 1 No. 1, pp. 1–24.

Foolmaun, R. K. and Ramjeeawon, T. (2013), 'Comparative life cycle assessment and social life cycle assessment of used polyethylene terephthalate (PET) bottles in Mauritius', *The International Journal of Life Cycle Assessment*, Vol. 18 No. 1, pp. 155–171.

Ford, H. (2008), *Mein Leben und Werk: Die Autobiografie,* Deltus.de, Borsdorf.

Forman, E. H. and Gass, S. I. (2001), 'The analytical hierarchy process – an exposition', *Operations Research*, Vol. 49 No. 4, pp. 469–487.

Forrester, J. W. (1971), 'Counterintuitive behavior of social systems', *Technology Review*, Vol. 73 No. 3, pp. 52–68.

Forrester, J. W. (1993), 'System dynamics and the lessons of 35 years. In A systems-based approach to policymaking', in Greene, K. B. de (Ed.), *A Systems-Based Approach to Policymaking*, Springer, Boston, pp. 199–240.

Forrester, J. W. (1994), 'System dynamics, systems thinking, and soft OR', *System dynamics review*, Vol. 10 No. 2-3, pp. 245–256.

Fortuna, L., Graziani, S., Rizzo, A. and Xibilia, M. G. (2007), *Soft Sensor for Monitoring and Control of Industrial Processes*, Springer, London.

Franklin, B. D., Shebl, N. A. and Barber, N. (2012), 'Failure mode and effects analysis: too little for too much?', *BMJ quality & safety*, Vol. 21 No. 7, pp. 607–611.

Freeman, R. E. (1984), *Strategic Management: A Stakeholder Approach*, Financial Times Prentice Hall, New Jersey.

Freiherr von Stackelberg, H. (1993), *Marktform und Gleichgewicht*, Schäffer Poeschel, Stuttgart.

Freimann, J. and Walther, M. (2012), 'Von der Effizienz zur Konsistenz?', in Hauff, M. von, Isenmann, R. and Müller-Christ, G. (Eds.), *Industrial Ecology Management: Nachhaltige Entwicklung durch Unternehmensverbünde*, Springer Gabler, Wiesbaden, pp. 187–204.

Friedman, M. and Schwartz, A. J. (2015), *A Monetary History of the United States, 1867-1960*, Princeton University Press, Princeton.

Frischknecht, R. (2013), 'Analyse und Beurteilung der Umweltverträglichkeit. Teil 2: Ökobilanzen (Life cycle assessment, LCA)', available at: *http://www.eco-bau.ch/resources/uploads/Bildungsinstitutionen/130208_Frischknecht_v1%200.pdf* (accessed 1 March 2017).

Frischknecht, R., Steiner, R. and Jungbluth, N. (2009), *Methode der ökologischen Knappheit – Ökofaktoren 2006: Methode für die Wirkungsabschätzung in Ökobilanzen, Umwelt-Wissen*, Bern.

Frischknecht, R., Wyss, F., Knöpfel, S. B., Lützkendorf, T. and Balouktsi, M. (2015), 'Cumulative energy demand in LCA: the energy harvested approach', *The International Journal of Life Cycle Assessment*, Vol. 20 No. 7, pp. 957–969.

Fritsche, O. (2015), *Biologie für Einsteiger: Prinzipien des Lebens verstehen,* 2nd ed., Springer Spektrum, Berlin.

Fritz, A. H. and Schulze, G. (2010), *Fertigungstechnik*, Springer, Berlin.

Frosch, R. A. and Gallopoulos, N. E. (1989), 'Strategies for manufacturing', *Scientific American*, Vol. 261 No. 3, 144–153.

Fujita, H., Shimizu, K., Hatsuda, R., Sugiyama, S. and Nagamori, E. (2011), 'Designing of a Si-MEMS device with an integrated skeletal muscle cell-based bio-actuator', *Biomedical Microdevices*, Vol. 13 No. 1, pp. 123–129.

Full, J., Miehe, R., Kiemel, S., Bauernhansl, T. and Sauer, A. (2019), 'The Biological Transformation of Energy Supply and Storage–Technologies and Scenarios for Biointelligent Value Creation', *Procedia Manufacturing*, Vol. 39, pp. 1204–1214.

Fullerton, J. (2015), *Regenerative Capitalism*, Greenwich, USA, available at: *https://capitalinstitute.org/wp-content/uploads/2015/04/2015-Regenerative-Capitalism-4-20-15-final.pdf.*

Gadd, K. and Delbrück, M. (2016), *TRIZ für Ingenieure: Theorie und Praxis des erfinderischen Problemlösens*, Wiley-VCH, Weinheim.

Gandenberger, C. (2011), 'Von der sozialen zur sozio-ökologischen Einbettung des Unternehmens', in Conrad, P. and Sydow, J. (Eds.), *Organisation und Umwelt*, Gabler Verlag, Wiesbaden, pp. 249–279.

Gao, B., Yang, Q., Zhao, X., Jin, G., Ma, Y. and Xu, F. (2016), '4D bioprinting for biomedical applications', *Trends in Biotechnology*, Vol. 34 No. 9, pp. 746–756.

Gao, L. and McCarthy, T.J. (2006), 'The "lotus effect" explained: two reasons why two length scales of topography are important', *Langmuir*, Vol. 22 No. 7, pp. 2966–2967.

Gardner, H. (1993), *Multiple Intelligences: The Theory in Practice*, Basic Books, New York.

Gardner, H. (2011), *Frames of Mind: The Theory of Multiple Intelligences,* 3rd ed., Basic Books, New York.

Gasiunas, G., Barrangou, R., Horvath, P. and Siksnys, V. (2012), 'Cas9–crRNA ribonucleoprotein complex mediates specific DNA cleavage for adaptive immunity in bacteria', *Proceedings of the National Academy of Sciences*, Vol. 109 No. 39, E2579-E2586.

Gausemeier, J., Plass, C. and Wenzelmann, C. (2014), *Zukunftsorientierte Unternehmensgestaltung,* 2nd ed., Hanser, München.

Gebhardt, A. (2013), *Generative Fertigungsverfahren – Additive Manufacturing und 3D Drucken für Prototyping – Tooling – Produktion,* 4th ed., Hanser, München.

Geissdoerfer, M., Savaget, P., Bocken, N.M. and Hultink, E.J. (2017), 'The Circular Economy – A new sustainability paradigm?', *Journal of Cleaner Production*, Vol. 143, pp. 757–768.

Geoffrey, S.K., Raven, J.E. and Schofield, M. (2001), *Die vorsokratischen Philosophen*, J.B. Metzler, Stuttgart.

Georgescu-Roegen, N. (1987), *The Entropy Law and the Economic Process in Retrospect, Schriftenreihe des IÖW*, Institut für Ökologische Wirtschaftsforschung, Berlin.

Georgescu-Roegen, N. (2014), *The Entropy Law and the Economic Process*, Reprint, Harvard University Press, Cambridge, USA.

Gerabek, W.E., Haage, B.D., Keil, G. and Wegner, W. (2004), *Enzyklopädie Medizingeschichte: Enzyklopädie Medizingeschichte*, De Gruyter, New York.

Gerard, M., Chaubey, A. and Chaubey, A., & Malhotra, B. (2002), 'Application of conducting polymers to biosensors', *Biosensors and Bioelectronics*, Vol. 17 No. 5, pp. 345–359.

Gerber, P.J., Steinfeld, H., Henderson, B., Mottet, A., Opio, C., Dijkman, J., Falcucci, A. and Tempio, G. (2013), *Tackling Climate Change Through Livestock: A Global Assessment of Emissions and Mitigation Opportunities*, Food and Agriculture Organization, Rom.

Gerber-Kreuzer, A. (2007), *Biotech – Biochip – Biogas: Faszinierende Berufswelt Life Sciences*, BW Bildung u. Wissen, Nürnberg.

Gesell, S. (2018), *Die Natürliche Wirtschaftsordnung Durch Freiland und Freigeld,* Nachdruck, Forgotten Books, London.

Giannakis, M. and Papadopoulos, T. (2016), 'Supply chain sustainability: A risk management approach', *International Journal of Production Economics*, Vol. 171, pp. 455–470.

Gibas, C. and Jambeck, P. (2001), *Einführung in die praktische Bioinformatik*, O'Reilly, Newton.

Gibson, I., Rosen, D.W. and Stucker, B. (2014), *Additive manufacturing technologies*, Springer, New York.

Giddings, G., Allison, G., Brooks, D. and Carter, A. (2000), 'Transgenic plants as factories for biopharmaceuticals', *Nature Biotechnology*, Vol. 18 No. 11, pp. 1151–1155.

Giesecke, M. (2006), *Der Buchdruck in der frühen Neuzeit.,* 4th ed., Suhrkamp, Frankfurt a.M.

Gigerenzer, G. (1999), *Simple heuristics that make us smart*, Oxford University Press, New York.

Gilles, E.D. (2004), 'Komplexität in Technik und Biologie', *Wissenschaft kreiert Technik*, Vol. 55.

Glassey, J., Montague, G.A., Ward, A.C. and Kara, B.V. (1994), 'Enhanced supervision of recombinant E. coli fermentation via artificial neural networks', *Process Biochemistry*, Vol. 29 No. 5, pp. 387–398.

Gleißner, W. (2004), *Future Value: 12 Module für eine strategische wertorientierte Unternehmensführung*, Gabler, Wiesbaden.

Gloria-Palermo, S. (1999), *Evolution of Austrian Economics: From Menger to Lachmann*, Routledge, London.

Goebel, C., Jacobsen, H.A., del Razo, V., Doblander, C., Rivera, J. and Ilg, J.e.a. (2014), 'Energieinformatik', *Wirtschaftsinformatik*, Vol. 56 No. 1, pp. 31–39.

Goguen, J. A. (1969), 'The logic of inexact concepts', *Synthese*, Vol. 19 No. 3-4, pp. 325–373.

Gogula, V., Wan, H.-D. and Kuriger, G. (2011), 'Impact of lean tools on energy consumption', *Revista S&T*, Vol. 19 No. 9, pp. 33–53.

Goldstein, J. (1999), 'Emergence as a Construct: History and Issues', *Emergence*, Vol. 1 No. 1, pp. 49–72.

Gömmel, R. (1998), *Die Entwicklung der Wirtschaft im Zeitalter des Merkantilismus 1620–1800: Enzyklopädie deutscher Geschichte, Band 46*, De Gruyter, München.

Goodfellow, I., Bengio, Y. and Courville, A. (2016), *Deep Learning*, MIT Press, Cambridge.

Goodwin, G. C. (2000), 'Predicting the performance of soft sensors as a route to low cost automation', *Annual Reviews in Control*, Vol. 24, pp. 55–66.

Görtz, H. D. and Brümmer, F. (2012), *Biologie für Ingenieure*, Spektrum Akademischer Verlag, Heidelberg.

Görz, G., Schneeberger, J. and Schmid, U. (2013), *Handbuch der Künstlichen Intelligenz,* 5th ed., Oldenbourg, Berlin.

Gossen, H. H. (2010), *Entwickelung der Gesetze des menschlichen Verkehrs und der daraus fliessenden Regeln für menschliches Handeln*, Nabu Press.

Gottfredson, L. S. (1997), 'Mainstream Science on Intelligence', *Intelligence*, Vol. 24, pp. 13–23.

Gottwald, F. T. and Krätzer, A. (2014), *Irrweg Bioökonomie. Kritik an einem totalitären Ansatz*, Suhrkamp, Berlin.

Götze, U. (2013), *Szenario-Technik in der strategischen Unternehmensplanung*, Springer, Berlin.

Gowdy, J. and Mesner, S. (1998), 'The evolution of Georgescu-Roegen's bioeconomics', *Review of Social Economy*, Vol. 56 No. 2, pp. 136–156.

Goyal, P., Rahman, Z. and Kazmi, A. A. (2015), 'Identification and prioritization of corporate sustainability practices using analytical hierarchy process', *Journal of Modelling in Management*, Vol. 10 No. 1, pp. 23–49.

Graedel, T. E., Barr, R. and Chandler, C. e.a. (2012), 'Methodology of Metal Criticality Determination', *Environmental Science & Technology*, Vol. 46 No. 2, pp. 1063–1070.

Greening, L. A., Greene, D. L. and Difiglio, C. (2000), 'Energy efficiency and consumption. The rebound effect – a survey', *Energy Policy*, Vol. 28, pp. 389–401.

Grefe, C. (2016), *Global Gardening. Bioökonomie neuer Raubbau oder Wirtschaftsform der Zukunft?*, Antje Kunstmann, München.

Greshake, T. (2017), *Methodik zur lebenszyklusweiten, umweltgerechten Produkt und Produktionsoptimierung,* Apprimus Wissenschaftsverlag, Aachen.

Grieves, M. W. (2014), 'Digital Twin: Manufacturing Excellence through Virtual Factory Replication', available at: *https://research.fit.edu/media/site-specif-ic/researchfitedu/camid/documents/1411.0_Digital_Twin_White_Paper_Dr_Grieves.pdf* (accessed 23 April 2020).

Griffith, L. G. (2000), 'Polymeric biomaterials', *Acta Materialia*, Vol. 48 No. 1, 263-277.

Griffith, L. G. and Naughton, G. (2002), 'Tissue engineering – current challenges and expanding opportunities', *Science*, Vol. 295 No. 5557, pp. 1009–1014.

Grinin, L. E., Grinin, A. L. and Korotayev, A. (2017), 'Forthcoming Kondratieff wave, Cybernetic Revolution, and global ageing', *Technological Forecasting and Social Change*, Vol. 115, pp. 52–68.

Grober, U. (2010), *Die Entdeckung der Nachhaltigkeit: Kulturgeschichte eines Begriffs,* 3rd ed., Kunstmann, München.

Gronau, N., Grum, M. and Bender, B. (2016), 'Determining the optimal level of autonomy in cyber-physical production systems', *IEEE 14th International Conference on Industrial Informatics*, pp. 1293–1299.

Guinée, J. B. (Ed.) (2004), *Handbook on Life Cycle Assessment: Operational Guide to the ISO Standards*, Kluwer Academic Publishers, New York.

Guinée, J. B., Heijungs, R. and Huppes, G. e.a. (2011), 'Life Cycle Assessment: Past, Present and Future', *Environmental Science & Technology*, Vol. 45 No. 1, pp. 90–96.

Guinée, J. B., Huppes, G. and Heijungs, R. (2001), 'Developing an LCA guide for decision support', *Environmental Management and Health*, Vol. 12 No. 3, pp. 301–311.

Günther, E. and Nowack, M. (2008), 'CO2-Management von Unternehmen', *uwf UmweltWirtschaftsForum*, Vol. 16 No. 2, pp. 49–51.

Günther, H.-O. and Tempelmeier, H. (2011), *Produktion und Logistik*, Springer, Berlin.

Guo, C., Dai, Z. D. and Sun, J. R. (2005), 'Current status and prospect of research on bio-robots', *Robot*, Vol. 27 No. 2, pp. 188–192.

Gustavsson, R. and Mandenius, C. F. (2013), 'Soft sensor control of metabolic fluxes in a recombinant Escherichia coli fed-batch cultivation producing green fluorescence protein', *Bioprocess and Biosystems Engineering,*, Vol. 36 No. 10, pp. 1375–1384.

Gutenberg, E. (1982), *Grundlagen der Betriebswirtschaftslehre: Die Produktion*, Vol. 1, 8th ed., Springer, Berlin.

Gutenberg, E. (1988), *Die Unternehmung als Gegenstand betriebswirtschaftlicher Theorie*, Topos Ruggell, Ruggell, Liechtenstein.

Gutmann, A. (2011), 'The ethics of synthetic biology: guiding principles for emerging technologies', *Hastings Center Report*, Vol. 41 No. 4, 17–22.

Guvendiren, M., Molde, J., Soares, R. M. and Kohn, J. (2016), 'Designing biomaterials for 3D printing', *ACS Biomaterials Science & Engineering*, Vol. 2 No. 10, pp. 1679–1693.

Haglund, D. G. (1984), 'Strategic minerals – A conceptual analysis', *Resources Policy*, Vol. 10 No. 3, pp. 146–152.

Hampicke, U. (1992), *Ökologische Ökonomie: Individuum und Natur in der Neoklassik · Natur in der ökonomischen Theorie: Teil 4*, VS Verlag für Sozialwissenschaften, Darmstadt.

Han, B. C. (2002), *Philosophie des Zen-Buddhismus*, Reclam, Ditzingen.

Han, C. and Lee, Y. H. (2002), 'Intelligent integrated plant operation system for Six Sigma', *Annual Reviews in Control*, Vol. 26 No. 1, pp. 27–43.

Hanashi, T., Yamazaki, T., Tanaka, H., Ikebukuro, K., Tsugawa, W. and Sode, K. (2014), 'The development of an autonomous self-powered bio-sensing actuator', *Sensors and Actuators B: Chemical*, No. 196, pp. 429–433.

Handfield, R., Walton, S. V., Sroufe, R. and Melnyk, S. A. (2002), 'Applying environmental criteria to supplier assessment: A study in the application of the Analytical Hierarchy Process', *European Journal of Operational Research*, Vol. 141 No. 1, pp. 70–87.

Hankel, M. (2017), 'Reference Architecture Model for Industry 4.0 and International Collaboration. OPC Day Finland', available at: *https://www.youtube.com/watch?v=_7XntF6K_0Y* (accessed 23 April 2020).

Hans-Böckler-Stiftung (2000), *Arbeit und Ökologie. Wege in eine nachhaltige Zukunft: Ergebnisse aus dem Verbundprojekt Arbeit und Ökologie*, Düsseldorf.

Harari, Y. N. (2014), *Sapiens: A Brief History of Humankind*, Harvill Secker, London.

Harari, Y. N. (2016), *Homo Deus: A Brief History of Tomorrow*, Harvill Secker, London.

Harary, F. and Batell, M. F. (1981), 'What is a system?', *Social Networks*, Vol. 3 No. 1, pp. 29–40.

Hartwick, J. (1977), 'Intergenerational Equity and the Investing of Rents from Exhaustible Resources', *American Economic Review*, Vol. 67 No. 5, pp. 972–974.

Harun, W. S.W., Kamariah, M. S.I.N., Muhamad, N., Ghani, S. A.C., Ahmad, F. and Mohamed, Z. (2018), 'A review of powder additive manufacturing processes for metallic biomaterials', *Powder Technology*, No. 327, pp. 128–151.

Hashim, J. (2007), 'Information communication technology (ICT) adoption among SME owners in Malaysia', *International Journal of Business and Information*, Vol. 2 No. 2, pp. 221–240.

Hatti-Kaul, R., Nilsson, L. J., Zhang, B., Rehnberg, N. and Lundmark, S. (2020), 'Designing biobased recyclable polymers for plastics', *Trends in Biotechnology*, Vol. 38 No. 1, pp. 50–67.

Hauschild, M. Z., Herrmann, C. and Kara, S. (2017), 'An integrated framework for life cycle engineering', *Procedia CIRP*, Vol. 61, pp. 2–9.

Hauschild, M. Z., Kara, S. and Røpke, I. (2020), 'Absolute sustainability: Challenges to life cycle engineering', *CIRP Annals*, Vol. 69, pp. 533–553.

Hauschildt, J. and Salomo, S. (2007), *Innovationsmanagement,* 4th ed., Vahlen, München.

Hawken, P., Lovins, A. and Lovins, L. H. (1999), *Natural Capitalism: Creating the Next Industrial Revolution*, Washington D. C., USA, available at: *http://www.usgbc.org/* (accessed 6 February 2020).

Hayden, O., Bindeus, R., Haderspöck, C., Mann, K. J., Wirl, B. and Dickert, F. L. (2003), 'Mass-sensitive detection of cells, viruses and enzymes with artificial receptors', *Sensors and Actuators B: Chemical*, Vol. 91 No. 1-3, pp. 316–319.

Hecker, F. T., Hussein, W. B., Paquet-Durand, O., Hussein, M. A. and Becker, T. (2013), 'A case study on using evolutionary algorithms to optimize bakery production planning', *Expert Systems with Applications*, Vol. 40 No. 17, pp. 6837–6847.

Hede, S., Ferreira, P. V., Lopes, M. N. and Rocha, L. A. (2015), 'TRIZ and the paradigms of social sustainability in product development endeavors', *Procedia Engineering*, Vol. 131, pp. 522–538.

Heertje, A. and Wenzel, H.-D. (2008), *Grundlagen der Volkswirtschaftslehre*, Springer, Berlin.

Heijungs, R. (Ed.) (1992), *Environmental Life Cycle Assessment of Products*, Multicopy, Leiden.

Heinen, E. (1969), 'Zum Wissenschaftsprogramm der entscheidungsorientierten Betriebswirtschaftslehre', *Zeitschrift für Betriebswirtschaft (ZfB)*, Vol. 39 No. 4, pp. 207–220.

Heinen, E. (1994), *Grundfragen der entscheidungsorientierten Betriebswirtschaftslehre*, Gabler Verlag, Wiesbaden.

Heinrich, G. (2007), *Allgemeine Systemanalyse*, Oldenbourg Wissenschaftsverlag, Oldenbourg.

Heiskanen, E. (2002), 'The institutional logic of life cycle thinking', *Journal of Cleaner Production*, Vol. 10 No. 5, pp. 427–437.

Helbig, H. (2018), *Welträtsel aus Sicht der modernen Wissenschaften: Emergenz in Natur, Gesellschaft, Psychologie,Technik und Religion*, Springer, Berlin.

Hench, L. L. and Wilson, J. (1984), 'Surface-active biomaterials', *Science*, Vol. 226 No. 4675, pp. 630–636.

Hendel, A., Bak, R. O., Clark, J. T., Kennedy, A. B., Ryan, D. E. and Roy, S. e.a. (2015), 'Chemically modified guide RNAs enhance CRISPR-Cas genome editing in human primary cells', *Nature Biotechnology*, Vol. 33 No. 9, pp. 985–989.

Hengstler, M., Enkel, E. and Duelli, S. (2016), 'Applied artificial intelligence and trust—The case of autonomous vehicles and medical assistance devices', *Technological Forecasting and Social Change*, Vol. 105, pp. 105–120.

Herb, R., Herb, T. and Kohnhauser, V. (2000), *TRIZ – Der systematische Weg zur Innovation*, Verlag Moderne Industrie, Landsberg/Lech.

Hering, E. and Schloske, A. (2019), *Fehlermöglichkeits- und Einflussanalyse: Methode zur vorbeugenden, systematischen Qualitätsplanung unter Risikogesichtspunkten*, Springer, Berlin.

Herlyn, W. (2011), *PPS im Automobilbau: Produktionsprogrammplanung und -steuerung von Fahrzeugen und Aggregaten*, Carl Hanser, München.

Hermann, M., Pentek, T. and Otto, B. (2016), 'Design principles for industrie 4.0 scenarios', *49th Hawaii international conference on system sciences*, pp. 3928–3937.

Herrmann, C., Blume, S., Kurle, D., Schmidt, C. and Thiede, S. (2015), 'The positive impact factory–transition from eco-efficiency to eco–effectiveness strategies in manufacturing', Procedia CIRP No. 29, pp. 19–27.

Herrmann, C., Cerdas, F., Abraham, T., Büth, L. and Mennenga, M. (2020), 'Biological transformation of manufacturing as a pathway towards environmental sustainability: Calling for systemic thinking', *CIRP Journal of Manufacturing Science and Technology*.

Herzwurm, G., Schockert, S. and Mellis, W. (2000), *Joint Requirements Engineering. QFD for Rapid Customer Focused Software and Internet-Development*, Vieweg, Braunschweig.

Hesser, F., Wohner, B., Meints, T., Stern, T. and Windsperger, A. (2017), 'Integration of LCA in R&D by applying the concept of payback period: case study of a modified multilayer wood parquet', *The International Journal of Life Cycle Assessment*, Vol. 22 No. 3, pp. 307–316.

Hetterich, J., Bonnemeier, S., Pritzke, M. and Georgiadis, A. (2012), 'Ecological sustainability–a customer requirement? Evidence from the automotive industry', *Journal of Environmental Planning and Management*, Vol. 55 No. 9, pp. 1111–1133.

Hinchey, M. G., Sterritt, R. and Rouff, C. (2007), 'Swarms and swarm intelligence', *Computer*, Vol. 40 No. 4, pp. 111–113.

Hirschberg, H. G. (1999), *Handbuch Verfahrenstechnik und Anlagenbau. Chemie, Technik, Wirtschaftlichkeit*, Springer, Berlin.

Hochrein, T. and Alig, I. (2011), *Prozessmesstechnik in der Kunststoffaufbereitung*, Vogel Business Media, Würzburg.

Hodgman, C. E. and Jewett, M. C. (2012), 'Cell-free synthetic biology: thinking outside the cell', *Metabolic Engineering*, Vol. 14 No. 3, pp. 261–269.

Hofmann, P. (2015), *Der kritische Faktor Mensch: Analyse und Würdigung einer Human-FMEA im Bereich des Qualitätsmanagements im Rahmen der Erstellung eines Maßnahmenkataloges zur Fehlerprävention*, Igel Verlag, Paderborn.

Hofstede, G. (2003), *Culture's Consequences: Comparing Values, Behaviors, Institutions and Organizations Across Nations*, 2nd ed., Sage, Los Angeles, USA.

Hoitsch, H. J. and Lingnau, V. (2004), *Kosten- und Erlösrechnung: Eine controllingorientierte Einführung*, 5th ed., Springer, Berlin.

Holub, H. W. (2010), *Eine Einführung in die Geschichte des ökonomischen Denkens: Band III: Physiokraten und Klassiker*, Vol. 3, LIT, Wien.

Honary, L. and Richter, E. (2011), *Biobased lubricants and greases: Technology and products*, John Wiley & Sons, Hoboken.

Hoorens, S., Ghez, J., Guerin, B., Schweppenstedde, D., Hellgren, T., Horvath, V., Graf, M., Janta, B., Drabble, S. and Kobzar, S. (2013), *Europe's Societal Challenges – An analysis of global societal trends to 2030 and their impact on the EU*, European Union, Santa Monica u. a.

Horkheimer, M. (2007), *Zur Kritik der instrumentellen Vernunft*, Fischer, Berlin.

Horn, J. L. and Cattell, R. B. (1966), 'Refinement and test of the theory of fluid and crystallized general intelligences', *Journal of Educational Psychology*, Vol. 57 No. 253–270.

Horne, R., Grant, T. and Verghese, K. (2009), *Life Cycle Assessment: Principle, Practice and Prospects*, CSIRO Publishing, Collingwood, Australia.

Horvath, P. and Barrangou, R. (2010), 'CRISPR/Cas, the immune system of bacteria and archaea', *Science*, Vol. 327 No. 5962, pp. 167–170.

Hotelling, H. (1931), 'The Economics of Exhaustible Resources', *Journal of Political Economy*, Vol. 39 No. 2, pp. 137–175.

Hoverstadt, P. (2009), *The fractal organization: Creating sustainable organizations with the viable system model*, Wiley, Hoboken.

Hozdić, E. (2015), 'Smart factory for industry 4.0: A review', *International Journal of Modern Manufacturing Technologies*, Vol. 7 No. 1, pp. 28–35.

Hua, A. H., Hsu, C.-W., Kuo, T.-C. and Wu, W.-C. (2009), 'Risk evaluation of green components to hazardous substance using FMEA and FAHP', *Expert Systems with Applications*, Vol. 36 No. 3, pp. 7142–7147.

Hubbell, J. A. (1995), 'Biomaterials in tissue engineering', *Biotechnology*, Vol. 13 No. 6, pp. 565–576.

Hubbell, J. A. (1999), 'Bioactive biomaterials', *Current opinion in biotechnology*, Vol. 10 No. 2, pp. 123–129.

Huber, J. (1995), 'Nachhaltige Entwicklung durch Suffizienz, Effizienz und Konsistenz', in Fritz, P., Huber, J. and Levi, H. W. (Eds.), *Nachhaltigkeit in naturwissenschaftlicher und sozialwisenschaftlicher Perspektive*, Wissenschaftliche Verlagsgesellschaft, Stuttgart.

Huber, J. (1998), *Vollgeld: Beschäftigung, Grundsicherung und weniger Staatsquote durch eine modernisierte Geldordnung*, Duncker & Humblot, Berlin.

Huber, J. (2011), *Allgemeine Umweltsoziologie*, 2nd ed., VS Verlag für Sozialwissenschaften, Wiesbaden.

Huijbregts, M. A., Hellweg, S., Frischknecht, R., Hendriks, H. W., Hungerbuhler, K. and Hendriks, A. J. (2010), 'Cumulative energy demand as predictor for the environmental burden of commodity production', *Environmental Science & Technology*, Vol. 44 No. 6, pp. 2189–2196.

Hülsmann, M. (2004), 'Bezugspunkte zwischen Strategischem Management und Nachhaltigkeit', in Hülsmann, M., Müller-Christ, G. and Haasis, H.-D. (Eds.), *Betriebswirtschaftslehre und Nachhaltigkeit*, Springer, Wiesbaden, pp. 25–72.

Humphreys, L. G. (1979), 'The construct of general intelligence', *Intelligence*, Vol. 3 No. 2, pp. 105–120.

Hungenberg, H. (2012), *Strategisches Management in Unternehmen: Ziele – Prozesse – Verfahren,* 7th ed., Springer Gabler, Berlin.

Hungerbühler, K., Ranke, J. and Mettier, T. (1999), 'Risikoanalyse chemischer Prozesse', *Chemische Produkte und Prozesse*, pp. 139–159.

Hunkeler, D. (2006), 'Societal LCA methodology and case study', *The International Journal of Life Cycle Assessment*, Vol. 11 No. 6, pp. 371–382.

Hunkeler, D., Lichtenvort, K. and Rebitzer, G. (2008), *Environmental life cycle costing*, Taylor & Francis, Milton Park.

Hunter, E. E., Brink, E. W., Steager, E. B. and Kumar, V. (2018), 'Toward soft micro bio robots for cellular and chemical delivery', *IEEE Robotics and Automation Letters*, Vol. 3 No. 3, pp. 1592–1599.

Hüther, M. (Ed.) (2006), *Klassiker der Ökonomie: Von Adam Smith bis Amartya Sen*, Bundeszentrale für politische Bildung, Bonn.

Hüther, M. and Wiggering, H. (1999), *Angemessenes Wachstum – Dauerhaft umweltgerechte Entwicklung: Konzeptionelle und instrumentelle Ansätze der beiden Sachverständigenräte, Ökonomisierung der Umweltpolitik. Beiträge zur volkswirtschaftlichen Umweltökonomik*, Berlin.

Ibarra, D., Ganzarain, J. and Igartua, J. I. (2018), 'Business model innovation through Industry 4.0: A review', *Procedia Manufacturing*, Vol. 22, pp. 4–10.

Illge, L. and Schwarze, R. (2006), 'A Matter of Opinion – How Ecological and Neoclassical Environmental Economists think about Sustainability and Economics', *DIW-Discussion Papers*, No. 619.

Imahori, H. (2004), 'Giant multiporphyrin arrays as artificial light-harvesting antennas', *The Journal of Physical Chemistry*, Vol. 108 No. 20, pp. 6130–6143.

Intergovernmental Panel on Climate Change (2014a), *Climate Change 2014: Synthesis Report. Contribution of Working Groups I, II and III to the Fifth Assessment Report of the Intergovernmental Panel on Climate Change*, Geneva, Switzerland.

Intergovernmental Panel on Climate Change (Ed.) (2014b), *Climate Change 2014 Mitigation of Climate Change: Working Group III Contribution to the Fifth Assessment Report of the Intergovernmental Panel on Climate Change*, Cambridge University Press, New York, USA.

Intergovernmental Panel on Climate Change (Ed.) (2018), *Summary for Policymakers. In: Global Warming of 1.5°C. An IPCC Special Report on the impacts of global warming of 1.5°C above preindustrial levels and related global greenhouse gas emission pathways, in the context of strengthening the global response to the threat of climate change, sustainable development, and efforts to eradicate poverty*, World Meteorological Organization, Geneva, Switzerland.

International Chemical Secretariat (2015), 'SIN List', available at: *http://chemsec.org/what-we-do/sin-list*.

International energy agency (2019), *World Energy Outlook 2019*, Paris, available at: *https://www.iea.org/reports/world-energy-outlook-2019* (accessed 17 May 2020).

Jang, J., Park, J. Y., Gao, G. and Cho, D. W. (2018), 'Biomaterials-based 3D cell printing for next-generation therapeutics and diagnostics', *Biomaterials*, Vol. 156, pp. 88–106.

Jarass, L., Obermair, G. M. and Voigt, W. (2009), *Windenergie: Zuverlässige Integration in die Energieversorgung,* 2nd ed., Springer, Berlin.

Jax, K. (2006), 'Ecological units: definitions and application', *The Quarterly Review of Biology*, Vol. 81 No. 3, pp. 237–258.

Jeuken, M. (1975), 'The biological and philosophical definitions of life', *Acta Biotheoretica*, Vol. 24, pp. 14–21.

Jevons, W. S. (1871), *The Theory of Political Economy*, Book on Demand Pod, Berlin.

Jevons, W. S. (1906), *Coal Question,* 3rd Reprint, Augustus M Kelley Pubs, New York.

Jianrong, C., Yuqing, M., Nongyue, H., Xiaohua, W. and Sijiao, L. (2004), 'Nano-technology and biosensors', *Biotechnology Advances*, Vol. 22 No. 7, pp. 505–518.

Jinek, M., Chylinski, K., Fonfara, I., Hauer, M., Doudna, J. A. and Charpentier, E. (2012), 'A Programmable Dual-RNA–Guided DNA Endonuclease in Adaptive Bacterial Immunity', *Science*, Vol. 337 No. 6096, pp. 816–821.

Johansson, G. and Sundin, E. (2014), 'Lean and green product development: two sides of the same coin?', *Journal of Cleaner Production*, Vol. 85, pp. 104–121.

Johnson, G., Scholes, K. and Whittington, R. (2011), *Strategisches Management – Eine Einführung: Analyse, Entscheidung und Umsetzung*, 9th ed., 978-3.868.940.565, München.

Jonas, H. (1993), *Das Prinzip Verantwortung: Versuch einer Ethik für die technologische Zivilisation,* 3rd ed., Suhrkamp, Frankfurt a. M.

Jørgensen, A., Finkbeiner, M., Jørgensen, M. S. and Hauschild, M. Z. (2010), 'Defining the baseline in social life cycle assessment', *The International Journal of Life Cycle Assessment*, Vol. 15 No. 4, pp. 376–384.

Jørgensen, A., Le Bocq, A., Nazarkina, L. and Hauschild, M. (2008), 'Methodologies for social life cycle assessment', *The International Journal of Life Cycle Assessment*, Vol. 13 No. 2, p. 96.

Jörissen, J., Kopfmüller, J. and Brandl, V. e.a. (1999), *Ein integratives Konzept nachhaltiger Entwicklung*, Karlsruhe, available at: *http://digbib.ubka.uni-karlsruhe.de/volltexte/fzk/6393/6393.pdf* (accessed 1 March 2017).

Joyce, A. and Paquin, R. L. (2016), 'The triple layered business model canvas: A tool to design more sustainable business models', *Journal of Cleaner Production*, Vol. 135, pp. 1474–1486.

Jozala, A. F., Lencastre-Novaes, L. C. de, Lopes, A. M., Carvalho Santos-Ebinuma, V. de, Mazzola, P. G. and Pessoa-Jr, A. e.a. (2016), 'Bacterial nanocellulose production and application: a 10-year overview', *Applied microbiology and biotechnology*, Vol. 100 No. 5, pp. 2063–2072.

June, C. H., O'Connor, R. S., Kawalekar, O. U., Ghassemi, S. and Milone, M. C. (2018), 'CAR T cell immunotherapy for human cancer', *Science*, Vol. 359 No. 6382, pp. 1361–1365.

Kacprzyk, J. and Pedrycz, W. (2015), *Handbook of Computational Intelligence*, Springer, Berlin.

Kagermann, H., Lukas, W. D. and Wahlster, W. (2011), 'Industrie 4.0: Mit dem Internet der Dinge auf dem Weg zur 4. industriellen Revolution', *VDI Nachrichten*, Vol. 13 No. 11, p. 2.

Kalhoff, J., Löwen, U., Manger, T. and Schneider, K. (2017), 'Open Source Projekt openAAS – mit offenen Standards Industrie 4.0 umsetzen', available at: *https://www.zvei.org/fileadmin/user_upload/Verband/Fachverbaende/Automation/Open-Source-ProjektopenAAS- mit-offenen-Standards-Industrie-4.0- umsetzen.pdf.*

Kamitaka, Y., Tsujimura, S. and Kano, K. (2009), 'High current density bioelectrolysis of D-fructose at fructose dehydrogenase-adsorbed and Ketjen black-modified electrodes without a mediator', *Chemistry Letters*, Vol. 36 No. 2, pp. 218–219.

Kamm, B. (2008), 'Biomasse-Wirtschaft und Bioraffinerie-Systeme. Internet-Zeitschrift des Leibniz-Instituts für interdisziplinäre Studien', *LIFIS ONLINE*, Vol. 16 No. 7, pp. 1–10.

Kamm, B. and Kamm, M. (2004), 'Biorefinery systems', *Chemical and Biochemical Engineering Quarterly*, Vol. 18 No. 1, pp. 1–7.

Kamp, H. and Partee, B. (1995), 'Prototype theory and compositionality', *Cognition*, Vol. 57 No. 2, pp. 129–191.

Kane, M. J., Hambrick, D. Z. and Conway, A. R.A. (2005), 'Working memory capacity and fluid intelligence are strongly related constructs: Comment on Ackerman, Beier, and Boyle', *Psychological Bulletin*, Vol. 131, pp. 66–71.

Kapp, K. W. (1971), *The Social Costs of Private Enterprise,* 2nd ed., Schocken Books, New York.

Kar, A. K. (2016), 'Bio inspired computing–a review of algorithms and scope of applications', *Expert Systems with Applications*, Vol. 59, pp. 20–32.

Karaboga, D. and Akay, B. (2009), 'A survey: algorithms simulating bee swarm intelligence', *Artificial Intelligence Review*, Vol. 31 No. 1-4, p. 61.

Karafyllis, N. C. (2012), 'Bios und Zoe [Version 1.0]', In: Kirchhoff, T.: Naturphilosophische Grundbegriffe, available at: *www.naturphilosophie.org* (accessed 20 April 2020).

Karanasiou, G. S., Papafaklis, M. I., Conway, C., Michalis, L. K., Tzafriri, R., Edelman, E. R. and Fotiadis, D. I. (2017), 'Stents: biomechanics, biomaterials, and insights from computational modeling', *Annals of Biomedical Engineering*, Vol. 45 No. 4, pp. 853–872.

Karg, S. R. and Kallio, P. T. (2009), 'The production of biopharmaceuticals in plant systems', *Biotechnology Advances*, Vol. 27 No. 6, pp. 879–894.

Katz, E. and Willner, I. (2003), 'A biofuel cell with electrochemically switchable and tunable power output', *Journal of the American Chemical Society*, Vol. 125 No. 22, pp. 6803–6813.

Katz, E., Willner, I. and Kotlyar, A. B. (1999), 'A non-compartmentalized glucose|O2 biofuel cell by bioengineered electrode surfaces', *Journal of Electroanalytical Chemistry*, Vol. 479 No. 1, pp. 64–68.

Keeso, A. (2014), 'Big data and environmental sustainability: a conversation starter', *Smith School of Enterprise and the Environment – Working Paper Series*, 14-04.

Kendel, H. (2007), 'Rezension zu Marktwirtschaft ohne Kapitalismus. Eine Neubewertung der Freiwirtschaftslehre aus wirtschaftsethischer Sicht', available at: *https://www.humane-wirtschaft.de/wp-content/uploads/2007/01/kendel_marktwirtschaft-ohne-kapitalismus.pdf* (accessed 11 January 2021).

Keoleian, G. A. and Menerey, D. (1994), 'Sustainable development by design: review of life cycle design and related approaches', *Air & Waste*, Vol. 44 No. 5, pp. 645–668.

Kern, W. (1992), *Industrielle Produktionswirtschaft*, Schäffer-Poeschel Verlag.

Kerr, L. D. (2010), 'The use of biologic agents in the geriatric population', *The Journal of Musculoskeletal Medicine*, Vol. 27 No. 5, p. 175.

Kesicki, F. and Ekins, P. (2012), 'Marginal abatement cost curves: a call for caution', *Climate Policy*, Vol. 12 No. 2, pp. 219–236.

Keynes, J. M. (2002), *Allgemeine Theorie der Beschäftigung, des Zinses und des Geldes*, 9th ed., Duncker & Humblot, Berlin.

Khalaf, M. A., Labib, A. A. and ElSayed, A. E. (2011), 'An Investigation Into The Relationship Between The Implementation of Lean Manufacturing and Energy Efficiency in Industrial Organizations', *Australian Journal of Basic and Applied Sciences*, Vol. 12 No. 5, pp. 1886–1896.

Khalil, H. (2014), *Nonlinear Control*, Pearson, London.

Khoo, Z. X., Teoh, J. E.M., Liu, Y., Chua, C. K. and Yang, S., An, J. et al. (2015), '3D printing of smart materials: A review on recent progresses in 4D printing', *Virtual and Physical Prototyping*, Vol. 10 No. 3, pp. 103–122.

Kicherer, A., Schaltegger, S., Tschochohei, H. and Ferreira Pozo, B. (2007), 'Eco-Efficiency. Combining Life Cycle Assessment and Life Cycle Costs via Normalization', *International Journal of Life Cycle Assessment*, Vol. 12 No. 7, pp. 537–543.

Kidwell, M. (2006), 'Lean Manufacturing and the Environment. Ignoring the 8th Deadly Waste leaves money on the table', *Association for Manufacturing Excellence*, Vol. 22 No. 6, pp. 13–18.

Kilic, T., Navaee, F., Stradolini, F., Renaud, P. and Carrara, S. (2018), 'Organs-on-chip monitoring: sensors and other strategies.', *Microphysiol. Syst.*, Vol. 2 No. 5.

Kim, D., Sakimoto, K. K., Hong, D. and Yang, P. (2015), 'Artificial photosynthesis for sustainable fuel and chemical production', *Angewandte Chemie International Edition*, Vol. 54 No. 11, pp. 3259–3266.

Kim, J. (2016), 'Emergence: Core ideas and issues', *Synthese*, Vol. 151 No. 3, pp. 547–559.

Kircher, M. (2012), 'The transition to a bio-economy: national perspectives', *Biofuels, Bioproducts and Biorefining*, Vol. 6 No. 3, pp. 240–245.

Kiridus-Göller, R. (2011), 'Bioware – Programmatik zur lebensfähigen Ökonomie', in Pinter, D. and Schubert, U. (Eds.), *Wirtschaft – Gesellschaft – Natur: Ansätze zu einem zukunftsfähigen Wirtschaften*, Festschrift für Prof. Dr. Eberhard K. Seifert, Metropolis, Weimar, pp. 223–236.

Kitano, H. (2004), '(2004). Biological robustness', *Nature Reviews Genetics*, Vol. 5 No. 11, pp. 826–837.

Klipp, E., Liebermeister, W., Wierling, C. and Kowald, A. (2016), *Systems biology: a textbook*, John Wiley & Sons, Hoboken.

Klöpffer, W. and Grahl, B. (2014), *Life cycle assessment (LCA): a guide to best practice*, 1st ed., Wiley-VCH, Weinheim.

Kloth, C. (2010), *Havariemanagement im Broadcast Engineering*, Vieweg+Teubner, Wiesbaden.

Koenen, G. (2017), *Die Farbe Rot: Ursprünge und Geschichte des Kommunismus*, 2nd ed., CH Beck.

Kolesky, D. B., Truby, R. L., Gladman, A. S., Busbee, T. A., Homan, K. A. and Lewis, J. A. (2014), '3D bio-printing of vascularized, hetero-geneous cell-laden tissue constructs', *Advanced Materials*, Vol. 26 No. 19, pp. 3124–3130.

Koon, D. W. (1998), 'Is polar bear hair fiber optic?', *Applied Optics*, Vol. 37 No. 15, pp. 3198–3200.

Kopfmüller, J. (2007), 'Auf dem Weg zu einem integrativen Nachhaltigkeitskonzept', *Ökologisches Wirtschaften*, No. 1, pp. 16–18.

Kopfmüller, J., Brandl, V. and Jörissen, J. e.a. (2001), *Nachhaltige Entwicklung integrativ betrachtet: Konstitutive Elemente, Regeln, Indikatoren, Global zukunftsfähige Entwicklung – Perspektiven für Deutschland*, Vol. 1, Edition Sigma, Berlin.

Koren, Y. (2010), *The Global Manufacturing Revolution: Product-Process-Business Integration and Reconfigurable Systems*, Wiley, Hoboken, NJ, USA.

Korpi, E. and Ala-Risku, T. (2008), 'Life cycle costing: a review of published case studies', *Managerial Auditing Journal*, Vol. 23 No. 3, pp. 240–261.

Koshland, J. and Daniel, E. (2002), 'The Seven Pillars of Life', *Science*, Vol. 295 No. 5563, pp. 2215–2216.

Kotler, P. (2010), 'The prosumer movement', in Blättel-Mink, B. and Hellmann, K.-U. (Eds.), *Prosumer Revisited*, VS Verlag für Sozialwissenschaften, Wiesbaden, pp. 51–60.

Kotler, P., Keller, K. L. and Bliemel, F. (2007), *Marketing Management – Strategien für wertschaffendes Handeln,* 12th ed., Person Education, München.

Kramer, M. O. (1960), 'Boundary layer stabilization by distributed damping', *Journal of the American Society for Naval Engineers*, Vol. 72 No. 1, pp. 25–34.

Kranzberg, M. (1986), 'Technology and History: Kranzberg's Laws', *Technology and Culture*, Vol. 27 No. 3, pp. 544–560.

Kreuzer, E., Lugtenburg, J. B., Meißner, H. G. and Truckenbrodt, A. (1994), *Industrieroboter: Technik, Berechnung und anwendungsorientierte Auslegung*, Springer, Berlin.

Kriegman, S., Blackiston, D., Levin, M. and Bongard, J. (2020), 'A scalable pipeline for designing reconfigurable organisms', *Proceedings of the National Academy of Sciences*.

Kristeller, P. O. (1974), *Humanismus und Renaissance I*, Wilhelm Fink Verlag, München.

Kritzinger, W., Karner, M., Traar, G., Henjes, J. and Sihn, W. (2018), 'Digital Twin in manufacturing: A categorical liter-ature review and classification', *IFAC-PapersOnLine*, Vol. 51 No. 11, pp. 1016–1022.

Krol, G. J. and Karpe, J. (1999), *Ökonomische Aspekte von Nachhaltigkeit: Die Umweltproblematik aus sozioökonomischer Sicht*, LIT, Münster.

Kruber, K. P. (2002), *Theoriegeschichte der Marktwirtschaft*, LIT, Münster.

Kruth, J. P., Leu, M. C. and Nakagawa, T. (1998), 'Progress in additive manufacturing and rapid prototyping', *CIRP Annals-Manufacturing Technology*, Vol. 47 No. 2, pp. 525–540.

Kuhlmann, T. and Sauer, A. (2016), 'Ultra-Efficiency Factory-Framework and Holistic Approach to Increase and Evaluate Efficiency and Sustainability', *Advanced Materials Research*, Vol. 1140 No. 481-488.

Kuhn, T. S. (1996a), *Die Struktur wissenschaftlicher Revolutionen,* 13th ed., Suhrkamp, Berlin.

Kuhn, T. S. (1996b), *Structure of Scientific Revolutions*, University of Chicago Press, Chicago.

Kumar, P., Chandrasekhar, K., Kumari, A., Sathiyamoorthi, E. and Kim, B. S. (2018), 'Electro-fermentation in aid of bioenergy and biopolymers', *Energies*, Vol. 11 No. 2, 343.

Küppers, U. (2004), 'Bionik und Wirtschaftlichkeit: Versuch einer Annäherung', *Biologie in unserer Zeit*, Vol. 34 No. 5, pp. 316–323.

Küppers, U. (2010), 'Systemisches Bionik-Management', *Wissenschaftsmanagement*, Vol. 1, pp. 37–42.

Küppers, U. and Tributsch, H. (2001), *Bionik der Verpackung: Verpacktes Leben – Verpackte Technik*, Wiley-VCH, Weinheim.

Kyllonen, P. C. and Christal, R. E. (1990), 'Reasoning ability is (little more than) working-memory capacity?!', *Intelligence*, Vol. 14, pp. 389–433.

Lachmann, W. (2004), *Volkswirtschaftslehre 2: Anwendungen*, 2nd ed., Springer, Berlin.

Laffont, J. J. and Tirole, J. (1993), 'A theory of incentives in procurement and regulation', *MIT Press*.

Lam, J. S.L. and Lai, K. H. (2015), 'Developing environmental sustainability by ANP-QFD approach: the case of shipping operations', *Journal of Cleaner Production*, Vol. 105, pp. 275–284.

Lamy, C. (2016), 'From hydrogen production by water electrolysis to its utilization in a PEM fuel cell or in a SO fuel cell: Some considerations on the energy efficiencies', *International Journal of Hydrogen Energy*, Vol. 41 No. 34, pp. 15.415–15.425.

Landes, D. S. (1986), *Der entfesselte Prometheus: Technologischer Wandel und industrielle Entwicklung in Westeuropa*, Kiepenheuer & Witsch GmbH, Köln.

Landherr, M., Schneider, U. and Bauernhansl, T. (2016), 'The Application Center Industrie 4.0-Industry-driven manufacturing, research and development', *Procedia CIRP*, No. 57, pp. 26–31.

Langenscheidt (Ed.) (1993), *Langenscheidt Taschenwörterbuch Altgriechisch*, Langenscheidt, Berlin.

Langenscheidt (Ed.) (2010), *Langenscheidt Universal-Wörterbuch Latein*, Langenscheidt, Berlin.

Längkvist, M., Karlsson, L. and Loutfy, A. (2014), 'A review of unsupervised feature learning and deep learning for time-series modeling', *Pattern Recognition Letters*, Vol. 42, pp. 11–24.

Lapitsky, Y., Zahir, T. and Shoichet, M. S. (2008), 'Modular biodegradable biomaterials from surfactant and polyelectrolyte mixtures', *Biomacromolecules*, Vol. 9 No. 1, pp. 166–174.

Latouche, S. (2015), *Es reicht! Abrechnung mit dem Wachstumswahn*, Oekom, München.

Laughlin, R. B. (2007), *Abschied von der Weltformel: Die Neuerfindung der Physik*, 4th ed., Piper, München.

Lavie, D. (2006), 'The competitive advantage of interconnected firms: An extension of the resource-based view', *Academy of management review*, Vol. 31 No. 3, pp. 638–658.

Le, H. P. (1998), 'Progress and trends in ink-jet printing technology', *Journal of Imaging Science and Technology*, Vol. 42 No. 1, pp. 49–62.

Lee, E. A. (2008), 'Cyber physical systems: Design challenges', *11th IEEE International Symposium on Object and Component-Oriented Real-Time Distributed Computing (ISORC)*, pp. 363–369.

Lee, J., Bagheri, B. and Kao, H. A. (2015), 'A cyber-physical systems architecture for industry 4.0-based manufacturing systems', *Manufacturing Letters*, Vol. 3, pp. 18–23.

Lee, J., Wu, F., Zhao, W., Ghaffari, M., Liao, L. and Siegel, D. (2014), 'Prognostics and health management design for rotary machinery systems – Reviews, methodology and applications', *Mechanical Systems and Signal Processing*, Vol. 42 No. 1-2, pp. 314–334.

Lee, J. Y., Buxton, G. A. and Balazs, A. C. (2004), 'Using nanoparticles to create self-healing composites', *The Journal of Chemical Physics*, Vol. 121 No. 11, pp. 5531–5540.

Lee, S. W., Lee, S. K. and Belcher, A. M. (2003), 'Virus-based alignment of inorganic, organic, and biological nanosized materials', *Advanced Materials*, Vol. 15 No. 9, pp. 689–692.

Lee, W. S., Grosh, D. L., Tillman, F. A. and Lie, C. H. (1985), 'Fault Tree Analysis, Methods, and Applications – A Review', *IEEE transactions on reliability*, Vol. 34 No. 3, pp. 194–203.

Leech, D., Kavanagh, P. and Schuhmann, W. (2012), 'Enzymatic fuel cells: Recent progress', *Electrochimica Acta*, Vol. 84, pp. 223–234.

Lemus, R. and Lal, R. (2005), 'Bioenergy crops and carbon sequestration', *Critical Reviews in Plant Sciences*, Vol. 24 No. 1, pp. 1–21.

Lentes, J., Mandel, J., Schliessmann, U., Blach, R., Hertwig, M. and Kuhlmann, T. (2017), 'Competitive and sustainable manufacturing by means of ultra-efficient factories in urban surroundings', *International Journal of Production Research*, Vol. 55 No. 2, pp. 480–491.

Lentes, J., Hertwig, M. (2019), 'Towards Ultra-Efficient Industrial Areas', *Procedia Manufacturing*, Vol. 39, pp. 804–813.

Lentini, R., Martín, N.Y. and Forlin, M. e.a. (2017), 'Two-Way Chemical Communication between Artificial and Natural Cells', *ACS Central Science*, Vol. 3 No. 2, pp. 117–123.

Lenz, R.W. and Marchessault, R.H. (2005), 'Bacterial polyesters: biosynthesis, biodegradable plastics and biotechnology', *Biomacromolecules*, Vol. 6 No. 1, pp. 1–8.

Lenzen, M. (2003), *Evolutionstheorien in den Natur- und Sozialwissenschaften*, Campus Verlag, Frankfurt a. M.

Leonard, A. (2009), 'The viable system model and its application to complex organizations', *Systemic practice and action research*, Vol. 22 No. 4, pp. 223–233.

Leontief, W.W. (1986), *Input-Output Economics*, 2nd ed., Oxford University Press, New York.

Li, R.Y.M., Chau, K.W. and Zeng, F.F. (2019), 'Ranking of risks for existing and new building works', *Sustainability*, Vol. 11 No. 10, p. 2863.

Li, Y.C., Zhang, Y.S., Akpek, A., Shin, S.R. and Khademhosseini, A. (2016), '4D bioprinting: the next-generation technology for biofabri-cation enabled by stimuli-responsive materials', *Biofabrication*, Vol. 9 No. 1, p. 12.001.

Liebig, M.A., Schmer, M.R., Vogel, K.P. and Mitchell, R.B. (2008), 'Soil carbon storage by switchgrass grown for bioenergy', *Bioenergy Research*, Vol. 1 No. 3-4, pp. 215–222.

Lindahl, M. (1999), 'E-FMEA – A new promising tool for efficient design for environment', *Proceedings First International Symposium on Environmentally Conscious Design and Inverse Manufacturing*, pp. 734–739.

Liravi, F. and Toyserkani, E. (2018), 'A hybrid additive manufacturing method for the fabrication of silicone bio-structures: 3D printing optimization and surface characterization', *Materials & Design*, No. 138, pp. 46–61.

Liu, Y., Rafailovich, M.H., Malal, R., Cohn, D. and Chidambaram, D. (2009), 'Engineering of bio-hybrid materials by electrospinning polymer-microbe fibers', *Proceedings of the National Academy of Sciences*, Vol. 106 No. 34, pp. 14.201–14.206.

Liu, Z., Feng, J. and Wang, J. (2020), 'Resource-Constrained Innovation Method for Sustainability: Application of Morphological Analysis and TRIZ Inventive Principles', *Sustainability*, Vol. 12 No. 3, pp. 1–23.

Locatelli, G. and Mancini, M. (2012), 'A framework for the selection of the right nuclear power plant', *International Journal of Production Research*, Vol. 50 No. 17, pp. 4753–4766.

Lorenz, A. (2000), 'Externe Kosten irgnorieren, antizipieren oder integrieren', paper presented at Expertenworkshop – Irgnorieren, antizipieren oder integrieren? Die Kosten der Umweltverschmutzung aus Unternehmenssicht, 8.6.2000, Witten.

Lozano, R. (2012), 'Towards a better understanding of sustainability into companies' systems: an analysis of voluntary corporate initiatives', *Journal of Cleaner Production*, Vol. 25, pp. 14–26.

Lu, D.Y., Lu, T.R. and Wu, H.Y. (2014), 'Personalized cancer therapy, a perspective', *Clinical Experimental Pharmacology*, Vol. 4 No. 2, p. 153.

Lu, Z., Chang, D., Ma, J., Huang, G., Cai, L. and Zhang, L. (2015), 'Behavior of metal ions in bioelectrochemical systems: A review', *Journal of Power Sources*, Vol. 275, pp. 243–260.

Luchsinger, R.H., Pedretti, A., Steingruber, P. and Pedretti, M. (2004), 'The new structural concept Tensairity: Basic principles', *Progress in Structural Engineering, Mechanics and Computation*, pp. 323–328.

Lud, S.Q., Nikolaides, M.G., Haase, I., Fischer, M. and Bausch, A.R. (2006), 'Field Effect of Screened Charges: Electrical Detection of Peptides and Proteins by a Thin-Film Resistor', *ChemPhysChem: a European Journal of Chemical Physics and Physical Chemistry*, Vol. 7 No. 2, pp. 379–384.

Lüdeke-Freund, F., Gold, S. and Bocken, N.M. (2019), 'A review and typology of circular economy business model patterns', *Journal of Industrial Ecology*, Vol. 23 No. 1, pp. 36–61.

Luhmann, N. (1996), *Social systems*, Stanford University Press, Stanford.

Lüthy, C., Murdoch, J.E. and Newman, W.R. (2001), *Late Medieval and Early Modern Corpuscular Matter Theory.*, Brill Academic Pub, Leiden.

Luttmann, R., Bitzer, G. and Hartkopf, J. (1994), 'Development of control strategies for high cell density cultivations', *Mathematics and Computers in Simulation*, Vol. 37 No. 2-3, pp. 153–164.

Luttmann, R., Bracewell, D. G., Cornelissen, G. and Gernaey, K. V. et al. (2012), 'Soft sensors in bioprocessing: a status report and recommendations', *Biotechnology Journal*, Vol. 7 No. 8, pp. 1040–1048.

Lv, Z., Zhou, Y., Han, S. T. and Roy, V. A.L. (2018), 'From biomaterial-based data storage to bio-inspired artificial synapse', *Materials Today*, Vol. 21 No. 5, pp. 537–552.

Lyle, J. T. (2008), *Regenerative Design for Sustainable Development*, John Wiley & Sons Ltd., Hoboken, New Jersey, USA.

Mahfouz, M. M., Piatek, A. and Stewart, C. N. (2014), 'Genome engineering via TALENs and CRISPR/Cas9 systems: challenges and perspectives', *Plant biotechnology journal*, Vol. 12 No. 8, pp. 1006–1014.

Mahner, M. and Bunge, M. (2000), *Philosophische Grundlagen der Biologie*, Springer, Berlin.

Maier, H. (2016), *Grundlagen der Robotik*, VDE VERLAG GmbH, Berlin.

Mainzer, K. (2016), *Künstliche Intelligenz – Wann übernehmen die Maschinen?*, Springer, Berlin.

Malik, F. (2006), *Strategie des Managements komplexer Systeme: Ein Beitrag zur Management-Kybernetik evolutionärer Systeme*, 9th ed., Haupt, Bern.

Malthus, T. (2007), *An Essay on the Principle of Population*, DOVER PUBN INC, Mineola, USA.

Man, J. C. de and Strandhagen, J. O. (2017), 'An Industry 4.0 research agenda for sustainable business models', *Procedia CIRP*, Vol. 63, pp. 721–726.

Mandenius, C. F. and Gustavsson, R. (2015), 'Mini-review: soft sensors as means for PAT in the manufacture of bio-therapeutics', *Journal of Chemical Technology & Biotechnology*, Vol. 90 No. 2, pp. 215–227.

Mandrycky, C., Wang, Z., Kim, K. and Kim, D. H. (2016), '3D bioprinting for engineering complex tissues', *Biotechnology Advances*, Vol. 34 No. 4, pp. 422–434.

Mangla, S. K., Govindan, K. and Luthra, S. (2017), 'Prioritizing the barriers to achieve sustainable consumption and production trends in supply chains using fuzzy Analytical Hierarchy Process', *Journal of Cleaner Production*, Vol. 151, pp. 509–525.

Mankiw, N. G. and Taylor, M. P. (2012), *Grundzüge der Volkswirtschaftslehre*, Schäffer-Poeschel, Stuttgart.

Mannix, L. H. and Stratton, J. A. (2005), 'Mind and Hand: The Birth of MIT', *MIT Press*, pp. 190–192.

Mansouri, S. S., Udugama, I. A., Cignitti, S., Mitic, A., Flores-Alsina, X. and Gernaey, K. V. (2017), 'Resource recovery from bio-based production processes: a future necessity?', *Current Opinion in Chemical Engineering*, Vol. 18, pp. 1–9.

Manyika, J., Sinclair, J., Dobbs, R., Strube, G. and Rassey, L. e.a. (2012), *Manufacturing the future: The next era of global growth and innovation*, San Francisco.

Marazzi, C. (2013), 'Bioökonomie und Biokapitalismus', in Borsò, V. and Cometa, M. (Eds.), *Die Kunst, das Leben zu »bewirtschaften«: Biós zwischen Politik, Ökonomie und Ästhetik*, transcript, Bielefeld, pp. 39–51.

Marcuse, H. (2004), *Der eindimensionale Mensch: Studien zur Ideologie der fortgeschrittenen Industriegesellschaft*, DTV, München.

Märkt, S. (2005), 'Marktwirtschaft und Freiwirtschaftslehre', *Zeitschrift für Wirtschafts- und Unternehmensethik*, Vol. 6 No. 2.

Marshalek, B., Lohman, D. F. and Snow, R. E. (1983), 'The complexity continuum in the radex and hierarchical models of intelligence', *Intelligence*, Vol. 7, pp. 107–127.

Martinez-Velazquez, R., Gamez, R. and Saddik, A. E. (2019), 'Cardio Twin: A Digital Twin of the human heart running on the edge', *IEEE International Symposium on Medical Measurements and Applications (MeMeA)*, pp. 1–6.

Marx, K. (2014), *Zur Kritik der politischen Ökonomie*, CreateSpace, Charleston, USA.

Marx, K. and Engels, F. (2017), *Das kommunistische Manifest*, CreateSpace Independent Publishing Platform, North Charleston, USA.

März, E. (1985), *Einführung in die Marxsche Theorie der wirtschaftlichen Entwicklung*, Baulino Verlag GmbH, Ulm.

Mastura, M. T., Sapuan, S. M. and Mansor, M. e.a. (2017), 'Environmentally conscious hybrid bio-composite material selection for automotive anti-roll bar', *The International Journal of Advanced Manufacturing Technology*, Vol. 89, pp. 2203–2219.

Masui, K., Sakao, T., Aizawa, S. and Inaba, A. (2002), 'Quality function deployment for environment (QFDE) to support design for environment (DFE)', American Society of Mechanical Engineers, *ASME 2002 International Design Engineering Technical Conferences and Computers and Information in Engineering Conference*, pp. 415–423.

Matten, D., Janzen, H. and Hahn, R. (2012), 'Die gesellschaftliche Verantwortung des Unternehmens. Eine Einführung', in Hahn, R. Janzen, H. and Matten, D. (Eds.), *Die gesellschaftliche Verantwortung des Unternehmens: Hintergründe, Schwerpunkte und Zukunftsperspektiven*, Schäffer-Poeschel, Stuttgart, pp. 1–17.

May, M. (2017), 'Companies in the cloud: digitizing lab operations', *Science*, Vol. 355 No. 6324, pp. 532–534.

McCaffrey, J. (2005), 'Test Run: The Analytic Hierarchy Process', *MSDN Magazine*, pp. 139–146.

McKay, C. P. (2004), 'What Is Life—and How Do We Search for It in Other Worlds?', *PLOS Biology*, Vol. 2 No. 9, e302.

McQuarrie, E. (2005), *The market research toolbox: a concise guide for beginners*, 2nd ed., Sage, New York.

Meadows, D. H., Meadows, D. L. and Randers, J. e.a. (1972), *The Limits to Growth*, Universe Books, New York.

Melhuish, C., Ieropoulos, I., Greenman, J. and Horsfield, I. (2006), 'Energetically autonomous robots: Food for thought', *Autonomous Robots*, Vol. 21 No. 3, pp. 187–198.

Mell, P. and Grance, T. (2011), *The NIST definition of cloud computing.*, Gaithersburg.

Mende, M. (1993), 'Vom Holz zur Kohle: Prozeßwärme und Dampfkraft', in Wengenroth, U. (Ed.), *Technik und Wirtschaft*, VDI, Düsseldorf, pp. 305–324.

Menger, C. (2006), *Grundsätze der Volkswirtschaftslehre*, VDM, Saarbrücken.

Meretz, W. (1963), 'Die Wabenzelle der Honigbiene', *The Bulletin of Mathematical Biophysics*, Vol. 25 No. 1, pp. 95–110.

Meyer, A. (2009), *Die Epoche der Aufklärung*, Akademie, Berlin.

Meyer-Abich, K. M. (2001), 'Nachhaltigkeit – ein kulturelles, bisher aber chancenloses Wirtschaftsziel', *Zeitschrift für Wirtschafts- und Unternehmensethik*, Vol. 3 No. 2, pp. 291–310.

Miehe, R. (2018), 'Methodik zur Quantifizierung der nachhaltigen Wertschöpfung von Produktionssystemen an der ökonomisch-ökologischen Schnittstelle anhand ausgewählter Umweltprobleme. Stuttgarter Beiträge zur Produktionsforschung', Dissertation, Industrielle Fertigung und Fabrikbetrieb, Universität Stuttgart, Stuttgart, 2018.

Miehe, R., Bauernhansl, T., Beckett, M. and Brecher, C. e.a. (2020a), 'The biological transformation of industrial manufacturing–Technologies, status and scenarios for a sustainable future of the German manufacturing industry', *Journal of Manufacturing Systems*, Vol. 54, pp. 50–61.

Miehe, R., Bauernhansl, T., Schwarz, O. and Traube, A. e.a. (2018), 'The biological transformation of the manufacturing industry–Envisioning biointelligent value adding', *Procedia CIRP*, Vol. 72, pp. 739–743.

Miehe, R., Bogdanov, I. and Sauer, A. (2017a), 'Eco Lean Sustainability Control Panel–A Tailored Indicator System for Day-to-Day Operations in Manufacturing Facilities', *Advances in Manufacturing Technology*, Vol. 6, pp. 395–400.

Miehe, R., Bogdanov, I., Schneider, R., Hirsch, M., Bauernhansl, T., Pawlik, E. and Horbal, R. (2016a), 'The Eco Lean method – A combined approach for low cost economic and ecologic optimization in the manufacturing industry', *Procedia CIRP*, Vol. 57, pp. 613–618.

Miehe, R., Buckreus, L., Kiemel, S., Sauer, A. and Bauernhansl, T. (2021a), 'A Conceptual Framework for Biointelligent Production—Calling for Systemic Life Cycle Thinking in Cellular Units', *Clean Technologies*, Vol. 3 No. 4, pp. 844–857.

Miehe, R., Fischer, E., Berndt, D. and Herzog, A. e.a. (2019a), 'Enabling bidirectional real time interaction between biological and technical systems: Structural basics of a control oriented modeling of biology-technology-interfaces', *Procedia CIRP*, Vol. 81, pp. 63–68.

Miehe, R., Full, J., Scholz, P., Demmer, A., Bauernhansl, T., Sauer, A. and Schuh, G. (2019b), 'The Biological Transformation of Industrial Manufacturing-Future Fields of Action in Bioinspired and Bio-based Production Technologies and Organization', *Procedia Manufacturing*, Vol. 39, pp. 737–744.

Miehe, R., Horbelt, J., Baumgarten, Y. and Bauernhansl, T. (2020b), 'Basic considerations for a digital twin of biointelligent systems: Applying technical design patterns to biological systems', *CIRP Journal of Manufacturing Science and Technology*.

Miehe, R., Mueller, S. and Schneider, R. e.a. (2015), 'Integrated Hazardous Materials Management: Combining Requirements from Various Environmental Legislations to Enable Effective Business Compliance Processes in Industries', *International Journal of Precision Engineering and Manufacturing-Green Technology*, Vol. 2 No. 3, pp. 289–298.

Miehe, R., Scheumann, R. and Jones, C. M. et al. (2016b), 'Regional carbon footprints of households: a German case study', *Environment, Development and Sustainability*, Vol. 18 No. 2, pp. 577–591.

Miehe, R., Schneider, R., Baaij, F. and Bauernhansl, T. (2016c), 'Criticality of material resources in industrial enterprises–Structural basics of an operational model', *Procedia CIRP*, Vol. 48, pp. 1–9.

Miehe, R., Stender, S., Hessberger, N., Mandel, J. and Sauer, A. (2017b), 'Improving Manufacturing Systems with Regard to the Concept of Ultra-Efficiency', *Advances in Manufacturing Technology*, No. 6, p. 321 ff.

Miehe, R., Waltersmann, L., Sauer, A. and Bauernhansl, T. (2021b), 'Sustainable production and the role of digital twins: basic reflections and perspectives', *Journal of Advanced Manufacturing and Processing*.

Miehe, R., Wiedenmann, M. and Bauernhansl, T. (2016d), 'Der Umsatz-Nachhaltigkeitsindex. Ein Benchmark für Fertigungsprozesse des deutschen Maschinen- und Anlagenbaus', *wt Werkstattstechnik online*, Vol. 106 No. 3, pp. 136–140.

Mill, J. S. (2006), *Utilitarismus*, Reclam, Leipzig.

Mill, J. S. (2014), *Schriften zur Politischen Ökonomie in fünf Bänden: Kleinere Schriften zur Politischen Ökonomie: 1825-1861*, Metropolis, Weimar.

Miller, J. G. (1978), *Living Systems*, McGraw-Hill, New York.

Milward, R., Gilless, C. and Wickes, G. (2013), 'Lean, Energy, and Savings: Energy Impacts of Lean Manufacturing', *ACEEE Summer Study on Energy Efficiency in Industry*, pp. 1–13.

Minteer, S. D., Liaw, B. Y. and Cooney, M. J. (2007), 'Enzyme-based biofuel cells.', *Current opinion in biotechnology*, Vol. 18 No. 3, pp. 228–234.

Mintzberg, H., Ahlstrand, B. and Lampel, J. (2008), *Strategy Safari: The complete guide through the wilds of strategic management,* 2nd ed., Pearson Education, London.

Mintzberg, H., Pascale, R. T., Goold, M. and Rumelt, R. P. (1996), 'The Honda effect revisited', *California Management Review*, Vol. 38 No. 4, pp. 77–117.

Mintzberg, H. and Waters, J. A. (1985), 'Of strategies, deliberate and emergent', *Strategic Management Journal*, Vol. 6 No. 3, pp. 257–272.

Mironov, V. (2003), 'Printing Technology to Produce Living Tissue', *Expert Opinion on Biological Therapy*, Vol. 3 No. 5, pp. 701–704.

Mironov, V., Reis, N. and Derby, B. (2006), 'Bioprinting: a beginning', *Tissue Engineering*, Vol. 12 No. 4, pp. 631–634.

Mises, L. (2005), *Theorie des Geldes und der Umlaufsmittel,* 2nd ed., Duncker & Humblot, Berlin.

Miyazaki, M., Honda, T., Yamaguchi, H., Briones, M. P.P. and Maeda, H. (2008), 'Enzymatic processing in microfluidic reactors. Biotechnology and Genetic', *Engineering Reviews*, Vol. 25 No. 1, pp. 405–428.

Mogas-Soldevila, L. and Oxman, N. (2015), 'Water-based engineering & fabrication: large-scale additive manufacturing of biomaterials', *MRS Online Proceedings Library Archive*, p. 1800.

Mohan, S. V., Nikhil, G. N., Chiranjeevi, P., Reddy, C. N., Rohit, M. V., Kumar, A. N. and Sarkar, O. (2016), 'Waste biorefinery models towards sustainable circular bioeconomy: critical review and future perspectives', *Bioresource Technology*, Vol. 215, pp. 2–12.

Mohr, H. (1990), 'Biologie und Ökonomik – Chancen für eine Interdisziplinarität', in Witt, U. (Ed.), *Studien zur Evolutorischen Ökonomik I*, Duncker & Humblot, Berlin, pp. 209–219.

Montgomery, D. C. (2012), *Design and Analysis of Experiments*, 8th ed., John Wiley & Sons, Hoboken.

Moore, C. M., Minteer, S. D. and Martin, R. S. (2005), 'Microchip-based ethanol/oxygen biofuel cell', *Lab on a Chip*, Vol. 5 No. 2, pp. 218–225.

Moreira, F., Alves, A. and Sousa, R. (2010), 'Towards eco-efficient lean production systems', in Bas, A. O., Franco, R. D. and Gasquet, P. G. (Eds.), *Balanced Automation Systems for Future Manufacturing Networks, July 21-23, Valencia, Spain*, Springer, Berlin, pp. 100–108.

Moreno-Vozmediano, R., Montero, R. S. and Llorente, I. M. (2012), 'Key challenges in cloud computing: Enabling the future internet of services', *IEEE Internet Computing*, Vol. 17 No. 4, pp. 18–25.

Morris, M., Schindehutte, M. and Allen, J. (2005), 'The entrepreneur's business model: toward a unified perspective', *Journal of business research*, Vol. 58 No. 6, 726-735.

Mortimer, C. E. and Müller, U. (2015), *Chemie: Das Basiswissen der Chemie*, 12th ed., Thieme, New York.

Moslemizadeh, A., Dehkordi, A. F., Barnaji, M. J., Naseri, M., Ravi, S. G. and Jahromi, E. K. (2016), 'Novel bio-based surfactant for chemical enhanced oil recovery in montmorillonite rich reservoirs: Adsorption behavior, interaction impact, and oil recovery studies', *Chemical Engineering Research and Design*, Vol. 109, pp. 18–31.

Mousses, S., Kiefer, J., Hoff, D. von and Trent, J. (2008), 'Using biointelligence to search the cancer genome: an epistemological perspective on knowledge recovery strategies to enable precision medical genomics', *Oncogene*, Vol. 27 No. 2, pp. 58–66.

Müller, A. H. (1922), *Die Elemente der Staatskunst*, Vol. 1, Berlin.

Müller, D. H. and Tietjen, T. (2003), *FMEA-Praxis*, 2nd ed., Hanser, München.

Müller, J. M. (2019), 'Business model innovation in small-and medium-sized enterprises', *Journal of Manufacturing Technology Management*, Vol. 30 No. 8, 1127-1142.

Müller-Armack, A. (1976), *Wirtschaftsordnung und Wirtschaftspolitik. Studien und Konzepte zur Sozialen Marktwirtschaft und zur Europäischen Integration*, Haupt, Bern.

Müller-Christ, G. (2010), *Nachhaltiges Management: Einführung in Ressourcenorientierung und widersprüchliche Managementrationalitäten*, Nomos, Baden-Baden.

Müller-Prothmann, T. and Dörr, N. (2011), *Innovationsmanagement. Strategien, Methoden und Werkzeuge für systematische Innovationsprozesse*, Hanser, München.

Munneke, M. A., Bakker, C. D., Goverde, E. A., Pasman, J. W. and Stegeman, D. F. (2018), 'On the electrode positioning for bipolar EMG recording of forearm extensor and flexor muscle activity after transcranial magnetic stimulation', *Journal of Electromyography and Kinesiology*, Vol. 40, pp. 23–31.

Murphy, S. V. and Atala, A. (2014), '3D bioprinting of tissues and organs', *Nature Biotechnology*, Vol. 32 No. 8, p. 773.

Nachtigall, W. (2005), *Biologisches Design: Systematischer Katalog für bionisches Gestalten*, Springer, Berlin.

Nachtigall, W. (2010), *Bionik als Wissenschaft: Erkennen – Abstrahieren – Umsetzen*, Springer, Berlin.

Nachtigall, W. (2013), *Bionik: Grundlagen und Beispiele für Ingenieure und Naturwissenschaftler*, 2nd ed., Springer, Berlin.

Nachtigall, W. and Wisser, A. (2015), *Bionics by Examples*, Springer, Berlin.

Nair, L. S. and Laurencin, C. T. (2007), 'Biodegradable polymers as biomaterials', *Progress in Polymer Science*, Vol. 32 No. 8-9, pp. 762–798.

Naplekov, I., Zheleznikov, I., Pashchenko, D., Kobysheva, P. and Moskvitina, A. e.a. (2018), 'Methods of computational modeling of coronary heart vessels for its digital twin', *MATEC Web of Conferences*, Vol. 172, pp. 1–6.

Narayanan, H., Sokolov, M., Morbidelli, M. and Butté, A. (2019), 'A new generation of predictive models: The added value of hybrid models for manufacturing processes of therapeutic proteins', *Biotechnology and Bioengineering*, Vol. 116 No. 10, pp. 2540–2549.

Nargund, S., Guenther, K. and Mauch, K. (2019), 'The Move toward Biopharma 4.0: Insilico Biotechnology develops 'smart' processes that benefit biomanufacturing through Digital Twins', *Genetic Engineering & Biotechnology News.*, Vol. 39 No. 6, pp. 53–55.

National Research Council of the National Academies (Ed.) (2007), *Minerals, Critical Minerals and The US Economy*, National Academies Press, Washington D. C., USA.

Navlakha, S. and Bar-Joseph, Z. (2011), 'Algorithms in nature: the convergence of systems biology and computational thinking', *Molecular Systems Biology*, Vol. 7 No. 1.

Nefeslioglu, H. A., Sezer, E. A., Gokceoglu, C. and Ayas, Z. (2013), 'A modified analytical hierarchy process (M-AHP) approach for decision support systems in natural hazard assessments', *Computers & Geosciences*, Vol. 59, pp. 1–8.

Negri, E., Fumagalli, L. and Macchi, M. (2017), 'A review of the roles of digital twin in cps-based production systems', *Procedia Manufacturing*, Vol. 11, pp. 939–948.

Neisser, U., Boodoo, G., Bouchard, T. J., J., Boykin, A. W., Brody, N., Ceci, S. J., Halpern, D. F., Loehlin, J. C., Perloff, R., Sternberg, R. J. and Urbina, S. (1996), 'Intelligence: Knowns and unknowns', *American Psychologist*, Vol. 51 No. 2, pp. 77–101.

Nertinger, S. (2015), *Carbon and Material Flow Cost Accounting: Ein integrierter Ansatz im Kontext nachhaltigen Erfolgs und Wirtschaftens*, Springer Gabler, Wiesbaden.

Neugebauer, R., Hippmann, S., Leis, M. and Landherr, M. (2016), 'Industrie 4.0 – Form the perspective of apllied research', *Procedia CIRP*, pp. 2–7.

Neumann, J. von and Morgenstern, O. (1973), *Spieltheorie und wirtschaftliches Verhalten*, Physica, Berlin.

Neumayer, E. (2003), *Weak versus strong sustainability: Exploring the limits of two opposing paradigms*, 2nd ed., Edward Elgar, Cheltenham, UK.

Newton, I. (1999), *Die mathematischen Prinzipien der Physik*, De Gruyter, Berlin.

Niederberger, M. and Renn, O. (2019), *Delphi-Verfahren in den Sozial- und Gesundheitswissenschaften: Konzept, Varianten und Anwendungsbeispiele*, Springer VS, Berlin.

Noireaux, V. and Libchaber, A. (2004), 'A vesicle bioreactor as a step toward an artificial cell assembly', *Proceedings of the National Academy of Sciences of the United States of America*, Vol. 101 No. 51, pp. 17.669–17.674.

Norris, G. A. (2001), 'Integrating life cycle cost analysis and LCA', *The International Journal of Life Cycle Assessment*, Vol. 6 No. 2, pp. 118–120.

Norton, B. G. and Toman, M. A. (1997), 'Sustainability. Ecological and Economic Perspectives', *Land Economics*, Vol. 73 No. 4, pp. 553–568.

Nurse, P. (2008), 'Life, logic and information', *Nature*, Vol. 454 No. 7203, pp. 424–426.

Nutzinger, H. G. and Wolfstetter, E. (2008), *Die Marxsche Theorie und ihre Kritik: Eine Textsammlung zur Kritik der Politischen Ökonomie*, Metropolis, Marburg.

Oertel, D. and Grunwald, A. (2006), *Potenziale und Anwendungsperspektiven der Bionik*, Vorstudie, *Arbeitsbericht*, available at: *http://www.itas.kit.edu/pub/v/2006/oegr06a.pdf* (accessed 23 September 2016).

Ohno, T. (2005), *Das Toyota-Produktionssystem*, Campus, Frankfurt a. M.

Olson, G. J., Brierley, J. A. and Brierley, C. L. (2003), 'Bioleaching review part B. Applied microbiology and biotechnology', Vol. 63 No. 3, pp. 249–257.

Organisation for Economic Co-Operation and Development (2001), 'Biotechnology, list based definition', available at: *https://stats.oecd.org/glossary/detail.asp?ID=218* (accessed 30 April 2020).

Osiro, L., Lima-Junior, F. R. and Carpinetti, L. C.R. (2018), 'A group decision model based on quality function deployment and hesitant fuzzy for selecting supply chain sustainability metrics', *Journal of Cleaner Production*, Vol. 183, pp. 964–978.

Osteroth, D. (1985), *Soda, Teer und Schwefelsäure: Der Weg zur Grosschemie*, Rowohlt, Hamburg.

Ott, K. and Döring, R. (2008), *Theorie und Praxis starker Nachhaltigkeit*, 2nd ed., Metropolis, Marburg.

Ozawa, T., Mickle, D. A., Weisel, R. D., Koyama, N., Wong, H., Ozawa, S. and Li, R. K. (2002), 'Histologic changes of nonbiodegradable and biodegradable biomaterials used to repair right ventricular heart defects in rats', *The Journal of Thoracic and Cardiovascular Surgery*, Vol. 124 No. 6, pp. 1157–1164.

Ozbolat, I. T. and Yu, Y. (2013), 'Bioprinting toward organ fabrication: challenges and future trends', *IEEE Transactions on Biomedical Engineering*, Vol. 60 No. 3, pp. 691–699.

Paech, N. (2012), *Befreiung vom Überfluss: Auf dem Weg in die Postwachstumsökonomie*, Oekom, München.

Pahl-Wost, C. (2007), 'The implications of complexity for integrated resource management', *Environmental Modelling and Software*, Vol. 22 No. 5, pp. 561–569.

Paninski, L. and Cunningham, J. P. (2018), 'Neural data science: accelerating the experiment-analysis-theory cycle in large-scale neuroscience', *Current Opinion in Neurobiology*, Vol. 50, pp. 232–241.

Panke, S. and Wubbolts, M. G. (2002), 'Enzyme technology and bioprocess engineering', *Current opinion in biotechnology*, Vol. 13 No. 2, pp. 111–116.

Pant, D., van Bogaert, G., Diels, L. and Vanbroekhoven, K. (2010), 'A review of the substrates used in microbial fuel cells (MFCs) for sustainable energy production', *Bioresource Technology*, Vol. 101 No. 6, pp. 1533–1543.

Pareto, V. (2007), *Ausgewählte Schriften*, Springer VS, Berlin.

Parkhey, P. and Mohan, S. V. (2019), 'Biosensing applications of microbial fuel cell: Approach toward miniaturization', *Microbial Electrochemical Technology*, pp. 977–997.

Parzinger, H. (2015), *Die Kinder des Prometheus. Eine Geschichte der Menschheit vor der Erfindung der Schrift*, 2nd ed., C. H.Beck, München.

Pastrana, E. (2011), 'Optogenetics. Controlling cell function with light', *Nature Methods*, Vol. 8 No. 1, pp. 24–25.

Patankar, N. A. (2004), 'Mimicking the lotus effect: influence of double roughness structures and slender pillars', *Langmuir*, Vol. 20 No. 19, pp. 8209–8213.

Patermann, C. and Aguilar, A. (2018), 'The origins of the bioeconomy in the European Union', *New biotechnology*, Vol. 40, pp. 20–24.

Pathak, A., Dastidar, M. G. and Sreekrishnan, T. R. (2009), 'Bioleaching of heavy metals from sewage sludge: A review', *Journal of Environmental Management*, Vol. 90 No. 8, pp. 2343–2353.

Patterson, M. G. (1996), 'What is energy efficiency? Concepts, indicators and methodological issues', *Energy Policy*, No. 5, pp. 377–390.

Paul, G. (1991), *CIM-Basiswissen für die Betriebspraxis: Für Unternehmer und Führungskräfte kleiner und mittlerer Unternehmen*, Vieweg+Teubner Verlag, Wiesbaden.

Pauli, G. (2012), *The Blue Economy: 10 Jahre – 100 Innovationen – 100 Millionen Jobs*, Konvergenta, Berlin.

Pavone, M., Arena, P., Fortuna, L., Frasca, M. and Patané, L. (2006), 'Climbing obstacle in bio-robots via CNN and adaptive attitude control', *International Journal of Circuit Theory and Applications*, Vol. 34 No. 1, pp. 109–125.

Peacocke, C. (2002), 'Three Principles of Rationalism', *European Journal of Philosophy*, Vol. 10, pp. 375–397.

Pearce, D. (2012), 'The Biointelligence Explosion – How recursively self-improving organic robots will modify their own source code and bootstrap our way to full-spectrum superintelligence', in Eden, A. H., Moor, J. H., Soraker, J. H. and Steinhart, E. (Eds.), *Singularity Hypotheses: A Scientific and Philosophical Assessment*, Springer, Berlin, pp. 199–236.

Pearce, D. W. (1988), 'Economics, Equity and Sustainable Development', *Futures*, Vol. 20 No. 6, pp. 598–605.

Peaucelle, J.-L. and Guthrie, C. (2015), *Henri Fayol, the Manager*, Routledge, London.

Pedretti, M. and Luscher, R. (2007), 'Tensairity-Patent – Eine pneumatische Tenso-Struktur', *Stahlbau*, Vol. 76 No. 5, pp. 314–319.

Peinado, V., Castell, P., García, L. and Fernández, Á. (2015), 'Effect of extrusion on the mechanical and rheological properties of a reinforced poly (lactic acid): reprocessing and recycling of biobased materials', *Materials*, Vol. 8 No. 10, pp. 7106–7117.

Peirce, C. S. and Jastrow, J. (1885), 'On Small Differences in Sensation', *Memoirs of the National Academy of Sciences*, Vol. 3, pp. 73–83.

Pennisi, E. (2013), 'The CRISPR craze', *Science*, Vol. 341 No. 6148, pp. 833–836.

Peter-André Alt (2007), *Aufklärung*, 3rd ed., J. B. Metzler, Stuttgart.

Petersen, B. (2017), 'Basics of genome editing technology and its application in livestock species', *Reproduction in Domestic Animals*, Vol. 52, pp. 4–13.

Petrini, F. M., Bumbasirevic, M., Valle, G., Ilic, V., Mijović, P. and Čvančara, P. e.a. (2019), 'Sensory feedback restoration in leg amputees improves walking speed, metabolic cost and phantom pain', *Nature medicine*, Vol. 25 No. 9, pp. 1356–1363.

Pfau, S. F., Hagens, J. E., Dankbaar, B. and Smits, A. J. (2014), 'Visions of sustainability in bioeconomy research', *Sustainability*, Vol. 6 No. 3, pp. 1222–1249.

Pfeffer, J. and Salancik, G. R. (2003), *The External Control of Organizations: A Resource Dependence Perspective*, Stanford University Press, Stanford, USA.

Pfeifer, R., Lungarella, M. and Iida, F. (2007), 'Self-organization, embodiment, and biologically inspired robotics', *Science*, Vol. 318 No. 5853, pp. 1088–1093.

Pfeiffer, W. and Weiß, E. (1994), *Lean Management: Grundlagen der Führung und Organisation lernender Unternehmen,* 2nd ed., Erich Schmidt Verlag GmbH & Co., Berlin.

Pfiffner, M. (2010), 'Five experiences with the viable system model. Five experiences with the viable system model', *Kybernetes: The International Journal of Systems & Cybernetics*, Vol. 39 No. 9-10, pp. 1615–1626.

Pham, T. H., Rabaey, K., Aelterman, P., Clauwaert, P., Schamphelaire, L. de, Boon, N. and Verstraete, W. (2006), 'Microbial fuel cells in relation to conventional anaerobic digestion technology', *Engineering in Life Sciences*, Vol. 6 No. 3, pp. 285–292.

Pigou, A. C. (1911), *Wealth and welfare*, Book on Demand.

Pillkahn, U. (2012), *Innovationen zwischen Planung und Zufall: Bausteine einer Theorie der bewussten Irritation*, Books on Demand, Norderstedt.

Plattform Industrie 4.0 (2019a), 'Die Verwaltungsschale im Detail – von der Idee zum implementierbaren Konzept', available at: *https://www.plattformi40.de/PI40/Redaktion/DE/Downloads/Publikation/verwaltungsschale- im-detail-pr%C3%A4sentation.html* (accessed 23 April 2020).

Plattform Industrie 4.0, V. D. (2019b), 'Diskussionspapier Verwaltungsschale in der Praxis: Wie definiere ich Teilmodelle, beispielhafte Teilmodelle und Interaktion zwischen Verwaltungsschalen (Version 1.0)', available at: *https://www.plattformi40.de/PI40/Redaktion/DE/Downloads/Publikation/2019-verwaltungsschale-in-der-praxis.html* (accessed 23 April 2020).

Plattform Industrie 4.0, Z. (2019c), 'Details of the Asset Administration Shell: Part 1 – The exchange of information between partners in the value chain of Industrie 4.0 (Version 2.0)', available at: *https://www.plattformi40.de/PI40/Redaktion/DE/Downloads/Publikation/Details-of-the-Asset-Administration-Shell-Part1.html* (accessed 23 April 2020).

Plattform Industrie 4.0, Z. (2019d), 'The asset administration shell: Implementing digital twins for use in Industrie 4.0: Making Industrie 4.0 components interoperable', available at: *https://www.plattformi40.de/PI40/Redaktion/DE/Downloads/Publikation/ VWSiD%20V2.0.html* (accessed 23 April 2020).

Poh, P. S., Chhaya, M. P., Wunner, F. M. and De-Juan-Pardo, E. M. et al. (2016), 'Polylactides in additive biomanufacturing', *Advanced Drug Delivery Reviews*, Vol. 107, pp. 228–246.

Polanyi, K. (1973), *The Great Transformation: Politische und ökonomische Ursprünge von Gesellschaften und Wirtschaftssystemen,* 12th ed., Suhrkamp, Berlin.

Popper, K. (2007), *Logik der Forschung*, Oldenbourg, München.

Porter, M. E. (1985), *Competitive Advantage: Creating and Sustaining Superior Performance*, The Free Press, New York.

Porter, M. E. (2008), 'The five competitive forces that shape strategy', *Harvard Business Review*, Vol. 86 No. 1, pp. 25–40.

Posten, C. and Walter, C. (2012), *Microalgal Biotechnology: Integration and Economy*, De Gruyter, Berlin.

Potočnik, J. (2005), *Transforming life sciences knowledge into new, sustainable, eco-efficient and competitive products. 2005.: SPEECH/05/513*, Brüssel, available at: *http://europa.eu/rapid/press-release_SPEECH-05-513_en.htm* (accessed 6 February 2020).

Potts, H. W., Anderson, J. E., Colligan, L., Leach, P., Davis, S. and Berman, J. (2014), 'Assessing the validity of prospective hazard analysis methods: a comparison of two techniques', *BMC health services research*, Vol. 14 No. 1, p. 41 ff.

Praetorius, F., Kick, B., Behler, K. L., Honemann, M. N., Weuster-Botz, D. and Dietz, H. (2017), 'Biotechnological mass production of DNA origami', *Nature*, Vol. 552 No. 7683, pp. 84–87.

Premack, D. and Woodruff, G. (1978), 'Does the chimpanzee have a theory of mind?', *Behavioral and Brain Sciences*, Vol. 1 No. 4, pp. 515–526.

Priddat, B. P. (2001), *Le concert universel: Die Physiokratie. Eine Transformationsphilosophie des 18. Jahrhunderts*, Metropolis, Marburg.

Priddat, B. P. (2005), *Unvollständige Akteure: Komplexer werdende Ökonomie*, VS Verlag für Sozialwissenschaften, Wiesbaden.

Ptak, R. (2013), *Vom Ordoliberalismus zur Sozialen Marktwirtschaft: Stationen des Neoliberalismus in Deutschland*, Springer, Berlin.

Pugh, D. S. and Hickson, D. J. (2007), *Writers on Organizations,* 6th ed., Penguin, London, UK.

Puglieri, F. N. and Ometto, A. R. (2011), 'Environmental and operational analysis of ecodesign methods based on QFD and FMEA', *Glocalized solutions for sustainability in manufacturing*, pp. 142–147.

Pulz, O. (2001), 'Photobioreactors: production systems for phototrophic microorganisms', *Applied microbiology and biotechnology*, Vol. 57 No. 3, pp. 287–293.

Qin, J., Liu, Y. and Grosvenor, R. (2016), 'A categorical framework of manufacturing for industry 4.0 and beyond', *Procedia CIRP*, Vol. 52, pp. 173–178.

Qin S. J., Yue H. and Dunia R. (1997), 'Self-validating inferential sensors with application to air emission monitoring', *Industrial and Engineering Chemistry Research*, Vol. 36 No. 5, pp. 1675–1685.

Quesnay, F. (1971), *Quesnay's Tableau Economique*, 3rd ed., MacMillan Publishing Company, New York.

Quinet, E. (2004), 'A meta-analysis of Western European external costs estimates', *Transportation Research Part D: Transport and Environment*, Vol. 9 No. 6, pp. 465–476.

Rahimi, M. and Weidner, M. (2002), 'Integrating design for environment (DfE) impact matrix into quality function deployment (QFD) process', *The Journal of Sustainable Product Design*, Vol. 2 No. 1-2, pp. 29–41.

Rajan, K. S. (2009), *Biokapitalismus: Werte im postgenomischen Zeitalter*, Suhrkamp, Frankfurt a. M.

Rajhans, A., Cheng, S.-W., Schmerl, B., Garlan, D. and Krogh, B. H. et al. (2009), 'An architectural approach to the design and analysis of cyber-physical systems', *Electronic Communications of the EASST*, Vol. 21, pp. 1–10.

Rajkumar, R., Lee, I., Sha, L. and Stankovic, J. (2010), 'Cyber-physical systems: The next computing revolution', *Design Automation Conference*, pp. 731–736.

Ramanavicius, A., Kausaite, A. and Ramanaviciene, A. (2005), 'Bio-fuel cell based on direct bioelectrocatalysis', *Biosensors and Bioelectronics*, Vol. 20 No. 10, pp. 1962–1967.

Ramani, K., Ramanujan, D. and Bernstein, W. Z. et al. (2010), 'Integrated sustainable life cycle design: a review', *Journal of Mechanical Design*, Vol. 132 No. 9.

Rampf, M., Speck, O., Speck, T. and Luchsinger, R. H. (2012), 'Structural and mechanical properties of flexible polyurethane foams cured under pressure', *Journal of Cellular Plastics*, Vol. 48 No. 1, pp. 53–69.

Ran, F. A., Hsu, P. D., Wright, J., Agarwala, V., Scott, D. A. and Zhang, F. (2013), 'Genome engineering using the CRISPR-Cas9 system', *Nature Protocols*, Vol. 8 No. 1, p. 2281.

Randek, J. and Mandenius, C. F. (2018), 'On-line soft sensing in upstream bioprocessing', *Critical Reviews in Biotechnology*, Vol. 38 No. 1, pp. 106–121.

Randers, J. (1980), *Elements of the System Dynamics Method*, Cambridge.

Randles, M., Lamb, D. and Taleb-Bendiab, A. (2010), 'A comparative study into distributed load balancing algorithms for cloud computing', *IEEE 24th International Conference on Advanced Information Networking and Applications Workshops*, pp. 551–556.

Rappaport, A. (1986), *Creating Shareholder Value: The New Standard for Business Performance*, The Free Press, New York.

Rappaport, A. (1998), *Shareholder Value: Ein Handbuch für Manager und Investoren,* 2nd ed., Schäffer-Poeschel, Stuttgart.

Rasmussen, S. (2012), *Production economics: the basic theory of production optimisation*, Springer Science & Business Media, Berlin.

Rathore, A. S. and Sofer, G. (2012), *Process validation in manufacturing of biopharmaceuticals*, Taylor & Francis, Milton Park.

Rausand, M. and Hoylan, A. (2003), *System Reliability Theory: Models, Statistical Methods, and Applications*, Wiley-Blackwell, Hoboken.

Rauschenberger, R. (2002), *Nachhaltiger Shareholder Value*, Haupt, Zürich.

Rawls, J. (1999), *A Theory of Justice*, 2nd ed., Harvard University Press, Cambridge.

Read, P. and Lermit, J. (2005), 'Bio-energy with carbon storage (BECS): a sequential decision approach to the threat of abrupt climate change', *Energy*, Vol. 30 No. 14, pp. 2654–2671.

Reap, J., Roman, F., Duncan, S. and Bras, B. (2008), 'A survey of unresolved problems in life cycle assessment', *The International Journal of Life Cycle Assessment*, Vol. 13 No. 5, pp. 374–388.

Rechberger, M. (2012), 'Selbstschärfende Schneidwerkzeuge für abrasive Schnittgüter: eine bionische Entwicklung', Dissertation, Universität Duisburg-Essen, Duisburg/Essen, 2012.

Rechenberg, I. (1973), *Evolutionsstrategie – Optimierung technischer Systeme nach Prinzipien der biologischen Information*, Friedrich Frommann Verlag, Stuttgart.

Reeves, G. A., Talavera, D. and Thornton, J. M. (2009), 'Genome and proteome annotation: organization, interpretation and integration', *Journal of the Royal Society Interface*, Vol. 6 No. 31, pp. 129–147.

Reichwald, R. and Piller, F. (2009), *Interaktive Wertschöpfung: Open Innovation, Individualisierung und neue Formen der Arbeitsteilung,* 2nd ed., Gabler, Wiesbaden.

Reif, W. E. and Dinkelacker, A. (1982), 'Hydrodynamics of the squamation in fast swimming sharks', *Neues Jahrbuch für Geologie und Paläontologie*, Vol. 164 No. 1-2, pp. 184–187.

Reinhart, G., Bengler, K., Dollinger, C., Intra, C. and Lock, C. e.a. (2017), 'Der Mensch in der Produktion von Morgen', in Reinhart G. (Ed.), *Handbuch Industrie 4.0. Geschäftsmodelle, Prozesse, Technik*, Hanser, München, pp. 50–88.

Ren, S., Li, X., Yuan, B., Li, D. and Chen, X. (2018), 'The effects of three types of environmental regulation on eco-efficiency: A cross-region analysis in China', *Journal of Cleaner Production*, Vol. 173, pp. 245–255.

Renneberg, R. and Süßbier, D. (2005), *Biotechnologie für Einsteiger*, Spektrum Akademischer Verlag, Heidelberg.

Rhiel, M., Ducommun, P., Bolzonella, I., Marison, I. and Stockar, U. von (2002), 'Real-time in situ monitoring of freely suspended and immobilized cell cultures based on mid-infrared spectroscopic measurements', *Biotechnology and Bioengineering*, Vol. 77 No. 2, pp. 174–185.

Rieder, J., Estler, M. and Kehle, F. (2014), 'Green Stream Analysis – Ökologisch Lean. Eine Methode zur ökologischen Optimierung der Wertschöpfungskette', *ZWF – Ökologische Optimierung*, Vol. 109 No. 12, pp. 948–951.

Ritzer, G., Dean, P. and Jurgenson, N. (2012), 'The coming of age of the prosumer', *American behavioral scientist*, Vol. 56 No. 4, pp. 379–398.

Roberts, E. B. (1978), *Managerial Applications of System Dynamics*, MIT Press, Cambridge.

Rockström, J., Steffen, W. and Noone, K. e.a. (2009a), 'A safe operating space for humanity', Feature, Vol. 461, pp. 472–475.

Rockström, J., Steffen, W. and Noone, K. e.a. (2009b), 'Planetary Boundaries: Exploring the Safe Operating Space for Humanity', *Ecology and Society*, Vol. 14 No. 2.

Rogal, J., Probst, C. and Loskill, P. (2017), 'Integration concepts for multi-organ chips: how to maintain flexibility?!', *Future Science OA*, Vol. 3 No. 2, FSO180.

Rogall, H. (2008), *Ökologische Ökonomie: Eine Einführung*, 2nd ed., VS Verlag für Sozialwissenschaften, Wiesbaden.

Röhm, M., Wiendahl, H.-H., Denner, T. and Schöllhammer, O. (2020), 'Ganzheitliche Produktionssysteme', in Bauernhansl, T. (Ed.), *Fabrikbetriebslehre 1: Management in der Produktion*, Springer Vieweg, Berlin, pp. 297–344.

Rohwerder, T., Gehrke, T., Kinzler, K. and Sand, W. (2003), 'Bioleaching review part A', *Applied microbiology and biotechnology*, Vol. 63 No. 3, pp. 239–248.

Ronkainen, N. J., Halsall, H. B. and Heineman, W. R. (2010), 'Electrochemical biosensors', *Chemical Society Reviews*, Vol. 39 No. 5, pp. 1747–1763.

Rosen, M. A. (2012), 'Engineering sustainability: A technical approach to sustainability', *Sustainability*, Vol. 4 No. 9, pp. 2270–2292.

Roser, M. and Ritchie, H. (2020), 'Technological Progress – Empirical Review', available at: *https://ourworldindata.org/technological-progress* (accessed 15 January 2020).

Roser, M., Ritchie, H. and Ortiz-Ospina, E. (2020), 'World Population Growth', available at: *https://ourworldindata.org/world-population-growth* (accessed 18 January 2020).

Roth, U. (1992), *Umweltkostenrechnung: Grundlagen und Konzeption aus betriebswirtschaftlicher Sicht, DUV: Wirtschaftswissenschaften*, Deutscher Universitäts-Verlag, Wiesbaden.

Rozendal, R. A., Hamelers, H. V., Rabaey, K., Keller, J. and Buisman, C. J. (2008), 'Towards practical implementation of bioelectrochemical wastewater treatment', *Trends in Biotechnology*, Vol. 26 No. 8, pp. 450–459.

Rudolph, T. (2009), *Modernes Handelsmanagement – Eine Einführung in die Handelslehre*, 2nd ed., Schäffer-Poeschel Verlag, Stuttgart.

Rüegg-Stürm, J. and Grand, S. (2015), *Das St. Galler Management-Modell*, 2nd ed., Haupt, Bern.

Russel, S. J. and Norvig, P. (2010), *Artificial Intelligence: A Modern Approach*, 3rd ed., Pearson, London.

Rüßmann, M., Lorenz, M., Gerbert, P., Waldner, M., Justus, J., Engel, P. and Harnisch, M. (2015), 'Industry 4.0: The future of productivity and growth in manufacturing industries', *Boston Consulting Group*, Vol. 9 No. 1, pp. 54–89.

Russo, D., Regazzoni, D. and Montecchi, T. (2011), 'Eco-design with TRIZ laws of evolution', *Innovative product design: 9th ETRIA World TRIZFuture Conference, Timisoara, Romania, 4–6 November 2009*, No. 9, pp. 311–322.

Russo, D., Rizzi, C. and Montelisciani, G. (2014), 'Inventive guidelines for a TRIZ-based eco-design matrix', *Journal of Cleaner Production*, Vol. 76, pp. 95–105.

Saaty, T. L. (1999), *Decision Making for Leaders: The Analytical Hierarchy Process for Decisions in a Complex World*, Bertrams, Norwich.

Saaty, T. L. and Peniwati, K. (2008), *Group Decision Making: Drawing out and Reconciling Differences*, RWS, Pittsburgh.

Sachs, W. (1993), 'Die vier E's: Merkposten für einen maß-vollen Wirtschaftsstil', *Politische Ökologie*, No. 33, pp. 69–72.

Santillan-Jimenez, E., Pace, R., Marques, S., Morgan, T., McKelphin, C., Mobley, J. and Crocker, M. (2016), 'Extraction, characterization, purification and catalytic upgrading of algae lipids to fuel-like hydrocarbons', *Fuel*, Vol. 180, pp. 668–678.

Santoro, C., Arbizzani, C., Erable, B. and Ieropoulos, I. (2017), 'Microbial fuel cells: from fundamentals to applications. A review', *Journal of Power Sources*, Vol. 356, pp. 225–244.

Sarkis, J. (2003), 'A strategic decision framework for green supply chain management', *Journal of Cleaner Production*, Vol. 11, pp. 397–409.

Satava, R. M. (2002), 'The bio-intelligence age: Surgery after the information age', *Journal of Gastrointestinal Surgery*, Vol. 6 No. 6, pp. 795–799.

Sato, T., Kano, T. and Ishiguro, A. (2011), 'A decentralized control scheme for an effective coordination of phasic and tonic control in a snake-like robot', *Bioinspiration & Biomimetics*, Vol. 7 No. 1, p. 16.005.

Scarff, M., Arnold, S. A., Harvey, L. M. and McNeil, B. (2006), 'Near infrared spectroscopy for bioprocess monitoring and control: current status and future trends', *Critical Reviews in Biotechnology*, Vol. 26 No. 1, pp. 17–39.

Schaefer, K. C. (1993), 'Die merkantilisitsche Wirtschaftspolitik', in Richard H. Tilly (Ed.), *Geschichte der Wirtschaftspolitik*, De Gruyter, Berlin.

Schäfer, B. (2018), 'Data Exchange in the Laboratory of the Future', available at: *https://analyticalscience.wiley.com/do/10.1002/gitlab.17.270/full/* (accessed 23 April 2020).

Schaltegger, S. (2004), 'Sustainability balanced scorecard', *Controlling*, Vol. 16 No. 8–9, pp. 511–516.

Schaltegger, S., Hansen, E. G. and Lüdeke-Freund, F. (2016), 'Business models for sustainability: Origins, present research, and future avenues', *Organization & Environment*, Vol. 29 No. 1, pp. 3–10.

Schaltegger, S., Kleiber, O. and Müller, J. (2003), 'Die Werkzeuge des Nachhaltigkeitsmanagements', in Linne, G. and Schwarz, M. (Eds.), *Handbuch Nachhaltige Entwicklung: Wie ist nachhaltiges Wirtschaften machbar?*, leske+budrich, Wiesbaden, pp. 332–341.

Schaltegger, S. and Sturm, A. (1999), *Ökologieorientierte Entscheidungen in Unternehmen: Ökologisches Rechnungswesen statt Ökobilanzierung: Notwendigkeit, Kriterien, Konzepte,* 3rd ed., Haupt, Bern.

Schaltegger, S. and Sturm, A. (2000), *Öko-Effizienz durch Öko-Controlling: Zur praktischen Umsetzung von EMAS und ISO 14.001*, Schäffer-Poeschel Verlag, Stuttgart.

Schatten, M. and Žugaj, M. (2011), 'Biomimetics in modern organizations–Laws or metaphors?', *Interdisciplinary Description of Complex Systems: INDECS*, Vol. 9 No. 1, pp. 39–55.

Schatzberg, E. (2006), "Technik' Comes to America: Changing Meanings of Technology Before 1930', *Technology and Culture*, Vol. 47 No. 3, pp. 486–512.

Schenk, H. O. (1991), *Marktwirtschaftslehre des Handels*, Gabler, Wiesbaden.

Schenk, M., Wirth, S. and Müller, E. (2014), *Fabrikplanung und Fabrikbetrieb: Methoden für die wandlungsfähige, vernetzte und ressourceneffiziente Fabrik*, 2nd ed., Springer Vieweg, Berlin.

Scheper, T., Hitzmann, B., Stärk, E., Ulber, R., Faurie, R., Sosnitza, P. and Reardon, K. F. (1999), 'Bioanalytics: detailed insight into bioprocesses', *Analytica Chimica Acta*, Vol. 400 No. 1-3, pp. 121–134.

Schiebahn, S., Grube, T., Robinius, M., Zhao, L., Otto, A. and Kumar, B. e.a. (2013), 'Power to gas', *Transition to Renewable Energy Systems*, pp. 813–848.

Schiele, N. R., Koppes, R. A., Corr, D. T., Ellison, K. S., Thompson, D. M. and Ligon, L. A. et al. (2009), 'Laser direct writing of combinatorial libraries of idealized cellular constructs: biomedical applications', *Applied Surface Science*, Vol. 255 No. 10, pp. 5444–5447.

Schloske, A. (2020), 'Innovation und Produktentwicklung', in Bauernhansl, T. (Ed.), *Fabrikbetriebslehre 1: Management in der Produktion*, Springer Vieweg, Berlin, pp. 67–104.

Schmidt-Bleek, F. (2014a), *Grüne Lügen. Nichts für die Umwelt, alles fürs Geschäft – wie Politik und Wirtschaft die Welt zugrunde richten*, Ludwig, München.

Schmidt-Bleek, F. (2014b), *Wieviel Umwelt braucht der Mensch? MIPS – das Maß für ökologisches Wirtschaften*, Birkhäuser, Basel.

Schneeweiß, C. (2002), *Einführung in die Produktionswirtschaft*, 8th ed., Springer, Berlin.

Schneider, M. P. (2006), 'Plant-oil-based lubricants and hydraulic fluids', *Journal of the Science of Food and Agriculture*, Vol. 86 No. 12, pp. 1769–1780.

Schneidewind, U. and Palzkill, A. (2011), *Suffizienz als Business Case: Nachhaltiges Ressourcenmanagement als Gegenstand einer transdisziplinären Betriebswirtschaftslehre, Impulse zur Wachstumswende*, Wuppertal.

Schönhoven, K. (2017), 'Soziale Gerechtigkeit als reformpolitische Perspektive. Staatsverständnis der sozialdemokratischen Gewerkschaftsbewegung 1848 bis 1914', in Ruck, M. (Ed.), *Gegner – Instrument – Partner: Gewerkschaftliche Staatsverständnisse vom Industrialismus bis zum Informationszeitalter*, Nomos, Baden-Baden, pp. 21–42.

Schreiner, M. (1996), *Umweltmanagement in 22 Lektionen*, 4th ed., Gabler Verlag, Wiesbaden.

Schröder, T. (2015), *Max Weber – Der Werturteilsstreit*, GRIN Verlag, München.

Schrödinger, E. (1967), *What is Life? Mind and Matter*, Cambridge University Press, Cambridge.

Schuh, G. (2011), 'Der Ordnungsrahmen Produktion und Management', in Schuh, G. and Klappert, S. (Eds.), *Technologiemanagement: Handbuch Produktion und Management 2*, Springer, Berlin, pp. 1–4.

Schuh, G., Boos, W., Kampker, A. and Gartzen, U. (2010), 'Strategie', in Schuh, G. and Kampker, A. (Eds.), *Strategie und Management produzierender Unternehmen: Handbuch Produktion und Management 1*, Springer, pp. 63–132.

Schüler, J. (2016), *Die Biotechnologie-Industrie: Ein Einführungs-, Übersichts-und Nachschlagewerk*, Springer, Berlin.

Schumacher, A., Erol, S. and Sihn, W. (2016), 'A maturity model for assessing Industry 4.0 readiness and maturity of manufacturing enterprises', *Procedia CIRP*, Vol. 52 No. 1, pp. 161–166.

Schumpeter, J. A. (2005), *Kapitalismus, Sozialismus und Demokratie*, 8th ed., UTB, Stuttgart.

Schumpeter, J. A. (2007), *Geschichte der ökonomischen Analyse*, 2 Bände, Neuauflage, Vandenhoeck & Ruprecht.

Schütz, H. and Bringezu, S. (2008), *Ressourcenverbrauch von Deutschland – aktuelle Kennzahlen und Begriffsbestimmungen: Erstellung eines Glossars zum "Ressourcenbegriff" und Berechnung von fehlenden Kennzahlen des Ressourcenverbrauchs für die weitere politische Analyse*, Dessau-Roßlau, available at: *http://www.umweltbundesamt.de/sites/default/files/medien/publikation/long/3426.pdf* (accessed 1 March 2017).

Schwaninger, M. (2006), 'Design for viable organizations: The diagnostic power of the viable system model', *Kybernetes: The International Journal of Systems & Cybernetics*, Vol. 35 No. 7-8, pp. 955–966.

Schwaninger, M. (2015), 'Organizing for sustainability: a cybernetic concept for sustainable renewal', *Kybernetes*, Vol. 44 No. 6/7, pp. 935–954.

Schwille, P. (2011), 'Bottom-up synthetic biology: engineering in a tinkerer's world', *Science*, Vol. 333 No. 6047, pp. 1252–1254.

Science Based Targets Initiative (2020), 'Science Based Targets', available at: *https://sciencebasedtargets.org/* (accessed 18 May 2020).

Sciortino, M., Watson, S. and Presnar, R. (2009), 'The Importance of Energy Efficiency in Lean Manufacturing: Declaring Energy the Ninth Waste', *ACEEE Summer Study on Energy Efficiency in Industry*, pp. 113–122.

Scott, J. C. (1977), *The moral economy of the peasant: Rebellion and subsistence in Southeast Asia*, Yale University Press, New Haven, Connecticut, USA.

Seif, A., Toro, C. and Akhtar, H. (2019), 'Implementing Industry 4.0 Asset Administrative Shells in Mini Factories', *Procedia Computer Science*, Vol. 159, pp. 495–504.

Selimović, Š., Dokmeci, M. R. and Khademhosseini, A. (2013), 'Organs-on-a-chip for drug discovery', *Current Opinion in Pharmacology*, Vol. 13 No. 5, pp. 829–833.

Sen, A. (1990), *Poverty And Famines: An Essay on Entitlement and Deprivation,* Reprint, Oxford University Press, Oxford.

Seol, Y. J., Kang, H. W., Lee, S. J., Atala, A. and Yoo, J. J. (2014), 'Bioprinting technology and its applications', *European Journal of Cardio-Thoracic Surgery*, Vol. 46 No. 3, pp. 342–348.

Settleworth, S. J. (2009), *Cognition, Evolution and Behavior,* 2nd ed., Oxford University Press, Oxford.

Shafto, M., Conroy, M., Doyle, R., Glaessgen, E. and Kemp, C. e.a. (2010), 'DRAFT Modeling, Simulation, Information Technology & Processing Roadmap: Technology Area 11.', available at: *https://www.nasa.gov/pdf/501321main_TA11-MSITP-DRAFT-Nov2010-A1.pdf* (accessed 23 April 2020).

Shah, D., Rust, R. T., Parasuraman, A., Staelin, R. and Day, G. S. (2006), 'The path to customer centricity', *Journal of service research*, Vol. 9 No. 2, pp. 113–124.

Shang, C., Yang, F., Huang, D. and Lyu, W. (2014), 'Data-driven soft sensor development based on deep learning technique', *Journal of Process Control*, Vol. 24, pp. 223–233.

Sharkh, M. A., Jammal, M., Shami, A. and Ouda, A. (2013), 'Resource allocation in a network-based cloud computing environment: design challenges', *IEEE Communications Magazine*, Vol. 51 No. 11, pp. 46–52.

Shebl, N. A., Franklin, B. D. and Barber, N. (2009), 'Is failure mode and effect analysis reliable?', *Journal of patient safety*, Vol. 5 No. 2, pp. 86–94.

Sheng, I. L. and Kok-Soo, T. (2010), 'Eco-efficient product design using theory of inventive problem solving (TRIZ) principles', *American Journal of Applied Sciences*, Vol. 7 No. 6, pp. 852–858.

Shin, H., Jo, S. and Mikos, A. G. (2003), 'Biomimetic materials for tissue engineering', *Biomaterials*, Vol. 24 No. 24, pp. 4353–4364.

Shuler, M. L. and Kargi, F. (2002), *Bioprocess Engineering: Basic Concepts,* 2nd ed., Prentice Hall, Upper Saddle River.

Siebert, H. (1982), *Ökonomische Theorie natürlicher Ressourcen*, Mohr Siebeck, Tübingen.

Simons, D. and Mason, R. (2003), 'Lean and Green: Doing More with Less', *ECR Journal*, Vol. 3 No. 1, pp. 84–91.

Simons, S., Abé, P. and Neser, S. (2017), 'Learning in the AutFab – The fully automated Industrie 4.0 learning factory of the University of Applied Sciences Darmstadt', *Procedia Manufacturing*, Vol. 9, pp. 81–88.

Sims, R. E., Mabee, W., Saddler, J. N. and Taylor, M. (2010), 'An over-view of second generation biofuel technologies', *Bioresource Technology*, Vol. 101 No. 6, pp. 1570–1580.

Sinemus, K. and Egelhofer, M. (2007), 'Transparent communication strategy on GMOs: Will it change public opinion?', *Biotechnology Journal: Healthcare Nutrition Technology*, Vol. 2 No. 9, pp. 1141–1146.

Singh, J., Kaushik, N. and Biswas, S. (2014), 'Bioreactors–technology & design analysis', *The Scitech Journal*, Vol. 1, pp. 28–36.

Singh, S., Ramakrishna, S. and Singh, R. (2017), 'Material issues in additive manufacturing: A review', *Journal of Manufacturing Processes*, Vol. 25, pp. 185–200.

Siontorou, C. G. and Batzias, F. A. (2014), 'A methodological combined framework for roadmapping biosensor research: A fault tree analysis approach within a strategic technology evaluation frame', *Critical Reviews in Biotechnology*, Vol. 34 No. 1, pp. 31–55.

Sitaram, R., Lee, S., Ruiz, S., Rana, M., Veit, R. and Birbaumer, N. (2011), 'Real-time support vector classification and feedback of multiple emotional brain states', *NeuroImage*, Vol. 56 No. 2, pp. 753–765.

Skalak, R. (1986), *Handbook of Bioengineering*, McGraw-Hill, New York.

Skansi, S. (2018), *Introduction to Deep Learning*, Springer, Berlin.

Sleutels, T. H., Lodder, R., Hamelers, H. V. and Buisman, C. J. (2009), 'Improved performance of porous bio-anodes in microbial electrolysis cells by enhancing mass and charge transport', *International Journal of Hydrogen Energy*, Vol. 34 No. 24, pp. 9655–9661.

Sleutels, T. H., Ter Heijne, A., Buisman, C. J. and Hamelers, H. V. (2012), 'Bioelectrochemical systems: an outlook for practical applications', *ChemSusChem*, Vol. 5 No. 6, pp. 1012–1019.

Smela, E., Kallenbach, M. and Holdenried, J. (1999), 'Electrochemically driven polypyrrole bilayers for moving and positioning bulk micromachined silicon plates', *Journal of Microelectromechanical Systems*, Vol. 8 No. 4, pp. 373–383.

Smith, A. (2013), *Der Wohlstand der Nationen*, Anaconda Verlag.

Söllner, F. (1997), 'Die Divergenz zwischen Zahlungs- und Akzeptanzbereitschaft bei der Bewertung von Umweltgütern', *Konjunkturpolitik*, Vol. 43, pp. 43–81.

Sommeregger, W., Sissolak, B., Kandra, K., Stosch, M. von, Mayer, M. and Striedner, G. (2017), 'Quality by control: Towards model predictive control of mammalian cell culture bioprocesses', *Biotechnology Journal*, Vol. 12 No. 7, p. 1.600.546.

Soroudi, A. and Jakubowicz, I. (2013), 'Recycling of bioplastics, their blends and biocomposites: A review', *European Polymer Journal*, Vol. 49 No. 10, pp. 2839–2858.

Sorrell, S. and Dimitropoulos, J. (2008), 'The rebound effect. Microeconomic definitions, limitations and extensions', *Ecological Economics*, Vol. 65, pp. 636–649.

Sousa-Zomer, T. T. and Miguel, P. (2017), 'A QFD-based approach to support sustainable product-service systems conceptual design', *The International Journal of Advanced Manufacturing Technology*, Vol. 88, pp. 701–717.

Spearman, C. (1904a), 'General intelligence objectively determined and measured', *American Journal of Psychology*, Vol. 15, pp. 201–293.

Spearman, C. (1904b), 'The proof and measurement of association between two things', *American Journal of Psychology*, Vol. 15, pp. 72–101.

Spearman, C. (1914), 'The theory of two factors', *Psychological Review*, Vol. 21 No. 2, p. 101.

Speck, O., Speck, D., Horn, R., Gantner, J. and Sedlbauer, K. P. (2017), 'Biomimetic bio-inspired biomorph sustainable? An attempt to classify and clarify biology-derived technical developments', *Bioinspiration & Biomimetics*, Vol. 12 No. 1, p. 11.004.

Spenko, M. J., Haynes, G. C., Saunders, J. A., Cutkosky, M. R., Rizzi, A. A., Full, R. J. and Koditschek, D. E. (2008), 'Biologically inspired climbing with a hexapedal robot', *Journal of Field Robotics*, Vol. 25 No. 4-5, pp. 223–242.

Spier, F. (1998), *Big History. Was die Geschichte im Innersten zusammenhält*, Primus in Wissenschaftliche Buchgesellschaft, Darmstadt.

Spur, G. (1998), *Technologie und Management: Zum Selbstverständnis der Technikwissenschaft*, Fachbuchverlag, Leipzig.

Sriram, G., Alberti, M., Dancik, Y., Wu, B., Wu, R. and Feng, Z. e.a. (2018), 'Full-thickness human skin-on-chip with enhanced epidermal morphogenesis and barrier function', *Materials Today*, Vol. 21 No. 4, pp. 326–340.

Stachowiak, H. (1973), *Allgemeine Modelltheorie*, Springer, Berlin.

Stachowiak, J. C., Richmond, D. L., Li, T. H., Brochard-Wyart, F. and Fletcher, D. A. (2009), 'Inkjet formation of unilamellar lipid vesicles for cell-like encapsulation', *Lab on a Chip*, Vol. 9 No. 14, pp. 2003–2009.

Stahel, W. (2010), *The Performance Economy*, 2nd ed., Palgrave Macmillan, London, UK.

Stamatis, D. H. (2003), *Failure mode and effect analysis: FMEA from theory to execution*, Quality Press, Milwaukee.

Statistisches Bundesamt (2019a), 'Forschung und Entwicklung – Interne Ausgaben für Forschung und Entwicklung nach Sektoren in Millionen Euro', available at: *https://www.destatis.de/DE/Themen/Gesellschaft-Umwelt/Bildung-Forschung-Kultur/Forschung-Entwicklung/Tabellen/forschung-entwicklung-sektoren.html*.

Statistisches Bundesamt (Destatis) (2019b), *Volkswirtschaftliche Gesamtrechnungen: Erste Ergebnisse zur gesamtwirtschaftlichen Lage 2018*, Wiesbaden.

Stead, J. G. and Stead, W. E. (2008), 'Sustainable strategic management: an evolutionary perspective', *International Journal of Sustainable Strategic Management*, Vol. 1 No. 1, pp. 62–81.

Steen, B. (1999), *A systematic approach to environmental priority strategies in product development (EPS): version 2000-general system characteristics*, Gothenburg.

Steffen, W., Crutzen, P. J. and McNeill, J. R. (2007), 'The Anthropocene: are humans now overwhelming the great forces of nature', *AMBIO: A Journal of the Human Environment*, Vol. 36 No. 8, pp. 614–622.

Steffen, W., Richardson, K., Rockström, J. and Cornell, S. E. et al. (2015), 'Planetary boundaries: Guiding human development on a changing planet', *Science*, Vol. 347 No. 6223, p. 1.259.855.

Steger, U. (1993), *Umweltmanagement: Erfahrungen und Instrumente einer umweltorientierten Unternehmensstrategie*, Gabler, Wiesbaden.

Steinfeld, H., Gerber, P., Wassenaar, T., Castel, V., Rosales, M. and Haan, C. de (2006), *Livestock's Long Shadow: Environmental Issues and Options*, Food & Agriculture Organization of the United Nations, New York.

Sterman, J. D. (2000), *Business Dynamics: Systems thinking and modeling for a complex world*, McGraw Hill, New York.

Stern, E. and Neubauer, A. (2016), 'Intelligenz: kein Mythos, sondern Realität', *Psychologische Rundschau*, Vol. 67 No. 1, pp. 15–27.

Stern, T. and Jaberg, H. (2007), *Erfolgreiches Innovationsmanagement*, 3rd ed., Gabler, Wiesbaden.

Sternberg, R. J. (1978), 'The theory of successful intelligence', *Review of General Psychology*, Vol. 3 No. 4, pp. 292–316.

Sternberg, R. J. (1982), *Handbook of Human Intelligence*, Cambridge University Press, Cambridge, UK.

Sternberg, R. J. (2003), 'A broad view of intelligence: The theory of successful intelligence', *Consulting Psychology Journal: Practice and Research*, Vol. 55 No. 3, pp. 139–154.

Steven, M. (1994), *Produktion und Umweltschutz: Ansatzpunkte für die Integration von Umweltschutzmaßnahmen in die Produktionstheorie, Beiträge zur betriebswirtschaftlichen Forschung*, Vol. 71, Gabler, Wiesbaden.

Stigler, S. M. (1978), 'Mathematical statistics in the early States', *Annals of Statistics*, Vol. 6 No. 2, pp. 239–265.

Stirnberg, S. (1999), 'Die Berücksichtigung der natürlichen Umwelt in Produktionsfaktorsystemen', Diplomarbeit, Ruhr-Universität Bochum, Bochum, 1999.

Stock, T. and Seliger, G. (2016), 'Opportunities of sustainable manufacturing in industry 4.0', *Procedia CIRP*, Vol. 40, pp. 536–541.

Stock, W. G. and Stock, M. (2013), *Handbook of Information Science*, De Gruyter Saur, München.

Stokes, R. (2006), 'Network-based strategy making for events tourism', *European Journal of Marketing*.

Stonier, T. (1997), *Information and meaning: an evolutionary perspective*, Springer, London.

Storhas, W. (2000), *Bioreaktoren und periphere Einrichtungen: Ein Leitfaden für die Hochschulausbildung, für Hersteller und Anwender*, Springer, Berlin.

Stürmer, R. and Breuer, M. (2006), 'Enzyme als Katalysatoren. Chemie und Biologie Hand in Hand', *Chemie in unserer Zeit*, Vol. 40 No. 2, pp. 104–111.

Su, L., Jia, W., Hou, C. and Lei, Y. (2011), 'Microbial biosensors: a review', *Biosensors and Bioelectronics*, Vol. 26 No. 5, pp. 1788–1799.

Suh, S. and Huppes, G. (2005), 'Methods for Life Cycle Inventory of a product', *Journal of Cleaner Production*, Vol. 13 No. 7, pp. 687–697.

Sun, F. T., Kuo, C., Cheng, H. T., Buthpitiya, S., Collins, P. and Griss, M. (2010), 'Activity-aware mental stress detection using physiological sensors', *International conference on Mobile computing, applications, and services*, pp. 282–301.

Sun, W., Luo, Z., Lee, J., Kim, H. J., Lee, K. and Tebon, P. e.a. (2019), 'Organ-on-a-Chip for Cancer and Immune Organs Modeling', *Advanced Healthcare Materials*, Vol. 8 No. 4, p. 1.801.363.

Sun, Y., Yan, H., Lu, C., Bie, R. and Thomas, P. (2012), 'A holistic approach to visualizing business models for the internet of things', *Communications in Mobile Computing*, Vol. 1 No. 1.

Suresh, P. V., Babar, A. K. and Raj, V. V. (1996), 'Uncertainty in fault tree analysis: A fuzzy approach', *Fuzzy sets and Systems*, Vol. 83 No. 2, pp. 135–141.

Süß, H.-M. and Beauducel, A. (2015), 'Modeling the construct validity of the Berlin Intelligence Structure Model', *Estudos de Psicologia*, Vol. 32 No. 1, pp. 13–25.

Suzumura, K., Funakoshi, K., Hoshino, T., Tsujimura, H. and Iwabuchi, K. e.a. (2011), 'A light-regulated bio-micro-actuator powered by transgenic Drosophila melanogaster muscle tissue', *IEEE 24th International Conference on Micro Electro Mechanical Systems*, pp. 149–152.

Syahir, A. Z., Zulkifli, N. W.M., Masjuki, H. H., Kalam, M. A., Alabdulkarem, A. and Gulzar, M. e.a. (2017), 'A review on bio-based lubricants and their applications', *Journal of Cleaner Production*, No. 168, pp. 997–1016.

Tabatabaee, H. A. and Kurth, T. L. (2017), 'Analytical investigation of the impact of a novel bio-based recycling agent on the colloidal stability of aged bitumen', *Road Materials and Pavement Design*, Vol. 18, pp. 131–140.

Taghizadegan, R. (2010), 'Cradle-to-cradle – die nächste Sau, die man durch das globale Dorf treibt?', *wirks Magazin für Zukunftskompetenz*, No. 1, pp. 21–26.

Taguchi, G. (1986), *Introduction to Quality Engineering. Designing Quality into Products and Processes*, Asian Productivity Organization, Tokyo.

Takahashi, N., Oertner, T. G., Hegemann, P. and Larkum, M. E. (2016), 'Active cortical dendrites modulate perception', *Science*, Vol. 354 No. 6319, pp. 1587–1590.

Tan, Y. and Zheng, Z. Y. (2013), 'Research advance in swarm robotics', *Defence Technology*, Vol. 9 No. 1, pp. 18–39.

Tandon, B., Magaz, A., Balint, R., Blaker, J. J. and Cartmell, S. H. (2018), 'Electroactive biomaterials: Vehicles for controlled delivery of therapeutic agents for drug delivery and tissue regeneration', *Advanced Drug Delivery Reviews*, Vol. 129, pp. 148–168.

Tantik, E. and Anderl, R. (2017), 'Potentials of the Asset Administration Shell of Industrie 4.0 for Service-Oriented Business Models', *Procedia CIRP*, Vol. 64, pp. 363–368.

Tao, J. and Yu, X. B. (2012), 'Hair flow sensors: from bio-inspiration to bio-mimicking – a review', *Smart Materials and Structures*, Vol. 21 No. 11, p. 113.001.

Tartarini, D. and Mele, E. (2016), 'Adult stem cell therapies for wound healing: biomaterials and computational models', *Frontiers in Bioengineering and Biotechnology*, Vol. 3, p. 206.

Tavassoly, I., Goldfarb, J. and Iyengar, R. (2018), 'Systems biology primer: the basic methods and approaches', *Essays in Biochemistry*, Vol. 62 No. 4, pp. 487–500.

Tay, K. M. and Lim, C. P. (2008), 'On the use of fuzzy inference techniques in assessment models: part II: industrial applications', *Fuzzy Optimization and Decision Making*, Vol. 7 No. 3, pp. 283–302.

Taylor, F. W. (2001), *The Principles of Scientific Management*, Dover Publications Inc., Mineola, USA.

Terninko, J. (1997), *Step by Step. Customer Driven Product Design,* 2nd ed., CRC Press, Cleveland.

Thabrew, L., Wiek, A. and Ries, R. (2009), 'Environmental decision making in multi-stakeholder contexts: applicability of life cycle thinking in development planning and implementation', *Journal of Cleaner Production*, Vol. 17 No. 1, pp. 67–76.

Thünen, J. H. von (1990), *Der isolirte Staat in Beziehung auf Landwirthschaft und Nationalökonomie,* 5th ed., Scientia, Amsterdam, Niederlande.

Thurestone, L. L. (2014), *The Nature of Intelligence*, Routledge, London.

Tibbits, S. (2014), '4D printing: multi-material shape change', *Architectural Design*, Vol. 84 No. 1, pp. 116–121.

Tietenberg, T. H. (1990), 'Economic instruments for environmental regulation', *Oxford review of economic policy*, Vol. 6 No. 1, pp. 17–33.

Tolman, E. C. (1948), 'Cognitive maps in rats and men', *Psychological Review*, Vol. 55, pp. 189–208.

Touyama, Y., Hoshino, T., Iwabuchi, K. and Morishima, K. (2009), 'Micro-encapsulation of bio-actuator using insect dorsal vessel', *2009 International Symposium on Micro-NanoMechatronics and Human Science*, pp. 644–649.

Trchounian, A. (2015), 'Mechanisms for hydrogen production by different bacteria during mixed-acid and photo-fermentation and perspectives of hydrogen production biotechnology', *Critical Reviews in Biotechnology*, Vol. 35 No. 1, pp. 103–113.

Trepl, L. (2005), *Allgemeine Ökologie: Band 1- Organismus und Umwelt*, Peter Lang GmbH, Bern.

Tributsch, H., Goslowsky, H., Küppers, U. and Wetzel, H. (1990), 'Light collection and solar sensing through the polar bear pelt', *Solar Energy Materials*, Vol. 21 No. 2-3, pp. 219–236.

Trifonov, E. N. (2012), 'Definition of Life: Navigation through Uncertainties', *Journal of Biomolecular Structure & Dynamics*, Vol. 29 No. 4, pp. 647–650.

Tuma, A., Reller, A. and Thorenz, A. e.a. (2014), *Nachhaltige Ressourcenstrategien in Unternehmen: Identifikation kritischer Rohstoffe und Erarbeitung von Handlungsempfehlungen zur Umsetzung einer ressourceneffizienten Produktion*, Augsburg, available at: *http://www.resource-lab.de/documents/Endbericht_DBU.pdf* (accessed 5 August 2015).

Turner, A. P.((2013), 'Biosensors: sense and sensibility', *Chemical Society Reviews*, Vol. 42 No. 8, pp. 3184–3196.

Turner, C., Rudnitskaya, A. and Legin, A. (2003), 'Monitoring batch fermentations with an electronic tongue', *Journal of Biotechnology*, Vol. 103 No. 1, pp. 87–91.

Ueda, K., Hatono, I., Fujii, N. and Vaario, J. (2001), 'Line-less production system using self-organization: A case study for BMS', *CIRP Annals*, Vol. 50 No. 1, pp. 319–322.

Uhlemann, T. H.-J., Lehmann, C. and Steinhilper, R. (2017a), 'The digital twin: Realizing the cyber-physical production system for industry 4.0', *Procedia CIRP*, Vol. 61, pp. 335–340.

Uhlemann, T. H.-J., Schock, C., Lehmann, C., Freiberger, S. and Steinhilper, R. (2017b), 'The digital twin: Demonstrating the potential of real time data acquisition in production systems', *Procedia Manufacturing*, Vol. 9, pp. 113–120.

Ulrich, H. (1970), *Die Unternehmung als produktives soziales System,* 2nd ed., Haupt, Bern.

Ulrich, P. and Fluri, E. (1995), *Management,* 7th ed., Stuttgart, UTB.

Ulrich, P. and Hill, W. (1976), 'Wissenschaftstheoretische Grundlagen der Betriebswirtschaftslehre – Teil I', *Wirtschaftswissenschaftliches Studium – Zeitschrift für Ausbildung und Hochschulkontakt*, Vol. 5 No. 7, pp. 304–309.

Umweltbundesamt (2012), *Ökonomische Bewertung von Umweltschäden: Methodenkonvention 2.0 zur Schätzung von Umweltkosten*, Dessau-Roßlau, available at: *http://www.umweltbundesamt.de/sites/default/files/medien/378/publikationen/uba_methodenkonvention_2.0_-_2012_gesamt.pdf* (accessed 1 March 2017).

Umweltbundesamt (2016), 'Prozessorientierte Basisdaten für Umweltmanagementsysteme', available at: www.probas.umweltbundesamt.de (accessed 10 February 2017).

United Nations (2015), *World Population Prospects: The 2015 Revision*, New York, USA, available at: *https://esa.un.org/unpd/wpp/Download/Standard/Population/* (accessed 15 January 2017).

United Nations Environment Programme (Ed.) (2011), *Decoupling natural resource use and environmental impacts from economic growth, A Report of the Working Group on Decoupling to the International Resource Panel*, United Nations Environment Programme, Paris.

United States Environmental Protection Agency (1974), 'Resource and Environmental Profile Analysis of nine beverage container alternatives'.

United States Environmental Protection Agency (2011), *LEAN, ENERGY & CLIMATE TOOLKIT: Achieving Process Excellence Through Energy Efficiency and Greenhouse Gas Reduction*, Washington D. C., USA.

Vahidi, A., Aliahmadi, A. and Teimoury, E. (2019), 'Researches status and trends of management cybernetics and viable system model', *Kybernetes*.

Vaidya, S., Ambad, P. and Bhosle, S. (2018), 'Industry 4.0–a glimpse', *Procedia Manufacturing*, Vol. 20, pp. 233–238.

van Ree, R. and Annevelink, B. (2007), 'Status Report Biorefinery'.

van Suntum, U. (2005), *Die unsichtbare Hand: Ökonomisches Denken gestern und heute*, 3rd ed., Springer, Berlin.

Váncza, J. and Monostori, L. (2017), 'Cyber-physical manufacturing in the light of Professor Kanji Ueda's legacy', *Procedia CIRP*, Vol. 63, pp. 631–638.

Vasconcelos, A., Freddi, G. and Cavaco-Paulo, A. (2008), 'Biodegradable materials based on silk fibroin and keratin', *Biomacromolecules*, Vol. 9 No. 4, pp. 1299–1305.

Vasiliadou, I. A., Berná, A., Manchon, C., Melero, J. A., Martinez, F., Esteve-Nuñez, A. and Puyol, D. (2018), 'Biological and Bioelectrochemical Systems for Hydrogen Production and Carbon Fixation Using Purple Phototrophic Bacteria', *Frontiers in Energy Research*, Vol. 6 No. 107.

Venkat, K. and Wakeland, W. (2006), 'Is Lean Necessarily Green?', *Proceedings of the 50th Annual Meeting of the ISSS-2006*.

Verband der Deutschen Automobilindustrie (2020), 'Special characteristics', available at: *https://www.vda-qmc-learning.de/module/glossar/glossardetails.php?id=292&letter=B&mode=&searchstr=&currentlanguage=de* (accessed 20 April 2020).

Verband Deutscher Ingenieure (2000), *VDI 3780 – Technikbewertung – Begriffe und Grundlage (engl. Technology assessment – Concepts and foundations)*, Beuth, Berlin.

Verband Deutscher Ingenieure (2019), *VDI 6220 Blatt 1: Bionik – Grundlagen, Konzeption und Strategie*, Beuth, Berlin.

Verein Deutscher Ingenieure (2012a), *Ganzheitliche Produktionssysteme – Grundlagen, Einführung und Bewertung,* No. VDI 2870, VDI, Berlin.

Verein Deutscher Ingenieure (2012b), *VDI 6220 Blatt 1: Bionik – Konzeption und Strategie – Abgrenzung zwischen bionischen und konventionellen Verfahren/Produkten*, Beuth, Berlin.

Verein Deutscher Ingenieure (2014), *Additive Fertigungsverfahren – Grundlagen, Begriffe, Verfahrensbeschreibungen,* No. VDI 3405, VDI.

Verrier, B., Rose, B. and Caillaud, E. e.a. (2014), 'Combining organizational performance with sustainable development issues: the Lean and Green project benchmarking repository', *Journal of Cleaner Production*, Vol. 85, pp. 83–93.

Vester, F. (1993), *Neuland des Denkens – Vom technokratischen zum kybernetischen Zeitalter*, DTV, München.

Vester, F. (2011), *Die Kunst vernetzt zu denken: Ideen und Werkzeuge für einen neuen Umgang mit Komplexität: Ein Bericht an den Club of Rome*, 8th ed., DTV, München.

Vilks, A. (1991), *Neoklassik, Gleichgewicht und Realität: Eine Untersuchung über die Grundlagen der Wirtschaftstheorie*, Heidelberg.

Villares, M., Işıldar, A., Beltran, A. M. and Guinee, J. (2016), 'Applying an ex-ante life cycle perspective to metal recovery from e-waste using bioleaching', *Journal of Cleaner Production*, Vol. 129, pp. 315–328.

Villares, M., Işıldar, A., van der Giesen, C. and Guinée, J. (2017), 'Does ex ante application enhance the usefulness of LCA? A case study on an emerging technology for metal recovery from e-waste', *The International Journal of Life Cycle Assessment*, Vol. 22 No. 10, pp. 1618–1633.

Vincent, J. F. (2009), 'Biomimetics – A Review', *Journal of Engineering in Medicine*, Vol. 223 No. 8, pp. 919–939.

Vincent, J. F., Bogatyreva, O. A., Bogatyrev, N. R., Bowyer, A. and Pahl, A. K. (2006), 'Biomimetics: its practice and theory', *Journal of the Royal Society Interface*, Vol. 9 No. 3, pp. 471–482.

Vinson, B. T., Sklare, S. C. and Chrisey, D. B. (2017), 'Laser-based cell printing techniques for additive biomanufacturing', *Current Opinion in Biomedical Engineering*, Vol. 2, pp. 14–21.

Vivien, F. D., Nieddu, M., Befort, N., Debref, R. and Giampietro, M. (2019), 'The hijacking of the bioeconomy', *Ecological Economics*, Vol. 159, pp. 189–197.

Vogt, K. (1980), 'Die Spiegeloptik des Flusskrebsauges', *Journal of Comparative Physiology*, Vol. 135 No. 1, pp. 1–19.

Vogtländer, J. G. and Bijma, A. (2000), 'The 'Virtual Pollution Prevention Costs 99'. A Single LCA-Based Indicator for Emissions', *International Journal of Life Cycle Assessment*, Vol. 5 No. 2, pp. 113–124.

Vogtländer, J. G., Brezet, H. C. and Hendriks, C. F. (2001), 'The Virtual Eco-Costs '99. A Single LCA-Based Indicator for Sustainability and the Eco-Costs-Value Ratio (EVR) Model for Economic Allocation', *International Journal of Life Cycle Assessment*, Vol. 6 No. 3, pp. 157–166.

Vogt-Schilb, A. and Hallegatte, S. (2014), 'Marginal abatement cost curves and the optimal timing of mitigation measures', *Energy Policy*, Vol. 66, pp. 645–653.

Voit, E. (2012), *A first course in systems biology*, Garland Science, New York.

Volkanovski, A., Čepin, M. and Mavko, B. (2009), 'Application of the fault tree analysis for assessment of power system reliability', *Reliability Engineering & System Safety*, Vol. 94 No. 6, pp. 1116–1127.

Vom Bruch, R. and Kaderas, B. (2002), *Wissenschaften und Wissenschaftspolitik: Bestandsaufnahmen zu Formationen, Brüchen und Kontinuitäten im Deutschland des 20. Jahrhunderts*, Franz Steiner Verlag, Stuttgart.

Wächter, M. (2018), *Kleine Entdeckungsgeschichte(n) der Chemie im Kontext von Zeitgeschichte und Naturwissenschaften*, Königshausen u. Neumann, Würzburg.

Wagner, C., Grothoff, J., Epple, U., Drath, R. and Malakuti, S. e.a. (2017), 'The role of the Industry 4.0 asset administration shell and the digital twin during the life cycle of a plant', *22nd IEEE International Conference on Emerging Technologies and Factory Automation (ETFA)*, pp. 1–8.

Waldner, J. B. (1992), *Principles of Computer-Integrated Manufacturing*, John Wiley & Sons Ltd., New Jersey.

Walras, L. (2003), *Elements of Pure Economics: Or the Theory of Social Wealth*, Routledge Chapman & Hall, New York.

Walsh, G. (2003), *Biopharmaceuticals: biochemistry and biotechnology*, 2nd ed., John Wiley & Sons, Hoboken.

Walsh, G. (2004), 'Second-generation biopharmaceuticals', *European Journal of Pharmaceutics and Biopharmaceutics*, Vol. 58 No. 2, pp. 185–196.

Walsh, G. (2005), 'Biopharmaceuticals: recent approvals and likely directions', *Trends in Biotechnology*, Vol. 231, pp. 553–558.

Waltersmann, L., Kiemel, S., Amann, Y. and Sauer, A. (2019a), 'Defining sector-specific guiding principles for initiating sustainability within companies', *Procedia CIRP*, Vol. 81, pp. 1142–1147.

Waltersmann, L., Kiemel, S., Bogdanov, I., Lettgen, J., Miehe, R., Sauer, A. and Mandel, J. (2019b), 'Benchmarking Holistic Optimization Potentials in the manufacturing Industry–A Concept to Derive Specific Sustainability Recommendations for Companies', *Procedia Manufacturing*, Vol. 39, pp. 685–694.

Wamba, S. F., Akter, S., Edwards, A., Chopin, G. and Gnanzou, D. (2015), 'How 'big data'can make big impact: Findings from a systematic review and a longitudinal case study', *International Journal of Production Economics*, Vol. 165, pp. 234–246.

Wander, J. D. and Rao, R. P. (2014), 'Brain–computer interfaces: a powerful tool for scientific inquiry', *Current Opinion in Neurobiology*, Vol. 25, pp. 70–75.

Wang, W., Li, G. and Huang, Y. (2009), 'Modeling of bubble expansion-induced cell mechanical profile in laser-assisted cell direct writing', *Journal of Manufacturing Science and Engineering*, Vol. 131 No. 5, 051.013-1-10.

Wang, Y., Jiang, X., Hu, C., Sun, T., Zeng, Z. and Cai, X. e.a. (2017), 'Optogenetic regulation of artificial microRNA improves H_2 production in green alga Chlamydomonas reinhardtii', *Biotechnology for Biofuels*, Vol. 10 No. 1, p. 257.

Wang, Y., Pham, D. T. and Ji, C. (2015), 'Self-healing composites: A review', *Cogent Engineering*, Vol. 2 No. 1, p. 1.075.686.

Wang, Y. M., Wang, Y. S. and Yang, Y. F. (2010), 'Understanding the determinants of RFID adoption in the manufacturing industry', *Technological Forecasting and Social Change*, Vol. 77 No. 5, pp. 803–815.

Wanieck, K. (2019), *Bionik für technische Produkte und Innovation: Ein Überblick für die Praxis*, Springer, Berlin.

Ward, D. J. (2014), 'The failure of marginal abatement cost curves in optimising a transition to a low carbon energy supply', *Energy Policy*, Vol. 73, pp. 820–822.

Warnecke, H.-J. (1992), *Die Fraktale Fabrik: Revolution der Unternehmenskultur*, Springer, Berlin.

Weber, I. (2007a), *Die Natur des Naturschutzes: Wie Naturkonzepte und Geschlechtskodierungen das Schützenswerte bestimmen*, Oekom, München.

Weber, M. (1992), 'Asketischer Protestantismus und kapitalistischer Geist', in Winckelmann, J. (Ed.), *Soziologie. Universalgeschichtliche Analysen: Politik*, Kröner, Stuttgart.

Weber, M. (2015), *Gesammelte Aufsätze zur Religionssoziologie. Band 1*, Severus Verlag, Hamburg.

Weber, W. (2007b), *Industrieroboter: Methoden der Steuerung und Regelung*, 2nd ed., Carl Hanser Verlag, München.

Wechsler, D. (1944), *The measurement of adult intelligence*, Williams & Wilkins, Baltimore.

Wegst, U. G.K. and Ashby, M. F. (2004), 'The mechanical efficiency of natural materials', *Philosophical Magazine*, Vol. 84 No. 21, pp. 2167–2186.

Weiblen, T. (2015), 'Opening up the business model: Business model innovation through collaboration', Dissertation, Universität St. Gallen, St. Gallen, 2015.

Weiser, M. (1991), 'The Computer for the 21st Century', *Scientific American*, pp. 94–104.

Weiss, M., Patel, M., Heilmeier, H. and Bringezu, S. (2007), 'Applying distance-to-target weighing methodology to evaluate the environmental performance of bio-based energy, fuels, and materials', *Resources, Conservation and Recycling*, Vol. 50 No. 3, pp. 260–281.

Weizsäcker, E. U. von and Wijkman, A. (2017), *Wir sind dran – Was wir ändern müssen, wenn wir bleiben wollen – Eine neue Aufklärung für eine volle Welt*, Gütersloher Verlagshaus, Gütersloh.

Welge, M. K., Al-Laham, A. and Eulerich, M. (2017), *Strategisches Management: Grundlagen-Prozess-Implementierung*, 7th ed., Springer, Berlin.

Wender, B. A., Foley, R. W. and Hottle, T. A. et al. (2014), 'Anticipatory lifecycle assessment for responsible research and innovation', *Journal of Responsible Innovation*, Vol. 1, pp. 200–207.

Werdich, M. (2012), *FMEA – Einführung und Moderation*, 2nd ed., Vieweg & Teubner, Wiesbaden.

Werfel, J., Petersen, K. and Nagpal, R. (2014), 'Designing collective behavior in a termite-inspired robot construction team', *Science*, Vol. 343 No. 6172, pp. 754–758.

Werner, H. J. (1989), *Geschichte der Freiwirtschaftsbewegung: 100 Jahre Kampf für eine Marktwirtschaft ohne Kapitalismus*, Waxmann, Münster.

Westkämper, E. (2006), *Einführung in die Organisation der Produktion*, Springer, Berlin.

Westkämper, E. and Warnecke, H.-J. (2010), *Einführung in die Fertigungstechnik*, Vieweg+Teubner, Wiesbaden.

Wicke, L. (1993), *Umweltökonomie: Eine praxisorientierte Einführung*, 4th ed., Vahlen, München.

Wiendahl, H.-P. (2010), *Betriebsorganisation für Ingenieure*, 7th ed., Hanser, München.

Wiendahl, H.-P., Reichardt, J. and Nyhuis, P. (2009), *Handbuch Fabrikplanung: Konzept, Gestaltung und Umsetzung wandlungsfähiger Produktionsstätten*, Hanser, München.

Wiener, N. (1961), *Cybernetics: Or Control and Communication in the Animal and the Machine*, 2nd ed., MIT Press Ltd., Cambridge.

Wikswo, J. P., Block III, F. E., Cliffel, D. E., Goodwin, C. R. and Marasco, C. C. et al. (2013), 'Engineering challenges for instrumenting and controlling integrated organ-on-chip systems', *IEEE Transactions on Bio-medical Engineering*, Vol. 60 No. 3, pp. 682–690.

Williams, S. C. and Deisseroth, K. (2013), 'Optogenetics', *Proceedings of the National Academy of Sciences*, Vol. 110 No. 41, p. 16.287.

Willke, G. (2006), *Kapitalismus*, Campus Verlag, Frankfurt a. M.

Wilms, F. (2006), *Szenariotechnik – Vom Umgang mit der Zukunft*, Haupt, Berlin.

Wilms, F. E.P. (2000), *Systemorientiertes Management*, Vahlen, München.

Wirth, R. (2003), 'Marktwirtschaft ohne Kapitalismus – Eine Neubewertung der Freiwirtschaftslehre aus wirtschaftsethischer Sicht. St. Galler Beiträge zur Wirtschaftsethik 34', Dissertation, Universität St. Gallen, St. Gallen, 2003.

Witt, U. (1999), 'Bioeconomics as economics from a Darwinian perspective', *Journal of Bioeconomics*, Vol. 1 No. 1, pp. 19–34.

Witzany, G. (2019), 'Communication as the Main Characteristic of Life', in Kolb, V. (Ed.), *Handbook of Astrobiology*, Taylor & Francis Ltd, London, pp. 91–105.

Wöhe, G. and Döring, U. (2010), *Einführung in die Allgemeine Betriebswirtschaftslehre*, 24th ed., Vahlen, München.

Woll, A. (2008), 'Kapitalismus', in May, H. (Ed.), *Lexikon der ökonomischen Bildung*, Oldenbourg Wissenschaftsverlag, München.

Wollschlaeger, M., Sauter, T. and Jasperneite, J. (2017), 'The future of industrial communication: Automation networks in the era of the internet of things and industry 4.0', *IEEE industrial electronics magazine*, Vol. 11 No. 1, pp. 17–27.

Wolpaw, J. R. (2007), 'Brain–computer interfaces as new brain output pathways', *The Journal of Physiology*, Vol. 579 No. 3, pp. 613–619.

Womack, J., Jones, D. and Roos, D. (1990), *The Machine That Changed the World: The Story of Lean Production*, Harper Collins, New York 1990; deutsche Übersetzung: J. Womack, D. Jones, D. Roos: *Die zweite Revolution in der Autoindustrie*, 4. Auflage. Campus, Frankfurt am Main 1992, Reprint, Westview Press Inc, Boulder.

Womack, J. P. and Jones, D. T. (2003), *Lean Thinking: Banish Waste and Create Wealth in Your Corporation,* 2nd ed., Free Press, New York.

Womack, J. P. and Jones, D. T. (2015), *Lean Solutions: How Companies and Customers Can Create Value and Wealth Together*, Free Press.

Wong, D., Beattie, E. E., Steager, E. B. and Kumar, V. (2013), 'Effect of surface interactions and geometry on the motion of micro bio robots', *Applied Physics Letters*, Vol. 103 No. 15, p. 153.707.

Woodward, D. G. (1997), 'Life cycle costing—theory, information acquisition and application', *International Journal of Project Management*, Vol. 15 No. 6, pp. 335–344.

World Commission on Environment and Development (1987), 'Our Common Future. Report of the World Commission on Environment and Development', available at: *www.un-documents.net/our-common-future.pdf* (accessed 10 July 2015).

World Society for the Protection of Animals (2008), *Eating our Future: The environmental impact of industrial animal agriculture*, London.

World Wide Fund For Nature (Ed.) (2014), *Living Planet Report 2014: Species and spaces, people and places*, WWF – World Wide Fund For Nature, Gland, Switzerland.

Xia, F., Yang, L. T., Wang, L. and Vinel, A. (2012), 'Internet of things', *International Journal of Communication Systems*, Vol. 25 No. 9, p. 1101.

Xiong, J. (2006), *Essential Bioinformatics*, Cambridge University Press, Cambridge.

Xu, X. (2012), 'From cloud computing to cloud manufacturing', *Robotics and Computer-Integrated Manufacturing*, Vol. 28 No. 1, pp. 75–86.

Yan, Y., Wang, X., Pan, Y., Liu, H., Cheng, J. and Xiong, Z. e.a. (2005), 'Fabrication of viable tissue-engineered constructs with 3D cell-assembly technique', *Biomaterials*, Vol. 26 No. 29, pp. 5864–5871.

Yang, G. H., Yeo, M., Koo, Y. W. and Kim, G. H. (2019), '4D bioprinting: technological advances in biofabrication', *Macromolecular Bioscience*, Vol. 19 No. 5, p. 1.800.441.

Yu, G., Chai, T. and Luo, X. (2011), 'Multiobjective production planning optimization using hybrid evolutionary algorithms for mineral processing', *IEEE Transactions on Evolutionary Computation*, Vol. 15 No. 4, pp. 487–514.

Zadeh, L. A. (1965), 'Fuzzy sets', *Information and control*, Vol. 8 No. 3, pp. 338–353.

Zadpoor, A. A. (2017), 'Design for additive bio-manufacturing: From patient-specific medical devices to rationally designed meta-biomaterials', *International Journal of Molecular Sciences*, Vol. 18 No. 8, p. 1607.

Zadpoor, A. A. and Malda, J. (2017), 'Additive manufacturing of biomaterials, tissues, and organs', *Annals of Biomedical Engineering*, Vol. 45, pp. 1–11.

Zahn, E. and Schmidt, U. (1996), *Produktionswirtschaft – Band 1: Grundlagen und operatives Produktionsmanagement*, UTB, Stuttgart.

Zainal, N. A., Zulkifli, N. W.M., Gulzar, M. and Masjuki, H. H. (2018), 'A review on the chemistry, production, and technological potential of bio-based lubricants', *Renewable and Sustainable Energy Reviews*, No. 82, pp. 80–102.

Zalasiewicz, J., Williams, M., Smith, A. and et al. (2008), 'Are we now living in the Anthropocene?', *GSA Today*, Vol. 18 No. 2, pp. 4–8.

Zander, T. O., Andreessen, L. M., Berg, A., Bleuel, M., Pawlitzki, J. and Zawallich, L. e.a. (2017), 'Evaluation of a dry EEG system for application of passive brain-computer interfaces in autonomous driving', *Frontiers in Human Neuroscience*, Vol. 11, p. 78.

Zander, T. O. and Kothe, C. (2011), 'Towards passive brain–computer interfaces: applying brain–computer interface technology to human–machine systems in general', *Journal of Neural Engineering*, Vol. 8 No. 2, p. 25.005.

Zelewski, S. (1999), 'Grundlagen', in Corsten, H. and Reiß, M. (Eds.), *Betriebswirtschaftslehre*, 3rd ed., R.Oldenburg Verlag, München, pp. 1–126.

Zencey, E. (2012), *The Other Road to Serfdom and the Path to Sustainable Democracy*, UPNE, Lebanon.

Zhang, B. and Radisic, M. (2017), 'Organ-on-a-chip devices advance to market', *Lab on a Chip*, Vol. 17 No. 14, pp. 2395–2420.

Zhang, D., del Rio-Chanona, E. A., Petsagkourakis, P. and Wagner, J. (2019), 'Hybrid physics-based and data-driven modeling for bioprocess online simulation and optimization', *Biotechnology and Bioengineering*, Vol. 116 No. 11, pp. 2919–2930.

Zhang, D., Xie, G., Yu, J. and Wang, L. (2007), 'Adaptive task assignment for multiple mobile robots via swarm intelligence approach', *Robotics and Autonomous Systems*, Vol. 55 No. 7, pp. 572–588.

Zhang, Q., Cheng, L. and Boutaba, R. (2010), 'Cloud computing: State-of-the-art and research challenges', *Journal of Internet Services and Applications*, Vol. 1 No. 1, pp. 7–18.

Zhang, Y., Lyu, F., Ge, J. and Liu, Z. (2014), 'Ink-jet printing an optimal multi-enzyme system', *Chemical Communications*, Vol. 50 No. 85, pp. 12.919–12.922.

Zhang, Y. S., Yue, K., Aleman, J., Mollazadeh-Moghaddam, K., Bakht, S. M. and Yang, J. e.a. (2017), '3D bioprinting for tissue and organ fabrication', *Annals of Biomedical Engineering*, Vol. 45 No. 1, pp. 148–163.

Zhao, R., Torley, P. and Halley, P. J. (2008), 'Emerging biodegradable materials: starch-and protein-based bio-nanocomposites', *Journal of Materials Science*, Vol. 43 No. 9, pp. 3058–3071.

Zhao, Y., Kankala, R. K., Wang, S. B. and Chen, A. Z. (2019), 'Multi-organs-on-chips: towards long-term biomedical investigations', *Molecules*, Vol. 24 No. 4, p. 675.

Zheng, H. and Shen, X. (2013), 'Double-acting sleeve muscle actuator for bio-robotic systems', *Actuators*, Vol. 2 No. 4, pp. 129–144.

Zhong, R. Y., Li, Z., Pang, L. Y., Pan, Y., Qu, T. and Huang, G. Q. (2013), 'RFID-enabled real-time advanced planning and scheduling shell for production decision making', *International Journal of Computer Integrated Manufacturing*, Vol. 26 No. 7, pp. 649–662.

Zhong, R. Y., Xu, X., Klotz, E. and Newman, S. T. (2017), 'Intelligent manufacturing in the context of industry 4.0: A review', *Engineering*, Vol. 3 No. 5, pp. 616–630.

Zhou, A., Liu, Z., Wang, S., Chen, E., Wei, Y. and Liu, W. e.a. (2019), 'Bio-electrolysis contribute to simultaneous bio-hydrogen recovery and phosphorus release from waste activated sludge assisted with prefermentation', *Energy*, Vol. 185, pp. 787–794.

Zhou, H., Li, X., Fan, T., Osterloh, F. E., Ding, J. and Sabio, E. M. et al. (2010), 'Artificial inorganic leafs for efficient photochemical hydrogen production inspired by natural photosynthesis', *Advanced Materials*, Vol. 22 No. 9, pp. 951–956.

Zhu, N., Zhang, L., Li, C. and Cai, C. (2003), 'Recycling of spent nickel–cadmium batteries based on bioleaching process', *Waste Management*, Vol. 23 No. 8, pp. 703–708.

Ziegler, B. (2008), *Geschichte des ökonomischen Denkens: Paradigmenwechsel in der Volkswirtschaftslehre,* 2nd ed., De Gruyter, Oldenbourg.

Ziessel, R. and Harriman, A. (2011), 'Artificial light-harvesting antennae: electronic energy transfer by way of molecular funnels', *Chemical Communications*, Vol. 47 No. 2, pp. 611–631.

Zitzler, E., Deb, K. and Thiele, L. (2000), 'Comparison of multiobjective evolutionary algorithms: Empirical results', *Evolutionary Computation*, Vol. 8 No. 2, pp. 173–195.

Zobel-Roos, S., Schmidt, A., Mestmäcker, F. and Mouellef, L., Ditz, R. et al. (2019), 'Accelerating Biologics Manufacturing by Modeling or: Is Approval under the QbD and PAT Approaches Demanded by Authorities Acceptable without a Digital-Twin?', *Processes*, Vol. 7 No. 2, 94 ff.

Zott, C., Amit, R. and Massa, L. (2011), 'The business model: recent developments and future research', *Journal of Management*, Vol. 37 No. 4, pp. 1019–1042.

Zulkifli, N. W.M., Azman, S. S.N., Kalam, M. A., Masjuki, H. H., Yunus, R. and Gulzar, M. (2016), 'Lubricity of bio-based lubricant derived from different chemically modified fatty acid methyl ester', *Tribology International*, No. 93, pp. 555–562.

Zündorf, I. (2006), *Der Preis der Marktwirtschaft: Staatliche Preispolitik und Lebensstandard in Westdeutschland 1948 bis 1963*, Franz Steiner Verlag, Stuttgart.

Zweig, J. (2016), 'Deep Neural Networks Go to the Movies', available at: *https://www.strong.io/blog/deep-neural-networks-go-to-the-movies* (accessed 23 April 2020).

Annex 1: The basics of strategic management

An elementary prerequisite for the development of BIS is the choice of a design path within the framework of existing strategic and technical options. The planning act of strategy development is an elementary component of the management of micro- and macroeconomic units such as states, companies and households. This section thus presents the basics of strategic management.

While the concept of strategy originally referred to the art of military command and/or the continuation of politics by other means (Clausewitz, 2010), today it is primarily understood as a pattern in the continuous stream of microeconomic decisions (Mintzberg and Waters, 1985; Porter, 1985; Mintzberg et al., 1996; Mintzberg et al., 2008).

1.1 The concept of strategy in corporate decision-making

The concept of strategy in the context of corporate decision-making was introduced around the middle of the 20th century in game theory to describe a complete plan that provides a correct choice for all conceivable moves of a player (Neumann and Morgenstern, 1973). In the 1960s, representatives of the Harvard School introduced the term into management theory (Welge et al., 2017). To date, however, no uniform interpretation of the concept of strategy has been established. Two approaches are generally distinguished: the classical understanding of strategy and the Mintzberg School (Welge et al., 2017). The former understands a strategy as a planned bundle of measures by a company to achieve its intended, longer-term goals. In this understanding, a strategy is the result of formal, rational planning that is clearly characterized by several features:

- Strategies consist of several interrelated individual decisions.
- Strategies are a hierarchical construct that require efficiency and organizational organization.
- Strategies make statements about the positioning of the company in the market, e. g., on the basis of strengths, weaknesses, opportunities and threats (SWOT).
- Strategies make statements about the allocation of resources within the framework of a concrete action plan.

The classical understanding of strategy, which is primarily characterized by the maxim of rationality, is widely criticized by the Mintzberg School (Welge et al., 2017). The fundamental insight is that strategies are not necessarily the result of formal, rational planning, but that there is a much broader spectrum of possible strategy types including planned strategies (essentially the classical understanding), strategies as a ploy (i. e., spontaneous, tactical measure to surprise one or more competitors), strategy as a pattern/emergent strategies (i. e., unintended strategy evolution resulting from the company's actions and decisions), strategy as a position (i. e., planned, unplanned, or spontaneous positioning of the company in relation to its environment) as well as strategy as a perspective (i. e., philosophy of management) (Hungenberg, 2012; Welge et al., 2017).[1)] Mintzberg et al. (1985; 1996; 2008) describe the practical implementation of strategies in three basic schemes: [1] intended (planned) strategies that are actually realized, [2] intended strategies that are not realized, and [3] realized strategies that were not intended to be realized. A realized strategy is therefore always the result of planned and unplanned strategies (Figure 1.1).

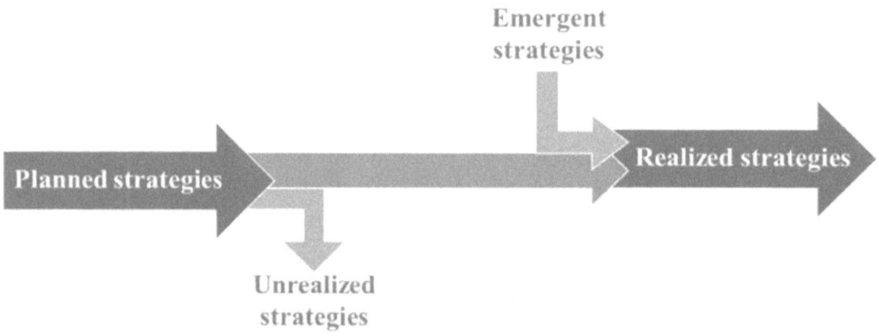

Figure 1.1 Basic patterns of observable corporate strategies according to Hungenberg (2012) and Welge et al. (2017)

[1] This enumeration is known as the 5P's of strategy.

1.2 Strategic management

The entirety of all goal-oriented actions, organization and planning with regard to this understanding of strategy is referred to as strategic management (Hungenberg, 2012). According to Hungenberg (2012), approaches of strategic management differ primarily in how they combine individual circumstances that shape the company's market position and resource base into objects of strategic management. The objects of strategic management include [1] the strategies, which are the focal point, [2] the structures, which are understood as basic rules for organizing the cooperation of employees in the company, and [3] the systems, which are needed to manage the company (e. g., information and incentive systems).

Decision-makers in strategic management generally operate in an environment of complex, often contradictory influencing factors. Hence, the development of corporate strategies is not trivial and requires a planned process that covers various activities. According to Hungenberg (2012), this process includes three stages: [1] strategic analysis, [2] strategy formulation and selection, and [3] strategy implementation (Figure 1.2).

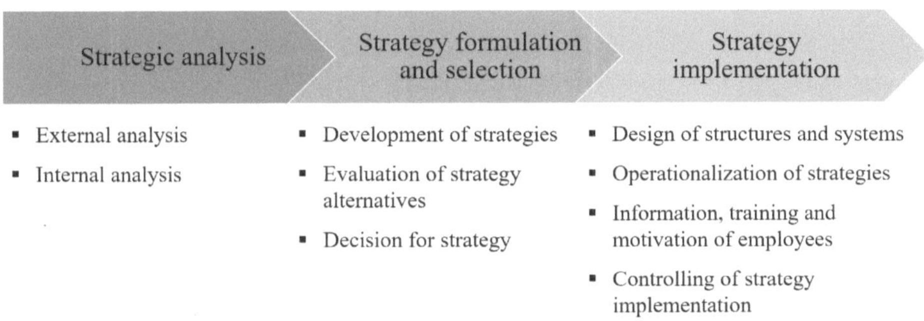

Figure 1.2 Process of strategic management according to Hungenberg (2012)

Basic elements of strategic management are vision, mission, and the development of basic goals (Gleißner, 2004; Hungenberg, 2012). The vision is the guiding idea of entrepreneurial activity, i. e., an imagination of what a company should look like in the future, in other words, a long-term picture of the company's goals (Johnson et al., 2011). A vision fulfils three functions (Hungenberg, 2012): [1] the identity function (describing a unique, unmistakable image of the future), [2] the identification function (showing employees a benefit or deeper meaning of their work), and [3] the mobilization function (encouraging companies to share the image of the future as a common goal). The mission is the written implementation of the vision in the form of corporate principles or guidelines. This is followed by the formulation of corporate goals. First and foremost, it is important to understand that companies do not have goals as such, but rather the individuals who work in the company. In economics, this issue is addressed in the so-called coalition theory (Hungenberg, 2012). Here, the company is understood as a coalition of individuals who perceive the enterprise as a vehicle for maximizing their own benefit. Strategic management must ensure that the corporate goals provide sufficient incentives for each individual in order to ensure a positive contribution to the company. The decision as to who is perceived as involved and/or affected individuals is a matter for management. The result is usually a compromise between the two extreme positions of the shareholder and stakeholder value approach (see Section 4.4.6.2). According to Doran (1981), corporate goals should be formulated according to the SMART principles, i. e., specific (target a specific area for improvement), measurable (quantify or at least suggest an indicator of progress), assignable (specify who will do it), realistic (state what results can realistically be achieved, given available resources), and time-related (specify when the result(s) can be achieved). The formulation of objectives within the framework of strategic management also serves to shape downstream decision-making processes, whose task is to identify approaches for achieving the objectives. Companies have a wide range of goals to set themselves. In scientific practice, little agreement exists on which goals should be the primary focus and which should not (Welge et al., 2017). Table 1.1 summarizes some of the possible goals of a company according to Ulrich and Fluri (1995).

Table 1.1 Potential corporate goal contents according to Ulrich and Fluri (1995)

Field	Potential goal contents
Market performance	▪ product quality ▪ product innovation ▪ product portfolio ▪ customer service
Market position	▪ revenue ▪ market share ▪ market valuation ▪ new markets
Profitability	▪ profit ▪ turnover ▪ return on total capital ▪ return on equity
Financing	▪ credit rating ▪ liquidity ▪ equity ▪ capital structure

Field	Potential goal contents
Power and prestige	▪ independence ▪ image ▪ political influence ▪ societal influence
Employees	▪ income and social security ▪ satisfaction ▪ social integration
Society and environment	▪ environmental protection ▪ mitigation of social costs ▪ non-commercial services for external stakeholder ▪ sponsoring

Generally, two levels of strategic management are distinguished: the corporate level and the business unit level (Hungenberg, 2012). A company that offers a comparatively homogeneous product group may only have one strategy. Larger companies in particular often exhibit a very inhomogeneous product range, often separated in different business units. In this case, the formulation of a strategy at the corporate level primarily serves the strategic orientation across all business units, while the business unit strategy determines the market position and the use of resources for a single product group.

1.3 Perspectives and methods of strategic management

Welge et al. (2017) emphasize that a corporate strategy provides information on four areas: [1] scope of the company's environmental relationships, [2] distinctive competencies, i. e., the company's relevant, necessary resources, [3] the company's competitive advantages, and [4] the synergy created by the strategic decision (Figure 1.3).

The resource basis is determined by the internal strength of a company, which results from the available core competencies. Examples of core competencies of a company are technological leadership, qualification of employees, existing network to suppliers, and customers or availability of information and data on issues relevant to competition. The market position results from the distinction against competitors on a market. Competitive advantages result from the core competencies and/or from the unique position of the company in the market. Consequently, strategic management distinguishes between the market and resource perspective (Gausemeier et al., 2014). Figure 1.4 illustrates the understanding according to Balzer et al. (2020).

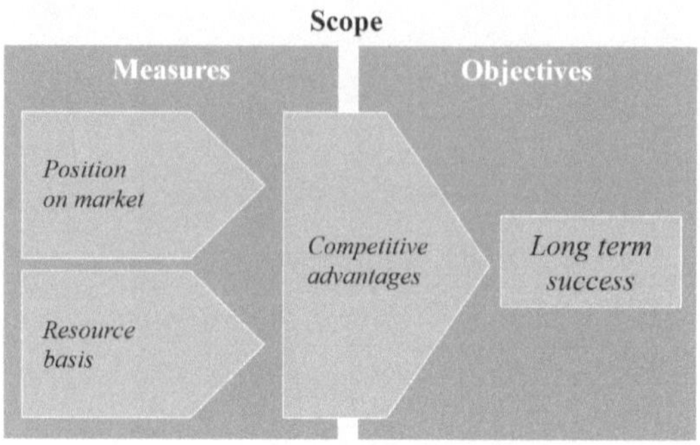

Figure 1.3 Basic understanding of strategy according to Hungenberg (2012) and Welge et al. (2017)

The market-based view attributes the company's success to the industry structure of the market in which the company operates. A company tends to be increasingly successful, the better it can adapt to changing conditions. Elementary methodological approaches of this view are the five forces model and the empirically supported analysis of Porter's competitive strategies (competitive advantages). Other methods commonly associated with the market-based view are PESTEL analysis[2] and the scenario technique (Annex 3). In turn, the resource-based view identifies the unique resources available in the company as the basis for success. Strategic resources can be either tangible or intangible. Material resources include raw materials, machinery, financial assets, real estate, hardware and software. Intangible resources include employees and their skills, patents, the organizational structure and processes, specific company values or cultural characteristics (Balzer et al., 2020). Basic methods of the resource-based view include VRIN, value chain and portfolio analysis, as well as the scenario technique (Annex 3). The following sections describe selected approaches of both perspectives in detail.

Figure 1.4 Market and resource perspective of strategic management according to Balzer et al. (2020)

2 The PESTEL method is used to analyze the company's operating environment. Thereby, six types of influencing factors are examined: political, economic, social, technological, ecological, and legal. The method is not further described in this work. Detailed information on the PESTEL analysis is provided by Johnson et al. (2011).

According to Porter (1985), the competitive advantages of a company result from the advantages of its suppliers and customers. Supplier advantages are differentiated into revenue advantages, i. e., the possibility of increasing the sales price, and cost advantages, i. e., the possibility of reducing production costs. Customer advantage is divided into benefit advantage (i. e., better product), and cost advantage (i. e., lower price). Furthermore, the question must be answered to what extent a company wants to compete with other companies. Porter (1985) distinguishes here between a broad and a narrow target range. The former can be an entire industry or the world market, while the latter is usually a niche or a segment of a larger market. The combination of strategic advantage and target range results in a matrix that allows a conclusion on three basic strategies. Cost leadership strategy focuses the individual economic actions on the minimization of costs. The core objective is to realize the lowest price in the competition for customers. Meanwhile, the differentiation strategy aims to increase customer benefit through unique product and/or service features. The customer should be inspired, e. g., by the exclusive nature or quality of the product. The choice of concentrating on market niches (e. g., regional markets or customer segments), which is possible in one form or another as a differentiation or cost leadership strategy, is based on the assumption of a better calculability and higher efficiency of the entrepreneurial production by reducing complexity. Figure 1.5 illustrates the concept of competitive strategies according to Porter (1985).

Figure 1.5 Concept of competitive advantages according to Porter (1985)

In contrast to the focus on customer advantages, supplier advantages result from the position of a company in the industry structure. Porter (2008) distinguishes five segments (forces model): new competitors, customers' bargaining power, suppliers' bargaining power, substitutes, and existing competitors. Figure 1.6 illustrates this so-called five-force model according to Porter (2008).

If these forces are strong in a sector, competition is not attractive. Accordingly, a company seeking market opportunities will likely find them in areas where certain forces are less developed. Then again, each force is interdependent and constantly changing (Johnson et al., 2011). The attractiveness of an industry can therefore change several times over time. Strategic management is thus responsible to constantly track latest developments in markets a company is involved in.[3] According to Johnson et al. (2011) and Dobbs (2014), various criteria should be examined when analyzing the five forces of a market or segment (Table 1.2).

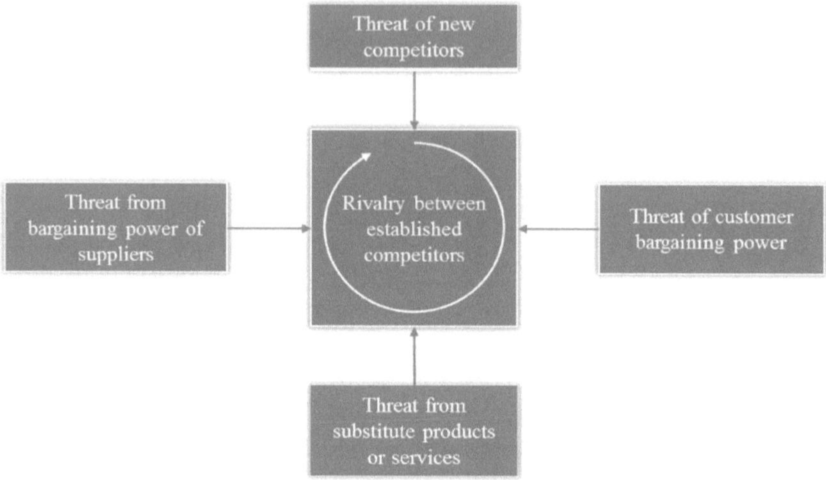

Figure 1.6 Five-force model according to Porter (2008)

According to Porter (2008), the application of the five forces method requires a procedure in four steps:

1. *Definition of relevant industry*, i. e., type of products, relation to other distinct industries, geographic scope of competition etc.
2. *Identification, grouping and description of participants and segments*, i. e., buyers and buyer groups, suppliers and supplier groups, competitors, potential substitutes, and potential entrants.
3. *Assessment of underlying drivers of each competitive force* to determine expression (strong vs. weak).
4. *Determination of overall industry structure and analysis for consistency*, i. e., level of profitability, controlling forces for profitability, better positioned, more-profitable players, recent and likely future changes in each force.

[3] For detailed discussion of market dynamics, hypercompetition, and competitive cycles, see Johnson et al. (2011).

Table 1.2 Analysis criteria of five forces method according to Johnson et al. (2011) and Dobbs (2014)[4]

Existing competitors	■ existing competitors ■ industry growth ■ fixed and/or storage costs ■ product differentiation ■ switching costs ■ strategic stakes ■ capacity expansion ■ exit barriers
New competitors	■ economies of scale ■ experience ■ capital requirements ■ access to suppliers and distribution channels ■ expected retaliation ■ government policy ■ competitive product differentiation
Customers' bargaining power	■ buyer orders ■ buyer information ■ buyer backward integration ■ industry products ■ buyer switching costs ■ overall buyer costs ■ buyer profitability ■ buyer product/service
Suppliers' bargaining power	■ supplier concentration ■ supplier volume/profit ■ supplier forward integration ■ supplier products ■ industry switching costs ■ supplier substitutes
Substitutes	■ price/indirect costs ■ buyer price sensitivity ■ performance ■ buyer switching costs ■ buyer profile

Common pitfalls in conducting the five forces analysis area too broad or narrow definition of the industry, confusing effect (price sensitivity) and cause (buyer econom-

4 Dobbs (2014) presents a five forces framework in form of a survey that can be applied to different use cases.

ics), equal attention to all forces rather than digging deep into the most important ones, application of static analysis that ignores industry trends (Porter, 2008).

In contrast to the market-based view, the resource-based view shifts the perspective to the inside of the company. Here, for example, the VRIN analysis is used as a moderation approach. Similar to the PESTEL and five forces analysis, it involves various fields that are examined on the basis of predefined criteria. VRIN is an acronym for value, rarity, inimitability, and non-substitutability (Johnson et al., 2011). According to Johnson et al. (2011), the following aspects are analyzed:

Value: Does the company have capabilities that provide a competitive advantage or for which customers are willing to pay?

Rarity: Does the company have capabilities that few or no competitors have?

Inimitability: Is it difficult for competitors to copy the existing capabilities?

Non-substitutability: Is the risk of substituting the existing capabilities low?

According to Schuh et al. (2010), portfolio analysis likewise complement the resource-based view. The analysis used in strategic management since the 1960s has led to a vast amount of different portfolio concepts. However, all concepts have in common that the variety of strategic success factors is limited to two dimensions. Not necessarily, but in many cases, these dimensions can be further differentiated into an external and an internal aspect (Schuh et al., 2010). The external dimension usually reflects the attractiveness of a market, whereas the internal dimension represents the strength of the company or business unit in relation to the competition. The external dimension of a portfolio cannot be influenced by the company, while the internal dimension can be controlled by the company. Portfolio concepts usually differ in the criteria used to capture the two dimensions. One of the most prominent portfolios is the market growth/market share portfolio (also BCG matrix), which was developed by the Boston Consulting Group in the late 1960s. Market growth is considered to be an expression of the external dimension, whereas relative market share represents the strength of a business (internal dimension). In order to ensure simplicity, the resulting four quadrants are referred to as cash cows, stars, question marks, and poor dogs. Cash cows have a high market share, but operate in markets that are in danger of stagnating or growing only marginally. Due to their good market position, business units in this section have a good cost position, which favors correspondingly high profits. Investments should only be made to a limited extent. The surpluses generated by these business units can be utilized by the company for other purposes. Business units that operate in growth markets and have a higher relative market share than their competitors are referred to as stars. The position taken in the market is attractive and strong. Question marks, on the other hand, are business units that have a comparatively low market share, while competing on markets with high growth. Competitors are already active in these markets and generally have a larger market share. For business areas of this type, it is not yet clear if they can survive in the market. Business areas with small market shares in low-growth or even stagnating markets are referred to as poor dogs. Investments in these areas are hardly promising. In addition, no great advantage is to be expected from investments due to the unfavorable market position. Companies should dispose of these business areas. Figure 1.7 illustrates the concept of the BCG matrix according to Schuh et al. (2010).

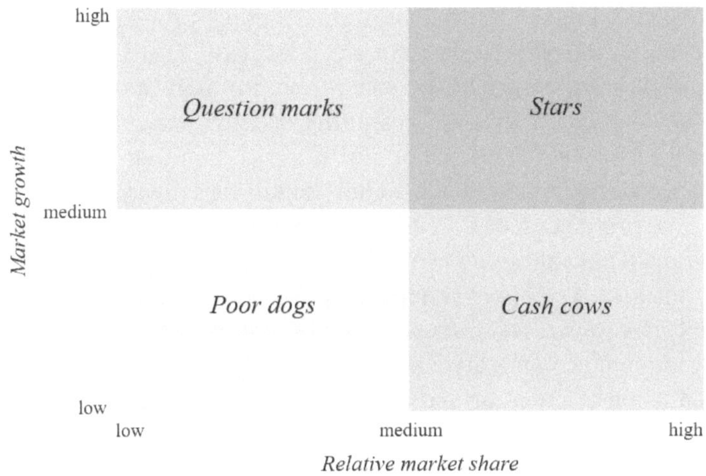

Figure 1.7 Market growth/market share portfolio (BCG matrix) according to Schuh et al. (2010)

A related approach comes from McKinsey and Company. Here, business strength is considered as an internal dimension. In contrast to the BCG matrix, it is not based on a single factor (relative market share), but on a number of different parameters. Any factors that are considered relevant in terms of their influence on business strength are selected. Examples of such factors include product quality, sales strength, research and development strength and financial strength. The external dimension may be represented by factors such as market size, market risk, market entry costs, competitive structure and demand situation. Among others, Schuh et al. (2010) and Udo-Imeh et al. (2012) provide further literature on portfolio analyses.

Another common method of strategic management that is closely related to the Portfolio analysis and combines market- and resource-based view is the SWOT analysis (Schuh et al., 2010). The term is an acronym for strengths, weaknesses, opportunities, and threats. The analysis takes an external perspective on opportunities and threats of a certain technology or market development as well as an internal view on relevant strengths and weaknesses. A comparison of the dimensions reveals whether the identified strengths and weaknesses of the company are relevant given an assumed development of the environment and offers the potential to address any approaching opportunities and risks. Section 6.6.2 provides an application example of the SWOT analysis. Further literature on the approach, procedures, and applications is provided by Johnson et al. (2011), Hungenberg (2012) and Welge et al. (2017).

Application example: a basic strategy portfolio for manufacturing enterprises in the context of biointelligence

Biointelligence offers new options for manufacturing companies, which promise different chances of success depending on core competencies, industry, and type of company. This raises the question of which fundamental strategic path a company might adopt. To visualize this, the creation of a portfolio is useful. Thereby, product and production can be distinguished as differentiating features (in this case both dimensions are internal). Both perspectives can in turn be divided into conventional and biointelligent products or production technologies. The resulting decision space comprises four squares: [A] the production of conventional products using conventional production technologies, [B] the production of biointelligent products using conventional production technologies, [C] the production of conventional products using biointelligent production technologies, and [D] the production of biointelligent products using biointelligent production technologies. A company that initially chooses to produce biointelligent products with its conventional technologies would follow a product-oriented primary strategy. If conventional products are produced using biointelligent technologies, this would be a production-oriented primary strategy. Should a company seek to be fully biointelligent, the respective other strategy is necessary within a two-phase approach. A company that directly chooses to produce biointelligent products via biointelligent technologies follows a single-phase basal strategy. Figure 1.8 illustrates the strategic decision-making scope of enterprises in the context of biointelligence.

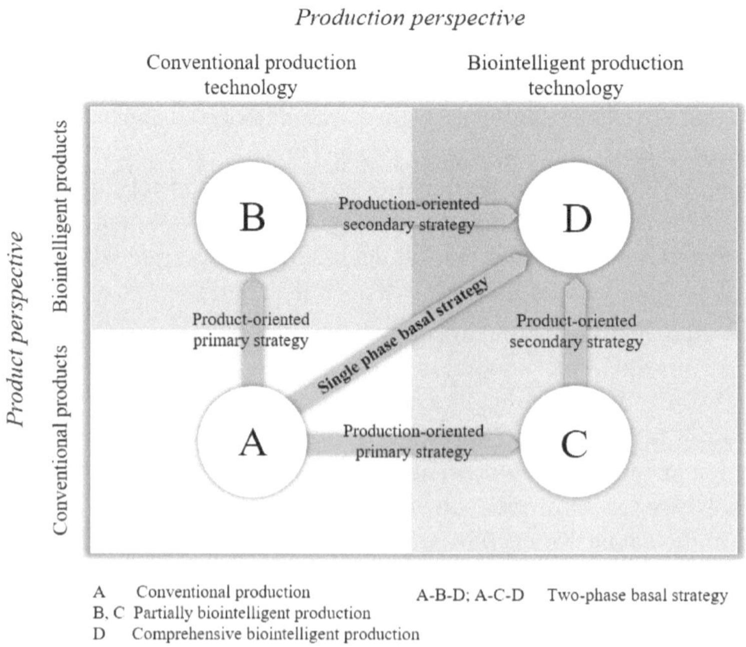

Figure 1.8 Strategic decision-making scope in the context of biointelligence [own illustration]

In any case, companies engaging in the field of biointelligence will have to consider current strengths, weaknesses, opportunities, and threats and may initially want to concentrate on basic technologies.

Annex 2: The basics of business models

A business model (BM) describes the functional nature of a company and its basic approach to generating profits (Morris et al., 2005; Zott et al., 2011; Burkhart et al., 2011). According to Morris et al. (2005), a sound BM addresses six key questions:

1. How do we create value?
2. Who do we create value for?
3. What is our key competence?
4. How do we competitively position ourselves?
5. How do we make money?
6. What are our time, scope, and size ambitions?

Thus, an elementary aspect of BM innovation is the search for success factors. These usually cover unique features in the supply and demand infrastructure as well as the value creation (Sun et al., 2012). Supply and demand infrastructure form a coherent structure. The supply side stimulates the demand side via specific products and services, which in turn responds by customer decisions. The supply infrastructure depends on the selection of key partners, resources, and activities. In turn, the demand infrastructure relies on the choice of distribution channels, customer relationships, and customer segments. Value proposition, revenue, and cost depict directly address customer aspirations. Figure 2.1 summarizes success factors of BMs based on Sun et al. (2012).

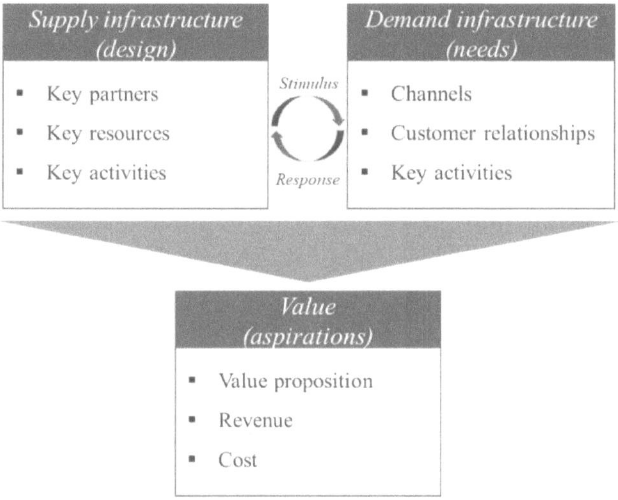

Figure 2.1 Success factors of business models [own illustration based on Sun et al. (2012)]

Marketing concept and sales channels are elementary components of a BM. Accordingly, BMs are generally oriented towards customer centricity (Shah et al., 2006, 2020). According to Burkhart et al. (2011), the emergence of the Internet in particular has led to a significant increase in the number of BMs. This trend continues with I4.0 (Man and Strandhagen, 2017). Thereby, especially the concept of collaborative BMs represents a promising path in the context of biointelligence. Weiblen (2015) presents fundamental considerations on the topic by developing a basic framework of BMs for network structures based on network configuration and customer centricity. This involves distinguishing between controlled, joint, and supported collaborative BMs (Figure 2.2).

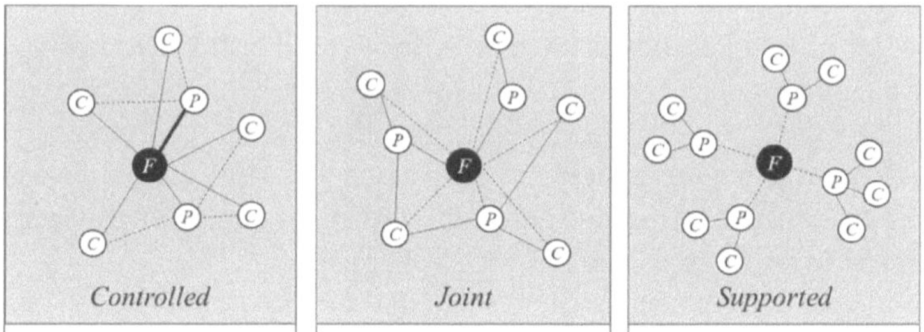

	Controlled	Joint	Supported
Customer centricity	High	Medium	Low
Network configuration	Few, strong ties	Medium amount and strength of ties	Many, weak ties, high shared vision

F = firm; C = customer; P = partner

Figure 2.2 Basic framework of BMs for network structures according to Weiblen (2015)

A number of approaches to sustainability have emerged under the heading of 'business models for sustainability' (Abdelkafi and Täuscher, 2016; Schaltegger et al., 2016). Among them, Abdelkafi and Täuscher (2016) present a systems dynamics approach to support the development of BMs. Another recent example is the work of Lüdeke-Freund et al. (2019), who present a framework for circular economy BMs by summarizing major BM dimensions in a morphological box. The framework presented in Figure 2.3 provides a decision-making horizon for businesses to differentiate themselves from contemporary competition.

Value proposition	*Products*	Repaired, refurbished, remanufactured or recycled products	Reusable or recyclable product	Products based on recycled waste	Long lasting product	Used product, components, materials or waste as production input	Reusable or recyclable production inputs			
	Services	Facilitating collaborations	Take back management	Customer education	Waste handling processes	Product-service based functions	Maintenance, repair, control	Product-service based results	Upgrading	Auxiliary services
Value delivery	*Target customers*	Quality conscious customers	Cost-conscious customers	Green customers	B2B customers	B2C suppliers	B2B suppliers	C2C suppliers		
	Value delivery process	Connecting suppliers and customer	Providing access to a products functionality	Providing product based services and results	Providing used products, components, materials or waste	Taking back used products, components, materials or waste	Sharing products, components or waste			
Value creation	*Partners and stakeholders*	Suppliers	Manufacturers	Retailers	Service providers	Public institutions	Collectors of products, components, materials or waste	Others (e.g. researchers)		
	Value creation process	Maintain or repairing products or components	Refurbishing products, components, materials or waste	Recycling of products, components, materials or waste	Upgrading or upcycling of products, components, materials or waste	Reselling of products, components, materials or waste	Winning back base materials	Using used products, components, materials or waste as input	Matching over- and under-capacities	Designing products, components or materials
Value capture	*Revenues*	Additional product revenues	Payments per unit of service	Payments for functions or results	Price premiums					
	Costs	Labor	Repair, maintenance, control	Waste handling, processing	Manufacturing	Resource inputs	Transportation logistics	Supply risks		

Figure 2.3 Morphological box of circular economy BM options according to Lüdeke-Freund et al. (2019)

Although these approaches provide valuable insights, they cannot be applied directly to biointelligence, as they neither adopt the technical basis nor the reorganization of the value creation system as a normative approach.

Annex 3: The basics of innovation management

3.1 Sustainable innovation paths and management

The success of a company strongly depends on the extent to which the products and services offered meet the requirements of the market (Stern and Jaberg, 2007; Brem, 2008). Responsible for this is innovation management, whose goal is the identification of market requirements as well as the medium to long-term development of new technologies (approx. 3 to 10 years). In addition to the process of the pure creative development of new concepts, innovation management focuses on operational exploitation (Hauschildt and Salomo, 2007; Müller-Prothmann and Dörr, 2011). Customer-specific solutions and variants required in the short term (0.5 to 3 years) are subsequently realized in product development. Figure 3.1 illustrates the difference between innovation management and product development in the operational practice according to Schloske (2020).

Innovation management tasks include futurology, trend analysis, initiation, planning, implementation and controlling of innovations. The process of innovation management is often portrayed as a funnel model, usually loosely based on the stage-gate process according to Cooper (2010). Figure 3.2 illustrates the funnel model of innovation management according to Schloske (2020).

Its starting points are the trends that are identified by market analyses. These trends are then examined for congruence with the corporate strategy in the form of a search field analysis in order to derive company-specific potentials. If a technology or a BM is identified as promising, ideas are generated using well-known creativity techniques, including brainstorming or the 6-3-5 method. Subsequently, the most promising approaches are transferred into partly competing pre-development projects, which are evaluated by an innovation committee after a certain period of time.

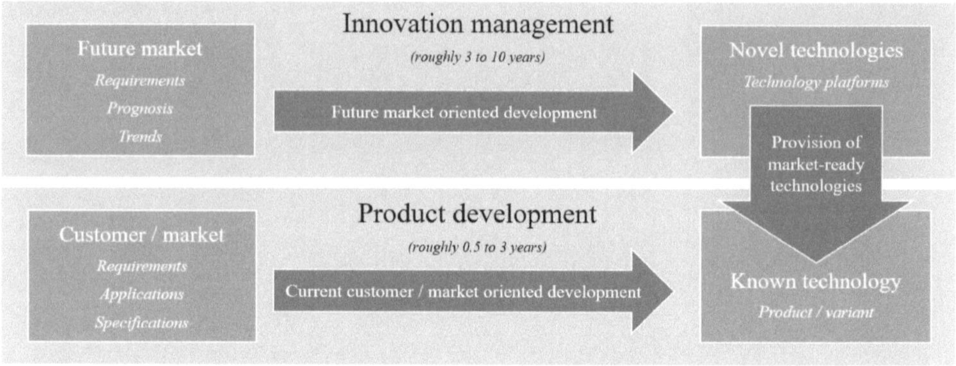

Figure 3.1 Difference between innovation management and product development according to Schloske (2020)

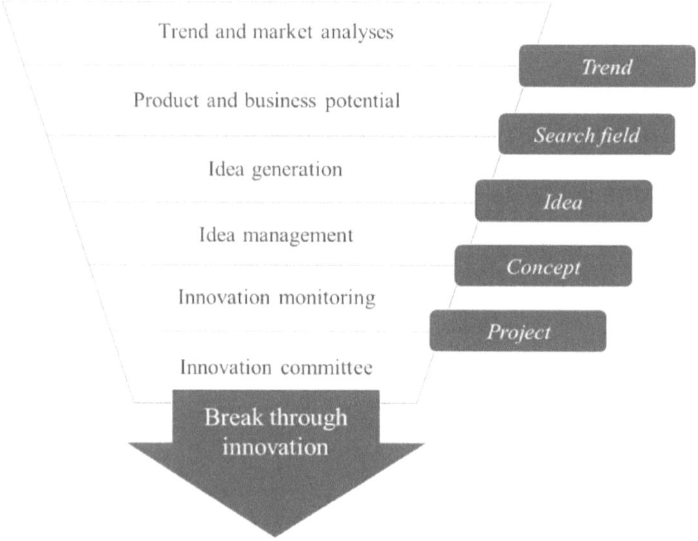

Figure 3.2 Funnel model of innovation management according to Schloske (2020)

3.2 Scenario technique

The development of scenarios is a substantial part of the investigation of trends and options. This is usually conducted using the scenario technique (Wilms, 2006; Fink and Siebe, 2011; Götze, 2013). Its purpose is to systematize strategic planning and forecasting in research, business, and politics. According to Gausemeier et al. (2014), a scenario is a generally comprehensible description of a possible situation in the future. In business practice, the method is used in different phases of innovation management to develop strategies and products for a possible future scenario. Thereby four steps are distinguished (Gausemeier et al., 2014):

1. *Scenario preparation* includes a definition and analysis of target areas and project objectives. Here, it is important to differentiate between the surroundings, which represent the future framework conditions, and the design field with possible configuration options.

The objective of the *scenario field analysis* is to define and describe the object of investigation, i. e., the scenario field. The problem is described by using influencing factors, so-called descriptors. A vehicle can be characterized by the descriptors mobility of society, petrol price and environmental legislation. In this step, the key factors are likewise identified by analyzing the cross-linking and relevance of the influencing factors with regard to their effect on the scenario field. Essentially, this is a comparison of the descriptors with the purpose of determining the respective relations. Basic methods for this are cross-linking tables, influence matrices, and the sensitivity model (Vester 2011).

2. Within the framework of the *development of projections*, alternative development possibilities for the identified key factors are elaborated on a predefined time axis.

3. Scenarios are generated from the individual future projections in *scenario building*. In this context, a scenario represents a consistent combination of several projections, commonly referred to as consistent worlds. Consistent worlds are collections of descriptors whose coexistence in the future appear plausible from today's perspective. A pairwise evaluation and analysis of the projections ensures the consistency of the scenarios formed. Within the scenario transfer, the effects of the scenarios on the design field are determined. Usually several scenarios are identified, as different values of the influencing factors are possible. A trend scenario is usually compared with two extreme scenarios (e. g., best case and worst case).

Supporting methods can be both quantitative and qualitative approaches. Figure 3.3 visualizes the phases of the scenario technique according to Gausemeier et al. (2014).

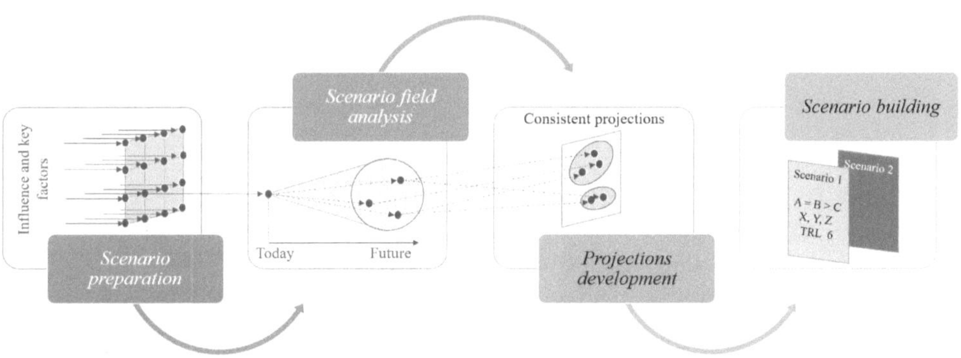

Figure 3.3 Stages of scenario technique according to Gausemeier et al. (2014)

3.3 Delphi method

The Delphi method seeks the opinions of various experts on a specific issue using a moderator with the aim of developing a solid consensus (Niederberger and Renn, 2019). The procedure of the Delphi method is usually divided into five steps. First, hypotheses are formulated in the instantiation to comprehensively describe the problem under investigation. The hypotheses can be developed by the moderator himself or by interviewing experts. Often a questionnaire is also developed in this step, which serves to test the hypotheses. In a second step, the first questioning of selected experts takes place. These experts are either presented with the hypotheses directly in order to comment them or to determine their opinion in the form of the questionnaire. The moderator then evaluates the feedback with regard to deviations in the opinion (aggregation) and presents this neutrally to the experts, who are then given the opportunity to express their opinion on the aggregated picture (second questioning of selected experts). The results of the second survey are consolidated and disseminated to a selected group of experts (consolidation and dissemination). Delphi analyses with and without discussion between the experts are distinguished (Schloske, 2020):

- In a *standard Delphi analysis*, the second survey of experts is conducted without further experts being involved. Consequently, a discussion between the interviewed experts is not possible.
- In contrast, the *broad Delphi analysis* informs the experts about the opinion of all other experts after the first survey. The experts involved have the opportunity to coordinate their opinions and initiate a discussion in order to develop a consensus on a specific issue. Usually, one or more workshops are organized for the second interview of the experts for the purpose of discussion, in which the respective experts participate collectively.

Often the Delphi method is used in combination with the scenario technique to increase the plausibility, consistency and/or acceptance of the developed scenarios by integrating different expert opinions.

Annex 4: The basics of responsible product development

Once innovation management has identified long-term trends and developed ideas for technical implementation, product development is responsible for adapting these to specific solutions and variants required in the short term (0.5 to 3 years). Various approaches exist, distinguishing similar phases. The most prominent approach is the V-model, an attempt borrowed from software development to describe the necessary activities in product development. According to Schloske (2020), the term V-model is derived from its form, which can be divided into the three areas specification (left side), product realization (basis) and verification (right side). Figure 4.1 illustrates the understanding according to Schloske (2020).

Starting point of the V-model is the requirements of the benefit system known from Chapter 4. From the initially loose set of customer requests, a specification sheet is developed in a first step and subsequently transferred into a technical specification sheet. The specification sheet thus represents a summary of the requirements for a product to be developed, while the technical specification sheet is the translation of these into engineering details. On the basis of the technical specification, the design department develops samples, prototypes, and products, which in turn are verified in the form of laboratory tests for compliance with the technical specifications and validated by tests with customer participation for compliance with the requirements of the benefit system. Between the moments of receiving the requirements of the benefit system (t0) and the moment of the product release to the benefit system (t1) a certain time span passes, which can strongly diverge depending on the industry. The decisive factor in bridging this time span is a certain flexibility in product development in order to be able to respond to any changes in customer requirements.

Usually, a transition from one stage of the V-model to the next is only realized after certain mandatory results have been achieved. The assessment of these partial goals is realized in so-called quality gates (Cooper, 2010), aiming at an early detection of deviations from the planned results. If one or more sub-goals are missed, an analysis of the causes and derivation of measures is necessary (Schloske, 2020).

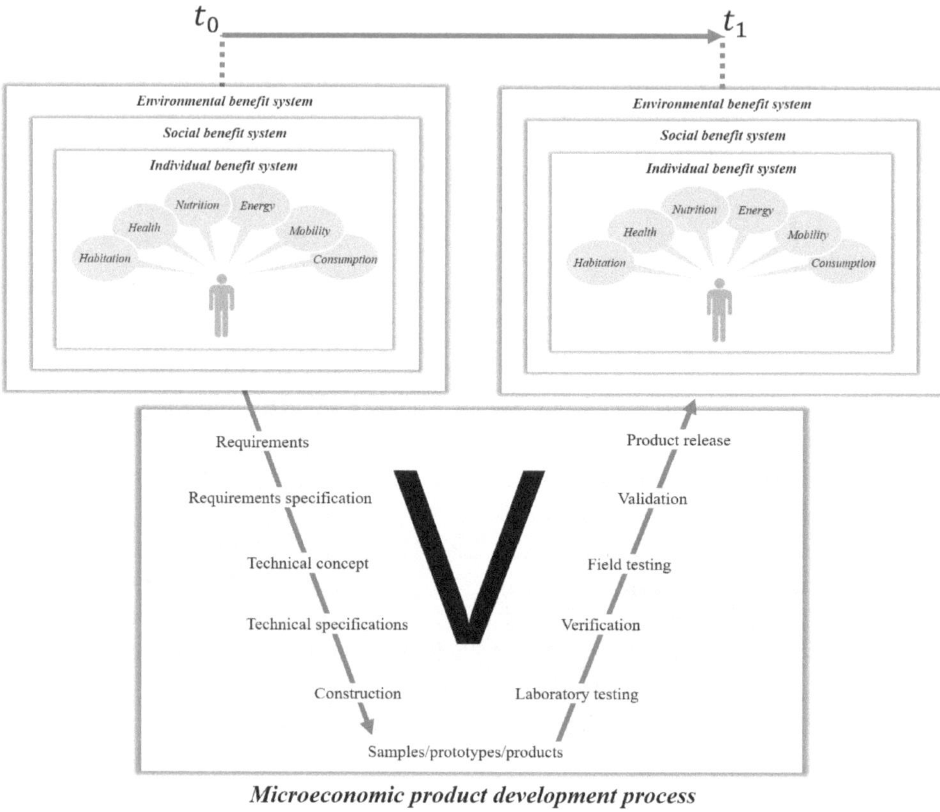

Figure 4.1 V-model of product development according to Schloske (2020)

The traditional product development process is supported by various methods, each of which fulfils a separate purpose (e. g., the efficient collection and promotion of customer requirements, ensuring product quality) and is used in different phases. According to Schloske (2020), two basic rules apply in practice when using methods to support the product development process. First, a method never fulfils an inherent purpose. Its use is only worthwhile if effective recommendations for action can be expected. Second, the effort of using a method must be kept as low as possible and duplication of work must be avoided.

Methods that seek to ensure sustainable product development are far less popular in business practice, yet a variety of concepts exist in scientific literature, e. g., design for environment (DFE), design for recycling (DFR), design for disassembly (DFD), design for sustainability (DFS) or life cycle design (LCD). Their aim is to design the product according to the concept of sustainability, which is a nearly impossible task due to the vague definition (see Section 3.1). However, several approaches exist that provide an orientation to approach the vision of sustainability iteratively. Table 4.1 summarizes examples of issues that have to be taken into account when thinking of designing a sustainable product (Keoleian and Menerey, 1994).

Table 4.1 Relevant issues in the context of sustainable product design according to Keolein and Menerey (1994)

Materials and energy	*Amount type*	Renewable vs. non-renewable
	Character	Virgin, reused, recycled
	Resource base	Location, scarcity, quality, restoration practices
	Impact caused by extraction and use	Material/energy used, residuals, ecosystem health, human health
Residuals	*Type*	Solid waste, air emissions, waterborne
	Characterization	Constituents, amount, concentration, toxicity, non-hazardous, radioactive
	Environmental fate	Containment, bioaccumulation, degradability, mobility/transport, treatment/disposal impacts
Ecological health	*Ecosystem stressors*	Physical, biological, chemical
	Impact categories	Diversity, sustainability, resilience to stressors, system structure and function, sensitive species
	Scale	Local, regional, global
Human health and safety	*Population at risk*	Workers, users, community
	Exposure routes	Inhalation, skin contact, ingestion, duration, and frequency
	Toxic character	Acute effects, chronic effects, morbidity/mortality
	Accidents	Type and frequency
	Nuisance effects	Noise, odors, visibility

The following sections summarize the design process according to the conventional V-model approach and complement aspects of sustainable design. In addition to the rigorous development of products with regard to individual, social and ecological benefit systems, emphasis must be placed on the greatest possible exploitation of degrees of freedom, which diminish significantly during the life cycle of a system. The costs of modification are inversely proportional to this. Figure 4.2 illustrates the qualitative curve of both effects according to Steen (1999).

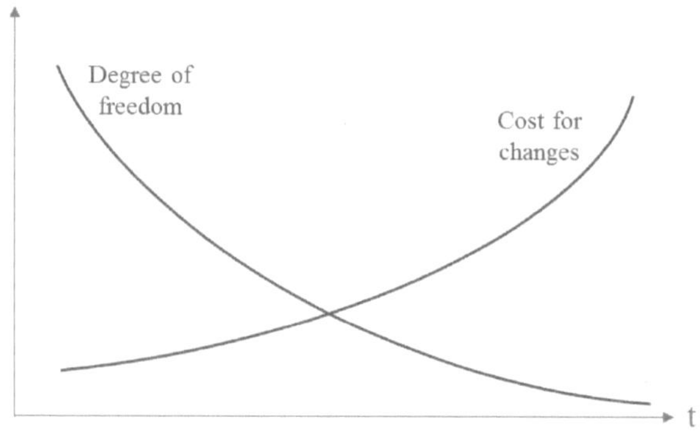

Figure 4.2 Qualitative development of degree of freedom and cost of changes during product life time according to Steen (1999)

4.1 Identifying demands from benefit systems (requirements specification)

In this work, this novel integral approach is expressed by the benefit system, which is designed to integrate the requirements of customers/markets, society and environment ab initio.

Individual benefit system

Identifying consumer requirements is an elementary task of classical market research. Its key instrument is the customer survey that aims to collect market-related information for product development for one or more target groups (McQuarrie, 2005; Berekoven et al., 2009). To guarantee a certain level of coherence, the use of questionnaires or interview guidelines is appropriate. This ensures that all participants of a survey deal with the same set of questions. A suitable structural framework for the questionnaire is the Kano model, which groups customer requirements into three classes: basic, functional, and excitement requirements (Schloske, 2020):

1. *Basic requirements* are defined as demands that a customer does not express verbally because he or she takes them for granted. Their fulfilment is not apparent to the customer. Yet, failure to meet them is immediately perceived by the customer and causes great dissatisfaction.

2. *Functional requirements* constitute all demands and objectives that are explicitly stated by a customer. The better these are fulfilled, the more satisfied a customer gets.

3. *Enthusiasm requirements* (unexpressed customer wishes) represent the third class. Customers are not yet able to express these requirements, as they are not known at this point. A fulfilment inspires him or her, as it represents an unexpected improvement. To excite customers for their products, companies constantly have to create new enthusiasm requirements.

Figure 4.3 illustrates the Kano model according to Schloske (2020).

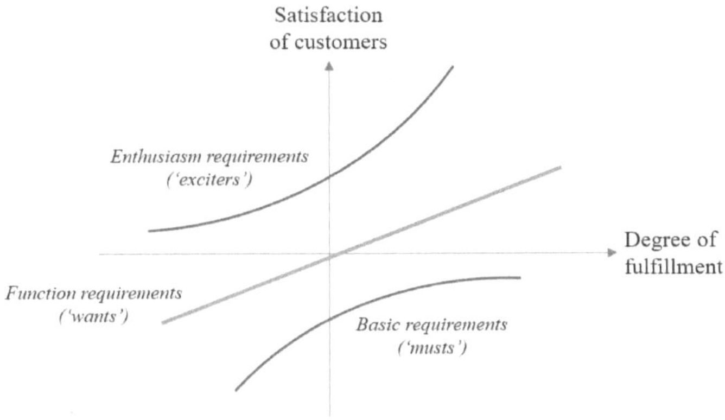

Figure 4.3 Kano model according to Schloske (2020)

According to Schloske (2020), a few basic rules have to be followed in designing a questionnaire. Open questions are suitable to identify not yet expressed desires or novel functionalities (enthusiasm requirements) of customers. In contrast, closed questions are used to assess existing customer requirements (function requirements) and should offer corresponding response options. Furthermore, it is useful to enable a simple prioritization with clear ranking of the expressed requirements. Basic requirements should only be questioned in special cases. As they are considered trivial from the customer's point of view, querying them unnecessarily extends the survey for the customer. Not only do they generally not provide new insights, they also draw the customer's attention to unimportant issues and lead to an increased risk of not identifying essential requirements.

Increasingly, it becomes crucial for companies to assess sustainability requirements of their customers at this point (Crittenden et al., 2011; Hetterich et al., 2012; Biju et al., 2015). These can be functional (e. g., energy consumption, recyclability, aluminum free) as well as enthusiastic (e. g., fairphone) requirements. Yet, from a holistic perspective, it has to be clear that these are based exclusively on subjective opinions that primarily serve to maximize personal benefit. Substantial insights for the social and ecological benefit system, which require a holistic view of the system dependencies, are thus not likely.

Social benefit system

In addition to the individual benefit creation for the customer, a biointelligent innovation requires meeting the needs of society. Societal requirements for entrepreneurial innovation result from widely accepted objectives and procedures, i. e., applicable regulations and existing standards, as well as from the demands of third parties affected by the entrepreneurial activity. The former is conducted within the framework of compliance management by identifying sales regions and thoroughly researching local legislations, among others in the areas of environment and health. Two approaches are common (Miehe et al., 2015). First, the acquisition of country warrants of apprehension from a specialized service provider. These services are generally expensive and limited to a single product. Second, a self-employed documentation of relevant legislations is by far less expensive and data sources are manifold. Research results should be documented in a legislation matrix that maps the sales regions and the names and requirements of the respective laws and standards. The matrix guarantees a comparatively effective overview of relevant legislations and enables a company to develop its distinct definition of a certain legal group according to relevant sales regions (e. g., EU-RoHS + Korea-RoHS + China-RoHS).

Identification of stakeholder requirements is usually part of environmental management or sustainability reporting, or can use established approaches in these areas, such as dialogs, platforms and questionnaires with previously identified stakeholders.

Environmental benefit system

The identification of requirements of the environmental benefit system requires a different approach, which is closely linked to corporate strategic management. According to Kuhlmann and Sauer (2016), Miehe et al. (2017b), Lentes et al. (2017), and Lentes and Hertwig (2019), the long-term goal must be ultra-efficiency, i. e., the complete avoidance of any harmful effects of an innovation and/or production on the environment through internal and external recycling and industrial-residential symbiosis. Since this requires a transformation process that cannot be implemented instantly, an iterative approach is necessary. The ecosystem perspective of biointelligence presupposes an orientation towards planetary boundaries. Yet, the concept presented by Rockström et al. (2009a, 2009b) outlines boundaries on the planetary level that are not easily transferable to the corporate level. This becomes possible through the science-based targets initiative, an association of several leading sustainability organizations, that transfers goals in key environmental problem categories to the company level on the basis of scientifically accepted methods (Science Based Targets Initiative, 2020). Waltersmann et al. (2019a; 2019b) present a similar approach in the context of ultra-efficiency. Based on a Delphi-supported scenario analysis, quantitative targets in key areas (energy, materials, emissions, personnel, and organization) are identified for different reference years (2025, 2035, etc.), to guide companies on their path to ultra-efficiency. Currently, these guidelines are available for four sectors (electrical and electronics, machinery and equipment, automotive and food) in Germany. Both approaches require a consideration of environmental objectives in the corporate strategy. A breakdown into individual areas (e. g., a single product or segment) can be realized by applying the sustainability balanced scorecard (Schaltegger, 2004).

Keolein and Menerey (1994) list several examples of environmental requirements for the case of a business telephone assuming a life cycle design (LCD) approach (Table 4.2).

Table 4.2 Examples of environmental requirements for the life cycle design of a business telephone according to Keolein and Menerey (1994)

	Manufacturing	Use/service	End-of-life
Product	▪ Materials should be recyclable (preferably on-site). ▪ Maximize use of recyclable material when environmentally preferable. ▪ Chose ozone-depleting-substance free components. ▪ Eliminate use of toxic materials.	▪ Extend useful life via modular design. ▪ Make product upgradable.	▪ Reuse parts. ▪ Standardize parts to facilitate remanufacturing. ▪ Design product and components to be easy recyclable, easy to disassemble. ▪ Housings should be shreddable.
Process	▪ Minimize process waste, incl. air emissions, liquid effluents, and hazardous/non-hazardous waste. ▪ Minimize resource and power consumption. ▪ Develop and meet corporate environmental goals.	▪ Ensure energy efficient operation.	▪ Maximize material recycling from components not reused. ▪ Minimize use of paints and solvents via service and reconditioning operations. ▪ Minimize waste (waste, incl. air emissions, liquid effluents and hazardous/non-hazardous waste) from refurbishment operations.
Distribution	▪ Minimize supplier packaging. ▪ Use recycled material for packaging when environmentally preferable. ▪ Reuse trays for parts in factory.	▪ Minimize packaging and number of products per package (if possible). ▪ Use recycled material. ▪ Minimize material variety for packaging.	▪ Use recyclable packaging. ▪ Use packing containing recycled material. ▪ Use reusable shipping containers.

4.2 Requirements design cube

Finally, all requirements from the three domains of the benefit system are to be documented in the specifications sheet that sets the basis for the next step. Here the conception of the three benefit systems as biointelligent design cube (BDC) provides assistance. The requirements from the single benefit systems are applied to the respective axes. The design cube can then, for example be, used to weight and/or analyze the interrelation of the requirements. At its core, it enables a simple visualization of variations in one or the other direction. Figure 4.4 illustrates the BDC qualitatively. Accordingly, the use of the BDC requires a three-dimensional logic in all following methods of the product development process.

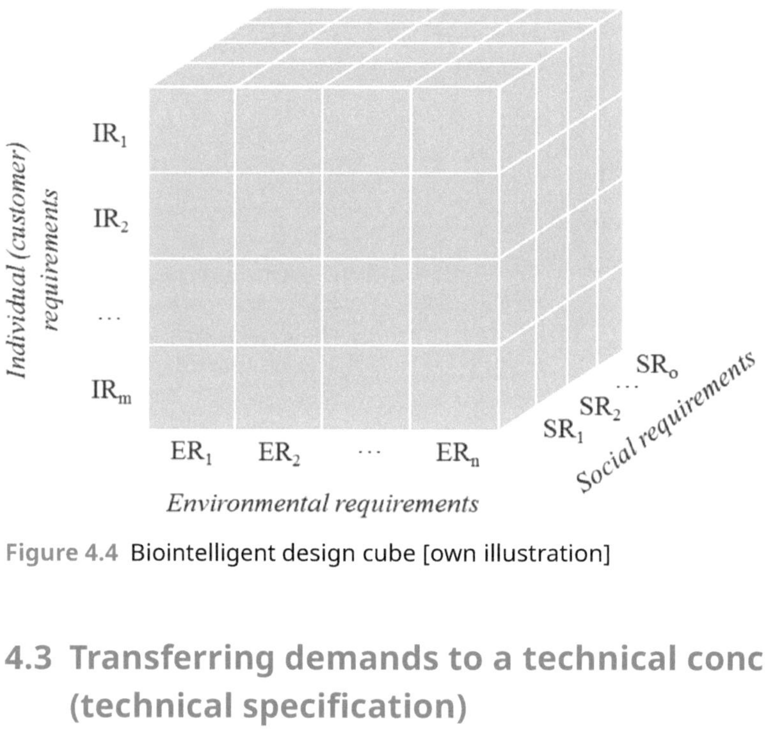

Figure 4.4 Biointelligent design cube [own illustration]

4.3 Transferring demands to a technical concept (technical specification)

Requirement specifications serve the development of a technical concept. Thus, a translation into 'technical language' is necessary. For this purpose, a so-called technical specification has to be developed. It represents the contractually binding technical description (specification) of the realization of the product. The preparation of the requirements specification is conducted by a contractor and is confirmed by the client. In the context of a customer-related development, the client is the customer himself, whereas in the context of a market-related development, the sales department or product management assumes the role of the client. The process of developing the technical specifications of a product can be supported by a several methods and tools.

Functional analysis as analytical hierarchy process (AHP)

A functional analysis describes the solution-neutral functional interaction in a planned system (Schloske, 2020). Essentially, this is an application of an analytical hierarchy process (AHP), which dates back to the works of Saaty (1999) from the 1970s onwards. The method seeks to structure and analyze various kinds of complex multi-party decisions (Saaty and Peniwati, 2008; Li et al., 2019), such as rankings, choice, prioritization, benchmarking, conflict resolution, and quality management (Forman and Gass, 2001). In addition to its use in product development, its generic approach enables an application to a multitude of fields, among others, covering the safety of nuclear reactors (Locatelli and Mancini, 2012), securing the quality of software systems (McCaffrey, 2005) or assessing the risk of cross-country petroleum pipelines (Dey, 2003). The aim is to quantify the weighting of decision criteria. For this purpose, the experience of individual experts is used to estimate the relative magnitudes of each factor by pairwise comparisons[1)]. In the context of product development, the factors are usually the required functions. Yet, a number of other factors (e. g., aesthetics, haptics) can be included as well. For a functional analysis, the required functions are first described regardless of their implementation and subsequently assigned to components. Thereupon, the pairwise comparison is conducted, by rating whether the considered criterion is less important (0), equally important (1) or more important (2) than the other. Similar to the interference matrix sketched above, both rows and columns of a two-dimensional matrix are labelled with the criteria. The diagonal of the matrix remains empty, since a criterion cannot be compared with itself. The row sums (cross sums) finally result in the weight of the respective criterion and thus allows a ranking. With a subsequent comparison of variants, the most suitable variant for solving a task can be identified by combining the advantages of individual variants. The weighted requirements from pair comparison are matched with the possible variants and evaluated. Figure 4.5 illustrates an example of a pair comparison matrix according to Schloske (2020).

AHP naturally facilitates decisions in the context of sustainability both within and beyond product development. Handfield et al. (2002) provide an example of AHP application for green supplier assessment. Others examples in the sustainability context are applications in natural hazard assessments (Nefeslioglu et al., 2013), identification and prioritization of corporate sustainability practices (Goyal et al., 2015), prioritizing barriers to achieve sustainable consumption and production trends in supply chains (Mangla et al., 2017).

[1] The concept developed in the early 20th century by Thurstone (2014) in the context of psychometry today serves as a basis for comparison or rank analysis in various fields.

Criterion (below) is more (= 2), equal (= 1) or less (= 0) important than criterion on the right	No chips in cavity	No chips in thread	No paint in thread	Simple manual assembly	Simple processing	Low material costs	Low assembly costs	Low manufacturing costs	Process-secure manufacturing	**Sum**
No chips in cavity		1	2	2	2	2	2	2	1	**14**
No chips in thread	1		2	2	2	2	2	2	1	**14**
No paint in thread	0	0		2	2	2	2	2	1	**11**
Simple manual assembly	0	0	0		1	1	2	2	0	**6**
Simple processing	0	0	0	1		2	2	2	0	**7**
Low material costs	0	0	0	1	0		0	1	0	**2**
Low assembly costs	0	0	0	0	0	2		1	0	**3**
Low manufacturing costs	0	0	0	0	0	1	1		0	**2**
Process-secure manufacturing	1	1	1	2	2	2	2	2		**13**

Figure 4.5 Example of a pair comparison matrix according to Schloske (2020)

Quality function deployment (QFD)

Another method to support the translation of requirements into technical specification is quality function deployment (QFD) (Schloske, 2020). The method is designed for customer and competition-oriented product development. It allows the systematic development of technical specifications based on prioritized requirement specifications (Taguchi, 1986; Terninko, 1997; Herzwurm et al., 2000; Brüggemann and Bremer, 2012). In addition to prioritization and customer orientation, the approach enables the consideration of competition. This ensures that the developers do not pursue individual goals and focus on ensuring to meet requirements. The application of the method is particularly suitable for market-related developments (e.g., consumer goods). A QFD comprises six steps:

1. Identification and prioritization of customer requirements
2. Comparison with the competition with regard to customer requirements
3. Definition of the product characteristics
4. Correlation between customer requirements and product features (HoQ matrix)

5. Technical competitive comparison
6. Mutual influence of product characteristics

According to Schloske (2020), the results of the six steps can be visualized in the so-called house of quality (Figure 4.6).

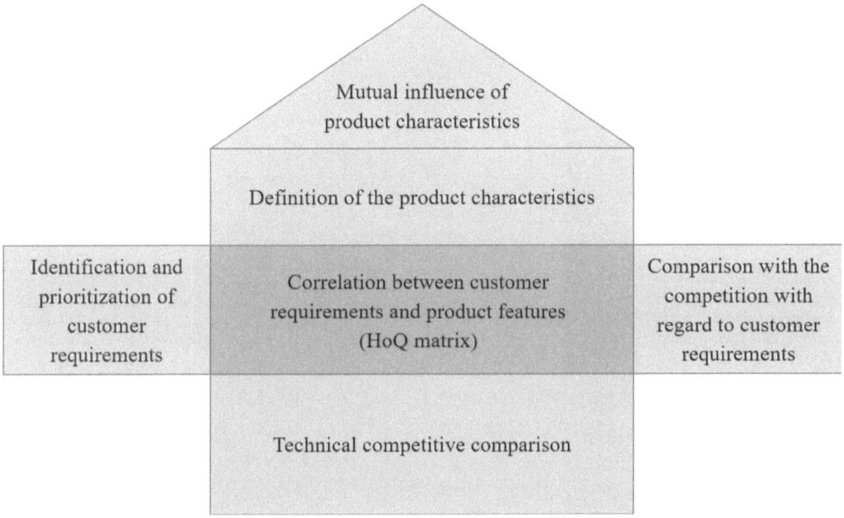

Figure 4.6 Quality function deployment according to Schloske (2020)

In the context of the development of sustainable technologies, QFD is, for instance, applied to transfer the requirements of stakeholders from the three sustainability dimensions into engineering metrics of products and services (Sousa-Zomer and Miguel, 2017), to select environmentally conscious (bio)materials (Mastura et al., 2017), to generally support the design for environment (DfE) process (Masui et al., 2002; Rahimi and Weidner, 2002), and to establish a green supply chain management (Dai and Blackhurst, 2012; Lam and Lai, 2015; Osiro et al., 2018). Essentially, these approaches simply use the generic framework of QFD to incorporate requirements from a sustainability or environmental perspective in addition to technical requirements.

Theory of inventive problem solving (TRIZ)

As sketched above in the context of biomimetics, the theory of inventive problem solving (TRIZ) is of great importance for the development of BIS as well. TRIZ is a complex toolbox based on the work of Altshuller (Altshuller, 1984; 1999). His work as a patent assessor enabled him to analyze patents. He noticed that patents were always created when seemingly insoluble contradictions were solved. This led him to develop a systematic approach which proposes 40 solution principles based on 39 technical parameters. Yet, no explicit solutions are given. Rather, it provides thought-provoking impulses, similar to creativity techniques that help to find a solution for a problem through abstraction. The approach thus focuses primarily on resolving contradictions within product development (e. g., conflicting requirements such as ‘stain should have 100 °C and stain must not be hotter than 80 °C). Figure 4.7 illustrates the TRIZ matrix according to Herb et al. (2000).

	1 Weight of a moving object	2 Weight of a static object	…	14 Firmness, strength	…	38 Degree of automation	39 Productivity
1 Weight of a moving object						*28, 26, 18, 32*	*32, 26, 24, 27*
2 Weight of a static object						*28, 26, 32, 10*	*28, 27, 12, 3*
…							
14 Firmness, strength	*1, 8, 40, 15*	*40, 26, 27, 1*		*9, 14, 17, 15*		*26, 2, 19*	*10, 28, 24, 32*
…							
38 Degree of automation	*28, 26, 18, 32*	*28, 26, 32, 10*					*5, 12, 32, 26*
39 Productivity	*32, 26, 24, 27*	*28, 27, 12, 3*				*5, 12, 32, 26*	

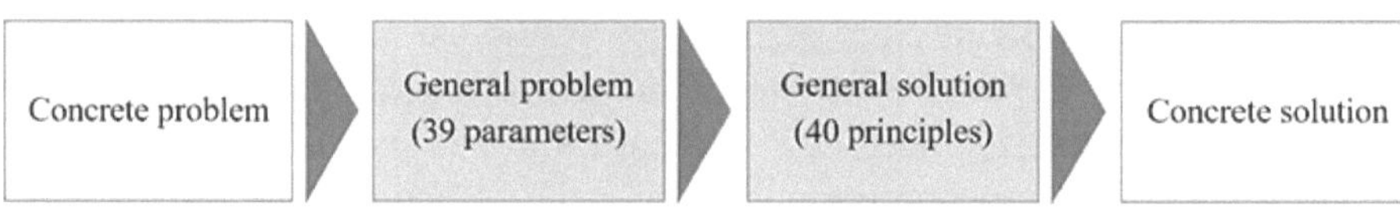

Figure 4.7 TRIZ matrix according to Herb et al. (2000)

The generic concept of the method and its inherent system-thinking approach enables an application in the context of sustainability. In addition to the above-mentioned application potential for the development of bioinspired systems (Nachtigall, 2010; Gadd and Delbrück, 2016), various authors use TRIZ as a tool to assess and innovate a technical system with regard to sustainability parameters (Russo et al., 2011; D’Anna and Cascini, 2011; Russo et al., 2014; Hede et al., 2015). Thereby the 40 solutions provided by TRIZ are in many cases already regarded as viable alternatives for a sustainable design, since they, for example, contribute to the reliability of the product. Sheng and Kok-Soo (2010) exemplify this within two case studies regarding an eco-efficient design of a water filter and a washing machine. Based on TRIZ, Liu et al. (2020) develop a resource-constrained innovation method (RCIM) that essentially serves to ensure a resource efficient product development. Among others, the work illustrates a correlation between resource-constrained innovation/principles and TRIZ principles (Table 4.3).

Table 4.3 TRIZ-based resource-constrained innovation method (RCIM) according to Liu et al. (2020)

Resource-constraint approach	TRIZ inventive principle	Resource-constrained principles
Partial optimization	▪ local quality ▪ asymmetry ▪ partial or excessive actions	*Local optimization principle*: Overcome a system's weak links and optimize the system state by improving local components.
Substitution and changing	▪ copying ▪ cheap short-living objects ▪ mechanics substitution ▪ porous materials ▪ color changes ▪ parameter changes ▪ phase transitions ▪ thermal expansion ▪ strong oxidants ▪ inert atmosphere ▪ composite materials	*Cheaper substitution principle*: Rtime and materials by replacing with cheaper materials and local services. *Theory substitution principle*: Optimize system efficiency by using interchange of the light field, sound field, force field, electric field, magnetic field or mechanic field.
Dynamic approach	▪ dynamics ▪ another dimension ▪ periodic action ▪ continuity of useful action ▪ skipping	*Re-use principle*: Make full use of waste materials and energy or parts to provide service for the system itself.
Friendliness and coordination	▪ spheroidality-curvature ▪ homogeneity	*Consistency principle*: Combine systematic characteristics, allocate system resources to maintain consistency, and achieve best system efficiency.

Finally, all technical specifications have to be documented in the technical specifications sheet. In case of an earlier application of the BDC, a three-dimensional logic is appropriate in this as well.

4.4 Construction and variant comparison

Once certain design principles that ideally meet the requirements of the individual, social, and ecological benefit system have been identified by applying the approaches described above, construction begins. A biointelligent design follows the approaches described in Chapter 4 and Chapter 5. However, in order to support the development process with respect to a holistic-systemic innovation, a wider view is needed. In addition to the TRIZ design principles that are occasionally supplemented by sustainability principles, various authors advocate an early application of LCA (see Section 6.2) to identify potential environmental impacts along the life cycle of a newly designed system (Steen, 1999; Rosen, 2012). Although a number of examples for early implementation exist (Wender et al., 2014; Villares et al., 2016; Villares et al., 2017; Hesser et al., 2017), ex ante applications prove to be challenging, as many determinants are unknown in this phase (Cucurachi et al., 2018). In this phase it is rather common to either use checklists with specific technical strategies in order to ensure that the requirements of one or more design alternatives are matched with the requirements from an individual, social or environmental perspective. In this context, Keolein and Menerey (1994) outline a few general strategies of a sustainable life cycle oriented design (incl. strategies for management) (Table 4.4).

Table 4.4 General strategies of a sustainable life cycle oriented design according to Keolein and Menerey (1994)

Product life extension	Extend useful life; ensure adaptability; facilitate service ability by simplifying maintenance and allowing reparations; enable remanufacturing and reuse
Material life extension	Use and specify recycled materials
Material selection	Substitute materials; reformulate product
Reduced material intensity	Conserve resources
Process management	Substitute processes; increase material and energy efficiency; improve process control, inventory, material handling, treatment and disposal
Efficient distribution	Chose efficient transportation; reduce packaging; use reusable packaging
Improved management practices	Use office material and equipment efficiently; phase out high-impact products; choose environmentally friendly suppliers

In the context of disassembly, Crowther (2005) presents a number of general design principles for improvement (Table 4.5).

Table 4.5 General design principles for improving disassembly according to Crowther (2005)

1. Use recycled and recyclable materials.
2. Minimize number and types of materials.
3. Avoid toxic and hazardous materials.
4. Avoid composite materials.
5. Avoid secondary finishes to materials (e. g., some coatings may contaminate base material).
6. Provide standard and permanent identification of material types.
7. Use mechanical connections rather than chemical.
8. Use modular design that is compatible with other materials and/or systems both dimensionally and functionally.
9. Use standard, simple, and 'low-tech' technologies.
10. Provide access to all parts and components.
11. Ensure simple handling and locating of components during assembly and disassembly.
12. Provide realistic tolerances to allow for maneuvering during disassembly.
13. Use minimum number and types of fasteners or connectors.
14. Design joints and connectors to withstand repeated use.
15. Allow for parallel disassembly rather than sequential disassembly.
16. Use lightweight materials and components.
17. Permanently identify points of disassembly and provide spare parts and on-site storage for them.
18. Retain all information on the systems and assembly and disassembly procedures.

Similar strategies and principles are proposed by a number of authors. A review of the scientific approaches as well as a classification along the different steps of a product life cycle design (design, manufacturing, supply chain, and end-of-life) is provided by Rahman et al. (2010). According to the respective design criteria, individual design alternatives may be easily compared in terms of fulfilment.

4.5 Risk and criticality analysis

Once the technical concept has been defined, a risk assessment is required to determine the technical risks of the product. For this purpose, potential failures have to be identified from the key functions of the product in order to evaluate their impact regarding the use phase. In addition, it is necessary to identify the chemical composition of the substances, materials, and components used and assess the risk regarding compliance with applicable substance regulations and associated specifics (especially exceptions and exclusions) in the previously identified sales regions. The requirement specifications should already contain information on the intended use of the product and its potential effects on society and the environment. Otherwise, a rough cause-effect tree regarding the impact of potential errors to the three benefit systems can be developed at this point (Figure 4.8).

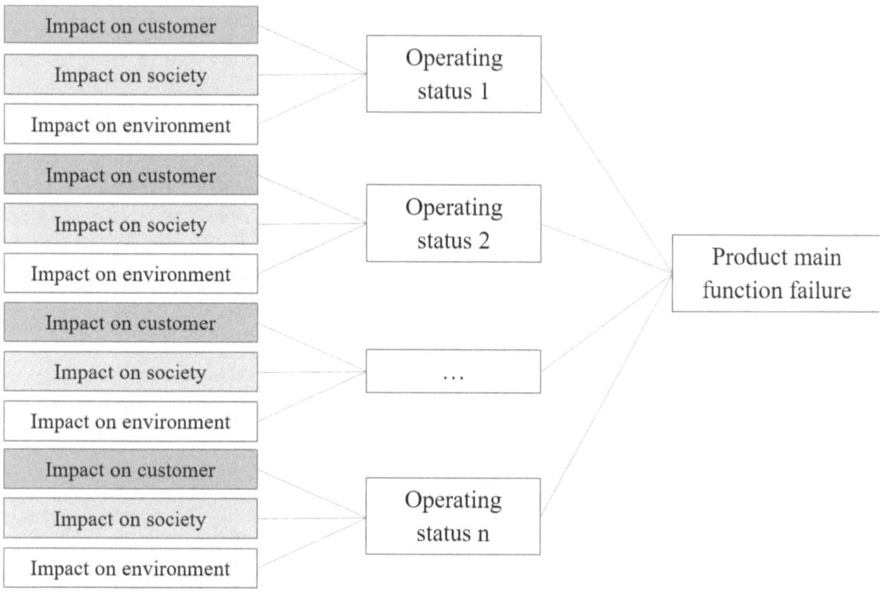

Figure 4.8 Extended initial visual assessment of the impact of potential errors [own illustration] based on Schloske (2020)

Based on the risk assessment, the functional and technical safety concept is developed. Supporting methods for this step include failure mode and effect analysis (FMEA), fault tree analysis as well as material and supplier evaluation.

Failure mode and effect analysis (FMEA)

The failure mode and effect analysis (FMEA) is a prominent method of corporate technical risk management that today is applied to a wide field of scientific disciplines (Rausand and Hoylan, 2003; Tay and Lim, 2008; Shebl et al., 2009; Franklin et al., 2012; Potts et al., 2014). The approach developed by NASA in the early 1960s to ensure the quality of the Apollo projects was quickly applied to other safety-critical fields such as aerospace and nuclear engineering. The application of FMEA in product development detects errors at an early stage and proactively avoids them through targeted planning of measures. It can be applied in different stages of product development (Figure 4.9).

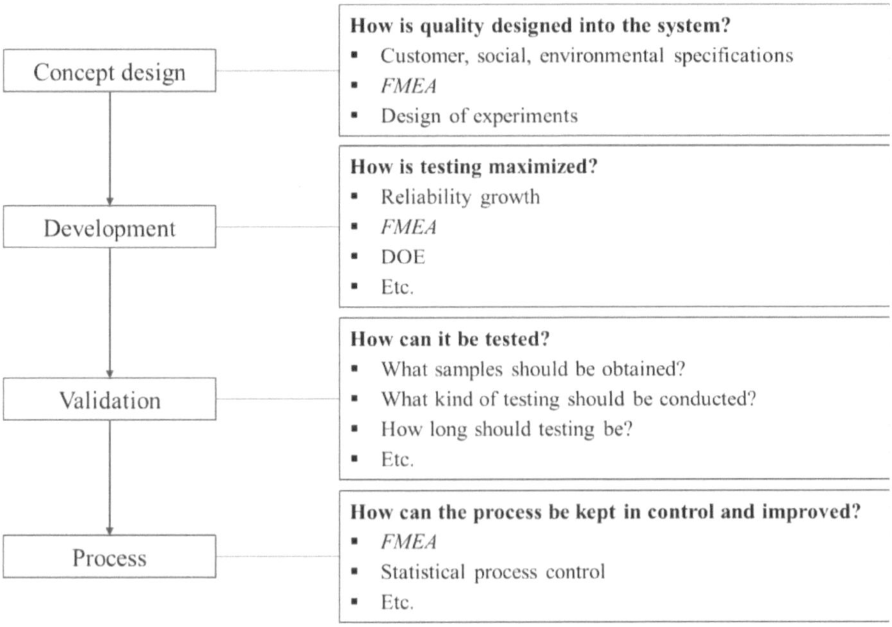

Figure 4.9 Integration of FMEA in different stages of product development according to Stamatis (2003)

FMEA ensures that products fulfil their intended function without any defects. FMEA thus contributes to meeting regulatory due diligence requirements when products are placed on the market. Today, FMEA is considered a standard method in many industries (Müller and Tietjen, 2003; Werdich, 2012). According to Werdich (2012), the approach comprises seven steps:

1. data collection (parts lists, specifications, regulations, benchmark results, lessons learned, QFD, etc.)
2. definition of FMEA scopes and depth of consideration (e.g., by block diagram and part function matrix)
3. functional analysis and linking of functions
4. structural analysis (recording and structuring of elements involved, assigning functions)
5. error analysis (assigning and linking malfunctions)
6. analysis of measures (documentation and evaluation of current status)
7. optimization (with further measures minimize risk and evaluate changed status)

Figure 4.10 illustrates the approach according to Wedrich (2012).

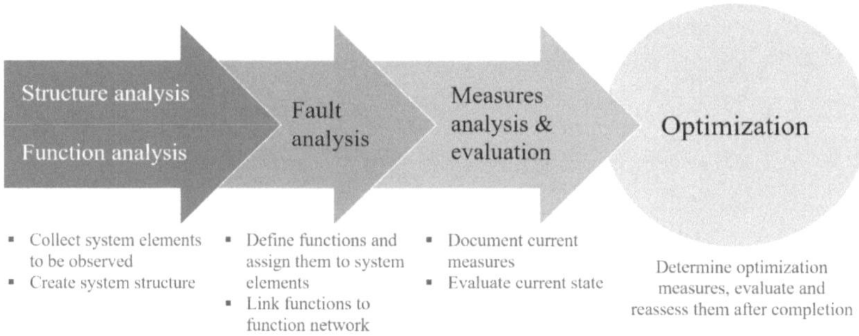

Figure 4.10 Phases of application of FMEA according to Werdich (2012)

According to Hering and Schloske (2019), three FMEA types are distinguished: system, design, and process. *System FMEA* (also product FMEA at system level) analyzes malfunctions in system components and interfaces as well as their effects on the overall system. Avoidance measures usually refer to conceptual safety measures of the system, while discovery measures generally relate to procedures that are applied during verification and/or validation of the system. Typical application fields for system FMEAs are interface examinations of complex systems. In turn, *design FMEAs* (also product FMEAs at component level) analyze whether the selected dimensions and tolerances as well as the selected materials are suitable for reliably fulfilling the required functions of a component over its life time. Avoidance measures generally concern design measures for the component, while discovery measures generally relate to procedures used in the verification and/or validation of the component and/or system. Finally, *process FMEA* analyzes malfunctions that may occur during production and/or assembly. The prevention measures generally refer to safeguards of the manufacturing and/or assembly process, while the discovery measures generally refer to measures to detect the defective manufactured products before they are delivered to the customer. Stamatis (2003) argues that once an FMEA has been conducted, it is never fully complete as long as the product remains on the market. Rather, it is to be understood as a dynamic tool that has to be updated constantly. Figure 4.11 summarizes the procedure and the embedding of FMEA in a dynamic, continuous development process according to Stamatis (2003).

Theoretically, the generic, systems-oriented approach of FMEA enables a transfer and/or linkage with various methods concerning different issues. For instance, the method offers considerable adaptability to integrate social and/or environmental aspects. Approaches to environmental FMEA (E-FMEA) attempt to detect environmental impacts (see Section 6.2) of technical or organizational actions. In this context, Lindahl (1999) presents an approach that links FMEA and LCA. Besides a systematic identification and evaluation of environmental impacts in all lifecycle stages of an investigated product, the approach aims to early detect possibilities for corrective and preventive actions to minimize the environmental burden. In contrast to a regular LCA (see Section 6.2) the approach is to be understood as a less detailed analysis of ecological impacts by only focusing on the most significant problem areas. According to Lindahl (1999), the approach comprises five steps, which are based on previously identified environmental requirements of a product or concept:

1. Identify a products/process life cycle phases (e. g., extraction, manufacturing, transport) as well as all linked activities/events in each phase.
2. Detect general environmental aspects in each stage (e. g., emission to atmosphere, contamination of land, waste).
3. Determine impact (e. g., GHG, AP, EP, ODP) each environmental aspect causes.
4. Evaluate environmental aspects and impacts that are to be regarded significant (e. g., from perspective of customer, society, supply chain, type of product) in three different criterions (S = controlling documents, I = public image, A = environmental consequences) by rating them from 1 to 3 depending regarding compliance previously identified environmental requirements of product or concept. The evaluation results in an environmental priority number (EPN) = S + I + A.
5. Identify, analyze, and prioritize improvement actions by starting with a listing of possible actions, analyze potential environmental benefit and rate them regarding on the effort in time, cost, and technical possibility needed for improving a product or concept (F = improvement).

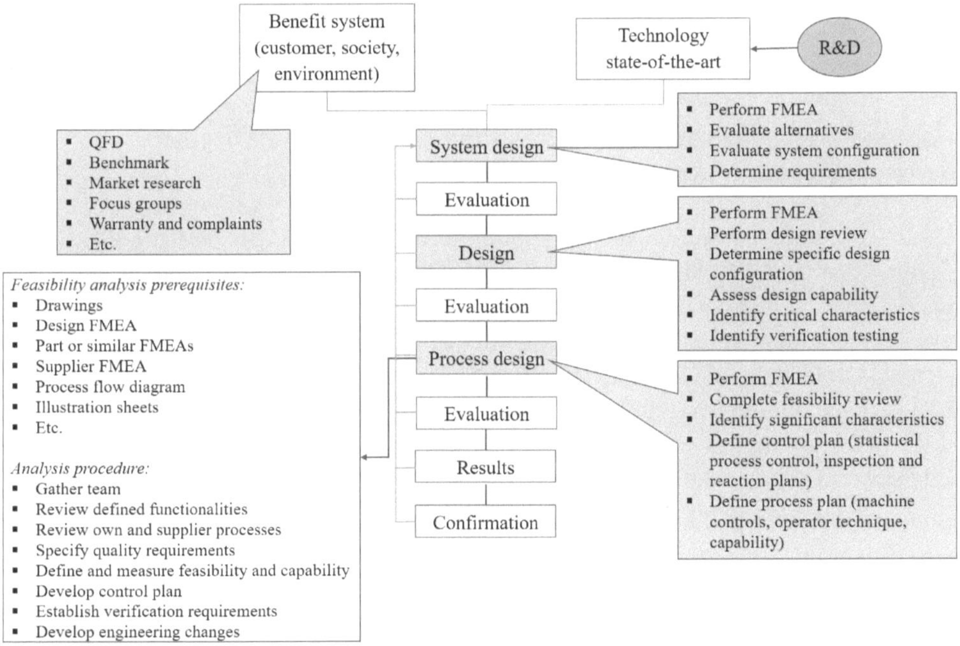

Figure 4.11 Procedure and embedding of FMEA in a dynamic, continuous development process according to Stamatis (2003)

Figure 4.12 exemplifies input and output figures of the E-FMEA approach as well as its application within the product development process according to Lindahl (1999).

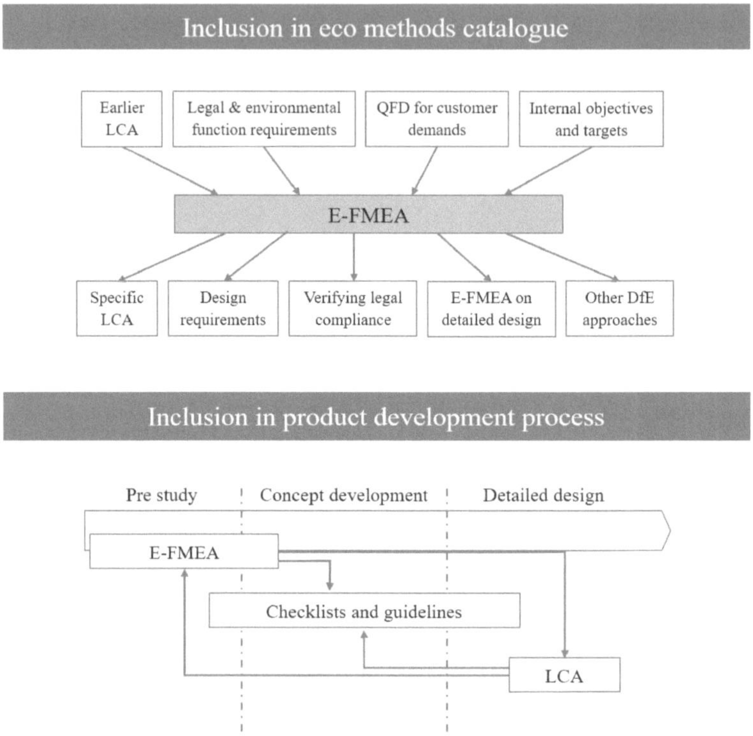

Figure 4.12 Input and output figures and application of E-FMEA in product development process according to Lindahl (1999)

Similar approaches of E-FMEA are, among others, documented by Greshake (2017), Giannakis et al. (2016) and Pulgieri et al. (2011). A related concept, yet from a social perspective, is human FMEA (H-FMEA). Based on the socio-technical systems approach, the aim of H-FMEA is the avoidance of product defects by a user-friendly design of the work system (Figure 4.13). While the procedure is similar to a conventional FMEA, the avoidance measures here focus primarily on aspects of work organization, work tasks, ergonomic working conditions, and personal performance requirements (Algedri and Frieling, 2015; Hofmann, 2015).

Given the high degree of expected complexity of BIS, FMEA represents an essential tool for supporting their development. Thereby, system FMEAs are particularly suitable for biointelligent product development, whereas process FMEAs serve to support the development of biointelligent production processes. To date few examples have been presented in the context of biomanufacturing (Rathore and Sofer, 2012).

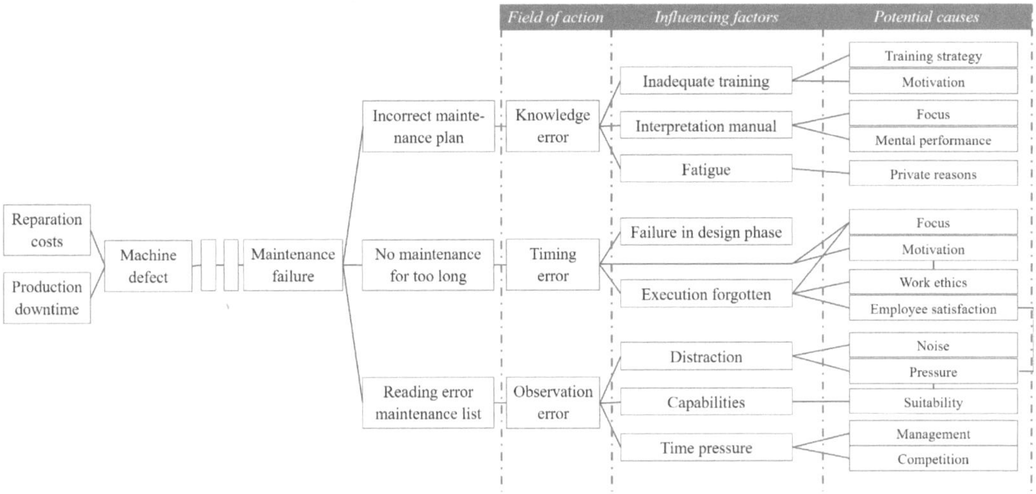

Figure 4.13 Example of industrial application of H-FMEA according to Algedri and Frieling (2015) and Hofmann (2015)

Fault tree analysis (FTA)

A further approach in this phase of product or concept engineering is fault tree analysis (FTA). The system engineering method relies on logical linkages between events in order to determine all possible causes that lead to a specified undesired event (Lee et al., 1985; Kloth, 2010). The concept addresses the problem that for new types of systems no immediate experience exists regarding the probability of failure and/or further risks at the system level. Accordingly, FTA intends to analyze the entire system in individual parts in order to be able to draw conclusions about the behavior of the entire system from the properties of the individual sub-systems, which are often better known. It starts with an undesired event. This can be the negation of a function or the non-fulfilment of a required product feature. Possible causes are then assigned to this top event. For each assigned cause, a check has to be conducted to determine whether it is linked with other assigned causes in an AND/OR operation (Schloske, 2020). The procedure continues similarly in the next levels. This establishes a chain of error-causes from the top event to the original cause (Kloth, 2010). The evaluation of the occurrence probabilities of individual results reveals a critical path that mostly leads to the top event. In order to establish the chain from top-event to causes, a uniform semantic exists (Figure 4.14).

Symbol	Name	Description
<Description>	Event description	Description of an event that describes an error/malfunction, system/component state, condition or action.
<Event> <ID>	Basic event	Fundamental event that cannot be developed further or whose development is not considered relevant.
<Event>	Undeveloped event	Event whose cause has not been further analysed or could not be investigated.
<Event>	House event	Event that describes an expected state (i.e. usually no fault). often serves as a technical tool to set certain states in a fault room by switching on or off.
<Event>	Conditional event	Event that expresses a condition, usually used as an input to a condition or PAND gate, can also be used to identify so-called systematic errors (e.g. of software)
<Event>	Dormant failure	Defect with special characteristics, which usually describes a dormant defect.

Figure 4.14 Symbols of fault tree analysis according to Edler et al. (2015)

The procedure of FTA can be qualitative or quantitative. While in the first case only the definition of the top event and the identification of all event combinations leading to the top event are necessary, the quantitative FTA follows four further steps: [1] determination and assignment of reliability parameters to each base event, [2] system modelling and calculation of the probability of occurrence of junctions and the top-event, [3] analysis of the dominant event combinations and contributions, and [4] development of optimization suggestions. The rules for the combination of the basic events are defined by Boolean algebra (Edler et al., 2015). Boolean formulas consist of a set of variables and the logical operators $\cap$ (AND, conjunction), $\cup$ (OR, disjunction) and $\neg$ (NONE, negation). A variable is always binary, i. e., it can only assume the values true and false. Regarding a fault tree, the events can be represented as variables whereas gates can be represented as logical operators. Given events from E_1 to E_n, the following applies to the gates:

- OR gates with inputs E_1 and E_2 correspond to $E_1 \cup E_2$.
- AND gates with inputs E_1 and E_2 correspond to $E_1 \cap E_2$.
- Decision gates with inputs E_1 to E_n and E_m as threshold: $(E_1 \cap \ldots \cap E_m) \cup (E_2 \cap \ldots \cap E_{m+1}) \cup \ldots \cup (E_{n-m} \cap \ldots \cap E_n)$
- Condition gate with event E and condition C: $E \cap C$
- XOR gates with E_1 and E_2 corresponds to $(E_1 \cap \neg E_2) \cup (\neg E_1 \cap E_2)$

- Non-gate of input event E_1 corresponds to $\neg E_1$
- PAND gates with inputs E_1 and E_2 results in $(E_1 \cap E_2) \cap P$ with $P = \{E_1 \; pre \; E_2\}$[2)]

Figure 4.15 exemplifies an FTA of a fictitious system according to Edler et al. (2015). Further literature on the procedure and the fields of application of FTA is provided by Lee et al. (1985), Suresh et al. (1996), and Kloth (2010), among others.

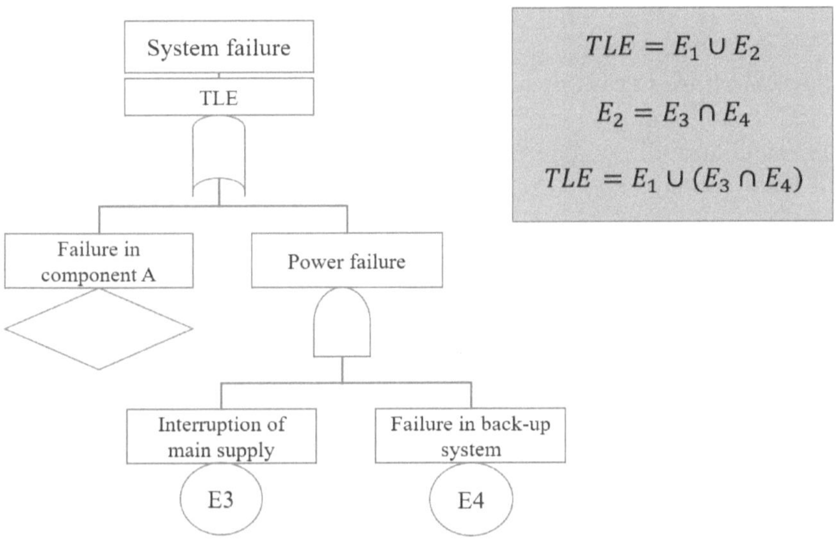

Figure 4.15 Example of FTA of fictitious system according to Edler et al. (2015)

A major advantage of FTA is the possibility to identify combinations of causes as well as failure probabilities and to display them clearly. It enables a system assessment with regard to operation and reliability. It also makes it possible to check to what extent a technical system can comply with specified error rates. As the method is often used in cases where severe damage to technologies, people, and the environment is likely to occur in case of a system failure, a variety of examples exist which illustrate how to contribute to a sustainable design. Applications in the chemical industry focus, for example, on avoidance of atmospheric dispersion of toxic substances, fires or explosions, as well as surface or ground water pollution (Hungerbühler et al., 1999). Another example is the application for assessment of power system reliability (Volkanovski et al., 2009).

The expected greater integration of biological systems in the context of biointelligence leads to two conclusions. On one hand, the overall system behavior of novel BIS is difficult to predict, since they may exhibit an increased degree of emergent behavior in addition to their mere novelty. Furthermore, essential biological relationships

2 Note: the PUND gate is described by an auxiliary construction via an explicit condition. It can only be described precisely with an extended logic in the predicates of the form 'A before B'.

may be unknown. On the other hand, it may be reasoned that the integration of biological systems into previously mechatronic and/or cyber-physical systems initially results in system behavior becoming increasingly unpredictable. The first case indicates an increased need for the use of FTAs in developing BIS, while the latter reveals that further development is necessary to ensure adequate use. In this context, Siontorou and Batzias (2014) explore the applicability of FTA in the context of biosensor systems by developing a methodological framework for R&D 'road mapping'. Thereby, biosensors based on ion selective field effect transistors are validated as a case study and applied prospectively, focusing on feasibility issues, using FTA. The results suggest that future trajectories are related to nanotechnology and in particular to nanofabrication and nanobioinformatics. Plus, it reveals that future research has to be conducted on the control of the natural process of self-assembly and the thermodynamics of bioelement-lipid interaction.

Substance and material risk assessment (SMRA)[3)]

Instead of ensuring functionality, compliance with relevant chemical and waste regulations requires a risk assessment of the suppliers and materials used at this stage of product development. Here the term risk refers to the probability of failing to be compliant. According to Miehe et al. (2015), this requires a four step approach. First, the status-quo of materials and substances has to be evaluated. Essentially, five feasible strategies for material evaluation exist: [1] experience values, [2] supplier declaration, [3] material declaration based on standards, [4] manufacturer information (e.g., material data sheet) or analysis (e.g., XRF screening) or [5] full material declaration. In case companies are not able to independently clarify materials applied on the basis of own research and/or testing, it is essential to establish communication with their suppliers. On this basis the identification of material risk can be conducted, loosely based on the steps of an LCA (Figure 4.16).

The definition of goal and scope first requires the grouping of material classes. As most companies already sort materials in internally agreed upon classes, this step is only mandatory if a classification has not yet been established. Yet, the usage of substance, material, and component types greatly differs between industries as well as within a single sector. IEC 62.474 and IEC 62.321-2 provide common classifications for the EEE sector. Particularly with regard to the increased use of biological substances and materials, practicable classification systems have to be developed and applied in different industries. Inventory analysis generally comprises two steps: initial evaluation of material classes and detailed evaluation of individual materials. The initial evaluation of material classes is conducted on the basis of empirical values. In the case of the EEE sector one option is to use IEC 62.321-2. Other sources might be analyses releases by government agencies. The initial rating serves to quickly identify high risk substances and materials. As the initial evaluation aims at defining recommendations for a de-

[3] This section is a summary of the findings from Miehe et al. (2015).

tailed rating, a categorization of material categories in three levels of urgency is advisable: high urgency (if many cases of non-compliance have been reported within this material class), medium urgency (if a justifiable number of cases of non-compliance has been reported within this material class) and low urgency (if no cases of non-compliance are known within this material class). Thereupon a detailed evaluation of each individual substance and material has to be executed. According to Miehe et al. (2016c), the following information is of capital importance: material/article number and denotation, frequency of use and/or quantity of application in different modules, use of standards/list of standards per product, confirmation of compliance with relevant regulations/use and duration of exemptions of regulations, as well as sources of information. Due to a high number of materials and substances in complex products, the focus is initially set on the previously identified high-risk material classes before all other classes are evaluated. Thereby, each material and component is analyzed in detail and apportioned in chemical components. Accepted methods for identification of chemical composition are technical standards, material declarations being linked to specific material data sheets, and analytical test results (e. g., X-ray fluorescence or wet chemistry analysis). The quantity of required documents should be based on the previously identified degree of urgency (L, M, H), e. g., one document suffices for L-materials, while at least two documents with one being a technical analysis should be gathered for a material of high urgency. This method is, however, associated with great efforts but allows complete coverage of all relevant legislations and allows long term validity due to its overall transparency. Against the background of the discussed exponential increase of environmental legislations only an integral material evaluation facilitates long term validation. The third step is impact assessment. Therefore, the previously identified risk rating is multiplied with the impact of non-compliance. Thereby, indicators such as frequency of use or quantity of application of a single material in different modules represent useful references for the impact of non-compliance. In order to visualize the risk and impact assessment, a risk matrix can be developed (Figure 4.17).

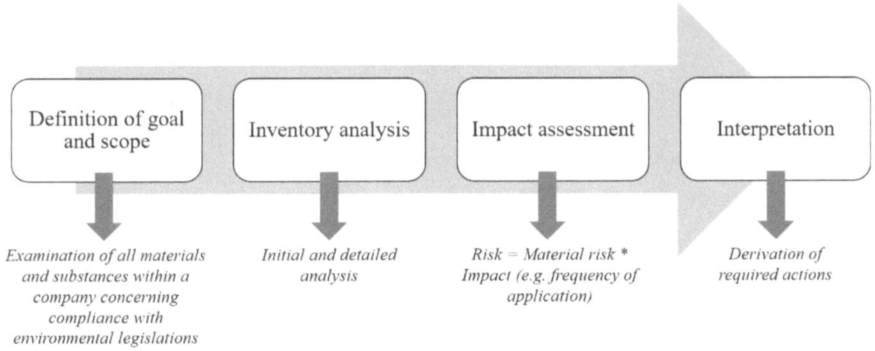

Figure 4.16 Four step approach to substance and material risk assessment according to Miehe et al. (2016c)

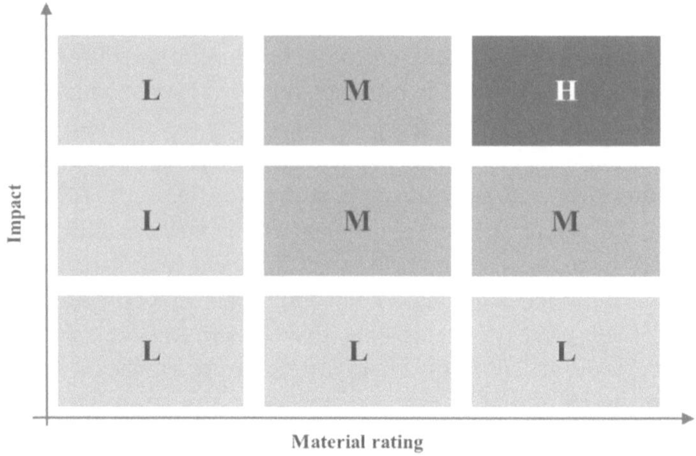

Figure 4.17 SMRA risk matrix according to Miehe et al. (2016c)

In a final step *specific actions* are derived for each of the categories. While for materials from low-risk categories a single supplier/manufacturer declaration is satisfactory, additional control sample analysis (e. g., XRF) are appropriate. Substitution or continuous analytical testing is advisable for materials and components from high-risk categories. The SMRA concept can easily be coupled with other approaches. Hua et al. (2009) present an example that uses the FMEA method described above for application in the context of SMRA.

Substance and material criticality assessment (SMCA)[4)]

A related analysis, which is particularly relevant especially in the context of biointelligence, is material criticality assessment. Instead of focusing on compliance with existing regulations (SMRA), SMCA aims to identify potentially critical factors that endanger the supply and/or use of a substance or material within a specific time horizon (Miehe et al., 2016c). Reasons for such factors can arise both in upstream and downstream stages of the supply chain. Depending on the specifics of the supply chain, various critical factors of each substance and material may occur. Each substance has certain applications in industrial products. While products become more and more complex, so do its components. With higher complexity of products comes higher complexity of supply chains.

The term substance and material criticality used here is closely related to the concept of resource criticality, yet should not be confused with it. In order to fully understand the unique approach of SMCA in an entrepreneurial environment, it is essential to explore this topic in depth. After early analyses were conducted in the first half of the 20th century in the context of both world wars to determine the strategic importance of certain minerals to warfare (Haglund, 1984), the term resource criticality has evolved

[4] This section is a summary of the findings from Miehe et al. (2016c).

to the standard designation for raw materials that exhibit great supply risks (Erdmann and Graedel, 2011). Today, resource criticality is commonly understood as the product of vulnerability and supply risk. The first study that provided an aggregation of both aspects was published by the US National Research Council of the National Academies (US NRC) (2007). The hereby presented analysis of supply risk and economic importance to the US economy was based on qualitative expert knowledge. As a tool for visualization, the US NRC used a Cartesian coordinate system, referred to as the criticality matrix. Table 4.6 illustrates the approach according to the US NRC (2007).

Table 4.6 Criticality matrix according to US NRC (2007)

		Supply risk		
		Low	*Medium*	*High*
Vulnerability	*Low*			
	Medium			
	High			

Subsequently, the approach was applied to various studies globally with only minor adjustments (European Commission, 2010a; Erdmann et al., 2011; Buijs et al., 2012; European Commission, 2014). While applying a similar approach, each study considers different criteria in order to quantify economic importance and supply risk. To include environmental aspects, Graedel et al. (2012) add a third dimension. Thereby, supply risk is derived from geologically economic, geopolitical, social, and regulatory aspects. In contrast, Tuma et al. (2014) divide the criticality analysis in economic, ecologic, and social aspects ab initio. Environmental aspects may be derived from impacts of resource use and/or damage to environmental conversation subjects, social aspects may include human rights, forced and child labor, corruption, discrimination etc. Yet, a simple transfer of these findings to corporate level is difficult, as company requirements may significantly differ from macro and meso level studies. For instance, a range of 100 years of a certain resource will not affect a company that acts in strategic planning intervals of 10 years max. Also, a general technological substitutability of a substance in a certain sector (e. g., automotive) does not provide evidence if a replacement is precluded for a certain product within a company. National level studies either calculate or estimate material flows. The degree of information regarding material/substance flows and transparency, however, declines considerably on the business level. Table 4.7 illustrates a comparison of national level studies and business requirements in order to identify intersections and gaps.

Clearly, product development decision-making in a company is not very well suited to existing resource criticality analysis at national level. This is where SMCA is applied. Thereby, substance and material criticality are understood as the risk of damage to a certain unit of analysis due to the utilization of certain substances and materials. Organizational damage hereby is expressed by a negative variation of factual and/or formal goals (e. g., failure of customer satisfaction via specific product development, denial of market access). The unit of analysis may either be a product, process, division or organization. The term criticality describes a violation of a certain limit of breach of one or more particular targets a company is willing to accept. Conversely: if a certain limit is violated, a substance or material becomes critical, e. g., a deviation of 5% of profit objective due to the price increase of a certain material. Similar to SMRA, SMCA is loosely based on ISO 14.040 and 14.044 (Figure 4.18).

Table 4.7 Comparison of national level studies and business requirements according to Miehe et al. (2016c)

	National level	Business level
Gaps	General scarcity of a resource depending on occurrence in earth's crust, technological capacity and degradation rate	Scarcity of a resource regarding strategy, product life span, time of stock-keeping, etc.
	General calculation and/or estimation of resource flows within a certain system (e. g., nation, sector)	Concrete resource transparency regarding certain products and processes
	General technological substitutability within a certain system (e. g., nation, sector)	Concrete substitutability regarding a certain technological application
	General market situation and development of a certain resource	Concrete supplier and customer market situation and development of a certain resource
Intersections	Adverse economic and/or social effects during life cycle of a certain resource	
	Concentration of reserves	
	Mine capacity	

Figure 4.18 Five step approach to substance and material criticality assessment according to Miehe et al. (2016c)

Definition of goal and scope: The first step sets the basis for an operational analysis. The aim of the approach is always exactly the same: to assess the danger to a company's essential performance requisition ('to provide a customer with the desired product in the required quality and time'). For use in product development the unit of analysis is always the product.[5] A reasonable approach is to use existing time frames a company has applied related to the unit of analysis, e.g., strategic horizon, planned product life span. Furthermore, three premier parameter of analysis that set the basis for a deployment of tailored indicators have to be differed: price, time and technology. The price represents the realized value of a product. As such, it is the value a customer is willing to pay. It should cover all previous value adding process steps as well as additional requirements (e.g., supporting processes, compliance with legislations and standards). The nature of this approach requires the assumption that possible environmental and/or social problems that occur during upstream processes and are regulated by law thus will be imparted to the customer. Time, however, depicts a different parameter of criticality assessment. Here, a company's position within the supply chain partly determines the short-term dependence on a certain substance or material. If a company for instance takes a spot at the end of a supply chain, it will thereinafter notice the effect of a potential change of supply. Technology, not least, depicts another source of criticality for product development. This aspect essentially covers functionality and substitutability. Non-supply of a certain resource might lead to impossibility of fulfilling certain technical functions. For a company it is necessary to develop a consistent weighting system of these three parameters in regard to its previously defined unit of analysis. For instance, the removal of direct access to a certain resource may initially lead to a higher price as well as a delay in delivery. The degree of criticality, however, increases significantly with substitutability. If a substitution is not possible, the fulfilment of a company's performance requisitions is hampered for long.

Inventory analysis and vulnerability assessment: The second step combines two aspects. First, an inventory analysis has to be conducted regarding the unit of analysis. Inventory, in this context, refers to the utilized materials and devices. The basis for this step is a classification of materials depending on the scope of analysis. For product development, the approach to classification described above in the context of SMRA can be applied. In a best possible scenario, the chemical composition for all used materials and devices can be identified. Realistically, information on chemical compound of devices is scarce. In this case the degree of information as well as empirical value may serve as alternatives. Therefore, sources of information gathering have to be determined. Similar to SMRA, these are standards, bill of materials, ERP-data, screenings (e.g., XRF, IR), expert knowledge, experience values, etc. The result of this step is an inventory table with the chemical composition of each relevant mate-

5 As the methodology can also be applied to other entities, other subjects are also possible (process, segment, company, etc.).

rial and device. Hereupon, the vulnerability of the analyzed unit can be determined. Based on the inventory table, an overall quantity and price of a certain resource can be calculated. In addition, an analysis of substitutability of the applied resources determines another source of vulnerability.

Supply risk assessment: The term supply risk describes the probability of supply shortfalls. In this context, a distinction into upstream and downstream system is required. The upstream system refers to the direct supply with the required substance and/or material. Their price, however, depict snap-shots. Potential future developments may not easily be derived from a simple price analysis. The upstream perspective thus solely focuses on the correlation of supply and demand, which determines the fulfilment of a company's performance requisitions. Hereby, two modes may be classified as critical to an organization. Either the offered substance or material on the market does not suffice to meet a company's demand regarding quantity and/or quality. In this mode, the manufacturing of a good is not possible. Then again, the quantity and/or quality of a substance or material a company has direct access to is subject to volatility. In this mode the manufacturing of a product is still possible although under inferior conditions. Ecological and/or social factors are not observed within the upstream system. It is assumed that possible changes will either lead to price adaptions within the supply chain or will be approached to the company from the downstream system via customer refusal or law.

- Mode 1 (no supply) has to focus on both primary and secondary sources of supply. The former is determined by the existence of sources in earth's crust and the existing technology. Possible indicators are static and dynamic range of feedstock as well as mine capacity. This again affects the type of manufacturing (e.g., coupled production) and the concentration of either sources or tiers within the upstream system. Here, especially uncertain events may cause a temporary stop of supply. As an indicator the Herfindahl-Hirschman-Index (HHI) may be adducted. The recycling rate of a resource represents an indicator for a secondary source of supply.
- Mode 2 (limited supply) requires a market analysis of supply and demand of each material. Hereby, an investigation of both supply and demand regarding its composition and development is necessary. Market composition of both may be differed in monopolistic (single supplier/consumer market), oligopolistic (few supplier/consumer market) and polypolistic (multi-supplier/-consumer market). The comparison of both determines the actual market type, as illustrated in Table 4.8 according to Stackelberg (1993).

Supply and demand development may be classified qualitatively as increasing, constant or decreasing. A comparison of both may be adducted as an indicator of critical market conditions. Table 4.9 illustrates the lineup of supply and demand development. Hereby, little (L), medium (M), and highly (H) critical market situations are differentiated.

Table 4.8 Market composition according to Stackelberg (1993)

		Supply		
		Polypolistic	*Oligopolistic*	*Monopolistic*
Demand	*Polypolistic*	Perfect competition	Supply oligopoly	Supply monopoly
	Oligopolistic	Demand oligopoly	Bilateral oligopoly	Limited supply monopoly
	Monopolistic	Demand monopoly	Limited demand monopoly	Bilateral monopoly

Table 4.9 Market development criticality according to Miehe et al. (2016c)

		Supply		
		Increase	*Constant*	*Decrease*
Demand	*Increase*	M	M	H
	Constant	L	L	M
	Decrease	L	L	M

Concerning a potential harmful development of supply and demand for a single product, the entire supply chain has to be analyzed regarding market type and development. In this context, a critical business situation occurs if a company's demand cannot be met by the supplied quantity and/or quality. The probability of occurrence of that condition increases with the degree of imbalance between supply and demand. Supply monopoly and oligopoly depict the highest form of criticality. Here, suppliers may easily establish an overly high prize or refuse access. A medium stage of criticality occurs in cases of perfect competition, bilateral oligopoly and bilateral monopoly, while demand monopoly and oligopoly may be classified as comparatively low risk.

As a supply stoppage of a certain resource may not only occur due to depletion, concentration, natural disaster etc., but also due to social issues, the downstream system analysis focusses on societal perception regarding ecological and/or social problems that may occur during the process of value adding, product usage, and end-of-life. The greatest critical factor that can occur in this system is that customers refuse purchase of a certain product. As an indication for this perspective an observation of medial reporting of potential harmful developments regarding the applied substances and materials is necessary. A solid approach is the development of so-called diffusion curves of environmental and/or social problems. Various authors indicate that every environmental and/or social problem follows this trend (Steger, 1993; Burschel et al., 2004). For companies the level of medial presence represents a possibility to derive future risks based on the utilization of certain resources. Therefore, the diffusion curve has to be grouped in different levels of criticality. The closer the medial presence approaches the end of the curve, the higher the criticality of a certain substance or material is. Figure 4.19 exemplifies a diffusion curve for a fictitious problem according to Steger (1993).

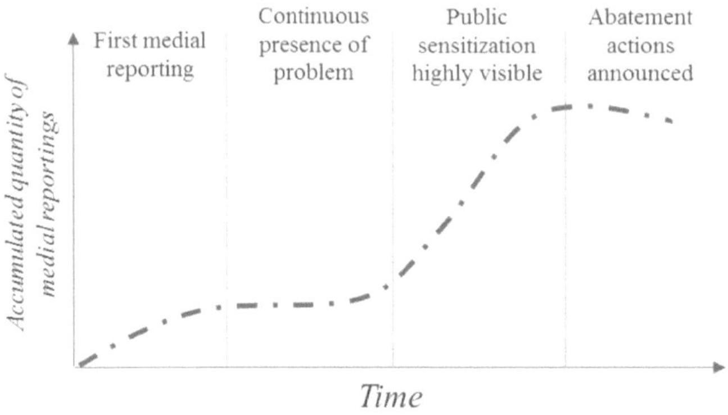

Figure 4.19 Diffusion curve of environmental problems according to Steger (1993)

Criticality assessment: Following the common definition (National Research Council of the National Academies, 2007), the criticality of a single substance or material (*i*) results from the product of its supply risk (*s*) and its vulnerability (*v*), i. e.

$$c_i = s_i \cdot v_i$$

The total criticality (*C*) of the unit of analysis hereupon may be calculated as the sum of all single resource criticality values, i. e.

$$C = \sum_{i=1}^{n} c_i$$

Derivation of improvement actions: Based on the previously defined limit of breach, it is now possible to determine which substance or material of a product is potentially critical. In cases of breach violation, improvement actions have to be conducted. As various aspects of criticality evaluation are based on estimations or assumptions, a sensitivity analysis is advisable in order to identify appropriate operating levers. Thereupon, a variety of actions exist in order to enhance the long-term resource criticality of socio-technical systems. To mention only a few, we suggest investments in R&D, changes of suppliers, insurance protection, installation of backup capacity, review of stock-keeping etc. Figure 4.20 summarizes the SMCA approach according to Miehe et al. (2016c).

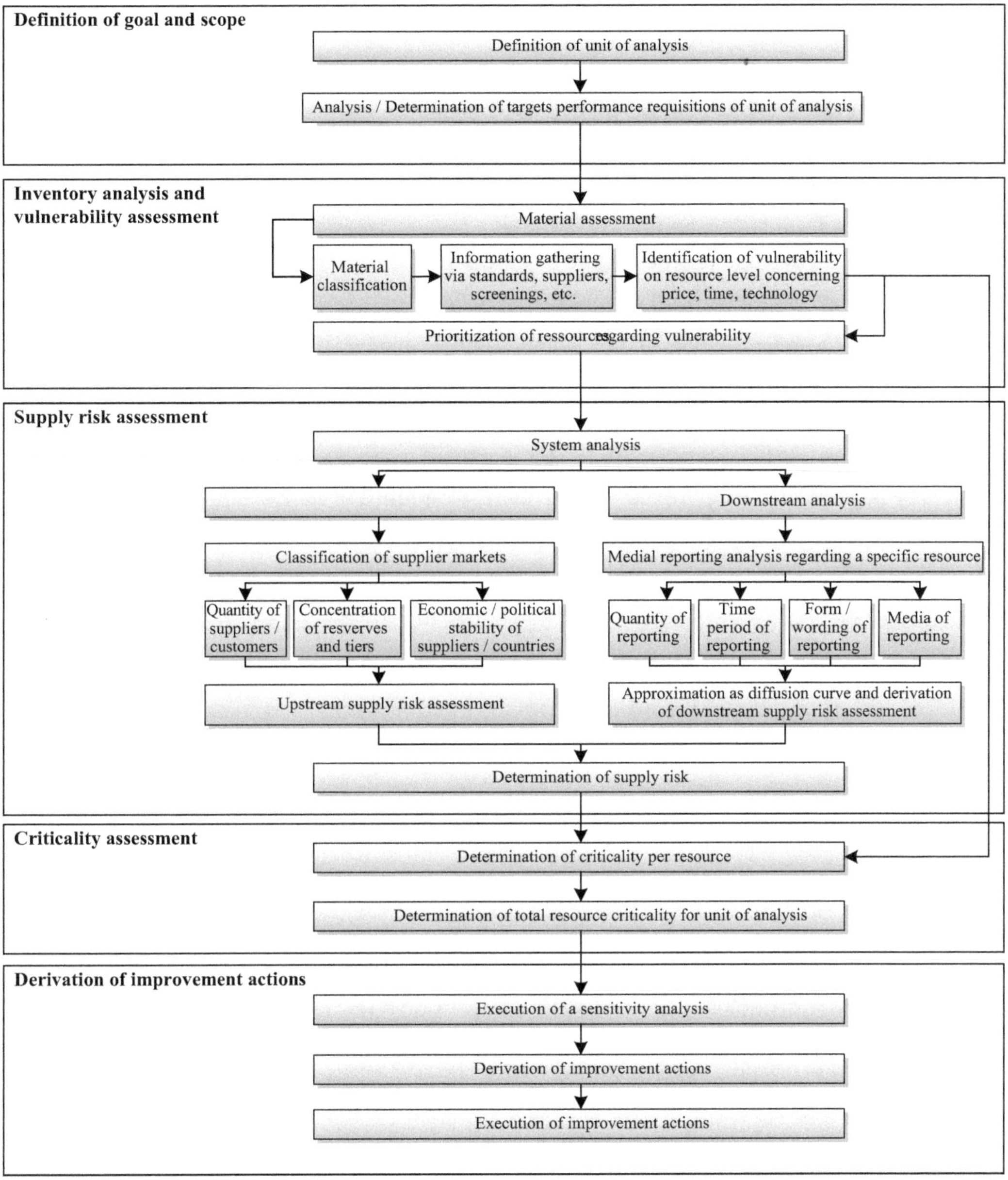

Figure 4.20 SMCA process according to Miehe et al. (2016c)

4.6 Prototypes and testing

Design of experiments (DOE)

Once a prototype has been produced, extensive testing is required before market release. In this context, DOE is applied to ensure that the experiments are planned statistically sound. The approach is based on the theory of statistical inference developed by Peirce at the end of the 19th century (Peirce and Jastrow, 1885; Stigler, 1978). In contrast to classical experimental design, where only one influencing factor is changed at a time, DOE seeks to change several influencing factors at the same time (Box et al., 2005; Montgomery, 2012; Dean and Voss, 2017). This significantly reduces the number of experiments to determine the relevant influencing parameters and the optimal parameter combination. Yet, the probability of misinterpreting results increases as well. According to Durakovic (2017), DOE is basically an application of linear regression that comprises seven steps for planning an experiment:

1. Define objectives, i. e., list problems that are to be investigated.
2. Define response variable, i. e., measurable outcome of experiment that is based on defined objectives.
3. Determine factors and levels, i. e., identify factors that may affect response variable (e. g., via fishbone diagram).
4. Determine appropriate experimental design type as well as number of test samples.
5. Perform experiments using design matrix.
6. Analyze data using statistical methods such as regression or ANOVA.[6)]
7. Develop practical conclusions and recommendations.

According to Durakovic (2017), selecting appropriate factors (factorial design) represents the most difficult part of DOE. Eldin (2011) describes the approach as follows: factorial design aims to identify the influences of all experimental variables (i. e., factors) and interaction effects on the response variable. If a combination of k factors is investigated at two levels, the factorial design consists of 2k experiments. Figure 4.21 exemplifies factorial designs for 2, 3 and 4 experimental variables according to Eldin (2011). The values of the factors are indicated by – (minus) for low level and + (plus) for high level. Also included is the value zero, a center where all variables are set to their mean value. Eldin (2011) recommends a number of 3 to 4 center experiments in the factorial design in order [1] to minimize the risk of nonlinear relationships in the middle of the intervals and [2] to determine confidence intervals by repetition. The

---

[6] Detailed descriptions of procedures for analysis of variance (ANOVA), linear regression as well as factorial design can be found in Durakovic (2017).

assumed reasonable variation to be studied determines what should be considered as – and + for each variable. In this way the size of the experimental space can be defined. For two and three variables, the experimental domain and the experimental design can be easily illustrated. The former represents the corners of a square, whereas a design with three variables results in a cube (Figure 4.22). More than three variables cause difficulties in geometric representation.

Two variables			Three variables				Four variables				
#	*Variables*		#	*Variables*			#	*Variables*			
	X_1	X_2		X_1	X_2	X_3		X_1	X_2	X_3	X_4
1	-	-	1	-	-	-	1	-	-	-	-
2	+	-	2	+	-	-	2	+	-	-	-
3	-	+	3	-	+	-	3	-	+	-	-
4	+	+	4	+	+	-	4	+	+	-	-
			5	-	-	+	5	-	-	+	-
			6	+	-	+	6	+	-	+	-
			7	-	+	+	7	-	+	+	-
			8	+	+	+	8	+	+	+	-
							9	-	-	-	+
							10	+	-	-	+
							11	-	+	-	+
							12	+	+	-	+
							13	-	-	+	+
							14	+	-	+	+
							15	-	+	+	+
							16	+	+	+	+

Figure 4.21 Examples of factorial designs for 2, 3 and 4 experimental variables according to Eldin (2011)

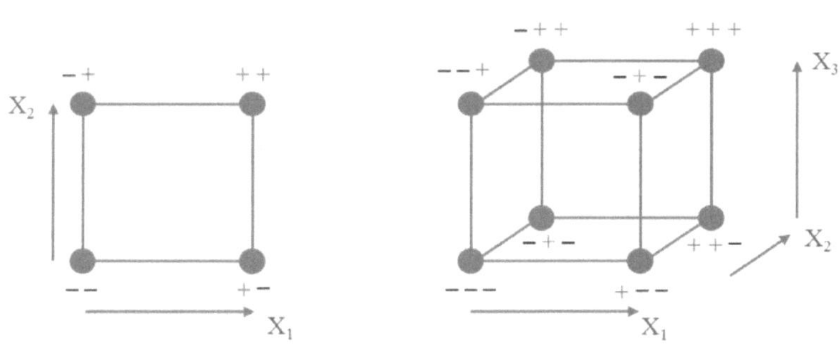

Figure 4.22 Comparison of geometric models for a factorial experimental design with two or three factors according to Eldin (2011)

The concept of special characteristics (SC)

Designs and prototypes need to be constantly reviewed for compliance with so-called special characteristics (SC) during the development process. The SC concept originates from the automotive industry. According to the German automotive industry association (VDA) (2020), the conventional interpretation of SC includes product or process features whose non-fulfilment or non-compliance with the requirements can have an impact on the following three criteria via plausible causal chains:

- SC S: *safety requirements*, whose violation can result in immediate danger to health or life. In the case of a vehicle examples of these are loss of road visibility, brake failure or uncontrolled drive.
- SC L: *legal and regulatory requirements* of the product relevant for approval. In the case of a vehicle examples of these are the locking system, vehicle emissions or recycling aspects.
- SC F: *functional requirements*. In the case of a vehicle examples of these are essential functional requirements such as 4F (form, fit, function, performance) or technical production requirements.

Like all conventional product development tools, the approach of the VDA (2020) focuses on the best possible fulfilment of the requirements of the consumer or market (individual benefit system). Simultaneously, requirements from the social benefit system, such as laws and standards, can be included in this interpretation of SC. However, the incorporation of requirements from the environment (ecological benefit system) requires additional thought. Future development processes should thus consider a further category of SC:

- SC E: *environmental requirements*, whose non-fulfilment might result in immediate and/or long-term negative effects to the environment. Requirements of this kind go beyond mere legal regulation and strive for the best possible technical design and the use of state-of-the-art environmental technology.

While decision-makers (e. g., designers and managers) may identify SC at an earlier stage of product development (e. g., in requirements or technical specification sheet), its assessment relies on methods such as AHP, FTA or FMEA. Simultaneously, fulfilment of SC requires a process of constant review, as proposed by Fitzgerald et al. (2007) (Figure 4.23).

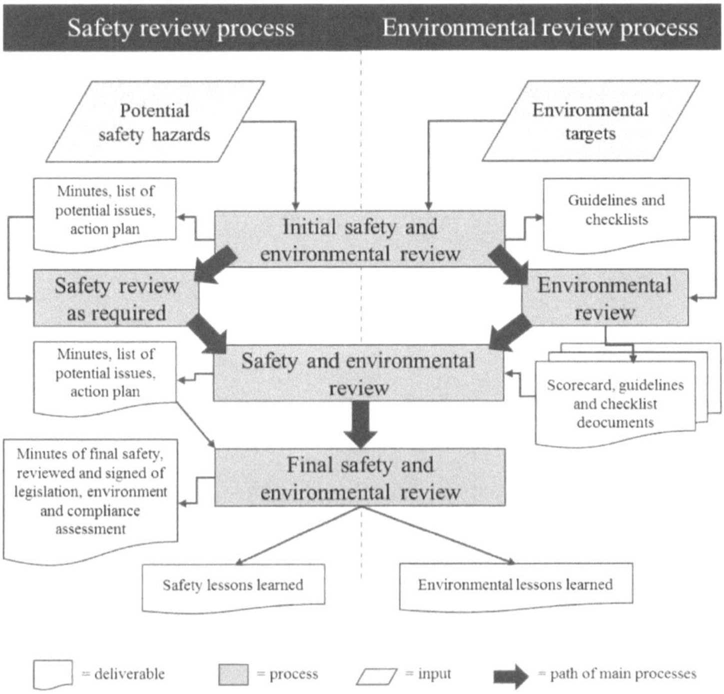

Figure 4.23 Combined safety and environmental SC review process according to Fitzgerald et al. (2007)

4.7 Verification and validation

Before market release, further steps of the V-model must be completed (Schloske, 2020). These are essentially verification and validation. First, it is necessary to verify the required specifications from the specifications book (verification). This usually involves laboratory testing. As soon as evidence of compliance with the technical specifications under laboratory conditions is provided, a validation under realistic conditions is conducted. Its purpose is to examine whether the requirements of the benefit systems (requirements specification sheet, e. g., in form of BDC) are met. While today customer testing suffices for a validation, the three dimensional BDC approach presented here requires the involvement of third parties that represent the societal and environmental benefit system. For instance, this could take place in form of a critical review by non-governmental organizations, consumer protection agencies, government bodies, etc. Extensive testing of this kind also enables the company to avoid unnecessary risks, which may appear after market-release and whose elimination could lead to disproportionately high costs. Once the requirements of the three benefit systems have been met by the developed product, market release can begin.

Annex 5: Examples of integrated production management

Apart from providing methodological support for the development of BIS, the management of BIS equally requires a transfer of the above outlined systems-oriented life-cycle thinking approach into industrial practice. Of vital importance here is the increasing integration of social and environmental aspects of corporate performance. The following sections thus summarize a number of selected existing integrative approaches of individual management concepts, which are of great importance for a future sustainable micro level economy.

5.1 The basics of management systems in manufacturing companies

As discussed in Chapter 2, the concept of production management is primarily based on Taylor's scientific management approach (Taylor, 2001), which led to the development of operational production science and factory management theory. According to Peaucelle (2015), early reflections on the structuring of the management originate from Fayol, who for the first time introduced the essential activities of management (planning, organizing, commanding, coordinating, and controlling), which are universally accepted today. Current industrial practice relies on a multitude of different separate management approaches exist, each addressing selected or all of these aspects from the perspective of a specific business challenge. Examples are strategic, innovation, production, quality, and environmental management. From the perspective of production, holistic production systems (HPS) essentially based on the example of the Toyota production system (TPS) are particularly important. However, the term 'holistic' is misleading in the context of this work, as these systems primarily focus on ensuring the most efficient and stable production processes possible. Figure 5.1 illustrates the range of selected management systems in the context of essential internal value creation processes.

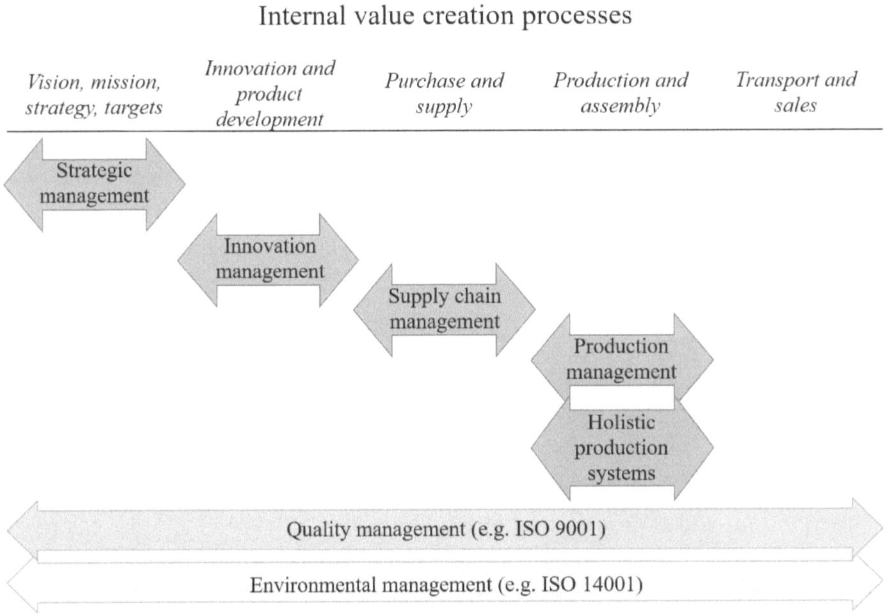

Figure 5.1 Range of selected management systems [own illustration]

While all of the approaches described here are equally relevant in the context of biointelligence, e. g., to ensure sufficient quality of the manufactured products, the here intended sustainability-oriented governance requires a greater focus on ecologically relevant concepts and approaches for their integration into conventional management. The following sections therefore primarily address environmental management as well as eco lean management (ELM). This, however, requires a basic review of lean management (LM) first.

5.2 Lean management (LM)

Today, LM is a wide-spread approach in the manufacturing industry. However, its roots date back about a century. While Europe benefited massively from the financial support of the USA in the period after WW2 in the course of the European recovery program (Marshall plan), Japan received far less support. Lacking financial support, Japanese companies were initially unable to keep up with western engineering. Firms thus needed alternative solutions. The Japanese government initially supported various sectors of the economy by almost completely excluding them from world trade. This guaranteed companies time to reposition themselves. The slow opening of the Japanese economy around the early second half of the 20th century quickly led to a Japanese economic boom, especially associated with one company: Toyota. Founded in 1937, the company is now considered one of the world's largest automobile manufacturers. However, even before the actual founding of the company, father Sakichi and son Kiichirō developed Toyoda spinning machines and looms. Besides the mere technological development, the Toyotas always tried to improve the production process. Thus, autonomous automation (jp. *Jidōka*) of production was already envisaged in 1924 as part of the development of an automated weaving machine. From the end of the 1920s, the family began to promote automobile production. In a Japanese automobile market that was poorly developed until the 1960s, the rise of the company began, which is closely associated with two names: Taiichi Ōno, the former production manager from 1950, and Eiji Toyoda, the president of the group between 1967 and 1982. While the latter is primarily known for his innovation-promoting corporate policy, the former is considered the inventor of the Toyota Production System (TPS), which is now known worldwide. After a further training trip to the USA with the then leading manufacturers Ford and General Motors, Ōno adapted essential principles of flow production to the Japanese way of doing business and gave its employees more freedom to exert influence. Subsequently, he developed various principles and systems (including kanban, just-in-time, Kaizen) that are now used worldwide. Ōnos creations led to an enormous increase in the company's production rates and quality standards.

For a long time, the reasons for Toyota's success remained hidden from its competitors, until Womack et al. (1990; 2003; 2015) published the essential principles of the Toyota production system under the synonym LM in the early 1990s. The supremacy of Toyota's production method was mainly the result of its lean production processes compared to western systems. While western car manufacturers produced items in stock, i. e., accumulated stocks of semi-finished and finished goods in order to be able to react quickly to a sudden increase in sales, Toyota aligned its system with the customer's wishes. Whenever European and U. S. companies had problems when the customer did not want exactly what had already been produced, Toyota was able to react flexibly to any changes in demand and wishes by reducing the throughput time of the produced goods to a minimum. The central lever of this approach is the avoidance of any activities for which the customer is not willing to pay. The avoidance of waste at Toyota was based on two pillars: just-in-time manufacturing design (pull principle) and autonomous automation (jp. *Jidōka*). Its foundation was continuous improvement (jp. *Kaizen*). Figure 5.2 illustrates the TPS according to Röhm et al. (2020).

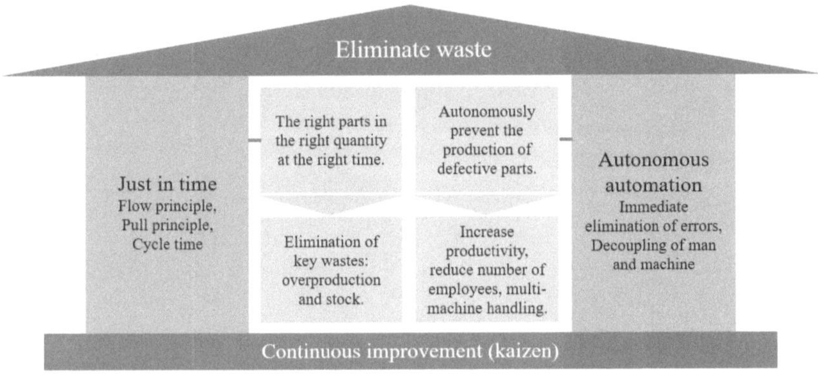

Figure 5.2 Toyota production system according to Röhm et al. (2020)

The discovery of the Toyota way resulted in a rethinking of entire industrial sectors. Today, LM is applied in almost every car manufacturer and supplier. Many of these companies (incl. Daimler, VW/Audi, Festo, Trumpf, ZF) have developed their own production system adopting the Toyota thinking. In addition, companies in other industries are increasingly adapting the approach as well.

In addition, the concept has been repeatedly adopted by the scientific community in order to gradually enhance its principles, methods and approaches. Today, the approach comprises a worldwide community of consultants and scientists as well as a variety of procedures, methods and best practice examples. In the following, only the very essence of lean management will be briefly summarized. Further basic literature on LM can be found in Womack et al. (1990; 2003; 2015) and Pfeiffer and Weiß (1994), among others. First, the perception of waste in production is decisive, i. e., the answer to the question for which activities the customer is not willing to pay. In classic LM, seven types of waste are distinguished: transport, inventory, motion, waiting, over-engineering, over-processing, and defects (TIMWOOD) (Womack and Jones, 2003). According to Pfeiffer and Weiß (1994) the same procedure is to be applied in any case: [1] analysis of current state of value adding of each activity, [2] comparison with target state, and [3] elimination/reduction of all non-value adding activities. This is also partly reflected in the plan-do-check-act principle (PDCA) of continuous improvement (jp. *Kaizen*), which was formulated in LM for the first time and is now included in almost every management principle. Its idea is that an operational improvement should always be implemented according to the same pattern: develop objectives and implementation approach (plan), execute implementation (do), monitor and measure implemented aspect regarding objectives (check), and continuously adapt and improve towards objective (act). The principle is designed as a circle, which is intended to start over and over again. Thereby, two dimensions are differed: improvement and performance. While performance describes the current value of an indicator, improvement is determined by its change within a certain interval $[t_n; t_{n+1}]$. Figure 5.3 illustrates the PDCA principle according to Miehe et al. (2017a).

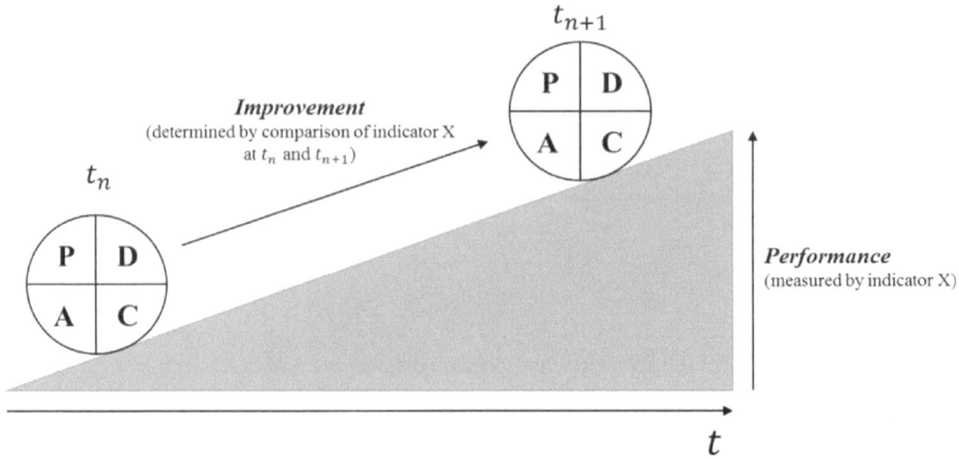

Figure 5.3 PDCA principle according to Miehe et al. (2017a)

The German standard VDI 2870 summarizes design principles of LM and methods for their implementation in a production environment (Table 5.1) (Verein Deutscher Ingenieure, 2012a).

Table 5.1 LM design principles and respective implementation methods according to VDI 2870 (Verein Deutscher Ingenieure, 2012a)

Design principles	Methods
Avoidance of waste	Chaku-chaku, low-cost-automation, total productive maintenance, value-stream-mapping and design
Standardization	6S, process standardization
Zero failure principle	5x why, 8D method, automation, Ishikawa diagram, short control cycles, poka yoke, Six Sigma, Statistische Prozessregelung, employee self-monitoring
Visual management	Andon, shopfloor management
Employee-oriented leadership	Hancho, target management
Flow principle	FIFO principle, one-piece-flow, rapid set-up, value stream planning, U-layout
Pull principle	Just-in-time/just-in-sequence, kanban, milk run, levelling, supermarket
Continuous improvement	Audit, benchmarking, cardboard engineering, PDCA

5.3 Environmental management (EM)

EM is defined as the management of the interaction and influence of production on the environment. According to the sustainability concept, its aim is to protect ecosystem services for future human generations (Pahl-Wost, 2007). EM seeks to identify and eliminate trade-offs that arise between a company's focus to satisfy needs and the immanent challenge to protect the environment. Today, environmental management systems (EMS) are applied in many companies. In order to ensure similar approaches across industries, various institutions have released guidelines and standards, among which ISO 14.001 and Eco-Management and Audit Scheme (EMAS) are the most popular. While environmental management according to ISO 14.001 is a pure management system, EMAS is a joint system of environmental management and environmental auditing. Table 5.2 summarizes the most significant differences of ISO 14.001 and EMAS, as it is outlined in the European eco-audit-regulation.

Table 5.2 Differences of EM based on ISO 14.001 and EMAS

	EMAS (EU eco-audit-regulation)	ISO 14001
Scope of application	Restricted to Europe	Valid worldwide
Proof	Validation (= state approval)	Certification (= private sector admission)
Financial support	Government funding possible	No government funding possible
Reporting	Environmental statement	No environmental statement
Scope of system	Location	Company
Marketing	Own logo	No logo

ISO 14.001 represents the most popular approach to EMS, while representing one of several management systems standardized by ISO, out of which ISO 9001 represents the baseline system. Accordingly, ISO 14.001 adopts its key elements such as continuous improvement (see LM), the concept of audits for performance tracking (internal or external), voluntary certification, the principles of documentation and the objective of standardizing procedures, tasks and roles (Baumast, 2013). Basic components of every EMS are: [1] environmental policy/strategy, [2] environmental objectives and program [3] indicators and ratios for measurement, [4] environmental audit, [5] control and corrective measures, [6] environmental reports as well as [7] stakeholder dialogue. Supporting standards and technical guidelines are supplemented within the ISO 14.000 standards program e. g., eco audit (ISO 14.010) or eco-labels and declarations (ISO 14.020). Similar to LM, PDCA represents the basis for continuous ecological improvement.

5.4 Eco Lean Management (ELM)

ELM seeks a combination of LM and EM.[1] With regard to the vision of a biointelligent enterprise, the approach thus illustrates an example of integrative management, which requires further development in the future. The following paragraphs aim to sensitize the reader for basic approaches and challenges of integrative sustainability-oriented management of production systems.

The aim of ELM is to specifically address environmentally relevant issues in day to-day operations at the shop floor level. It is to be understood as an integral system of principles, procedures and methods for a combined economic and ecologic low-cost improvement of manufacturing environments. Thereby, the term low-cost improvement is defined as an organizational enhancement of the manufacturing process in small steps (Kaizen) via minor investments (Miehe et al., 2016a). A concrete cap of investment is not intended. Yet, the payback period of such measures should not transcend three years. Improvement measures may thus be classified as low cost if either one of the following criteria is met: [1] an improvement can be executed without additional costs (costs that accrue in any case, e. g., employee salary, are not considered) or [2] the payback period of an investment necessary for improvement does not exceed three years.

The concept of ELM is primarily based on the works of Miehe et al. (2016a; 2017a), which provide the basis for the following sections. Yet, the idea of using LM to improve industrial environmental performance is not new. Initial concepts have been presented under the term Lean&Green by Bergmiller and McCright (2009), Khalaf et al. (2011), Diaz-Elsayed et al. (2013), Verrier et al. (2014). For instance, the US EPA (2011) presents a toolkit that among others estimates environmental impact of lean waste reduction, identifies lean approaches as potentially beneficial and demands ecologically oriented Kaizen shop floor workshops. Other authors focus on certain lean related methods and aim at extending its application to environmentally related aspects. Hereby, especially the method of value stream mapping depicts a foundation of advancement. Among others, Erlach and Westkämper (2009), Faulkner et al. (2012), Erlach and Sheehan (2014), and Rieder et al. (2014) add various ecology related aspects to the analysis, e. g., resource use, energy consumption, CO_2 emissions. Figure 5.4 illustrates the concept of a resource-oriented value stream mapping approach according to Faulkner et al. (2012) and Rieder et al. (2014).

[1] This section is a summary of the findings of Miehe et al. (2016a; 2017a).

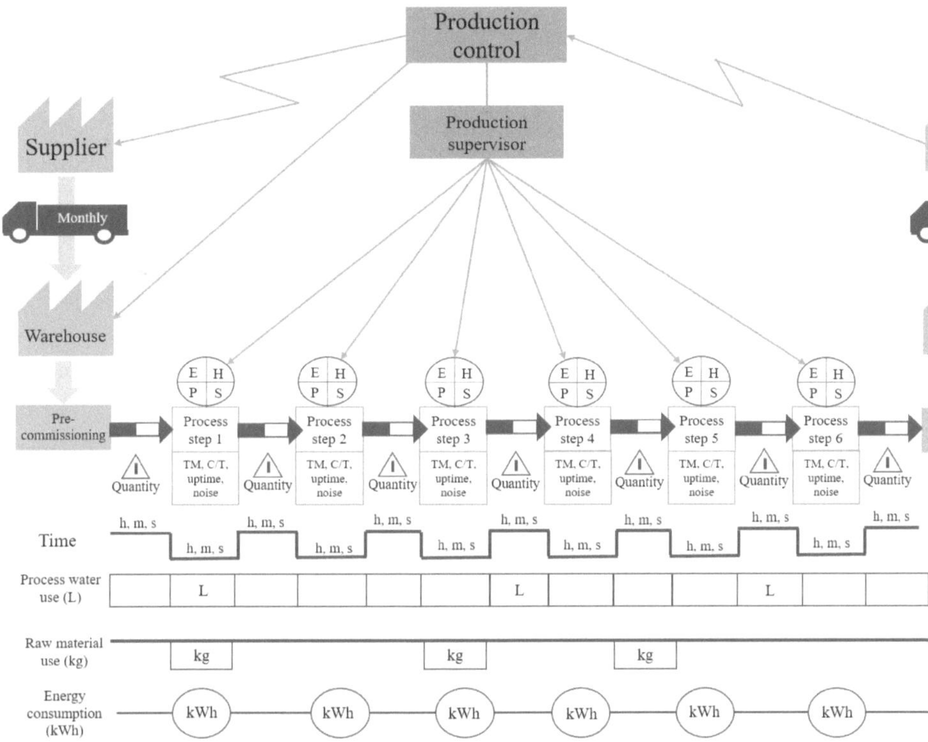

Figure 5.4 Resource-oriented value stream mapping according to Faulkner et al. (2012) and Rieder et al. (2014)

Yet, the transfer to an environmentally friendly value stream design (VSD) is rather vague. Chiarini (2014) provides an overview of the environmental impact of various Lean related methods in the context of European motorcycle component manufacturers. Hereby, value stream mapping, 5S, cellular manufacturing, and TPM are identified as beneficial, while single minute exchange of die (SMED) is not found to add benefit to the environment. The overall assumption of these studies that a reduction of lean waste categories ultimately leads to environmental benefits, however, is rarely questioned. In fact, few studies exist that provide an extension of traditional lean waste categories by adding environmentally related wastage, e.g., energy consumption, emissions (Kidwell, 2006; Sciortino et al., 2009; Milward et al., 2013), yet these definitions lack of scientific foundation. Venkat and Wakeland (2006) show that especially frequent replenishment and long distances of delivery of JIT/JIS strategies have negative influence to the environment.

Basic reflections on the ecology of LM

When understanding the term ecology as a synonym for environmental conservation, the idea of ecology significantly differs from traditional lean thinking. First, the goal of LM is to increase customer satisfaction (Pfeiffer and Weiß, 1994). Therefore, it intends to eliminate processes that do not generate value for the customer. In this context, reduction of resource consumption remains a side effect of the application of LM methods (Kidwell, 2006; Gogula et al., 2011). A solely ecology-oriented approach in turn would result in not using natural resources at all (Lozano, 2012). The mandatory requirement of LM – to use resources in order to satisfy the customer – is thus inconsistent with environmental conservation. In terms of its goal both approaches are thus fundamentally different (Venkat and Wakeland, 2006). Second, the scope of both approaches has to be distinguished. While LM represents a process-orientated approach (Pfeiffer and Weiß, 1994), environmental conservation seeks an improvement of products and processes (Moreira et al., 2010). LM mainly reduces development and/or lead time but never questions substances or materials that are used to produce the product itself (Johansson and Sundin, 2014). Environment-oriented improvement, in turn, demand an orientation towards environmental impact (United States Environmental Protection Agency, 2011). Therefore, different system boundaries are possible, e. g., cradle-to-grave, cradle-to-gate, gate-to-gate, gate-to-consumer. Then, as outlined above, LM is labelled a holistic approach due to its focus on process improvements within the supply chain and the company (Pfeiffer and Weiß, 1994; Womack and Jones, 2015). Yet, its focus ends when the product is handed out to the customer (Womack and Jones, 2015). In contrast, ELM relies on life-cycle thinking, considering both upstream and downstream stages, e. g., usage, end-of-life (Sarkis, 2003). Finally, LM focusses on low-cost problem solving (Pfeiffer and Weiß, 1994), which mainly results in advancements of process stability and lead time reduction (Simons and Mason, 2003). In contrast, EMS related improvements generally results in a reduction of resource use and impact during the entire life-cycle (Sarkis, 2003). Concluding, LM and EMS exhibit a variety of differences that prevent a pragmatic combination of both. A leading philosophy is required. As an environmental focus would lead to significant restrictions to economic activities, ELM relies on LM as the leading philosophy.

The Eco Lean Management approach

With LM as the leading philosophy, the overall target hence remains the satisfaction of customer needs and may be summarized as follows: to provide a customer with the desired product in the required quality and time. In contrast to traditional LM, ELM perceives customer requirements not as exclusively direct expressed demands, but as indirectly implied. It is assumed that a customer does not intend to destroy the environment when buying a product. Instead, she or he is relying on the company to do its best to prevent environmental harm. While traditional LM focusses on time and stability, ELM thus adds resource efficiency as the third key measure. Improvement measures shall solely be implemented if the resource efficiency constraint is fulfilled. For instance, a measure that improves time but downgrades resource efficiency should not be implemented. Figure 5.5 illustrates the relation according to Miehe et al. (2016a).

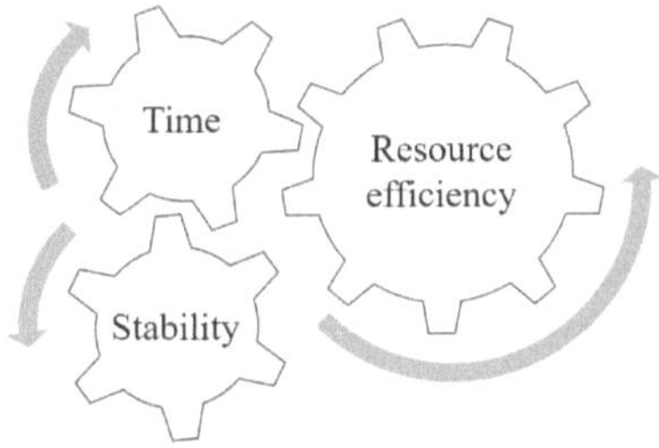

Figure 5.5 Key figures of ELM according to Miehe et al. (2016a)

This understanding directly affects the interpretation of value-adding activities and waste within a production process. By summarizing all activities for which the customer is willing to pay and substances/materials that remain in the product as value-added resources, three major types of waste are differed (Miehe et al., 2016a):

1. *Lost resources*: activities a customer is not willing to pay for and substances/materials that do not remain in the product. This category may be further distinguished according to traditional LM in TIMWOOD plus material waste, energy and emissions.
2. *Additional resources*: substances/materials that are added to the product but do not affect its functionality (e.g., packaging).
3. *Critical resources*: substances/materials that pose a risk to the company of failing to meet its performance requirement.

Organizational implementation of ELM requires the deployment of certain principles by both managers and employees that are direct micro-level implementations of the principles outlined in Section 3.6. Here, mental and operational premises are distinguished. The former serves as a mind-set for decision-makers, while the latter requires implementation in actual day-to-day operations. Table 5.3 summarizes the ELM premises according to Miehe et al. (2017a).

Table 5.3 Eco lean management premises according to Miehe et al. (2017a)

Mental premises	Operational premises
1. Continuity and consistency of actions 2. Avoidance of waste 3. Systemic life cycle thinking and acting 4. Striving for economic and ecological perfection 5. Awareness of social responsibility 6. Participation orientation 7. Material before capital orientation 8. Process oriented thinking and acting	1. Determine/define value from the perspective of customers, society, and environment. 2. Define goals from customer, society, and environment perspective and establish indicators to measure success. 3. Create transparency about all substances and materials used. 4. Identify value stream. 5. Implement flow principle. 6. Introduce pull principle. 7. Minimize resource consumption.

Based on to the PDCA principle, the ELM optimization procedure follows four standard stages, as illustrated in Figure 5.6.

Figure 5.6 ELM optimization procedure according to Miehe et al. (2016a)

The first step aims to achieve transparency with little effort. Its goal is to initially identify obvious problem areas. Therefore, an advanced Gemba[2] walk has to be conducted on the shop floor. In contrast to the traditional approach, additional aspects have to be examined, e.g., loss of resources. A plant layout has proven to be a helpful tool in this step in order to roughly delineate facility installations, breakdown of divisions, work flows and problem areas. Selective interrogation with facility employees provides additional information, e.g., machine performance at different load levels, problems with a certain machine. A secondary part of orbit view is a data analysis regarding ecological performance. Therefore, the utilized resources have to be classified in type and quantity of consumption over a previously defined period of time. Suitable sources among others are energy and material bills, supplier contracts, and standard reports (e.g., ISO 9001, ISO 14.001 and the like). In a final step, consumption data have to be allocated to certain divisions or products. If a direct assignment of consumption is not possible, allocation rates have to be developed. After initially identifying problem areas, a prioritization has to be made in order to detect major and minor preferences. Therefore, various methods exist, among others ranging from simple scoring to pairwise comparison or AHP. Given initial solutions have been identified already, a weighting regarding their impact on waste reduction and expected effort for implementation represents a pragmatic approach (Figure 5.7).

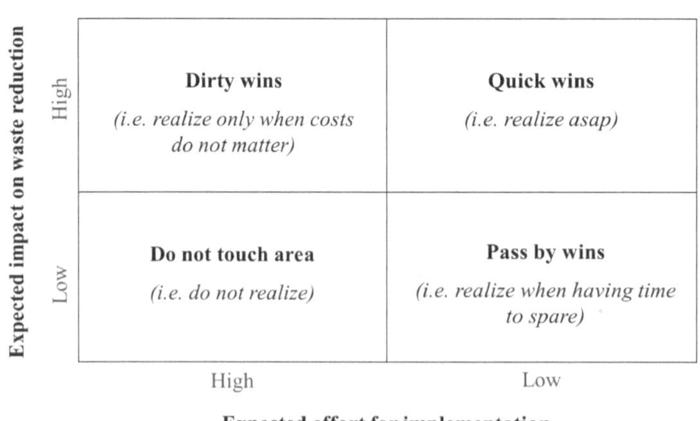

Figure 5.7 Exemplified selection matrix according to Miehe et al. (2016a)

[2] Gemba refers to the standard approach of LM to personally visit a certain site in order to gain a better understanding of the situation. A common phrasing is 'Go to Gemba'.

After the decision to focus on a particular problem, its root cause has to be identified. This step focuses on achieving the highest possible level of transparency. In a final step, improvement measures have to be implemented according to the A3-Report problem solving procedure in eight steps (Figure 5.8).

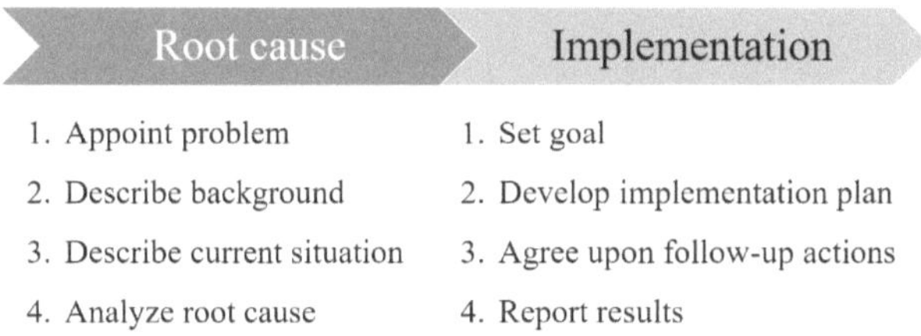

Figure 5.8 ELM root cause and implementations approach according to Miehe et al. (2016a)

Eco Lean Sustainability Control Panel (EL-SCP)

ELM requires constant controlling of continuous improvement in order to maintain economic and ecological enhancement of business processes. Therefore, Miehe et al. (2017a) present a tailored indicator system, referred to as the Eco Lean Sustainability Control Panel (EL-SCP). Here, term indicator is understood as a condensed figure aiming to quantify a specific issue. Indicators may either be absolute or relative values. Absolute figures result from measurement (e. g., material/electricity consumption) or calculation (e. g., sum, deviation, mean value), while relative figures are ratios of two separate indicators. The formulation of an indicator has to have a certain purpose that generally is deducted from corporate strategy. In doing so, EL-SCP differs between three major levels of operational implementation: [1] the company and/or facility location itself, [2] intersector level and [3] intrasector level. Figure 5.9 illustrates the approach according to Miehe et al. (2017a).

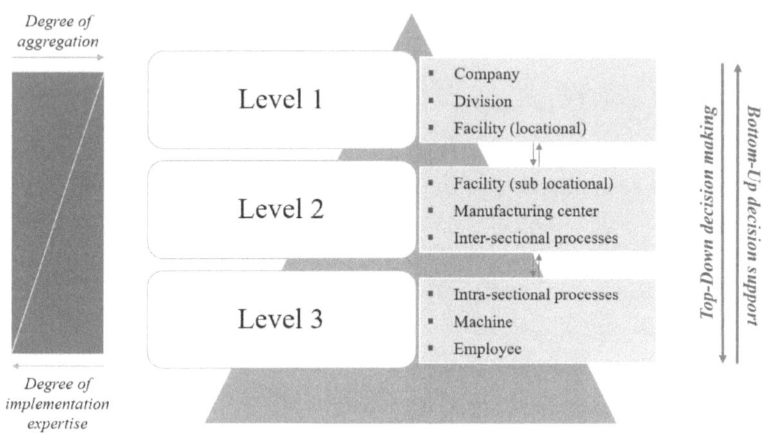

Figure 5.9 Perspectives of EL-SCP according to Miehe et al. (2017a)

According to the basic ELM approach, EL-SCP combines economic and ecologic indicators. Thereby, ecologic performance is determined by the quantity of substances entering and leaving the system, its impact on the environment, and its criticality to the company. Table 5.4 has summarized the 18 aggregated indicators of EL-SCP according to Miehe et al. (2017a).

Table 5.4 EL-SCP indicators according to Miehe et al. (2017a)

Ecologic indicators[3]	Economic indicators
1. **TMI** – Total material input [quantity by weight] 2. **TEU** – Total energy use [quantity by power input] 3. **TPO1** – Total product output [quantity by weight] 4. **LMO** – Lost material output [quantity by weight] 5. **TCMU** – Total critical material in use [quantity by weight] 6. **TEO** – Total emission output [quantity by particles] 7. **TMF** – Total material footprint [impact] 8. **LMF** – Lost material footprint [impact] 9. **TEF** – Total emission footprint [impact] 10. **ORE** – Overall resource efficiency [maturity] 11. **ODIC** – Overall degree of internal circular flow [%]	1. **ACT** – Average cycle time per product [min.] 2. **TPO2** – Total product output [quantity by no.] 3. **TMC** – Total manufacturing costs [€] 4. **TOP** – Total output performance [%] 5. **ADR** – Average defect ratio [%] 6. **OTD** – On time delivery [%] 7. **OEE** – Overall equipment effectiveness [%]

As the economic indicators applied here are commonly implemented in manufacturing companies, no further description is required. Yet, the main ecologic indicators demand for specific explanation. The quantity related indicators (*QRI = TMI, TEU, TPO1, LMO, TCMU, TEO*) are thus defined as the sum of all measures with similar characteristics, i. e.:

$$QRI = \sum_{i=1}^{n} q_i$$

As EL-SCP seeks to be easily applicable to industry standards, a simple sum of all materials entering a system is reasonable. The *total critical material in use* (*TCMU*) indicator refers to the specific waste category. Its calculation is, however, slightly more complex taking into account different types of supply and demand markets as well as the diffusion curve of environmental problems.

The *impact related indicators* (*IRI = TMF, LMF, TEF*) depict special cases of accounting. Quantity based EL-SCP indicators rather focus on internally relevant issues such as

[3] The user of Eco Lean SCP is given free choice of system boarders for accounting of each indicator as the approach presented here does not refer to a particular point of time but serves as an instrument for improvement over time.

materials and energy that again lead to significant costs. Impact accounting in contrast takes into account the external perspective (see Section 6.2). In order to simplify impact accounting for daily execution (especial for shopfloor management), EL-SCP focusses on three indicators:

- The formulation of the *total material footprint* (*TMF*) is derived from equivalence model approaches of LCA, e. g., CML. As such, it is defined as the sum of the *natural material footprint* (*NMF*) and the *synthetic material footprint* (*SMF*) , i. e.:

$$TMF = NMF + SMF = \left(\sum_{i=1}^{I} \left[\frac{NCR_i}{TQ_i} \right] \cdot R_i \right) + \left(\sum_{i=1}^{I} \left[\frac{SCR_i}{TPV_i} \right] \cdot \frac{\sum_{i=1}^{I} [TQ_i \cdot R_i]}{TQ} \right)$$

NCR_i Consumption rate of natural material i

TQ_i Total quantity of natural material i in earth's crust

R_i Regeneration ratio of natural material i in earth's crust

SCR_i Consumption rate of synthetical material i

TPV_i Total (global) production rate of synthetic material i

GAR Global average regeneration ratio

TQ_i Total quantity of natural material i in earth's crust

TQ Total quantity of earth's crust

- *Lost material footprint* (*LMF*) is simply an adaptation related to a different quantity allocation.

- The *total emission footprint* (*TEF*) is derived from the equivalence model approach of common LCA as well. In comparison to the simple sum of single emissions via quantity of particles in the TEO indicator, TEF adds up the results of characterization models (e. g., IPCC GWP100) by weight. As such, it may be understood as a simple adaptation of a quantity related indicator, i. e.:

$$TEF = \sum_{i=1}^{n} q_i = x \text{ kg } CO_2\text{e} + x \text{ kg } SO_2\text{e} + x \text{ kgR11e} + \ldots$$

- The overall resource efficiency may be defined with regard to two perspectives, internal and external. The former (ORE_I) is defined as the ratio of *TPO* and *TPI*, i. e.:

$$ORE_I = \frac{TPO}{TPI}$$

- The latter (ORE_E) is defined as the ratio of *TPO* and the sum of *TMF* and *TEF* under disregard of units, i. e.:

$$ORE_I = \frac{TPO}{TMF + TEF}$$

As the indicators sketched above are not directly measurable, Figure 5.10 illustrates the relation between measurable key figures and EL-SCP indicators.

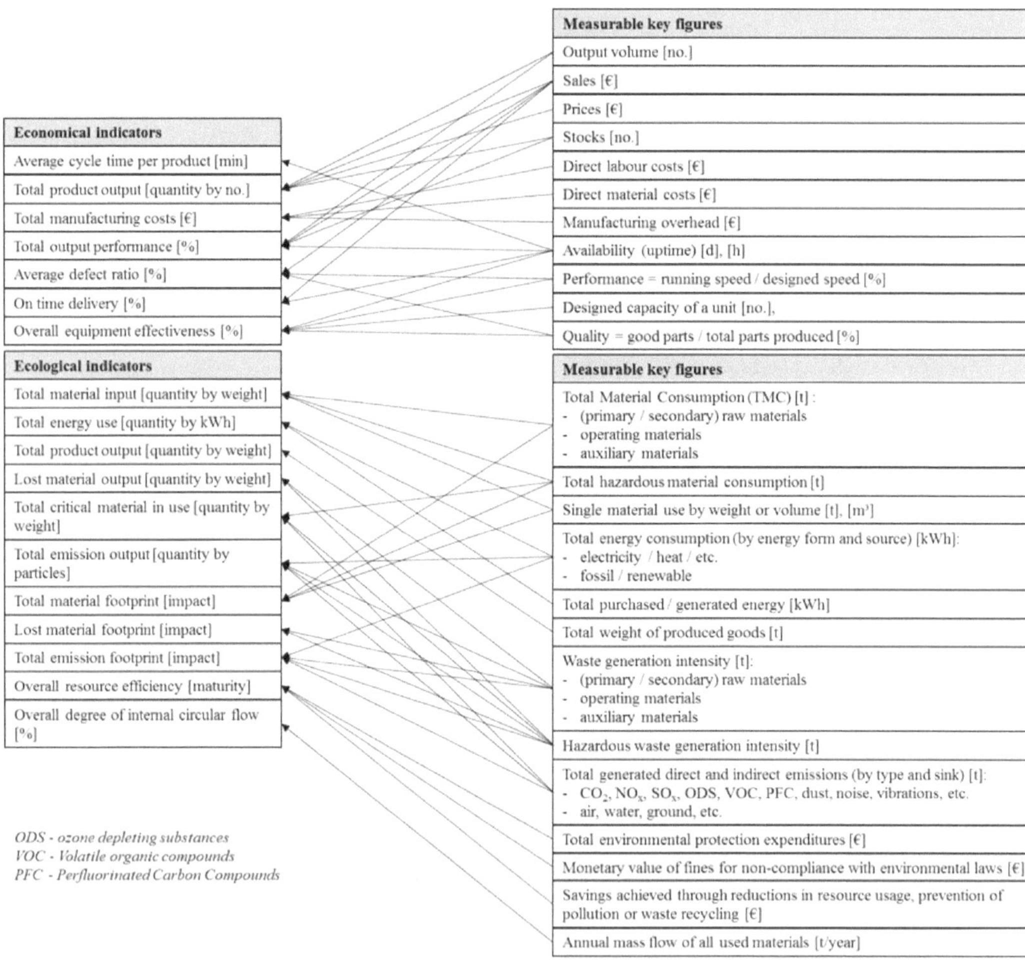

Figure 5.10 Relation between measurable key figures and EL-SCP indicators according to Miehe et al. (2017a)

Yet, the mere quantification of EL-SCP indicators does not easily enable simultaneous economic and ecologic optimization, as the interdependencies between indicators of the two dimensions are still unknown. Thus EL-SCP requires an analysis of correlations, which represents a test of great complexity. While simple scoring, pairwise comparison or AHP, involving selected experts, may be used. Miehe et al. (2017a) present a tailored approach to develop a correlation matrix by solely focusing on obvious interactions in four steps: [1] define target state for each indicator, [2] assume target state of a single indicator to be fulfilled, [3] ask experts within or outside the company for specific grading of impact on all other either economic or ecologic indicators, and [4] summarize results. Figure 5.11 exemplifies a correlation matrix of ecologic and economic indicators according to Miehe et al. (2017a).

		Economic Indicators						
		Average cycle time per product	Total product output	Total manu-facturing costs	Total output performance	Average defect ratio	On time delivery	Overall equip-ment effective-ness
Ecologic Indicators	Target state	Min.	Max.	Min.	Max.	Min.	Max.	Max.
Total material input	Min.	0 / 0	- / -	+ / 0	0 / +	0 / +	0 / 0	0 / +
Total energy use	Min.	0 / +	- / -	+ / 0	- / -	0 / 0	0 / 0	+ / +
Total product output	Max.	+ / +	+ / +	- / -	+ / +	0 / 0	+ / 0	+ / +
Lost material output	Min.	0 / 0	+ / +	+ / 0	+ / 0	+ / +	+ / 0	+ / 0
Total critical material in use	Min.	0 / 0	0 / 0	+ / 0	+ / 0	+ / +	0 / 0	0 / 0
Total emission output	Min.	0 / +	0 / 0	0 / 0	0 / 0	0 / 0	0 / 0	0 / +
Total material footprint	Min.	0 / 0	0 / 0	+ / 0	+ / -	0 / +	0 / 0	0 / +
Lost material footprint	Min.	0 / 0	0 / 0	+ / 0	+ / +	+ / +	0 / 0	0 / +
Total emission footprint	Min.	0 / +	0 / 0	0 / 0	0 / 0	0 / 0	0 / 0	0 / +
Overall resource efficiency	Max.	+ / +	0 / 0	+ / 0	+ / +	0 / +	+ / +	0 / +
Overall degree of internal circular flow	Max.	0 / 0	0 / 0	+ / 0	0 / 0	0 / 0	0 / 0	0 / 0

Figure 5.11 Correlation matrix of ecologic and economic indicators according to Miehe et al. (2017a)

In order to implement EL-SCP sufficiently, a reasonable organizational implementation is essential. While measurable figures such as time, number of parts or weight of materials may be raised periodically, a system for both documentation and cross-linking is required. Therefore, a report template should be designed in a company's ERP and/or MES system linking figure allocation, correlation matrix and at best an automatic measurement at the machine level.

Results from case study at small manufacturer of special machinery in Germany

ELM was first implemented as a whole at a small local manufacturer of special machinery. Thereby, EL- SCP was applied in order to quantify performance and improvement of different enhancement actions. In this particular case four optimization measures were selected leading to both significant economic and ecologic gains:

1. Implementing 5S at the workplace reduced searching time by roughly 40% and increased overall resource efficiency by 53% leading to cost savings of around €167,000 p. a.
2. Developing and introducing a tailored production planning and inventory system increased cycle time of a sample product by 36%. Simultaneously this measure reduced over ordering by roughly 21%. In total, improving production planning and inventory resulted in cost savings of €91,000 p. a.
3. Redesigning machine arrangement included an adaptation to three main requirements: short distances, job shop work profiles of employees, and waste heat recovery via specific combination of heat exchanger and pump. Besides an improvement of waste heat recovery of roughly 100%, employee productivity was boosted by around 18% leading to cost savings of €66,000 p. a.
4. Developing solution for circular flow of cooling liquid by separating pollutants from used cooling agents via oil skimmer technology increased physical life time by around 400%. Although a skimmer had to be purchased for each machine, annual savings accounted for €23,000.

5.5 Systemic management according to the concept of ultra-efficiency

Especially in the context of the implicit interpretation of biointelligence, the previous approaches are not enough. As a focal point for companies, the goal should be holistic, systemic management towards a positive impact production. A prominent example of a positive impact factory is the concept of ultra-efficiency, among others presented by Kuhlmann and Sauer (2016), Miehe et al. (2017b), Lentes et al. (2017), and Lentes and Hertwig (2019). The concept seeks to combine the efficiency and consistency strategy of sustainability on a factory and management level in order to establish a production system that not only has no negative impact on its environment, but realizes a positive impact in five essential fields of action (material, energy, emissions, staff, organization). The goal of management must be the development of an integrative optimum in order to optimize production within the framework of the life cycle of the factory and its produced products, the satisfaction of needs of the organization and its staff as well as the symbiotic embedding in its surroundings, both from an aesthetic and a social perspective. Table 5.5 summarizes the main objectives for systemic management of production according to the concept of ultra-efficiency.

Table 5.5 Main objectives for systemic management of production according to the concept of ultra-efficiency [own illustration]

Field of action	Main objectives
Material	The ultra-efficient use of materials strives for a full circular economy: it includes reused or recycled use of residual and waste materials as well as products or parts after their use. Ultra-efficient productions systems thus only use materials that are completely free of critical and harmful substances, that completely originate from local resources, and that are fully recyclable. This applies to internal small material cycles as well as to finished products.
Energy	Regenerative energies are used with maximum efficiency and effectiveness, and energy demand is optimally coordinated with the aim of minimizing it as far as possible (energy recycling). Ultra-efficient productions systems thus exclusively use renewable energy and strive to generating more energy than consumed (plus-energy factory).
Emissions	Ultra-efficient production systems do not allow any emissions to accompany their production. If they are unavoidable, they should be fully recycled in production. The vision is not only to avoid emissions and waste in the value creation process, but also to develop a company that reduces emissions and waste (emission sink factory).
Staff	People are far more than just workers. Ultra-efficiency is about achieving production goals, and generating a positive contribution to adjacent companies, employees, and neighbors. Ultra-efficient production systems ensure that employees enjoy their work as much as possible and integrate the factory perfectly into their living environment. Both white-collar and blue-collar employees work at and create personalized, innovative workplaces and models that effectively promote performance. Working in the ultra-efficient factory does not make sick, but healthier.
Organization	An ultra-efficient organization provides structures and rules that enable measures to be introduced and implemented quickly and smoothly. Ultra-efficient production systems operate as flexible, changeable, and resilient as possible. They establish transparency as a guiding principle in corporate culture and maximize employee identification with the organization.

The main objectives listed here serve as a guide for future decision-makers. The essential task of management is to identify and resolve conflicting goals, to avoid frictional losses due to divisional thinking and to provide the necessary framework for achieving the objectives. This again relies on systems-oriented life cycle thinking in cellular units as well as on the above outlined approaches (e.g., AHP). In a similar concept, Herrmann et al. (2015) distinguish between Dos and Don'ts when designing and managing a production system (Table 5.6).

Table 5.6 Dos and Don'ts of ultra-efficient production systems according to Herrmann et al. (2015)

Dimension	Category	Don'ts	Dos
Environ-mental	Energy and resource demand	High demand for natural resources/materials	Use of secondary materials, renewable energies (energy flexibility)
	Waste	High quantities of residual waste	Reuse, recycling, conversion of industrial or residential waste and sewage water
	Particle emissions	Air and land pollution, PM emissions peril human health	Emission sink, e. g., for air and water pollutants
	Noise emissions	High noise level	Factory as a noise sink
	Biodiversity	Destruction of bio-diversity	Creation of new habitats for animals and environmental regeneration
	Traffic	Congestion	Development of transport infrastructure, short dis-tances for workers
	Risks	Source of danger	Source of domestic emer-gency supplies, factory as shelter
Social	Worker satisfaction	Lack of motivation and development perspectives	Place for life-long learning and personal development
	Work Life Balance	Workers adapt to factory rhythm	Factory adapts to the work-ers' capabilities, place for recreational activity
	Working conditions	Harmful manual labor and mental stress	Factory as a fitness studio
	Automation	Substitutes human workers	Safeguards human health
	Openness	Factory as a black box	Factory is open for custom-ers and residents
	Individuality of products	Anonymous mass production	Individualized products
	Architecture	Factory spoils landscape, no acceptance among residents	Factory as architectural attraction, harmonic inte-gration into surrounding

Annex 6: A SWOT analysis of the German industry

The development of BIS largely depends on the existing strengths, weaknesses, opportunities, and threats (SWOT). According to Miehe et al. (2020a), these were compiled from a general perspective for the German industry as part of the BIOTRAIN survey.[1] The survey includes the opinion of 123 experts that were questioned in regard to 35 predefined SWOT-hypotheses within four perspectives and 14 focus topics in the framework of a Delphi-supported scenario-based research design (Figure 6.1).

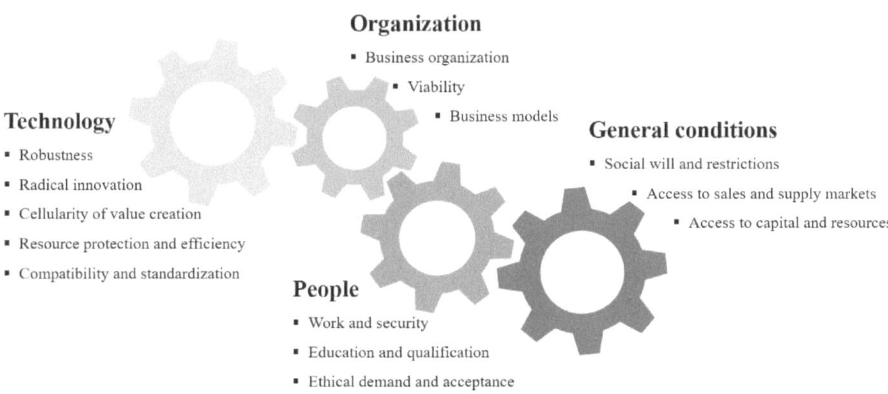

Figure 6.1 Framework of Delphi-supported scenario-based research design of BIOTRAIN survey according to Miehe et al. (2020a)

The SWOT results are summarized below according to Miehe et al. (2020a). It should be noted that this is an assessment at a given point in time, which is intended to provide information on the existing potentials and threads of biointelligent innovations. Future changes are likely (e. g., in the context of the Corona crisis).

[1] This section is a summary of the findings of Miehe et al. (2020a).

6.1 Strengths

The German manufacturing industry possesses seven key strengths regarding biointelligence. First, German basic research in general is well positioned in an international comparison. Relevant fields of research such as biomimetics, biotechnology and bioeconomy already resulted in a variety of innovations, e. g., lotus effect in surface coatings. In an international comparison, research in Germany is competitive in these fields. In addition to existing research, small and medium-sized enterprises (SMEs) form the basis for agility and the versatility of German industry. The existing SME landscape is globally renowned for its reliability, tradition, and quality. In addition, many large German corporations increasingly gain versatility by acquiring startups. Overall, the German ability to transform its economy is at a good state.

In any case, the label 'Made in Germany' is a recognized for quality worldwide. Products that are developed and manufactured in Germany or at least produced under German standards abroad are considered robust, functional, powerful, and durable. This positive association strengthens the reputation of German goods abroad and forms one of the central foundations for Germany's success as an export nation in international trade. In addition, the label is more and more associated with sustainability. Another strength is seen in the country's privileged position in international trade, as a comprehensive access to both sales and procurement markets depicts key strategic aspects of a company's success. The perception of Germany in this context is strong due to its good foreign policy relations and its economic strength. Besides having access to various global trade agreements that ensure largely free trade, Germany has direct access to one of the world's largest domestic markets, the Schengen area. The German political and economic situation are viewed as comparatively stable overall due to a generally consolidated democracy, comprehensive social security, a de-escalative foreign policy and a stable currency. Moreover, the German government aims to maintain its production location in the long term by providing tailored supporting measures, which especially relieves industries in times of crisis. Other strengths are standardization and occupational safety measures as well as environmental awareness and ethical standards. The introduction of novel value creation structures, new working conditions and work processes can only be achieved safely and efficiently by standardized guidelines. Germany occupies a strong international position in this field, as local standards are considered to be of extremely high quality and trustworthy worldwide. Hence, they often form the basis for international standards. Beyond that, Germany is viewed as a pioneer in environmental protection and sustainability. German companies occupy leading positions in environmental engineering worldwide. The nation's society exhibits a comparatively high sensitivity for environment related issues, which in terms of sustainability may have a catalytic effect on certain relevant topics. Finally, the diversity and quality of the German education and training system is in a good state, as it offers a large number of disciplines as well as numerous forms of training, mostly free of charge. The ability of Germany to create demand-oriented new training and study programs is satisfactory. The training portfolio ranges from applied training in companies to deep scientific studies at universities. All in all, interdisciplinary study programs and the dual education system form a sound basis for an adaptable, multidisciplinary education, which is considered necessary.

6.2 Weaknesses

In contrast, six key weaknesses of the German manufacturing industry exist. First, the German mentality is regarded as risk averse. Conservative attitudes are a widespread phenomenon in large parts of society, business, and politics leading to manifold impacts: restrictive legislations, little venture capital due to risk aversion of investors etc. Especially the latter is regarded as critical. Without a greater availability of venture capital, the conversion of many inventions to innovations will not succeed. Related to risk aversion, the German society is regarded as extremely critical towards new and unknown technologies. The risks of innovations is often weighted higher than its opportunities. This fact especially applies to fields such as biotechnology and AI that are crucial to biointelligence in an explicit sense. Especially in these fields regulations in Germany are comparatively strict, which impedes free action and rapid progress. Accordingly, the scientific freedom in genetic engineering is more limited than in comparable industrialized countries. Germany is exceeding the already strict EU rules and prohibiting, for example, research on embryos in the field of genetic engineering. As a result, the potential for innovation in Germany is severely hampered. This again leads to a low attractiveness of the location for leading experts and research institutions in the respective fields. High costs of German research and development depict another major obstacle of innovation as a result of high safety demands. German companies are thus often forced to acquire expensive technologies from abroad. German companies are predominantly viewed as hierarchically organized and highly static, a fact that depicts the exact opposite of a biological, decentralized and dynamic system, which represents the target state of biointelligence. Furthermore, hierarchical structures are regarded to lead to a concentration of knowledge on individuals or groups. The ability of the German manufacturing industry to implement new technologies quickly is thus perceived as improvable. As a resource-poor country, Germany is heavily dependent on imports from abroad. This applies to both energy and materials. A dependency, especially with regard to the energy supply, reduces the foreign policy scope of action of the federal government extremely and represents an obstacle to achieving political goals. In addition, the attractiveness of Germany as a business location is directly reduced by higher energy and raw material prices. Another weakness is seen in the limited development of digital systems and

BMs. In an international comparison, only a few Internet-based start-ups from Germany have been able to establish themselves internationally.

6.3 Opportunities

The survey revealed seven opportunities to the German manufacturing industry. First, in order to maintain the attractiveness of the industrial location, biointelligence is regarded as an outstanding chance. This especially applies to the fields of bio and information technology. In both fields Germany has recently lost ground against the USA and China. In the course of the transition, manufacturing companies now have a chance to catch up. Related to consolidation of the industrial location, biointelligence may have an increasing impact on the labor market and on ways how work is done. Complex, self-optimizing systems can be monitored by well-trained specialists. The likelihood of new jobs being created is high. Contents and condition of work are as likely to change, as the emerging decentralized production systems require more responsibility and design possibilities of each individual employee. Digital and biological know-how will increasingly determine job market chances for employees. Another opportunity is seen in the regionalization and resource efficiency, as biointelligent technological innovations offer the potential to increase the efficiency of value creation and close loops. New organizational BMs and biological structures may enable a higher degree of employee participation and more efficient work processes. Bioinspired structures, mechanisms and principles such as swarm intelligence create new competencies directly within a company. These can be adapted, collected and managed decentrally to the respective circumstances. Decentralized, networked or autonomous systems are considered to be extremely efficient and dynamically adaptable. Then again, the current dispersion of biotechnological systems in industrial production is expandable. These systems are going to diffuse strongly in the near future. A comprehensive standardization thus represents a promising instrument in order to address the prevalent societal scepticism. In addition, biointelligence increases process adaptability and robustness, by creating more stable and multifunctional processes that deliver a higher performance. The speed and adaptability of the product development process may as well be improved. Stable, BIS structures facilitate a more agile adaptability and performance of product development and production cycles. Not least, biointelligence is viewed to offer great potential for radical innovation, as it aims to combine new ways of thinking and disciplines.

6.4 Threats

In contrast, five threats to the German manufacturing industry exist. First, legal regulations are a strong influencing factor for the success or failure of innovation. Especially radical innovations require a high degree of freedom. Restrictive legislations are a major obstacle. On the systems level they might lead to a decline of Germany's innovation potential. The degree of regulative limitations is determined by social will. Unclear added value, e. g., due to bad marketing combined with unclear economic benefits and short-term thinking are major risk factors, as they might negatively influence social perception. The greatest threat of bioinspired, biobased or biohybrid technologies is, however, a possible loss of control over the processes and its results. While early standardization might pave the path to greater controllability, it poses a risk for the robustness of competing technical solutions and the existing innovation potential. Another threat is new dangers at the workplace. The implementation of biological organisms into production may result in new and still largely unknown hazard potentials.

Figure 6.2 summarizes the strengths and weaknesses of the German manufacturing industry according to Miehe et al. (2020a).

Strengths

- Strong research and SME landscape as well as innovative global players
- "Made in Germany"
- Privileged position in international trade
- Crisis security and stability
- Standardization and occupational safety measures
- Environmental awareness and ethical standards
- Diversity and quality of education and training system

Weaknesses

- Risk aversion and conservatism leading to incremental advancements
- Lack of acceptance and inadequate marketing of new technologies
- Restrictive legislations
- Hierarchical structures inhibit dynamic viability
- Dependence on raw materials and energy
- Limited development of digital systems and business models

Opportunities

- Preservation of the industrial location
- New jobs and conditions
- Regionalization and resource efficiency
- Bioinspired structures enabling cellularity of value creation
- Standardization as a tool for greater acceptance and security
- Process adaptability and robustness
- Basic requirements for innovations and digital business models

Threats

- Restrictive legislations
- Social will
- Standardization as critical success factor
- Controllability of autonomous systems
- New dangers at the workplace
- Protectionist tendencies and resource dependence

Figure 6.2 SWOT-matrix of current status of German manufacturing industry according to Miehe et al. (2020a)

Annex 7: An LCE perspective on the biological transformation

This section briefly summarizes a recent study conducted by Herrmann et al. (2020) as an example of a life cycle thinking approach towards BT. Herein, the authors analyze pathways towards environmental sustainability via BT of manufacturing. Herrmann et al. (2020) correctly emphasize the potential risks against the background of overly optimistic expectations by evaluating a variety of technological options. After the assignment of the technology strategies to the development modes presented by Miehe et al. (2018; 2019b; 2020a), life cycle impacts and benefits of each strategy are qualitatively compared. Figure 7.1 summarizes the results according to Herrmann et al. (2020).

The technology strategies presented here must however be viewed with caution. In the context of this work, they are intended to serve as examples for the reader from a contemporary perspective of production science and not as a comprehensive portrayal of the potential of biointelligence and/or BT. On the one hand, they originate from the research of individual authors. On the other hand, the technologies chosen show that the time horizon of consideration was obviously too short to adequately depict the transformative potential of biointelligence.

System level	*Description*	*Environmental relevance*	*Inspiration*	*Integration*	*Intelligence*
Product / component	Microbial-based metal working fluids	- Improved tribological properties leading to prolonged tool life time - Self-healing may extend working fluid life time - Substitution of mineral / chemical oil-based additives		■	
Product / component	Rodent tooth as blueprint for industrial blades	- Self-sharpening for reduced cutting force - Self-healing for prolonged tool life time	■		
Product / component	Mimicry of lotus effect for heat exchangers in food industry	- Antifouling effect for extended service time - Increased thermal conductivity over time - Reduced demand for cleaning agents	■		
Product / component	Filamentus bacteria as electric cable	- Possible for metal based electric cables leading to reduced demand for electric conductive materials (e.g. copper)		■	
Product / component	Bioinspired machine tool design	- Improved machine flexibility and performance lead to increased energy efficiency	■		
Process / machine	Process to manufacture balsa wood inspired hierarchical structures	- Increased material yield ratio - Potentially increased energy demand	■		
Process / machine	Using microbes for battery recycling	- Increased energy efficiency - Reduced chemical pollution and emissions		■	
Process / machine	Using lotus effect as tool property to process adhesive materials	- Relinquishment from lubricants (often mineral oil based) - Reduced demand for cleaning agents	■		

Figure 7.1 Examples for BT of manufacturing at different systems level with regard to environmental performance according to Herrmann et al. (2020)

	Using microbes as tools for micro machining	- Substitution of energy and material intensive processes - Reduced chemical pollution			
	Using microbes for polymer recycling	- Substitution of energy and working material intensive processes - Reduced pollution / production of environmentally benign monomers			
	Using microbes for polymer synthesis	- CO_2 sequestration - Reduced energy consumption due to photosynthetic microorganisms			
Process chain / factory	Bioinspired manufacturing control system	- Potentially increased operational efficiency leading to resource efficiency			
	Urban controlled and symbiotic food production system	- Efficient land use - Production of nutrients - Avoid released agrochemical - (Increased) energy demand			Implicit
	Bee's foraging behavior as concept for shop floor control	- Potentially increased operational efficiency leading to increased yield ratio - Reduced waste - Reduced energy demand			
	Green facades as sustainable approach towards climate change	- Building energy efficiency - Higher air quality - CO_2 sequestration			
	Algae based power generation within the façade	- Eco-efficient energy generation - CO_2 sequestration			
Production network	Industrial symbiosis as concept for collective resource utilisation	- Reduction of waste and resource consumption via collaborative resource usage and supply			
	Symbiotic water system for urban circularity	- Reduction of organic waste - Cleaning of waste water and air - Production of nutrients and minerals with low energy consumption			Implicit
	Bioinspired capacity control	- Reduction of work-in-progress with less unfinished goods in the manufacturing process including labor, raw materials, and overhead			
	Energy flexible production systems as autonomous entities	- Integration of renewable energy into manufacturing - Potential integration of battery storages as buffer			Implicit

Figure 7.1 Examples for BT of manufacturing at different systems level with regard to environmental performance according to Herrmann et al. (2020) (*continued*)

Index